Advances in Neurobiology

Michael J. O'Donovan • Mélanie Falgairolle
Editors

Vertebrate Motoneurons

Springer

Editors
Michael J. O'Donovan
NINDS, NIH
Bethesda, MD, USA

Mélanie Falgairolle
NCCIH, NINDS, NIH
Bethesda, MD, USA

ISSN 2190-5215 ISSN 2190-5223 (electronic)
Advances in Neurobiology
ISBN 978-3-031-07166-9 ISBN 978-3-031-07167-6 (eBook)
https://doi.org/10.1007/978-3-031-07167-6

This Springer imprint is published by the registered company Springer Nature Switzerland AG
The registered company address is: Gewerbestrasse 11, 6330 Cham, Switzerland

Dedicated to Robert E. Burke, MD.
A mentor and friend.

Preface

The purpose of this volume is to provide an overview of current developments in motoneuron function, organization, and susceptibility to disease. Despite being one of the most thoroughly examined neurons in neuroscience, their recent study has yielded new and unexpected discoveries that have provided the impetus for this book. The volume has been written in a didactic style that should be useful to students as well as medical and scientific professionals who are interested in the latest discoveries about these remarkable cells.

The book is organized into three parts. The first deals with the molecular and physiological development of motoneurons. This is a particularly challenging problem because motoneurons are a remarkably diverse neuronal class. Not only are they specialized to innervate every skeletal muscle in the body, but within an individual muscle at least three distinct classes of motoneuron can be distinguished according to the properties of the muscle fibers they innervate. Furthermore, motoneurons can innervate exclusively the extrafusal muscle fibers that comprise the bulk of the muscle (α-motoneurons), both the extrafusal fibers and the intrafusal fibers that are a component of sensory muscle spindles (β-motoneurons), and a group that innervates exclusively intrafusal muscle fibers (γ-motoneurons). Despite this enormous complexity, substantial progress is being made understanding the gene networks regulating motoneuron differentiation, pathfinding, and central connectivity as discussed in chapter "Establishing the Molecular and Functional Diversity of Spinal Motoneurons". This chapter argues that studies of motoneuron development have pioneered our understanding of the differentiation and development of other neuronal types and describes how the expression of Hox genes in both motoneurons and spinal interneurons might underlie the complexity and specificity of their synaptic interactions.

A unique aspect of neural development is the depolarizing nature of the inhibitory neurotransmitters, because of elevated neuronal intracellular chloride. This renders developing networks hyperexcitable and is crucial in the generation of spontaneous neural activity that is essential for many aspects of developing networks. Chapter "Chloride Homeostasis in Developing Motoneurons" summarizes the role of cation-chloride co-transporters in regulating intracellular chloride and the significance of chloride homeostasis in the development and pathophysiology of motoneurons, and details the changes in the function and distribution of these co-transporters during the switch of GABA- and glycinergic neurotransmission from functionally excitatory to inhibitory. Chapter "Normal Development and Pathology of Motoneurons: Anatomy, Electrophysiological Properties, Firing Patterns and Circuit Connectivity" describes the development of the electrophysiological and anatomical properties of motoneurons and how these are expressed in the context of the neural circuits in which motoneurons are embedded. This latter point is particularly important for motoneuron disease where early abnormalities of spinal interneurons and proprioceptive afferents are critical determinants of motoneuron pathology. Chapter "Homeostatic Regulation of Motoneuron Properties in Development" describes the critical role homeostatic processes play in maintaining patterns of developing motoneuron activity in response to external perturbations and to the natural changes in connectivity and size that motoneurons experience as they mature.

The next part of the book focuses on the connectivity and function of mature motoneurons. Chapter "Homeostatic Plasticity of the Mammalian Neuromuscular Junction" starts with a discussion of homeostatic plasticity at the neuromuscular junction. The experimental benefits of studying this process at this comparatively simple synapse are discussed and the various triggers initiating the plastic changes are described. These properties are likely to be crucially engaged in neuromuscular disease and may ameliorate and compensate the concomitant loss of motoneuron function. The next chapter (chapter "Diversity of Mammalian Motoneurons and Motor Units") describes in detail the various classes of motoneuron that innervate a single muscle, including α, β, and γ-motoneurons, each of which can be further divided into multiple subclasses based on their properties. Particular consideration is given to the different classes of α-motoneuron that are specialized to innervate the different types of skeletal muscle fiber. We then examine the synaptic connections of motoneurons within the spinal cord. Until recently, it was assumed that in mammals, motoneurons only innervated inhibitory Renshaw cells intraspinally. However, the new work described in chapter "Synaptic Projections of Motoneurons Within the Spinal Cord" has revealed that motoneurons form powerful chemical synapses with each other and with several different types of excitatory interneuron. These findings are among those that have led to a reconsideration of the role motoneurons play in the genesis of motor behaviors.

For many years, the way motoneurons were recruited into voluntary and reflex movements was believed to be regulated by the "size principle" that postulated that motoneurons were recruited according to their size. The new work reviewed in chapter "Recruitment of Motoneurons" shows that the recruitment process is much more complicated than originally hypothesized and depends upon the repertoire of ionic conductances in the motoneuron membrane and the organization of the interneuronal presynaptic inputs to motoneurons. Ultimately, the behavior of motoneurons is determined by their passive, transition, and active membrane properties.

These are considered comprehensively in chapter "Electrical Properties of Adult Mammalian Motoneurons" which also documents their role in determining motoneuron output and describes their modulation by C-boutons and dendritic inward currents. The chapter also emphasizes the value of comparative studies of humans and animals in understanding the function of motoneurons. This discussion is extended into chapter "The Cellular Basis for the Generation of Firing Patterns in Human Motor Units" that describes the properties of motoneurons in humans and the way that active membrane conductances influence how motoneurons fire and are recruited during voluntary contractions. The chapter also describes the use of computer modelling to "reverse engineer" the firing pattern of motoneurons to infer features of their synaptic drive that cannot otherwise be studied in humans.

The view of motoneurons as the output elements of the nervous system has been the dominant paradigm for interpreting their organization and function for the last 100 years. However, new findings discussed in chapter "Motoneuronal Regulation of Central Pattern Generator and Network Function" show that motoneurons are actively involved in generating and modulating motor behavior. This work has largely focused on the causal role of motoneuron outputs in central pattern generator function across several species. Motoneurons are now believed to contribute directly to the operation of several different central pattern generators including those for quadrupedal locomotion, swimming, and vocalization. The final chapter (chapter "Extraocular Motoneurons and Neurotrophism") in this part deals with a specialized motoneuronal population that regulates eye movements and describes their firing behavior during normal eye movements and during the vestibulo-ocular reflex. This chapter also describes an unusual class of motoneuron that innervates a subset of oculomotor muscle fibers with multiple synaptic contacts along the muscle fiber and highlights the importance of neurotrophic factors in regulating the firing properties and synaptic inputs of oculomotor motoneurons.

The final part of the book deals with motoneuron diseases and highlights the progress being made in understanding the molecular and physiological basis of these diseases, focusing on amyotrophic lateral sclerosis (ALS). Chapter "Motoneuron Diseases" surveys motoneuron diseases by considering the genetic and environmental factors contributing to the disease and the degree to which the pathology can be considered cell autonomous or as the result of interactions of motoneurons with the circuits in which they reside. Chapter "Electrical and Morphological Properties of Developing Motoneurons in Postnatal Mice and Early Abnormalities in SOD1 Transgenic Mice" describes the use of various mouse models of ALS to show that changes in motoneuronal morphology, excitability, and sensorimotor function can be detected very early in development well before the denervation and motor dysfunction that characterizes the disease. The final chapter (chapter "From Physiological Properties to Selective Vulnerability of Motor Units in Amyotrophic Lateral Sclerosis") discusses the selective susceptibility of the different motoneuron types to the disease process and postulates that the early reduced firing of the most susceptible motoneurons leads to alterations in multiple activity-dependent processes that ultimately lead to their demise. Collectively, these new findings indicate that motoneuronal degenerative diseases are not exclusively diseases of the motoneuron because they exhibit pathologies that are present in spinal interneurons, sensory afferents, and glial cells. This broadened perspective offers hope for the development of new therapies for these devastating ailments.

Bethesda, MD, USA

Michael J. O'Donovan
Mélanie Falgairolle

Contents

Part I
Motoneuron Development

Establishing the Molecular and Functional Diversity of Spinal Motoneurons

Jeremy S. Dasen

Abstract Spinal motoneurons are a remarkably diverse class of neurons responsible for facilitating a broad range of motor behaviors and autonomic functions. Studies of motoneuron differentiation have provided fundamental insights into the developmental mechanisms of neuronal diversification, and have illuminated principles of neural fate specification that operate throughout the central nervous system. Because of their relative anatomical simplicity and accessibility, motoneurons have provided a tractable model system to address multiple facets of neural development, including early patterning, neuronal migration, axon guidance, and synaptic specificity. Beyond their roles in providing direct communication between central circuits and muscle, recent studies have revealed that motoneuron subtype-specific programs also play important roles in determining the central connectivity and function of motor circuits. Cross-species comparative analyses have provided novel insights into how evolutionary changes in subtype specification programs may have contributed to adaptive changes in locomotor behaviors. This chapter focusses on the gene regulatory networks governing spinal motoneuron specification, and how studies of spinal motoneurons have informed our understanding of the basic mechanisms of neuronal specification and spinal circuit assembly.

Keywords Motoneuron · Development · Spinal circuit · Neural patterning · Transcription factor

J. S. Dasen (✉)
NYU Neuroscience Institute, Department of Neuroscience and Physiology, NYU School of Medicine, New York, NY, USA
e-mail: Jeremy.Dasen@nyulangone.org

M. J. O'Donovan, M. Falgairolle (eds.), *Vertebrate Motoneurons*, Advances in Neurobiology 28, https://doi.org/10.1007/978-3-031-07167-6_1

1 Introduction

Although making up only a small fraction of all the neurons in the vertebrate central nervous system, motoneurons play crucial roles in facilitating nearly all aspects of animal behavior. Motoneurons within the spinal cord and brainstem are responsible for controlling basic motor functions including walking, breathing, posture, and balance. To accomplish their specific functions, motoneurons acquire specific features during embryonic development, including subtype-specific gene expression profiles, peripheral target specificity, and central connectivity with local and descending premotor networks. Compared to other neuronal classes, motoneurons have been relatively accessible to explore basic mechanisms of neuronal diversification and synaptic specificity. This is in part because the cell bodies of motoneurons targeting a specific peripheral target are organized into anatomically distinct columnar and pool groups. This relatively simple organization has enabled exploration in determining the relationship between the molecular profile and synaptic specificity of specific motoneuron subtypes.

In this chapter I describe studies that have illuminated the genetic programs governing vertebrate spinal motoneuron differentiation, focusing on the operation of gene regulatory networks that contribute to motoneuron subtype diversity, central organization, and connectivity. I will focus predominantly on the role of transcription factors (TFs), as they are primary targets of early signaling that pattern the neural tube, and play key roles in specifying both class- and subtype-specific features of motoneurons. I will describe some of the key targets of transcription factors that determine motoneuron central organization and peripheral target specificity. The evolutionary origins of motoneurons subtypes will be discussed, as well as plausible mechanisms through which these programs have been modified in vertebrates that have adapted specialized locomotor strategies. The role of motoneuron subtype identity in circuit assembly will be explored, particularly in relation to the local connections established between motoneurons, sensory neurons, and spinal interneurons. The overall theme is to highlight how studies on a single neuronal class has enabled multi-scale analyses of the development and function of motor circuits.

2 Anatomical and Functional Diversity of Spinal Motoneurons

At the time of their birth, all motoneurons possess a set of core features that distinguish them from other neuronal classes, but subsequently diversify into dozens of molecularly and anatomically specialized subtypes. Motoneurons subtypes can be differentiated on the basis of their anatomical position, functional properties, specificity of their connections, and molecular profiles. Historically, retrograde labeling of motoneuron through tracer injection into different muscle groups have allowed characterization subtype identities on the basis of their organization within the spinal cord (Jessell et al. 2011; Landmesser 2001). These studies revealed that motoneurons targeting the same or related groups of peripheral targets are organized centrally into discrete clusters termed motor columns and motor pools (Fig. 1). Groups of motoneurons that target the same peripheral target can be further differentiated on the type of muscle fiber they innervate (i.e. fast versus slow, alpha versus gamma).

A major organizational feature of spinal motoneurons is their clustering into columns longitudinally arrayed along the rostrocaudal (head to tail) axis (Landmesser 1978b; Romanes 1941). Neurons within a column innervate a common set of target tissues, which includes both somatic muscles and visceral ganglia (Fig. 1a). Motoneurons targeting limb muscles are referred to as lateral motor column (LMC) neurons, indicating their position relative to medial motor column neuron (MMC) neurons, which target dorsal axial muscle. LMC neurons are generated specifically in brachial and lumbar levels of the spinal cord and innervate forelimb and hindlimb muscles, respectively, while MMC neurons are present at each segmental level. At thoracic levels visceral preganglionic column (PGC) neurons, which derive from the same precursors as somatic motoneurons, innervate sympathetic ganglia (Fig. 1c) (Prasad and Hollyday 1991). Visceral neurons sharing genetic signatures with thoracic PGC neurons are also present in sacral segments (Espinosa-Medina et al. 2016). Motoneurons of the hypaxial motor column (HMC) innervate intercostal and abdominal wall muscles, and are found predominantly in thoracic segments (Gutman et al. 1993; Prasad and Hollyday 1991). In mammals, phrenic motoneurons (PMC) are generated at rostral cervical levels and targets the diaphragm muscle. Each of these columnar groups form within the spinal cord shortly after motoneurons are generated (between E10.5 and E12.5 in mouse), and this basic organizational pattern persists into adulthood.

Additional layers of somatotopic organization of motoneurons are present within a single motor column (Fig. 1b). At brachial and lumbar levels, the LMC segregates into two molecularly and anatomically distinct divisional subtypes: a medial division (LMC_m) which contains motoneurons that project axons ventrally in the limb, and a lateral division (LMC_l) that projects dorsally (Landmesser 1978a; Tosney and Landmesser 1985a, b). Dorsally projecting LMC_l axons typically innervate limb extensor muscles, whereas LMC_m axons innervate flexor muscles. The organization of LMC neurons into columnar and divisional groups is present in all tetrapod classes that have been examined, as well as some species of cartilaginous fish (Jung et al. 2014, 2018).

A third level of motoneuron organization is evident in the diversification of LMC neurons into motor pools (Fig. 1d, e). A motor pool is defined as the population of motoneurons that innervates a single muscle (Hollyday et al. 1977; Landmesser 1978b; Romanes 1942). Motor pools occupy distinct positions within the spinal cord, and are invariant among animals of the same species (Hollyday and Jacobson 1990; Landmesser 1978b). While both brachial and lumbar LMC neurons share a common columnar and divisional organization, the motor pools of these two populations are distinct, reflecting differences in the anatomy of forelimb and hindlimb

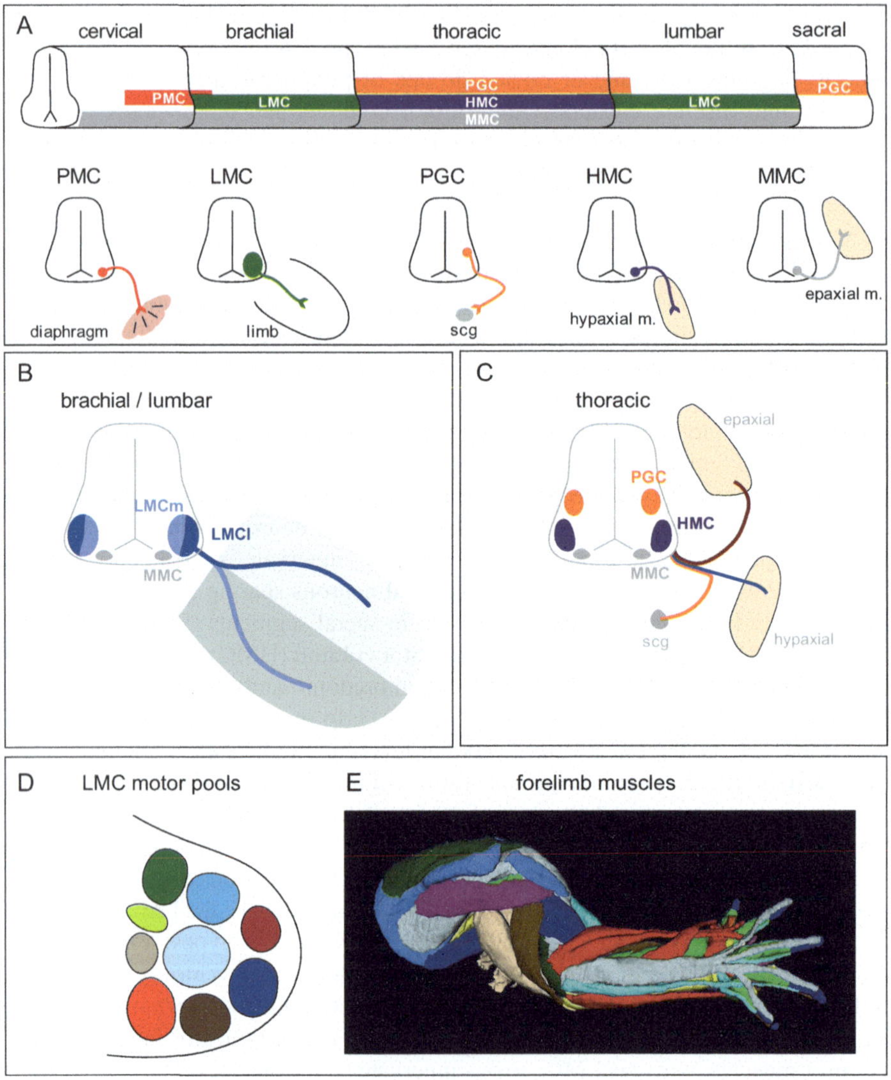

Fig. 1 Anatomical diversity of spinal motoneurons. (**a**) Organization of spinal motor columns and their peripheral targets. Phrenic motor column (PMC) neurons are found at cervical levels and target the diaphragm. Lateral motor column (LMC) neurons are generated at brachial and lumbar levels an innervate limb muscle. Preganglionic motor column (PGC) neurons are located in thoracic and sacral segments and target sympathetic chain ganglia (scg). Hypaxial motor column (HMC) neurons target ventral hypaxial muscle (e.g. intercostal and abdominal muscles). Medial motor column (MMC) neurons target dorsal epaxial muscles. (**b**) Motoneuron columnar organization at limb levels. Divisional identities of LMC neurons and initial trajectories are shown. LMC_l neurons target the dorsal limb compartment and LMC_m neurons project ventrally. (**c**) Organization of motor columns in thoracic segments. Motor axon projections of PGC, HMC, and MMC neurons are shown. PGC neurons are also observed in cervical, rostral lumbar and at lumbar segment L6. (**d**) Motor pool organization. Each motor pool targets a specific limb muscle. (**e**) Forelimb muscles of an E12.5 mouse embryo. (Credit: Muscles and tendons of mouse forelimb. NIMR, MRC. Attribution-NonCommercial 4.0 International (CC BY-NC 4.0))

muscles (Hollyday 1980; Hollyday and Jacobson 1990). Nevertheless, motor pools of both brachial and lumbar LMC neurons are somatotopically organized. At both segmental levels, motor pools positioned rostrally in the LMC typically innervate rostral and proximal limb muscles while caudal motor pools project to more caudal and distal muscles (Catela et al. 2016; Hollyday and Jacobson 1990; Landmesser 1978b; Vanderhorst and Holstege 1997).

Why motoneurons organize into defined columnar and pool clusters is not well understood. As discuss later, one argument in the formation of motoneuron topographic maps is to facilitate connections between motoneuron subtypes with premotor interneurons and sensory neurons (Jessell et al. 2011). Motoneuron columnar and pool organization may also facilitate the coordinated rhythmic firing of motoneurons, through gap junctions that are transiently formed during embryogenesis (Chang et al. 1999; Fulton et al. 1980). This rhythmic bursting activity of motoneurons has been shown to be important during the selection of peripheral trajectories by motor axons (Hanson and Landmesser 2004, 2006)

Additional layers of motoneuron diversification are also present within a single motor pool (reviewed in Kanning et al. 2010). Motoneurons targeting skeletal muscle can synapse on either extrafusal fibers (alpha-motoneuron), intrafusal fibers (gamma-motoneuron), or both (beta-motoneuron). Alpha motoneurons innervate extrafusal muscle fibers, have large cell bodies, and are the main drivers of muscle contraction. Gamma motoneurons innervate intrafusal fibers associated with muscle spindles, while beta-motoneuron target both intrafusal and extrafusal fibers. Alpha-motoneurons can be further classified based on whether they innervate fast-fatigable, fast fatigue-resistant, or slow-fatigable muscle fiber types. The appearance of these functional characteristics of motoneurons appears to emerge in later development, and multiple functional subtypes can be present within a single motoneuron pool. Thus, while motoneurons represent only a small fraction of neurons within the CNS, they display remarkable degree of anatomical and functional diversity. The diversification of motoneurons is governed by genetic programs deployed during embryonic development.

3 Specification of Spinal Motoneuron Class Identity

Like many neuronal classes, motoneurons acquire their fates as a consequence of developmental programs operating during the early stages of embryonic development. Shortly after neural induction, when ectoderm acquires a neuronal fate, the neural plate closes to form the neural tube. Gradients of secreted signaling molecules act on progenitors within the neural tube to specify positional fates along both the dorsoventral and rostrocaudal axes. Along the dorsoventral axis neural progenitors acquire unique molecular identities through the actions of multiple secreted patterning molecules (morphogens). Sonic hedgehog (Shh) is secreted ventrally from the notochord and floor plate, while bone morphogenetic protein (BMP) and wingless (Wnt) signaling originates from the dorsal roof plate and surface ectoderm

(Fig. 2a) (Jessell 2000; Sagner and Briscoe 2019). Graded Shh signaling is essential for the specification of motoneuron progenitors, as well as multiple classes of ventral spinal interneurons. RA from the paraxial mesoderm, and fibroblast growth factors (FGFs) from the caudal mesoderm and tail bud also contribute to the specification of motoneuron progenitors (Novitch et al. 2003). These spatially-graded morphogens establish unique populations of progenitors in the ventral spinal cord through inducing specific patterns of transcription factor expression (Fig. 2a). Molecularly distinct progenitor domains give to a single or multiple classes of postmitotic neurons, including four interneuron classes (termed V0, V1, V2, V3) and spinal motoneurons.

Shh acts via the Gli (glioma-associated oncogene) family of transcription factors to regulate the expression profile of two classes of homeodomain TFs, termed Class I and Class II (Shirasaki and Pfaff 2002). Class II proteins are induced in progenitors by differing levels of Shh, while Class I TFs are repressed by Shh and expressed more dorsally (Briscoe et al. 2000). Gradients of Shh along the dorsoverntal axis are

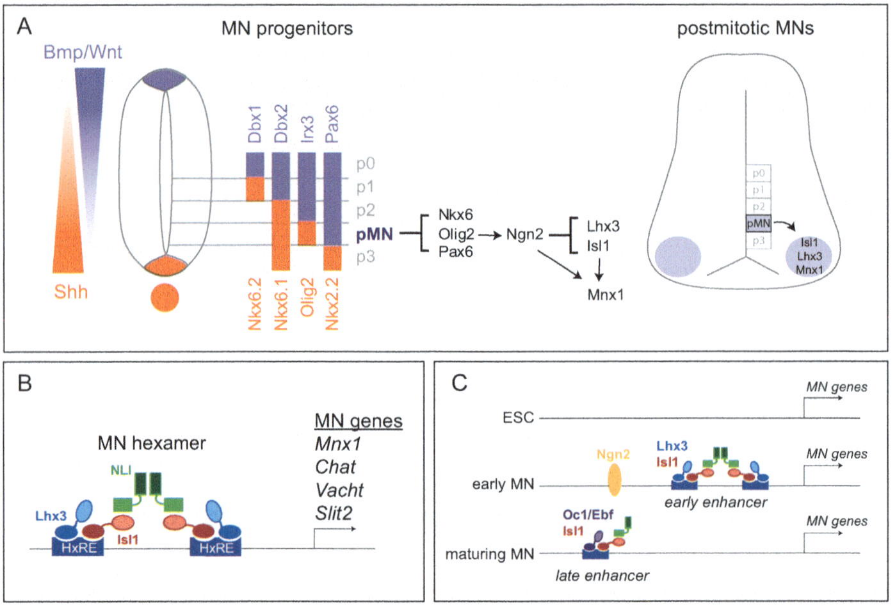

Fig. 2 Specification of core molecular feature of spinal motoneurons. (**a**) Early patterning of neural tube progenitors and motoneuron specification. BMP/Wnt and Shh gradients induce transcription factors in the neural tube. Motoneuron progenitors (pMN) and four ventral interneuron classes (p0, p1, p2, and p3) are shown. Olig2 promotes maintained expression of Neurogenin2 (Ngn2) which coordinately regulates motoneuron gene expression with Isl1 and Lhx3. Postmitotic motoneurons initially express Isl1, Lhx3, and Mnx1. (**b**) Transcriptional regulation of core motoneuron (MN) genes by Isl1, Lhx3, and NLI. Isl1, Lhx3 and NLI bind to hexamer response elements (HxRE) to regulate core motoneuron genes. (**c**) Enhancer switching during motoneuron maturation. Early motoneuron genes are coordinately regulated by Ngn2 and Isl1/Lhx3, later switching to enhancers containing Isl1/Oc or Isl1/Ebf motifs

interpreted by progenitors, in part, through spatial differences in the level of Gli activator and Gli repressor activity (Ribes and Briscoe 2009). In the absence of Shh, Gli repressor activity dominates, while Shh promotes ventral fates through promoting Gli activator function.

The initial patterns of Class I and Class II proteins are further fine-tuned and maintained through selective cross-repressive interaction between pairs of transcription factors (Balaskas et al. 2012). A consequence of these intrinsic transcriptional repressive networks is to establish sharp boundaries between progenitor domains and ensure the production of unique classes of postmitotic neurons (Fig. 2a). Spinal motoneuron progenitors (pMNs) are defined by the domain of Olig2, Pax6, and Nkx6.1 co-expression. Because cross-repressive interactions are critical in establishing the boundaries between progenitor domains, loss of one transcription factor often causes factors expressed in adjacent domains to become derepressed, leading to a switch in neuronal class fates (Ericson et al. 1997; Lanuza et al. 2004; Vallstedt et al. 2001). For example, mutation in the *Olig2* gene in mice leads to a loss of spinal motoneurons and a ventral expansion of V2 interneurons (Zhou and Anderson 2002).

Transcription factors expressed by progenitors play a critical role in setting up the regulatory landscape necessary for the emergence of general features of postmitotic motoneurons. Olig2 promotes expression of Neurogenin2 (Ngn2), and both Olig2 and Ngn2 operate to promote distinct aspects of motoneuron maturation (Mizuguchi et al. 2001; Novitch et al. 2001). Olig2 functions predominantly as a transcriptional repressor, where it establishes the pMN boundary. Olig2 also indirectly upregulates Ngn2 by repressing a repressor of *Ngn2* gene expression (Sagner et al. 2018). Ngn2 interacts with retinoic acid receptors (RARs) and histone modifying complexes to establish the chromatin environment critical for activation of pan-neuronal as well as motoneuron-restricted genes (Lee et al. 2009). Ngn2 also synergizes with the postmitotic motoneuron determinants Lhx3 and Isl1 to promote expression of motoneuron-restricted genes, including the transcription factor Mnx1 (also known as Hb9) (Lee and Pfaff 2003; Ma et al. 2008). Olig2 can counteract the function of Ngn2 in promoting motoneuron differentiation, suggesting the relative balance between Olig2 and Ngn2 expression serves as a gate for timing the activation of motoneuron-specific gene expression (Lee et al. 2005).

Like spinal progenitors, postmitotic neuronal classes can be defined by the specific transcription factors they express and can be further categorized on the basis of their connectivity pattern, neurotransmitter systems, and intrinsic physiological properties (Jessell 2000; Shirasaki and Pfaff 2002). Spinal motoneurons express a set of common early postmitotic determinants (Mnx1, Isl1, and Lhx3), extend axons outside the CNS, are cholinergic, and co-release glutamate (Mentis et al. 2005; Nishimaru et al. 2005). Lhx3, Isl1, Mnx1 play instructive roles in motoneuron specification, as misexpression of these factors, either alone or combination, can direct dorsal spinal interneurons to a motoneuron fate (Tanabe et al. 1998; Thaler et al. 2002). Genetic analyses in mice indicate that Lhx3 (and its paralog Lhx4), Isl1 (and Isl2), and Mnx1 are all essential for multiple aspects of motoneuron differentiation and peripheral connectivity (Arber et al. 1999; Liang et al. 2011; Sharma et al.

1998; Thaler et al. 1999, 2004). Isl1 and Lhx3 bind to the Lim-domain cofactor Lbd1 to promote spinal motoneuron identity by forming a hexameric complex consisting of two Lhx3:Isl1 dimers and NLI proteins (Fig. 2b) (Lee and Pfaff 2003). The interaction of these factors also prevents motoneurons from expressing genes associated with alternate spinal interneuron fates (Lee et al. 2008).

Because motoneurons represent only a small fraction of all the neurons in the spinal cord, it has historically been challenging to use biochemical approaches to determine the sets of genes that are directly regulated by motoneuron-restricted TFs. The ability to direct embryonic stem (ES) cells to a motoneuron fate has been an invaluable tool to determine the molecular mechanisms by which TFs regulate gene expression. ES cells can be differentiated to spinal motoneurons through treating with combinations of Shh agonist and RA, or through direct lineage programming by inducible expression of Ngn2, Lhx3, and Isl1 (Lee et al. 2012; Mazzoni et al. 2013a; Wichterle et al. 2002). Using these approaches, it has been demonstrated that the Lhx3-Isl1 complex directly binds and regulates a battery of genes associated with core motoneuron features, including other class-restricted transcription factors (i.e. Mnx1), cholinergic synthesis pathway genes, axon guidance receptors, and adhesion molecules (Fig. 2b).

Although the core TFs Isl1, Lhx3, and Mnx1 are expressed by all early born spinal motoneurons, they subsequently become restricted to distinct columnar, divisional, and pool subtypes. For example, while Lhx3 is required for the differentiation of all spinal motoneurons, it is subsequently restricted to MMC neurons which innervate dorsal axial muscle (Sharma et al. 1998). Thus, factors that regulate genes essential for motoneuron class identity are only transiently expressed, raising the question of how these programs are maintained after differentiation. Studies in ES-derived motoneurons indicate that gene enhancers occupied by Lhx3 and Isl1 are only transiently engaged during differentiation, subsequently switching to enhancers containing binding sites for Isl1, Onecut, and Ebf1 TFs (Rhee et al. 2016; Velasco et al. 2017). Maintenance of core motoneuron features therefore appears to depend on the switch from early enhancers bound by Lhx3-Isl1 to distinct enhancers that maintain expression in postmitotic motoneurons (Fig. 2c).

4 Early Patterning of Spinal Motoneurons Along the Rostrocaudal Axis

While all motoneurons share certain features, such as expression of genes encoding cholinergic synthesis and release pathways, they diversify into hundreds of subtypes. Motoneuron subtype identities have been traditionally defined by the specificity of their peripheral connections with muscle (Dasen and Jessell 2009; Landmesser 2001). Thus, efforts to define the developmental programs contributing to motoneuron subtype diversity have largely focused on the molecular profiles of neurons targeting specific muscle groups. The acquisition of muscle-specific fates relies on patterning along the rostrocaudal axis. In addition to their early roles in establishing motoneuron progenitors, RA and FGFs determine the rostrocaudal positional identities of spinal progenitors (Bel-Vialar et al. 2002; Liu et al. 2001). RA and FGFs form opposing concentration gradients, where RA secreted from the paraxial mesoderm and somites acts as a rostralizing signal, while FGFs patterns more caudal levels (Fig. 3a).

RA and FGFs confer rostrocaudal positional identities to motoneurons and other neuronal classes through controlling the temporal and spatial expression of *Hox* family transcription factors (Bel-Vialar et al. 2002; Dasen et al. 2003; Liu et al. 2001). *Hox* genes are a large family of chromosomally arrayed genes comprising 39 genes encoded by 4 clusters (*HoxA*, *HoxB*, *HoxC*, and *HoxD*), most of which are expressed in the spinal cord and hindbrain (Parker and Krumlauf 2020; Philippidou and Dasen 2013). In general, *Hox1-Hox5* paralog group genes are expressed in the hindbrain while *Hox4-Hox13* genes are detected in the spinal cord. Unlike the sequential expression of different sets of TFs in motoneuron progenitors (Olig2, Pax6, Nkx6) and early postmitotic motoneurons (Mnx1, Lhx3, Isl1), *Hox* genes are expressed during both phases of motoneuron differentiation.

Expression of individual *Hox* genes directly correlates with its position within a chromosomal cluster, a principle termed spatial colinearity (Kmita and Duboule 2003). *Hox* genes positioned at the 3′ end of a cluster are expressed at rostral levels, while 5′ genes are expressed at more caudal regions of the neuroaxis (Fig. 3a). The initiation of *Hox* gene expression occurs during axis extension, as stem cell-like populations emerge from the node (the organizer for gastrulation in the vertebrate embryo), and generate neuronal and non-neuronal progenitors. This early rostrocaudal patterning step appears to be initiated prior to neurogenesis (Metzis et al. 2018). Growth of the tail bud is associated with the progressive removal of repressive chromatin marks from *Hox* loci, and the appearance of chromatin marks indicative of gene activation (Soshnikova and Duboule 2009).

RA has an important role in patterning *Hox* expression in the spinal cord where it regulates expression of genes associated with cervical/brachial levels (*Hox4-Hox6* genes), in conjunction with other signaling systems (Liu et al. 2001). RA acts through retinoic acid receptors (RARs) to directly activate *Hox1-Hox5* paralogs (Mazzoni et al. 2013b). FGF signaling has a prominent role in establishing the patterns of *Hox4-Hox10* gene expression in the spinal cord and increasing levels of FGF induces *Hox* genes with a progressively more posterior character (Liu et al. 2001; Peljto et al. 2010). Artificially increasing FGF signaling in vivo induces a rostral shift of *Hox* expression and transforms the identities of brachial motoneuron subtypes to a thoracic fate (Bel-Vialar et al. 2002; Dasen et al. 2003, 2005). The effects of FGF signaling in the neural tube are mediated through Cdx homeodomain proteins, as FGF can induce *Cdx* expression, and Cdx proteins can directly activate expression of caudal *Hox* genes (Bel-Vialar et al. 2002; Mazzoni et al. 2013b). FGFs also works in concert with other signaling systems to orchestrate patterns of Hox expression in the spinal cord. At rostral levels FGF acts with RA to establish expression of *Hox6-Hox8* genes in brachial motoneurons. At more posterior levels, FGFs functions with secreted Growth differentiation factor 11 (Gdf11) to initiate

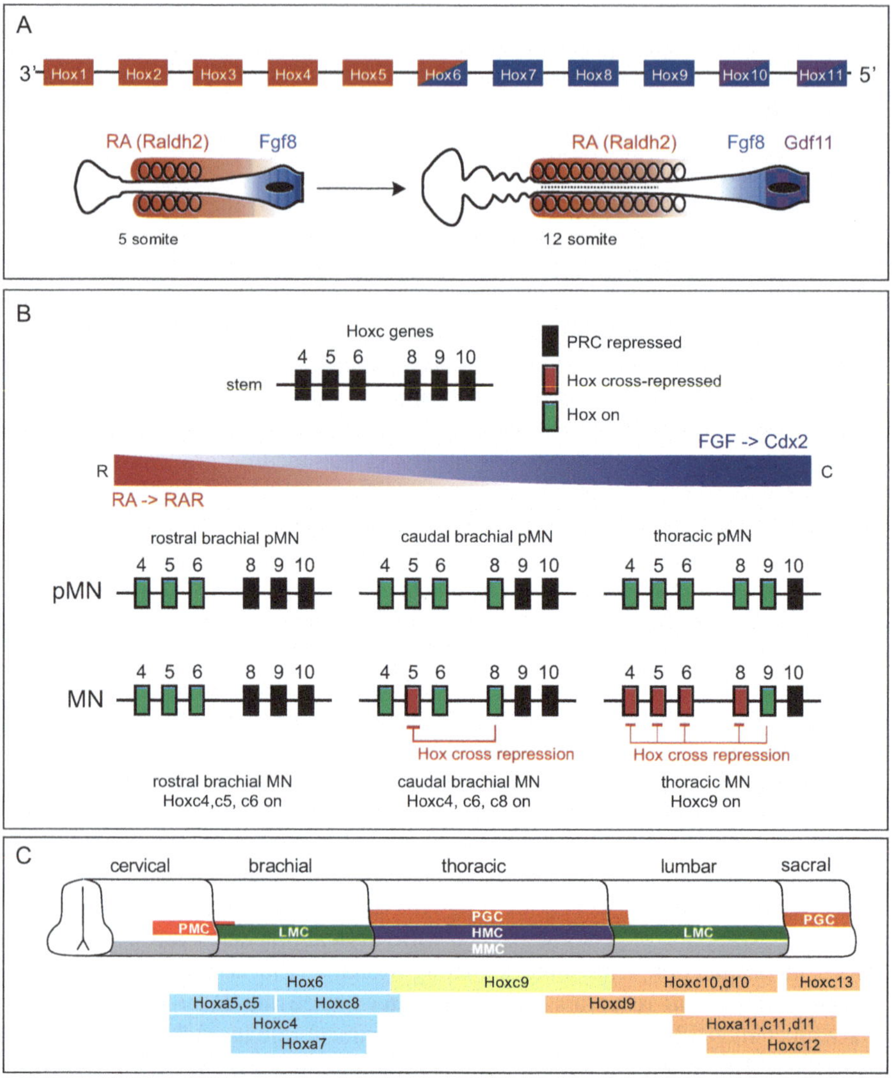

Fig. 3 **Regulation of *Hox* gene expression along the rostrocaudal axis.** (**a**) Patterning of *Hox* gene expression by RA, FGF, and Gdf11. Chromosomally arrayed *Hox* genes are activated by gradients of signaling molecules. Color coding of *Hox* genes denotes paralog groups regulated by indicated morphogens. RA induces *Hox1-Hox5* genes, FGF *Hox6-Hox9*, and Gdf11/FGF8 *Hox10-Hox11* genes. (**b**) Regulation of *Hox* gene expression by Polycomb repressive complexes (PRCs) and Hox cross-repression in brachial and thoracic segments. A subset of *HoxC* cluster genes is shown, which are silenced in stem cells by PRCs (indicated in black). In progenitors, RA (through RAR) and FGFs (through Cdx2) remove PRC marks sequentially from *Hox* genes, activating their expression (indicated in green). In early postmitotic motoneurons, Hox cross-repressive interactions establish caudal *Hox* boundaries (indicated in red). (**c**) Summary of *Hox* expression in motoneurons. Hox patterns and lumbar and sacral segments are based on mRNA expression and have not been confirmed with protein analyses

expression of *Hox10* genes which define lumbar fates (Liu et al. 2001). Wnt signaling also functions in rostrocaudal patterning and modulates the responsiveness of progenitors to RA and FGF (Nordstrom et al. 2006). Subsequent to these early patterning events, the rostral boundaries of *Hox* gene expression are maintained through differentiation by the action of the Polycomb group family of repressor proteins (Fig. 3b) (Golden and Dasen 2012; Mazzoni et al. 2013b; Narendra et al. 2015).

The pattern of *Hox* genes induced by morphogens in spinal progenitors is characterized by specific rostral boundaries, with caudal boundaries that often extend to the tail bud. Thus, there is initially extensive overlap in *Hox* mRNA expression in caudal neural progenitors. By the time of motoneuron differentiation posterior boundaries become apparent. These caudal boundaries are established through cross-repressive interaction between *Hox* genes (Fig. 3b) (Dasen et al. 2003, 2005). The molecular mechanisms mediating Hox cross-repressive interactions have been studied in detail for the Hoxc9 protein, which is expressed by thoracic motoneurons where it represses brachial *Hox4-Hox8* genes (Jung et al. 2010). In *Hoxc9* mutants, *Hox4-Hox8* paralog genes are derepressed at thoracic levels leading a transformation of thoracic motoneurons to a brachial LMC fate (Jung et al. 2010). In addition, within a specific segmental level, cross-repressive interactions amongst *Hox* genes contribute to the diversification of motor pools (Catela et al. 2016; Dasen et al. 2005).

Each of the segmentally-restricted motoneuron columnar subtypes can be defined by a specific pattern of postmitotic Hox expression (Fig. 3c). At cervical levels, Hox5 proteins are expressed by phrenic motor column neurons (Philippidou et al. 2012). Hox6 and Hox10 proteins mark brachial (forelimb-innervating) and lumbar (hindlimb) LMC neurons, respectively (Dasen et al. 2003, 2008; Lacombe et al. 2013; Rousso et al. 2008; Shah et al. 2004; Wu et al. 2008). Intervening limb-level LMC neurons, the *Hoxc9* gene is essential for the appearance of thoracic neuronal fates, including preganglionic column (PGC) neurons (Jung et al. 2010). Expression of an additional two dozen *Hox* genes within the LMC defines the molecular identity of motor pools (Dasen et al. 2005). The next section describes the roles Hox proteins in determining the molecular profiles and peripheral target specificity of motoneuron subtypes.

5 Developmental Mechanisms of Motoneuron Diversification

Patterning systems along the rostrocaudal axis act to define the expression profiles of *Hox* genes in motoneurons. While *Hox* gene expression is initiated in progenitors, they continue to be expressed by early postmitotic motoneurons, where they play important roles in generating segmentally-restricted subtypes. Although how dorsoventral and rostrocaudal patterning systems are integrated during development is poorly understood, it appears that Hox proteins operate in conjunction with early motoneuron determinants (e.g. Mnx1, Isl1) to generate segment-specific subtypes. In parallel with these programs, Hox-independent pathways are also deployed to specify the identities of motoneuron subtypes innervating axial muscle.

Hox Genes and the Specification of LMC Neurons and Divisional Subtypes

Because of their critical roles in facilitating limb movement, the specification of LMC neurons has been thoroughly examined, both in terms of the genetic programs that specify their subtype identities and mechanisms of muscle-target specificity (Fig. 4a). As with other neuronal classes, the diversification of LMC neurons involves a hierarchical program that progressively refines the molecular profiles of motoneurons and shapes the specificity of limb innervation. Early in development, Hox transcription factors ensure that LMC neurons are generated in register with the developing limb buds. As LMC neurons further mature, a combination of both motoneuron-intrinsic and limb-derived cues operate to generate the diversity of

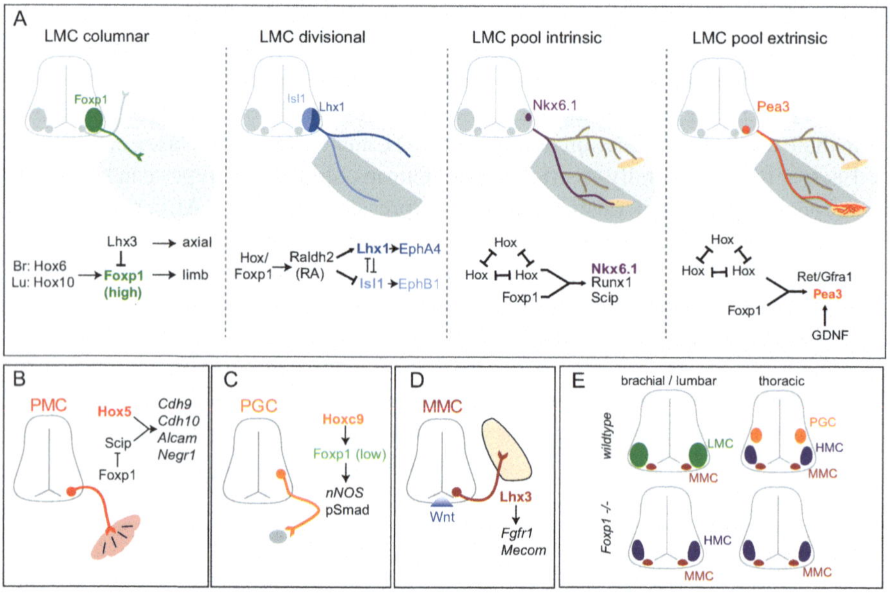

Fig. 4 Diversification of spinal motoneuron subtype identities. (**a**) Transcriptional regulation of LMC diversification. Hox proteins expressed at limb levels promote high expression of Foxp1. Brachial (Br) LMC neurons express *Hox6* genes, while lumbar (Lu) LMC neurons express *Hox10* genes. Foxp1 is required for Raldh2 expression, which contributes to the specification of LMC divisional identities. Isl1 and Lhx1 promote expression of Eph receptors within LMC divisions. Hox proteins specify motor pool fates by regulating expression of downstream transcription factors (e.g. Nkx6.1, Scip, Runx1). Limb-derived cues are also required to induce other pool-specific factors such as Pea3 and Etv1. (**b**) PMC development. Hox5 proteins act in conjunction with Scip to promote PMC-restricted gene expression. Foxp1 represses Scip expression. (**c**) PGC specification. Hoxc9 promotes a low level of Foxp1 expression. Foxp1 is required for the appearance of markers for PGC fate, including nNOS, and pSmad1/5/8. (**d**) MMC neuron development. Wnt signaling promotes maintained expression of Lhx3 in MMC neurons. Lhx3 regulates expression of Fgfr1 and the MMC-restricted marker Mecom. (**e**) Consequences of *Foxp1* mutation. In *Foxp1* mutants presumptive brachial/lumbar LMC and thoracic PGC neurons acquire an HMC identity

LMC subtypes necessary to innervate the dozens of muscles present in a typical tetrapod limb.

Hox proteins confer subtype-specific features of motoneurons, in part, through regulating expression of other transcription factors. A key target of Hox proteins in spinal motoneurons is the transcription factor *Foxp1*, which is expressed by LMC neurons, and at reduced level by thoracic and sacral PGC neurons (Dasen et al. 2008; Espinosa-Medina et al. 2016; Rousso et al. 2008). The pattern of Foxp1 in motoneurons is determined by Hox proteins expressed at specific segmental levels. Hoxc6 and Hoxc10 promote high levels of Foxp1 in brachial and lumbar LMC neurons, respectively, while Hoxc9 sets low Foxp1 levels in thoracic PGC neurons. Misexpression studies in chick embryos indicate that Hoxc6 and Hoxd10 can convert thoracic motoneurons to an LMC fate, as measured by ectopic expression of high Foxp1 levels and other LMC molecular markers (Dasen et al. 2008). Hox proteins therefore appear to act in the early phases on LMC differentiation to induce expression of genes that regulate subsequent steps of diversification.

While expression of individual *Hox* genes correlates with the positioning brachial and lumbar LMC neurons, there is significant redundancy among *Hox* genes during LMC specification. Although *Hox6* genes are expressed by the majority of brachial LMC neurons, in mice lacking all *Hox6* paralog genes (*Hoxc6*, *Hoxa6,* and *Hoxb6*) LMC neurons are still generated, although in reduced numbers (Lacombe et al. 2013). Misexpression studies in chick indicate that multiple *Hox5-Hox8* paralogs can confer an LMC identity to thoracic motoneurons, suggesting LMC specification involves the activities of multiple redundant Hox inputs. Consistent with this idea, combined mutation of the *HoxA* and *HoxC* gene clusters abolishes the specification of brachial LMC neurons (Jung et al. 2014). Thus, the early columnar identity of LMC neurons is determined through redundant Hox inputs that regulate the pattern of Foxp1 expression.

Foxp1 is essential for specifying nearly all early molecular features of LMC neurons. Genetic depletion of *Foxp1* in mice does not preclude the expression of general motoneuron features, such as expression of Mnx1 and cholinergic synthesis pathway genes, or affect the pattern of Hox expression (Dasen et al. 2008; Rousso et al. 2008). In *Foxp1* mutants, prospective LMC neurons fail to express early embryonic markers of motoneuron columnar, divisional and pool identities. In the absence of Foxp1, presumptive LMC and PGC neurons express markers of ventrally-projecting HMC neurons (Fig. 4c). These observations suggest that the HMC subtype represents an ancestral population from which LMC neurons and other segmentally-restricted subtypes evolved.

Foxp1 is also required for the expression of the retinoic acid synthetic enzyme Raldh2, which is expressed by brachial and lumbar LMC neurons. Raldh2 expression in motoneurons creates a local neuronal source of RA, and is involved in specifying LMC divisional identities (Sockanathan and Jessell 1998). Neurons within the LMC differentiate into to a lateral division (LMC_l) which targets muscle in the dorsal limb compartment, and a medial division (LMC_m) that targets ventral muscles (Fig. 4a). Each of these divisional groups are characterized by selective expression of Lim homeodomain (HD) proteins: LMC_l neurons express Lhx1, while

LMC_m neurons express Isl1 (Tsuchida et al. 1994). Expression of Lhx1 in LMC_l neurons is regulated by local RA signaling from Raldh2$^+$ LMC neurons, as well as RA from the paraxial mesoderm, and is consolidated through cross-repressive interactions between Lhx1 and Isl1 (Ji et al. 2006; Kania et al. 2000; Sockanathan and Jessell 1998). The activities of Lim HD factors in LMC_l and LMC_m also play important roles in determining the selection of axonal trajectories within the limb. Lhx1 promotes the dorsal projection of LMC_l neurons through regulating expression of the guidance receptor EphA4, which repels axons from ephrin-expressing ventral limb mesenchyme (Eberhart et al. 2002; Kania and Jessell 2003). Lim HD protein mediated regulation of Eph receptor/ephrin interactions are also involved in determining the ventral projection of LMC_m axons (Luria et al. 2008).

Intrinsic Programs of Motoneuron Pool Specification

Despite the converging actions of multiple redundant *Hox* genes to specify LMC fate, individual *Hox* genes are required to diversify LMC neurons into motor pools targeting specific limb muscles. Of the 39 *Hox* genes present in mammalian genomes, approximately half are expressed by spinal motoneurons, with the majority being expressed in LMC neurons (Dasen et al. 2005; Lacombe et al. 2013). Brachial LMC neurons express *Hox4-Hox9* paralogs, while lumbar LMC neurons express *Hox9-Hox12* genes. The profile of Hox expression in LMC motor pools is established through repressive interactions between *Hox* genes in motoneurons, contributing to both their rostrocaudal and intrasegmental organization. For example, repressive interactions between Hoxc5 and Hoxc8 establish the boundary between rostral and caudal brachial LMC motor pools, while a network of interactions between Hox4-Hox9 paralogs shape motor pool intrasegmental diversity (Catela et al. 2016; Dasen et al. 2005; Mendelsohn et al. 2017).

As with the specification of motoneuron columnar identity, Hox proteins determine motor pool diversity through regulating expression of downstream transcription factors. For example, the ETS domain TF Pea3 (also known as Etv4), the Pou domain protein Scip (Pou3f1), and Nkx6.1 are all expressed by pools of LMC neurons targeting specific limb muscles. Expression of these pool-restricted factors requires specific combinations of Hox proteins. Both Hoxc8 and Hoxc6 are required for the specification of the brachial motor pools defined by Pea3 expression, while only Hoxc8 is required for expression of Scip. In both *Hoxc6* and *Hoxc8* mutants there is a depletion in Pea3+ motoneurons and a severe reduction in the arborization of the muscle normally targeted by this pool (Catela et al. 2016; Lacombe et al. 2013; Tiret et al. 1998). Analyses of mice lacking the motor pool restricted TF Nkx6.1, demonstrate that downstream targets of Hox proteins are critical for muscle target specificity (De Marco Garcia and Jessell 2008). The motor pool-restricted expression of *Pea3*, *Scip*, and *Nkx6.1* is also lost in *Foxp1* mutants, indicating that Hox factors work in conjunction with Foxp1 to specify the molecular identities of motor pools. Foxp1 therefore acts both as a target and an accessory factor of Hox proteins to specify motoneuron pool fates.

Collectively, studies of *Hox* gene function indicate varying degrees of redundancy at the level of motoneuron columnar identity, but highly specific roles during motor pool specification. Multiple *Hox* genes promote the specification of LMC identity, through regulation of *Foxp1* expression, while individual *Hox* genes specify the molecular profiles and connectivity of motor pools. Because *Hox* expression is dynamic during the course of motoneuron pool differentiation, one possible purpose of *Hox* redundancy is to ensure LMC neurons maintain *Foxp1* expression in circumstances where individual *Hox* genes are downregulated. For example, in caudal cervical LMC neurons expression of Hoxc6 is downregulated in a subset of $Foxp1^+$ LMC neurons, while expression of Hoxc8 is maintained.

Target-Dependent Regulation of Motoneuron Pool Identities

While early features motor pool identity are controlled through cell-intrinsic Hox networks, the maturation of motoneurons also depends on cues from peripheral targets. Expression of the ETS transcription factors Pea3 and Etv1 in LMC pools depends on neurotrophic signals provided by the limb mesenchyme and muscle (Haase et al. 2002; Lin et al. 1998). These signals appear to be permissive rather than instructive and not all motoneurons are competent to respond to neurotrophin signaling. In explants of spinal cord treated with glial-derived neurotrophic factor (GDNF), Pea3 is induced in a pattern approximating the normal distribution observed *in vivo* and is confined to the caudal LMC segments that normally express Pea3 (Haase et al. 2002). Ectopic expression of Hoxc8 in rostral LMC neurons can expand the domain of Pea3 expression (Dasen et al. 2005), suggesting that Hoxc8 activity defines the region in which LMC neurons are competent to respond to GDNF. Consistent with this model, in *Hoxc8* mutants motoneurons fail to fully activate Pea3 expression (Catela et al. 2016; Vermot et al. 2005). Hoxc8 promotes the ability of motoneurons to respond to GDNF by regulating expression of the Ret and Gfra1 receptors, which are required to transduce GDNF signaling (Catela et al. 2016).

These observations indicate that target-derived cues contribute to the programming of motor pool fates. Expression of the target-induced factor Pea3 is critical for multiple aspects of differentiation, including the clustering of motoneurons into pools, muscle-specific patterns of axonal innervation, and sensory-motor connectivity (Livet et al. 2002; Vrieseling and Arber 2006). Motor pool specification therefore appears to unfold in two main phases: an early Hox/Foxp1-dependent phase that confers aspects of motoneuron molecular identity involved in the selection of target muscle connectivity (Landmesser 2001; Milner and Landmesser 1999), and a later phase, that is associated with ETS gene expression and the clustering of motoneurons within the LMC (Livet et al. 2002; Price et al. 2002).

Development of Phrenic Motor Column (PMC) Neurons

Breathing is an essential motor behavior which relies on rhythmic activation of phrenic motor column (PMC) neurons. PMC neurons are unique to mammals and are located in rostral cervical segments of the spinal cord. As with other segmentally-restricted motoneuron subtypes, PMC neuron development relies on *Hox* function to deploy subtype-specific gene regulatory programs (Fig. 4b). Two Hox5 paralogs, Hoxa5 and Hoxc5 are expressed by PMC neurons, and in the absence of *Hox5* function in mice, animals perish at birth due to respiratory failure (Philippidou et al. 2012). In *Hox5* mutants, motor axons extend to the diaphragm, but fail to arborize within the muscle, and PMC neurons eventually are lost due to programmed cell death. Hox5 proteins appear to act in conjunction with another transcription factor, Scip, to deploy PMC-specific gene programs in motoneurons (Machado et al. 2014; Philippidou et al. 2012).

Studies of *Hox5* gene function in PMC development have also provided insights into the development of spinal respiratory circuits required for breathing. Hox5 proteins are required to establish the stereotypical dendritic organization of PMC neurons, and *Hox5* mutants are characterized by an increased crossing of dendrites to the contralateral spinal cord (Vagnozzi et al. 2020). Hox5 proteins also regulate expression of cadherins in PMC neurons, and loss of cadherin function leads to similar defects in dendritic morphology. *Hox5* mutants are characterized by marked reduction in the number of inhibitory inputs, possibly a consequence of changes in the molecular profiles or dendritic morphology of PMC neurons. This loss of inhibitory inputs onto PMC neurons likely contributes to the respiratory defects observed in mice with reduced *Hox5* function (Vagnozzi et al. 2020).

Development and Diversity of Preganglionic Column (PGC) Neurons

In addition to muscle-innervating somatic motoneurons, the spinal cord contains an additional class of neurons that derive from the same progenitor domain as motoneurons and target peripheral ganglia in the sympathetic nervous system (Fritzsch et al. 2017). Although not directly involved in skeletal muscle contraction, preganglionic column (PGC) neurons share multiple features with somatic motoneurons, including expression of Lim HD factors, cholinergic synthesis genes, and projection of axons outside the CNS. Spinal PGC neurons are located predominantly in thoracic, rostral lumbar, and sacral segments and innervate sympathetic chain ganglia. PGC neurons are born in the ventral spinal cord near somatic motoneurons, migrating to a dorsomedial position in chick (forming the "Column of Terni"), while in mouse they migrate to a mediolateral position. In both species, embryonic PGC neurons can be distinguished from somatic motoneurons by selective phospho-Smad1/5/8 immunoreactivity. In mice, PGC neurons are also marked by expression of neuronal nitric oxidase synthase (nNOS).

Genetic studies indicate that like LMC neurons, PGC neuron molecular identities are determined by Hox transcription factor activity, which regulates the pattern of Foxp1 expression (Fig. 4c). Mutation in either *Hoxc9* of *Foxp1* leads to a loss of PGC marker expression, and sympathetic chain ganglia lack innervation (Dasen et al. 2008; Jung et al. 2010; Rousso et al. 2008). Hoxc9 functions to generate neurons that express low levels of Foxp1, and manipulations that reduce *Foxp1* levels in vivo can generate ectopic PGC neurons. For example, misexpression of Hoxc9 in brachial LMC neurons, reduces Foxp1 expression to low levels and converts LMC neurons to a PGC fate. Sacral PGC neurons also express *Foxp1* (Espinosa-Medina et al. 2016), and likely depend on sacral-level *Hox* genes for their specification.

While studies of visceral motoneuron specification have focused on a few known molecular markers, more recent studies have revealed that, like LMC neurons, PGC neurons are molecularly highly diverse. Single cell RNA sequencing studies on adult motoneurons have revealed up to 16 distinct subtypes, characterized by selective expression of neuropeptides and other neuromodulatory proteins (Alkaslasi et al. 2021; Blum et al. 2021). Interestingly, these studies also reveal that PGC neurons are not exclusive to thoracic and upper lumbar segments, but are also detected in cervical spinal cord.

Specification Hypaxial (HMC) and Medial Motor Column (MMC) Neurons

In contrast to the Hox-dependent programs that determine the identities of segmentally-restricted subtypes, motoneurons innervating axial muscles can be present throughout the spinal cord, and do not appear to directly rely on Hox protein function for their specification. Axial motoneurons can be broadly divided into two columnar subtypes, depending on whether they innervate dorsal epaxial or ventral hypaxial muscle. Dorsal epaxial muscles are associated with posture and balance, and are innervated by neurons in the medial motor column (MMC). Hypaxial muscles reside ventrally, include abdominal and intercostal muscles, and are innervated by HMC neurons. Similar to limb motoneurons, both MMC and HMC neurons also appear diversify into muscle-specific pools (Gutman et al. 1993; Smith and Hollyday 1983). Although the genetic programs that specify muscle-specific subtypes of axial motoneurons are not understood, recent studies have begun to define a few markers that may delineate MMC and HMC pools (Catela et al. 2019; Hanley et al. 2016).

MMC and HMC neurons can be distinguished by the absence of Foxp1 expression and maintained expression of factors involved in early motoneuron fate, such as Isl1, Lhx3, and Mnx1 (Dasen et al. 2008; Tsuchida et al. 1994). MMC neurons maintain expression of Lhx3, whereas HMC neurons, like most other columnar subtypes, downregulate Lhx3 as they differentiate. The maintained expression of

Lhx3 in MMC neurons appears to be regulated by Wnt signaling originating from the floorplate (Fig. 4d) (Agalliu et al. 2009). Elevation of Wnt signaling in vivo promotes the generation of MMC neurons at the expense of other motoneuron subtypes, while combined genetic removal of *Wnt* genes depletes MMC numbers. MMC neurons can also be defined expression of the Prdm family transcription factor Mecom, and Mecom expression is regulated by Lhx3 and Mnx1 (Hanley et al. 2016). Although Mecom is one of the few TFs selectively expressed by MMC neurons, its function is currently unknown.

Lhx3 is expressed by the precursors of all spinal motoneurons, and is required for the differentiation of most subtypes, confounding attempts to understand its specific function during MMC development (Sharma et al. 1998). Gain of function studies show that Lhx3 can convert Hox-dependent LMC and PGC neurons to an MMC fate and redirect motor axons dorsally to epaxial muscle. Lhx3 therefore has a dominant and instructive role in determining the identity and trajectories of MMC axons (Sharma et al. 2000). The projection of MMC neurons has been shown to depend on target-derived chemoattractant signaling emanating from a somite-derived structure, the dermomyotome (Tosney 1987, 1988). The dermamyotome expresses FGFs which act on the axons of MMC neurons, which selectively express FGF receptor 1 (Fgfr1) (Shirasaki et al. 2006). Mutation of *Ffgr1* in mice causes defects in the peripheral trajectory of MMC axons. In addition, misexpression of Lhx3 leads to ectopic expression of *Fgfr1* in non-MMC motoneurons and causes limb motor axons to gain sensitivity to FGFs (Shirasaki et al. 2006).

Studies in ES cell-derived motoneurons suggest that additional signaling pathways act in conjunction with Wnt signaling to promote MMC specification (Tan et al. 2016). Inhibition of Notch signaling in ES-cell derived motoneurons promotes the specification of HMC neurons at the expense of MMC neurons, suggesting that Wnt4/5 and Notch cooperate to specify MMC identity. These observations also raise the possibility that signaling between newly born motoneuron may play a role in subtype differentiation.

In contrast to MMC neurons, little is known about the developmental programs that specify HMC fate. Although trunk-level HMC neurons can be defined by expression of specific transcription factor combinations (Mnx1^{+}, Isl1/2^{+}, Lhx3^{-}, Foxp1^{-}), the intrinsic determinants of HMC identity are currently unknown. There are only a few known genetic manipulations that disrupt the specification of HMC neurons *in vivo*. For example, mutation of the *Hoxc9* gene, which is expressed broadly in thoracic segments, causes a derepression of multiple *Hox4-Hox8* genes and a conversion of presumptive HMC neurons to an LMC fate. By contrast, in *Foxp1* mutants the number of HMC expands throughout the spinal cord. It is possible that HMC neurons may represent a motoneuron ground state, through which instructive factors like Lhx3 and Hox proteins can specify subtype identities.

Establishing Motoneuron Functional Subtype Diversity

While studies of motoneuron diversification have largely focused on the developmental programs involved in establishing subtypes defined by their peripheral targets, motoneurons also acquire additional features related to the type of muscle fiber innervated (Kanning et al. 2010; Stifani 2014). Motoneurons can be classified as alpha, beta, or gamma depending on their innervation of intrafusal versus extrafusal fibers. Alpha motoneurons can also be classified as fast or slow, depending on the type muscle fiber a motoneuron innervates. Thus, even within a single motor pool, there are additional layers of diversification that contribute to motoneuron functional features.

Studies have revealed some of the pathways and molecular signatures that specify motoneuron functional subtype identities. Gamma motoneurons are marked by expression of *Wnt7a*, and expression of this marker depends on the presence of muscle spindles (Ashrafi et al. 2012). This observation suggests spindle-derived cues are involved in gamma motoneuron specification. The delta-like homolog Dlk-1 has been shown to promote the biophysical signatures of fast motoneurons, by activating expression of the potassium channel subunit Kcng4 (Muller et al. 2014). A slow muscle-fiber type specific cue also appears to promote the expression of molecular signatures of slow motoneurons over fast (Chakkalakal et al. 2010). Recent molecular profiling studies of adult motoneurons have also revealed additional molecular markers that distinguish between multiple motoneuron functional subtypes (Alkaslasi et al. 2021; Blum et al. 2021), opening up a means to resolve the developmental mechanisms through which these features are specified.

6 Evolution of Spinal Motoneuron Organization and Function

Animals exhibit a remarkable diversity in the types of motor behaviors they display. Evolutionary changes in motor behaviors allows animal to optimize the means through which they seek out mates, find food, and escape predators. Cross-species comparative analyses of the molecular and anatomical features of motoneuron subtypes can provide insights into how locomotor circuits may have evolved (Fetcho 1992; Jung and Dasen 2015; Murakami and Tanaka 2011). Even within a single species, genetic analyses of transcription factor function can provide insights into the origins of specific neuronal subtypes. For example, analyses of mice lacking *Foxp1* suggest that limb-innervating LMC neurons evolved from an HMC-like precursor that acquired sensitivity to Hox transcription factor activity (Dasen et al. 2008; Rousso et al. 2008).

Origins of Tetrapod Limb Motoneurons

The ability to travel on land was a key step in the evolution of vertebrates, and many of the circuit elements necessary for limb-based locomotion are contained within the spinal cord (Arber 2012; Goulding 2009; Grillner 2006; Kiehn 2016). It is thought that tetrapod locomotion evolved through a gradual transformation of a spinal network used to activate axial muscles during undulatory swimming, to a more localized circuitry dedicated to coordinating muscles in the limbs (Grillner and Jessell 2009; Jung and Dasen 2015). This idea is supported by the observation that contemporary representatives of intermediate steps in tetrapod evolution, including amphibians and reptiles, often display locomotor behaviors that are a composite of these two fundamentally distinct forms of locomotion (Chevallier et al. 2008).

However, in certain fish species forward propulsion is driven by oscillatory waves of muscle contraction within the fins, and in some cases, fins can be used to generate bipedal-like locomotor gaits (Holst and Bone 1993; Koester and Spirito 2003; Rosenberger 2001). In the little skate *Leucoraja erinacea,* the pelvic fins are used in locomotor behaviors displaying hallmark features of ambulatory locomotion, including reciprocal extension and flexion of the fins that alternate between the left and right sides of the spinal cord (Lucifora and Vassallo 2002; Macesic and Kajiura 2010). Skates generate this behavior using motoneuron subtypes that are remarkably similar to those found in tetrapods. Both pectoral and pelvic fin-innervating motoneurons of skates express Foxp1, and the Lim HD code essential for selective innervation of flexor and extensor muscles (Jung et al. 2018). Because the common ancestor of skates and tetrapods was the common ancestor to all vertebrates with paired appendages, these observations indicate that LMC neurons originated in ancient fin-bearing species (Fig. 5a).

Hox Genes and the Evolution of Motoneuron Segmental Organization and Diversity

Comparisons of the Hox-regulatory networks between vertebrates displaying different locomotor behaviors can provide mechanistic insights into how nervous systems may have evolved. Snakes evolved from limb-bearing reptiles that have repressed limb development and have reverted to an axial muscle-based form of locomotion (Leal and Cohn 2018). In African house and corn snakes, expression of the LMC determinants Foxp1 and Raldh2 are largely absent from the spinal cord, while axial MMC and HMC neurons are present throughout most spinal segments (Jung et al. 2014). This reversion of motoneurons to an almost entirely axial system is associated with a loss of brachial *Hox* gene expression from motoneurons and an expansion in the domain of *Hoxc9* expression within the spinal cord (Fig. 5b, c). As in other tetrapods, Hoxc9 likely acts in snakes by repressing anterior *Hox* genes and

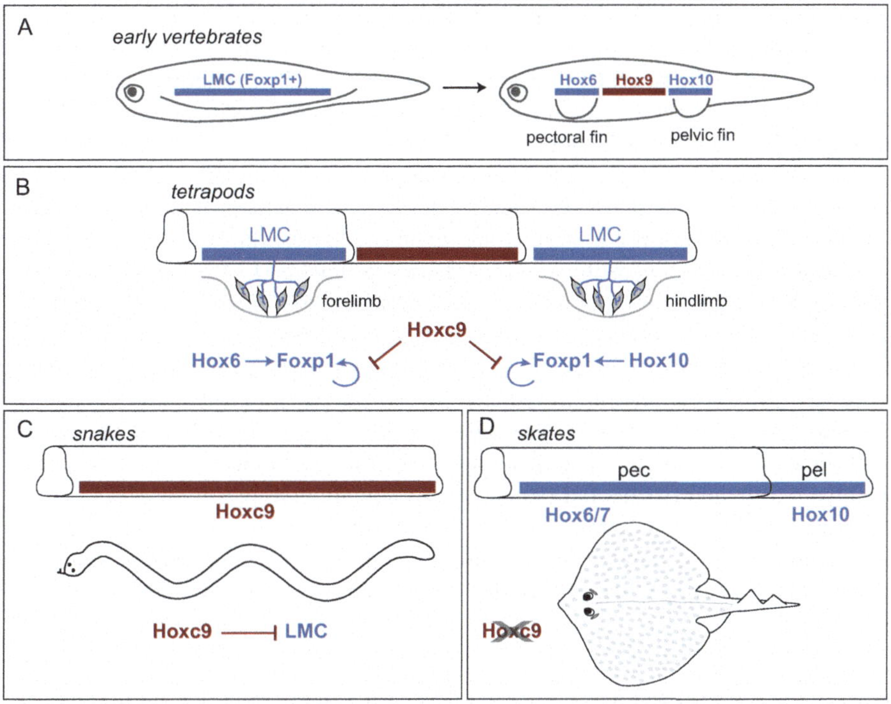

Fig. 5 Evolution of motoneuron diversity. (**a**) Model indicating that ancestral fin-bearing fish had LMC-like neurons, characterized by Foxp1 expression. This population may have extended along the length of the trunk. Hoxc9 activity was involved in generating separate pectoral and pelvic fin-innervating LMC populations, through repressing Foxp1 expression. (**b**) Regulatory interactions between Hox and Foxp1 in tetrapods. Limb level Hox proteins promote a high level of Foxp1 expression, which is maintained through Foxp1 autoregulation. Hoxc9 acts by blocking the ability of Foxp1 to positively autoregulate its own expression. (**c**) Evolutionary modifications in Hox-regulated organization of motor columns. In snakes, which have suppressed limb development programs, expression of Hoxc9 extends throughout most segments, and LMC neurons are not detected. (**d**) In skates, pectoral LMC neurons have expanded, likely as consequence of a natural deletion in the *HoxC* cluster, which removes the *Hoxc9* gene

Foxp1. Interestingly, this repressive activity is present in *Hoxc9* proteins of both terrestrial and aquatic vertebrate species (Jung et al. 2014), suggesting that Hox-dependent patterning is an ancestral strategy through which spinal motoneurons are organized.

In contrast to the loss of LMC neurons in snakes, in the little skate there is an extended domain of LMC neurons, spanning continuously from the pectoral fin-innervating population to the caudal pelvic domain. Skates also have a natural deletion in the *Hoxc9* gene (King et al. 2011), which likely contributed to the expanded domain of pectoral fin-innervating motoneurons (Fig. 5d). The pectoral and pelvic LMC neurons of skate also express Hox proteins that are analogous to those of tetrapods (Fig. 5d). These observations are consistent with a model in which the ancestral LMC population was present throughout the spinal cord, with Hoxc9 activity being important for the generation of separate rostral (pectoral/forelimb) and caudal (pelvic/lumbar) LMC populations (Fig. 5a). Thus, modulation in the pattern of Hox expression is associated with evolutionary changes in motoneuron organization and locomotor behaviors.

The appearance of digits represents an important evolutionary modification of the limb, enabling fine motor skills and dexterity. The genetic programs that specify forelimb digit-innervating motoneurons involves a divergent Hox-dependent program that is deployed near the boundary between brachial LMC and thoracic subtypes. Digit motoneurons can be distinguished from other brachial motor pools by low levels of *Hoxc9* expression, as well as digit motor pool restricted markers (e.g. *Cpne4*, *Fign*) (Mendelsohn et al. 2017). Although high levels Hoxc9 normally represses LMC specification, at reduced levels it appears to only repress Raldh2, allowing for the generation of motoneurons that express Foxp1, and the caudal LMC determinant Hoxc8. Specification of forelimb digit-innervating motoneurons requires Hoxc8 and Hoxc9 and evasion of RA signaling, as elevation of RA signaling represses digit motoneuron specification. Analysis of digit motoneuron specification have revealed how changes in the levels and activities of Hox proteins can contribute to the evolutionary diversification of neuronal subtypes.

Divergence of Axial Motoneuron Specification Programs in Fish

In most species of fish, forward propulsion is governed by motoneurons that innervate axial muscle (Fetcho 1987; Romer and Parsons 1977). Although used for distinct motor behaviors, the axial motoneurons of fish and tetrapods share common early developmental programs, including reliance on Mnx1 and Lim HD proteins for the specification of motoneuron class identity. Despite sharing some common specification programs, zebrafish has diverged significantly from mammals, particularly in the programs specifying the subtype identities of axial muscle-innervating motoneurons.

In contrast to the motoneurons of tetrapods, axial motoneurons of zebrafish lack a clear somatotopic organization that relates cell body position to muscle target specificity. Despite the absence of an obvious somatotopic organization, zebrafish axial motoneurons are functionally organized along the dorsoventral axis. This organization is associated with how motoneurons are recruited at specific swimming speeds. Axial motoneurons active during slow swimming speed reside ventrally in the spinal cord, motoneurons recruited at fast swimming speeds are located dorsally, while motoneurons involved in intermediate speeds are positioned between fast and slow motoneurons (Ampatzis et al. 2013; Liu and Westerfield 1988; McLean et al. 2007).

Unlike mammals and birds, where all motoneurons are generated during a single wave of neurogenesis, zebrafish motoneurons are generated over two waves, termed primary and secondary. Studies of motoneuron specification in zebrafish have largely focused on the four primary motoneuron types: the dorsal rostral primary (dRoP), ventral rostral primary (vRoP), caudal primary (CaP), and middle primary (MiP) neurons. dRoP and MiP motoneurons are similar to MMC neurons, in that they project to muscles located dorsal to the horizontal myoseptum, while CaP and vRoP project ventrally. However, unlike tetrapod MMC and HMC neurons, zebrafish primary motoneuron types cannot be distinguished by differential expression of Lhx3. Nevertheless, disruption of the core determinants Lhx3/4, Isl1/2, and the Mnx1 causes defects in primary motoneuron specification and connectivity. For example, loss of Lhx3/4 leads to motoneurons with hybrid motoneuron/interneuron fates (Seredick et al. 2014), while loss of Mnx proteins affects the specification of MiP motoneurons (Seredick et al. 2012).

Little is known about the specification of the later-born and more numerous secondary motoneurons (Myers et al. 1986). Secondary axial motoneurons are the most diverse, comprising at least 16 distinct subtypes, as classified by electrophysiological properties and morphological features (Bello-Rojas et al. 2019; Menelaou and McLean 2012). Although markers including Isl1, Isl2, and Mnx proteins can differentiate primary motoneuron subtypes, these factors are dynamically expressed and cannot distinguish secondary subtypes throughout development (Appel et al. 1995; Hutchinson et al. 2007; Seredick et al. 2012). Although secondary motoneurons make up the majority of subtypes in zebrafish, and are thought to be more similar to mammalian motoneurons, very little is known about the intrinsic TF codes responsible for their differentiation (Beattie et al. 1997).

Both primary and secondary motoneuron subtypes can be differentiated based on several criteria, such as birthdate, soma size, position, presence or absence of intraspinal or intermyotomal collaterals, and firing properties (Bello-Rojas et al. 2019; Menelaou and McLean 2012). There are three distinct types of firing patterns expressed by embryonic zebrafish axial motoneurons, tonic, chattering, and burst firing. Tonic firing patterns are specific to primary motoneurons, while chattering and burst firing patterns are specific to secondary motoneurons. Each secondary motoneuron subtype has a different distribution of these two firing patterns. While the distinct physiologic and anatomic features of secondary motoneurons have been well-characterized, it is yet unknown whether they reflect the operation of motoneuron-intrinsic genetic programs acting during development. Given that axial motoneurons are organized by functional attributes (fast, intermediate, and slow), as opposed to muscle-target specificity, and are present throughout the rostrocaudal axis, it is unlikely that their diversification relies on Hox-dependent differentiation programs. As with the differentiation of tetrapod MMC and HMC neurons, zebrafish motoneuron patterning may depend on signaling from the floorplate, or through cell-cell interactions between newly born neurons.

7 Assembly of Proprioceptive Sensory-Motor Circuits

Coordinate motor behavior depends on the establishment of central connections between motoneuron subtypes and premotor spinal and supraspinal circuits. While significant advances have been made in deciphering the mechanisms of motoneuron diversification and peripheral target specificity, the programs which determine the central connectivity of motoneurons are less well defined. Recent technological advancements in transsynaptic labeling methods, as well as the ability to selectively manipulate motoneuron subtype identities, have enabled investigation into the mechanisms of central synaptic specificity in spinal motor circuits. These studies provide evidence that motoneurons are not passive participants of motor commands, but can play instructive roles in shaping connectivity and activity of motor networks.

One of the most thoroughly studied circuits within the spinal cord is the monosynaptic stretch-reflex circuit, consisting of a limb muscle, a pool of alpha-motoneurons, and type Ia proprioceptive sensory neurons (pSNs) with stretch-sensing mechanoreceptor endings embedded within muscle spindles (Fig. 6a) (Imai and Yoshida 2018; Tuthill and Azim 2018). Despite its apparent simplicity, each motor pool receives selective inputs from pSNs that target the same peripheral muscle, suggesting a requirement for programs within pSNs to distinguish between different motoneuron subtypes. During development, pSN axons navigate through the spinal cord, preferentially contacting motoneuron pools innervating the same peripheral target, while avoiding motoneurons of functionally antagonist muscles (Eccles et al. 1957; Mears and Frank 1997). These connections are remarkably selective, as a single pSN establishes monosynaptic connections with each of the ~50–300 neurons within a motor pool that supplies the same muscle target (Mendell and Henneman 1968). Connections between pSNs and motoneurons appear to be established independent of patterned neural activity (Mendelsohn et al. 2015; Mendelson and Frank 1991), suggesting pSN-motoneuron matching relies on genetic programs operating during neural development. An important and still largely unanswered question is whether pSNs acquire molecular features that allow them to recognize specific motoneuron subtypes.

Fig. 6 (continued) and have central connections to alpha motoneurons that target the same muscle. A central connection between a flexor motoneuron and pSN is shown. Flexor pSNs synapse onto to flexor motoneurons, but not extensor motoneurons. (**b**) Consequences of *Foxp1* mutation on sensory-motor connections. In *Foxp1* mutants, motor pool organization is disrupted. Individual motoneurons can still target limb muscles their position is disorganized in *Foxp1* mutants. Nevertheless, pSN still project to the same dorsoventral domain in the ventral spinal cord. (**c**) Effect of *Hoxc9* mutation on sensory-motor connections. In *Hoxc9* mutants, ectopic LMC neurons are generated in thoracic segments and innervate hypaxial muscle. In *Hoxc9* mutants, limb-innervating pSNs project to, and form synapses on thoracic LMC neurons. (**d**) Coordinate *Hox* function in proprioceptive sensory-motoneuron connectivity. The same *Hox* genes expressed in motoneurons are selectively expressed by pSNs. Summary of Hoxa5 and Hoxc8 in brachial LMC neurons is shown in the top panels. After conditional deletion of *Hoxc8* from pSNs, flexor pSNs form connections with extensor motoneurons

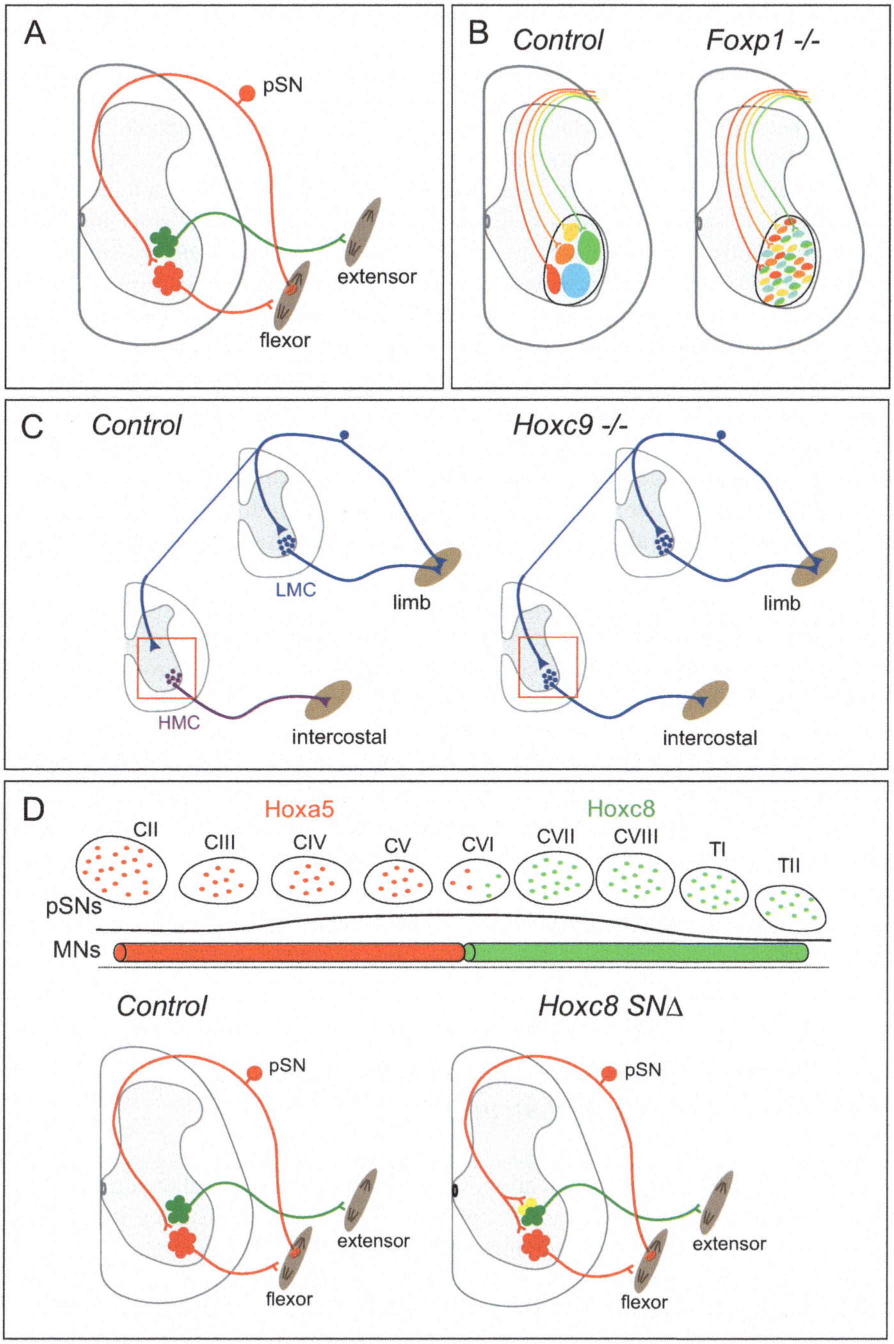

Fig. 6 Assembly of proprioceptive sensory-motor circuits. (**a**) Connections in the monosynaptic stretch reflex circuit. Type Ia proprioceptive sensory neurons (pSNs) innervate muscle spindles

Specification of Proprioceptor Sensory Neuron Class Identities

Similar to the diversification of spinal motoneurons, pSNs advance through hierarchical genetic programs in which expression of specific genes coincides with the acquisition of specialized characteristics, including peripheral target specificity, central projection pattern, and molecular identity (Dasen 2009; Lallemend and Ernfors 2012). Sensory neurons generated at spinal levels derive from migratory neural crest cells which coalesce to form dorsal root ganglia (DRG) (Butler and Bronner 2015). Most sensory neurons co-express the homeodomain transcription factors Isl1 and Brn3a, which are necessary for deployment of pan-sensory neuron genetic programs (Dykes et al. 2011).

Proprioceptive sensory neurons can be discriminated from other DRG sensory neuron classes by expression of the transcription factors Runx3 and Etv1, the neurotrophin receptor *Nrtk3*, and the calcium binding protein Parvalbumin (PV). Genetic studies in mice indicate that *Runx3* and *Etv1* are essential for establishing and maintaining core features of pSN identity, including their survival and ability to extend central axons towards motoneurons (Arber et al. 2000; Inoue et al. 2002).

While the transcriptional programs governing features common to all pSNs have been characterized, understanding later developmental facets of pSN-motoneuron circuit assembly, such as central connectivity with motoneurons, has been particularly challenging. In contrast to the topographic organization of motoneurons, pSNs are intermixed with other DRG sensory classes, with no clear organizational pattern (Honig et al. 1998; Jessell et al. 2011). A dearth of molecular markers for more nuanced neuronal features has made it challenging to characterize how pSNs and other sensory modalities further diversify into specific subtypes. One particular gap in our understanding is how the specificity of central connections between pSNs and motoneurons of the same muscle is achieved, since pSN axons must distinguish between multiple potential postsynaptic targets within the ventral spinal cord.

Target-Derived Signals Regulate pSN Subtype Identity and Connectivity with Motoneurons

In contrast to motoneuron specification, where early developmental features emerge largely independent of peripheral cues, pSN subtype specification relies on extrinsic signals provided by limb mesenchyme and muscle (Arber 2012; Sharma et al. 2020; Wu et al. 2019). Expression of the neurotrophin receptor *Nrtk3* renders pSNs sensitive to peripheral neurotrophin-3 (*Ntf3*) signaling, and both *Nrtk3* and *Ntf3* are essential for the differentiation and survival of pSNs (Chen et al. 2003). *Ntf3*/*Nrtk3* signaling regulates expression of *Etv1* and *Runx3*, and muscle-by-muscle differences in the level of *Ntf3* expression appear to contribute to pSN subtype identities (de Nooij et al. 2013; Patel et al. 2003; Wang et al. 2019). In addition, signals originating from the limb mesenchyme can trigger expression of genes that mark muscle-specific pSN subtypes (Poliak et al. 2016). Although certain features common to all pSNs have been shown to be limb-independent (Chen et al. 2002), a predominant mechanism of pSN fate specification appears to be through muscle-specific cues.

Target-derived signals also regulate aspects of motoneuron pool identity, and can contribute to the specificity of connections between motoneurons and pSNs. Expression of the transcription factor *Pea3* is induced in brachial and lumbar LMC neurons by limb-derived neurotrophins. In the absence of *Pea3* function, pSNs target inappropriate motoneuron subtypes (Livet et al. 2002; Vrieseling and Arber 2006). Pea3 regulates expression of the guidance receptor ligand *Sema3e*, and in *Sema3e* mutants, pSNs target inappropriate motoneurons (Fukuhara et al. 2013; Pecho-Vrieseling et al. 2009). Although target-induced expression of guidance determinants is one strategy for controlling specificity in reflex circuits, they appear to operate in only a limited number of motoneuron pools.

Roles of Motoneuron Subtype Identity in Sensory-Motor Circuit Assembly

While studies of Pea3 function indicate that motoneuron identity can influence sensory -motor connectivity, some aspects appear to be motoneuron-independent. Mutation in the Hox-dependent transcription factor *Foxp1* strips LMC neurons of motor pool-specification programs, and while motoneurons are still able to innervate muscle, motor axons select targets in a stochastic manner (Dasen et al. 2008; Rousso et al. 2008). Nevertheless, pSNs project to the appropriate dorsoventral position within the ventral spinal cord and synapse with motoneurons (Surmeli et al. 2011). These observations suggest that pSNs can project to specific regions within the ventral spinal cord in the absence of motoneuron subtype-specific cues (Fig. 6b). An interpretation of these findings is that one function of motoneuron topographic organization is to ensure the alignment of sensory central afferents with the corresponding motoneuron pools. Additional studies on the connections established between pSNs and spinal interneurons reinforce the idea that target cell position is involved in establishing synaptic specificity in spinal circuits (Bikoff et al. 2016; Tripodi et al. 2011).

Studies of *Foxp1* mutants suggest that motoneuron-derived cues do not play an instructive role in determining the dorsoventral targeting of pSN within the ventral spinal cord. However, because motoneurons are stripped of all LMC-specific features, it leaves open the possibility that pSNs are capable recognizing molecular differences among LMC neurons. In particular, the targeting of a pSN to its appropriate motor pool presumably also relies on the recognition of differences in motoneurons along the rostrocaudal axis. One way to test this would be to extend the rostrocaudal domain of motoneuron pools, and assess the impact on pSN-motoneuron connectivity. Mutation in the *Hoxc9* gene transforms thoracic motoneurons to a limb-level LMC fate, and extends the domain of Hoxc8⁺ motor pools into thoracic segments. In *Hoxc9* mutants, transformed thoracic motoneurons receive input from forelimb-innervating pSNs (Fig. 6c) (Baek et al. 2017). Thus, while the dorsoventral targeting of pSNs in the spinal cord may be motoneuron independent, the selection of termination zones along the rostrocaudal axis appears to depend on recognition of Hox-mediated differences in motoneuron subtype identity.

Evidence for Coordinate Regulation of pSN-Motoneuron Connectivity by Hox Genes

While studies suggest that Hox-dependent programs are essential in motoneurons for aspects of sensory-motor specificity, whether pSNs acquire muscle-specific molecular features similar to those of motoneurons is not fully known. *Hox* expression in pSNs also displays segmentally-restricted patterns, paralleling the rostrocaudal profiles of *Hox* genes in motoneurons (Shin et al. 2020). Proprioceptive sensory neurons may therefore use similar diversification programs as motoneurons. For example, the *Hoxc8* gene is expressed by LMC neurons targeting distal forelimb muscle, and is also expressed by pSNs targeting similar muscle groups (Fig. 6d). Expression of *Hox* genes in sensory neurons also does not appear to require limb-derived cues, as limb ablation experiments in chick do not affect Hox expression in pSNs (Shin et al. 2020). Hox activity in pSNs may therefore operate to provide a limb-independent mechanism of neuronal diversification.

Genetic studies in mice reveal that Hoxc8 activity in pSNs contributes to sensory-motor synaptic specificity. Deletion of *Hoxc8* selectively from SNs does not affect neuronal survival or prohibit the ability of pSNs to target the correct forelimb muscle. However, in *Hoxc8* sensory mutants, flexor pSNs make inappropriate connections with extensor motoneurons (Shin et al. 2020). Similar defects in sensory-motor matching is observed when *Hoxc8* is deleted from motoneurons. These observations suggest that the coordinate activity of *Hox* genes in pSNs and motoneurons, acting in conjunction with muscle-derived cues, contributes to synaptic specify in sensory-motor circuits. The specificity of pSN-motoneuron connections has been shown to correlate with the relative approach angles between pSN central axons and motoneuron dendrites (Balaskas et al. 2019). Sensory axons approach the dendrites of motoneurons targeting the same muscle at small angles, relative to the approach angle to dendrites of other motoneuron subtypes, suggesting the angle of axo-dendritic approach contributes to connection specificity. Loss of pSN-motoneuron alignment may account for the sensory-motor mismatch observed in both motoneuron and pSN-restricted *Hoxc8* mutants.

Centrally, pSNs establish connections with a variety of postsynaptic targets, including local spinal and projection interneurons that relay proprioceptive information to the brain (Bermingham et al. 2001; Bikoff et al. 2016; Koch et al. 2017; Tripodi et al. 2011; Yuengert et al. 2015). The same *Hox* genes expressed by pSNs and motoneurons are also expressed by multiple classes of spinal interneurons, suggesting Hox proteins play a broader role in shaping synaptic specificity within the spinal cord. Consistent with this idea, both long ascending spinocerebellar and local-circuit spinal interneurons require *Hox* function to generate molecularly distinct subtypes at cervical and thoracic levels (Baek et al. 2019; Sweeney et al. 2018). These observations suggest that the same *Hox* gene can act in multiple neuronal classes during development, implying a coherent molecular strategy for wiring the circuits essential for motor control.

8 Motoneurons as Regulators of Circuit Assembly and Function

Dedicated circuits contained within the vertebrate brainstem and spinal cord are capable of generating rhythmic patterns of motoneuron activation essential for basic motor functions such as walking and breathing. In the circuits controlling locomotion, the rhythm and pattern of motoneuron activity is facilitated by central pattern generators (CPGs) composed of multiple classes of excitatory and inhibitory spinal interneurons (Grillner 2006; Grillner and Jessell 2009). In tetrapods, spinal locomotor CPGs coordinate alternating patterns of motoneuron firing across the left and right side of the spinal cord (L-R CPG) as well as reciprocal activation extensor and flexor motoneurons (E-F CPG). The spinal interneurons required for the pattern of CPG output in fictive locomotor preparations are well-defined (Goulding 2009; Kiehn 2016). Left-right alternation relies on V0-class interneurons which cross the midline and inhibit the contralateral CPG (Lanuza et al. 2004; Talpalar et al. 2013). Coordination of extensor-flexor motoneuron activity requires V1 and V2b class inhibitory interneurons (Britz et al. 2015; Zhang et al. 2014). Several additional classes of genetically defined spinal interneurons are essential to determine the robustness of CPG output as well as modulating motoneuron firing at different locomotor speeds (Kiehn 2016).

While the identity and function of the spinal interneurons involved in locomotor CPGs have been elucidated, how these premotor populations engage specific motoneuron subtypes is less well understood. Because a single motoneuron can receive inputs from a variety of presynaptic neuronal classes, and there are dozens of distinct motoneuron subtypes, it has been challenging to determine whether there are rules that govern connectivity between spinal motoneuron subtypes and locomotor CPG interneurons. One possible solution could be through generating interneurons of a similar diversity to that of motoneurons. Evidence in support of this idea has come from recent studies on the diversification of spinal interneuron subtypes, revealing a remarkable degree of molecular heterogeneity within a single interneuron class (Bikoff et al. 2016; Griener et al. 2015; Hayashi et al. 2018).

Evidence suggesting a role for motoneuron subtype identity in the assembly of locomotor circuits has emerged from studies examining the distribution of interneurons that directly synapse onto motoneurons. The development of transynaptic tracing methods has enabled the mapping the distribution of inputs onto motoneurons (Esposito et al. 2014; Stepien et al. 2010). Analyses of the patterns of premotor connectivity have revealed marked differences in the types and distribution of interneurons that motoneuron subtypes engage. LMC neurons receive inputs that are largely biased towards spinal inhibitory interneurons localized ipsilaterally (Fig. 7a) (Goetz et al. 2015). In contrast, motoneurons projecting to axial muscles, including those located in the MMC and HMC, receive inputs that are evenly distributed across both sides the spinal cord, with a bias towards contralateral populations. Although interpretation of these findings is constrained by the inherent inefficiency of transsynaptic labeling methods, these observations suggest there are differences in the distribution of premotor interneurons that specific motoneuron columnar subtypes are connected with.

Differences in the distribution of premotor interneurons are also for observed for inputs onto motor pools that occupy specific regions within the LMC. Motoneurons situated ventromedially receive a greater proportion of inputs from contralateral premotor interneurons than motoneurons projecting to distal limb muscles, which are dorsolaterally positioned (Goetz et al. 2015). A suggested possible mechanism for these biases is through the configuration of motoneuron dendrites. MMC neurons, which are positioned medially, have dendrites which extend across the midline and can captures a greater proportion of input originating from contralateral interneuron populations. Recent studies also suggest a bias in the types of local interneurons that synapse onto motoneurons projecting to extensor and flexor muscles, which could similarly reflect differences in motoneuron position and/or dendritic morphology (Britz et al. 2015). Given that motor pool-restricted factors such as *Pea3* are known influence the pattern of motoneuron dendritic morphology (Vrieseling and Arber 2006), inputs from premotor interneurons could be shaped by both the positional and architectural features of motoneurons.

Motoneuron subtype identity has also been shown to influence activity of locomotor CPGs. The role of motoneurons subtype identity has been examined using mouse mutant lines in which motoneuron identities have been genetically transformed (Hinckley et al. 2015; Machado et al. 2015). Removal of *Foxp1* from motoneurons leads to an LMC to HMC transformation, but these motoneurons still target limb muscles. In motoneuron-specific *Foxp1* mutants, basic features of the CPG are retained, including normal rhythmic bursting patterns and left-right alternation. By contrast the output pattern from the E-F CPG is disrupted, reverting to a predominantly flexor-like state (Hinckley et al. 2015; Machado et al. 2015). In *Foxp1* mutants, transformed LMC populations are also shifted to a more ventromedial position. This raised the possibility that the defects in *Foxp1* mutants are due to changes in premotor connectivity as a consequence of altered motoneuron position. This idea was tested by misexpressing Lhx3 neurons in all motoneurons, a

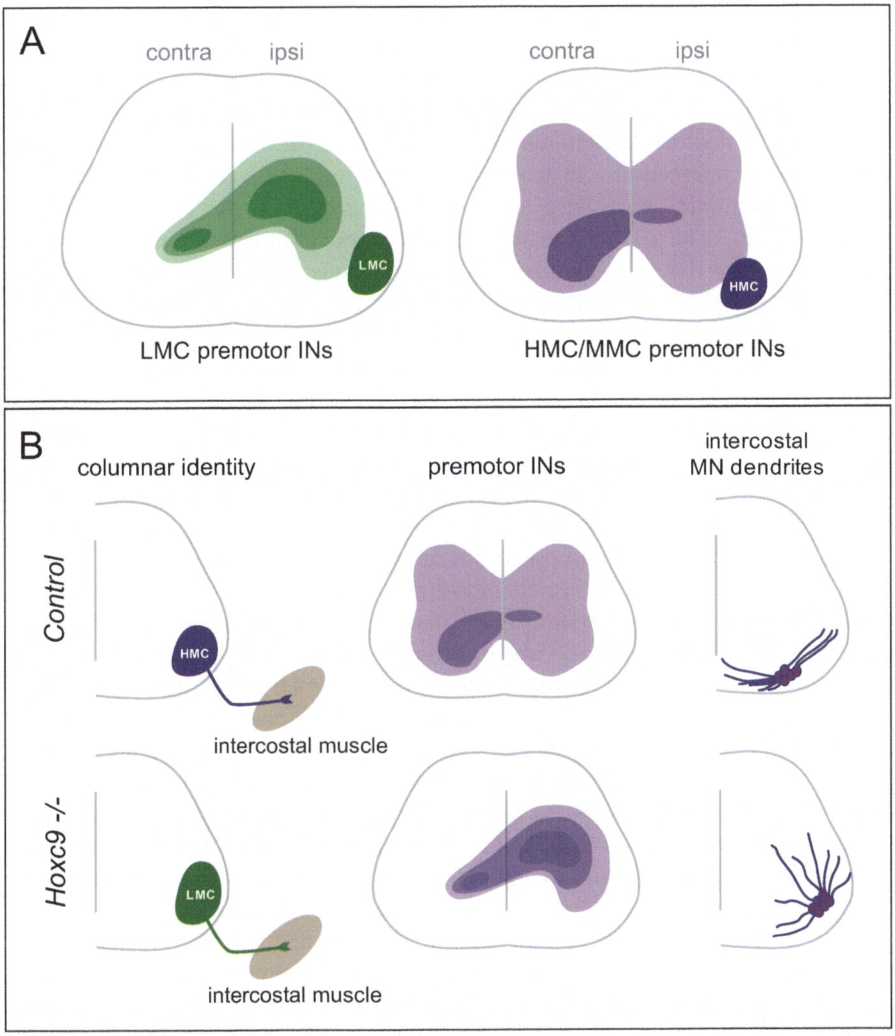

Fig. 7 **Motoneurons in the development of locomotor circuits.** (**a**) Pattern of premotor inputs onto LMC and MMC/HMC subtypes. LMC premotor interneurons are localized predominantly on the ipsilateral side of the spinal cord, whereas interneurons synapsing onto MMC and HMC neurons are equally distributed on both sides of the cord. Darker color shading indicates higher density of premotor inputs. (**b**) Effect of *Hoxc9* mutation on the distribution of premotor inputs. In *Hoxc9* mutants, thoracic HMC neurons are converted to an LMC fate. Ectopic LMC neurons project to intercostal muscle in *Hoxc9* mutants and dendritic morphology of thoracic motoneurons is altered. Thoracic motoneurons receive a higher distribution of inputs from ipsilateral premotor interneurons in *Hoxc9* mutants

manipulation that suppresses LMC differentiation and promotes an MMC identity (Dasen et al. 2008; Sharma et al. 2000). Many of these ectopic MMC neurons occupy the same position as LMC neurons, but still display a predominantly flexor-like motoneuron bursting pattern (Hinckley et al. 2015). These observations suggest that position alone is not sufficient to determine the E-F pattern of motoneurons firing.

In addition to determining the pattern of output from E-F CPGs, motoneurons have also been demonstrated to retrogradely influence the firing properties of excitatory premotor interneurons. In zebrafish, motoneurons involved in slow, intermediate, a fast swimming have been shown to be connected to interneurons via gap junctions (Song et al. 2016). A major excitatory input to motoneurons derives from V2a interneurons that form monosynaptic connections with motoneurons, and V2a interneurons are key components of the rhythm-generating networks that drive zebrafish locomotion (Ampatzis et al. 2014; El Manira 2014). Interestingly, gap junctions occur not only between V2a and motoneurons, but also between groups of V2a neurons and groups of motoneurons. Hyperpolarization of motoneurons increases the firing threshold and decreases the firing frequency of V2a interneurons. Conversely, depolarization of motoneurons decreases V2a thresholds and increases their firing frequency. Backward propagation of electrical signals from motoneurons can a have important influence on the activities of CPG circuits, by setting the firing threshold and properties of excitatory V2a interneurons. In mammals, motoneurons have also been demonstrated to provide feedback to regulate the CPG rhythm in fictive locomotor assays, but not through gap junctions between motoneurons and interneurons (Falgairolle et al. 2017).

While studies show that motoneurons play an instructive role in defining functional properties of locomotor networks, the role of motoneuron identity in spinal circuit assembly has remained unclear. Similar to their roles in shaping the specificity between pSNs and motoneurons, Hox-dependent programs also can regulate the patterns of connectivity between spinal interneurons and motoneuron subtypes. After global or motoneuron-restricted removal of *Hoxc9*, which transforms thoracic HMC and PGC neurons to an LMC fate, the pattern of premotor inputs to these fate-transformed thoracic motoneurons is markedly altered (Baek et al. 2017). The distribution of intercostal premotor interneurons shifts to an ipsilateral bias, similar to patterns normally observed for limb-innervating LMC motoneurons (Goetz et al. 2015). These changes do not appear to be due to peripheral signals provided by the limb, indicating motoneuron-intrinsic programs regulate connectivity between spinal interneurons and columnar subtypes. Mutation in *Hoxc9* also alters the settling position and dendritic architecture of transformed motoneuron populations (Fig. 7b). The changes in interneuron premotor distributions in *Hoxc9* mutants therefore could reflect changes in both the molecular profile and somatodendritic features of motoneurons. These results are consistent with the view that aspects motor circuit assembly relies on both positional and molecular recognition cues (Jessell et al. 2011).

9 Conclusions

Studies of spinal motoneuron development have provided fundamental insights into the molecular programs underlying the diversification and synaptic specificity of a single neuronal class. Work on motoneurons has enabled exploration into basic questions in developmental neurobiology, such as how multipotent cells interpret morphogen gradients, how gene regulatory networks are deployed during differentiation, and how subtype specific gene programs determine neuronal synaptic specificity. These studies have revealed molecular mechanism that are applicable to other classes of neurons in the CNS. Further cross-species analysis, taking advantage of the relative simplicity of the neuromuscular system of skates, or the divergent functional organization of motoneurons in zebrafish, should yield novel insights into the mechanisms of motoneuron differentiation, and how motoneuron identities shape circuit assembly and function.

While modern molecular approaches have enabled unprecedented resolution in the characterization of subtype-specific gene expression patterns, there remain significant gaps in our understanding of how molecular features of motoneurons contribute to their connections and functional properties. One challenge is that motoneurons receive synaptic input from a daunting array of neuronal classes, including excitatory and inhibitory spinal interneurons, proprioceptive sensory neurons, neuromodulatory systems, and descending cortical inputs. While the mechanisms of premotor synaptic specificity are still not fully resolved, the ability to genetically manipulate motoneuron differentiation programs provides an entry point to further resolve how motor circuits are assembled during development. Recent findings reveal that the same *Hox* gene expressed by motoneurons are also required for the diversification of interneurons and proprioceptive sensory neurons, suggest that the use of a common group of regulatory factors may be an important mechanism regulating synaptic specificity in spinal circuits.

References

Agalliu D, Takada S, Agalliu I, McMahon AP, Jessell TM (2009) Motor neurons with axial muscle projections specified by Wnt4/5 signaling. Neuron 61:708–720

Alkaslasi MR, Piccus ZE, Hareendran S, Silberberg H, Chen L, Zhang Y, Petros TJ, Le Pichon CE (2021) Single nucleus RNA-sequencing defines unexpected diversity of cholinergic neuron types in the adult mouse spinal cord. Nat Commun 12:2471

Ampatzis K, Song J, Ausborn J, El Manira A (2013) Pattern of innervation and recruitment of different classes of motoneurons in adult zebrafish. J Neurosci 33:10875–10886

Ampatzis K, Song J, Ausborn J, El Manira A (2014) Separate microcircuit modules of distinct v2a interneurons and motoneurons control the speed of locomotion. Neuron 83:934–943

Appel B, Korzh V, Glasgow E, Thor S, Edlund T, Dawid IB, Eisen JS (1995) Motoneuron fate specification revealed by patterned LIM homeobox gene expression in embryonic zebrafish. Development 121:4117–4125

Arber S (2012) Motor circuits in action: specification, connectivity, and function. Neuron 74:975–989

Arber S, Han B, Mendelsohn M, Smith M, Jessell TM, Sockanathan S (1999) Requirement for the homeobox gene Hb9 in the consolidation of motor neuron identity. Neuron 23:659–674

Arber S, Ladle DR, Lin JH, Frank E, Jessell TM (2000) ETS gene Er81 controls the formation of functional connections between group Ia sensory afferents and motor neurons. Cell 101:485–498

Ashrafi S, Lalancette-Hebert M, Friese A, Sigrist M, Arber S, Shneider NA, Kaltschmidt JA (2012) Wnt7A identifies embryonic gamma-motor neurons and reveals early postnatal dependence of gamma-motor neurons on a muscle spindle-derived signal. J Neurosci 32:8725–8731

Baek M, Pivetta C, Liu JP, Arber S, Dasen JS (2017) Columnar-intrinsic cues shape premotor input specificity in locomotor circuits. Cell Rep 21:867–877

Baek M, Menon V, Jessell TM, Hantman AW, Dasen JS (2019) Molecular logic of spinocerebellar tract neuron diversity and connectivity. Cell Rep 27(2620–2635):e2624

Balaskas N, Ribeiro A, Panovska J, Dessaud E, Sasai N, Page KM, Briscoe J, Ribes V (2012) Gene regulatory logic for reading the Sonic Hedgehog signaling gradient in the vertebrate neural tube. Cell 148:273–284

Balaskas N, Abbott LF, Jessell TM, Ng D (2019) Positional strategies for connection specificity and synaptic organization in spinal sensory-motor circuits. Neuron *102*:1143

Beattie CE, Hatta K, Halpern ME, Liu HB, Eisen JS, Kimmel CB (1997) Temporal separation in the specification of primary and secondary motoneurons in zebrafish. Dev Biol 187:171–182

Bello-Rojas S, Istrate AE, Kishore S, McLean DL (2019) Central and peripheral innervation patterns of defined axial motor units in larval zebrafish. J Comp Neurol 527:2557–2572

Bel-Vialar S, Itasaki N, Krumlauf R (2002) Initiating Hox gene expression: in the early chick neural tube differential sensitivity to FGF and RA signaling subdivides the HoxB genes in two distinct groups. Development 129:5103–5115

Bermingham NA, Hassan BA, Wang VY, Fernandez M, Banfi S, Bellen HJ, Fritzsch B, Zoghbi HY (2001) Proprioceptor pathway development is dependent on MATH1. Neuron 30:411–422

Bikoff JB, Gabitto MI, Rivard AF, Drobac E, Machado TA, Miri A, Brenner-Morton S, Famojure E, Diaz C, Alvarez FJ et al (2016) Spinal inhibitory interneuron diversity delineates variant motor microcircuits. Cell 165:207–219

Blum JA, Klemm S, Shadrach JL, Guttenplan KA, Nakayama L, Kathiria A, Hoang PT, Gautier O, Kaltschmidt JA, Greenleaf WJ et al (2021) Single-cell transcriptomic analysis of the adult mouse spinal cord reveals molecular diversity of autonomic and skeletal motor neurons. Nat Neurosci 24:572–583

Briscoe J, Pierani A, Jessell TM, Ericson J (2000) A homeodomain protein code specifies progenitor cell identity and neuronal fate in the ventral neural tube. Cell 101:435–445

Britz O, Zhang J, Grossmann KS, Dyck J, Kim JC, Dymecki S, Gosgnach S, Goulding M (2015) A genetically defined asymmetry underlies the inhibitory control of flexor-extensor locomotor movements. elife 4:e13038

Butler SJ, Bronner ME (2015) From classical to current: analyzing peripheral nervous system and spinal cord lineage and fate. Dev Biol 398:135–146

Catela C, Shin MM, Lee DH, Liu JP, Dasen JS (2016) Hox proteins coordinate motor neuron differentiation and connectivity programs through Ret/Gfralpha genes. Cell Rep 14:1901–1915

Catela C, Correa E, Wen K, Aburas J, Croci L, Consalez GG, Kratsios P (2019) An ancient role for collier/Olf/Ebf (COE)-type transcription factors in axial motor neuron development. Neural Dev 14:2

Chakkalakal JV, Nishimune H, Ruas JL, Spiegelman BM, Sanes JR (2010) Retrograde influence of muscle fibers on their innervation revealed by a novel marker for slow motoneurons. Development 137:3489–3499

Chang Q, Gonzalez M, Pinter MJ, Balice-Gordon RJ (1999) Gap junctional coupling and patterns of connexin expression among neonatal rat lumbar spinal motor neurons. J Neurosci 19:10813–10828

Chen HH, Yip JW, Stewart AF, Frank E (2002) Differential expression of a transcription regulatory factor, the LIM domain only 4 protein Lmo4, in muscle sensory neurons. Development 129:4879–4889
Chen HH, Hippenmeyer S, Arber S, Frank E (2003) Development of the monosynaptic stretch reflex circuit. Curr Opin Neurobiol 13:96–102
Chevallier S, Jan Ijspeert A, Ryczko D, Nagy F, Cabelguen JM (2008) Organisation of the spinal central pattern generators for locomotion in the salamander: biology and modelling. Brain Res Rev 57:147–161
Dasen JS (2009) Transcriptional networks in the early development of sensory-motor circuits. Curr Top Dev Biol 87:119–148
Dasen JS, Jessell TM (2009) Hox networks and the origins of motor neuron diversity. Curr Top Dev Biol 88:169–200
Dasen JS, Liu JP, Jessell TM (2003) Motor neuron columnar fate imposed by sequential phases of Hox-c activity. Nature 425:926–933
Dasen JS, Tice BC, Brenner-Morton S, Jessell TM (2005) A Hox regulatory network establishes motor neuron pool identity and target-muscle connectivity. Cell 123:477–491
Dasen JS, De Camilli A, Wang B, Tucker PW, Jessell TM (2008) Hox repertoires for motor neuron diversity and connectivity gated by a single accessory factor, FoxP1. Cell 134:304–316
De Marco Garcia NV, Jessell TM (2008) Early motor neuron pool identity and muscle nerve trajectory defined by postmitotic restrictions in Nkx6.1 activity. Neuron 57:217–231
de Nooij JC, Doobar S, Jessell TM (2013) Etv1 inactivation reveals proprioceptor subclasses that reflect the level of NT3 expression in muscle targets. Neuron 77:1055–1068
Dykes IM, Tempest L, Lee SI, Turner EE (2011) Brn3a and Islet1 act epistatically to regulate the gene expression program of sensory differentiation. J Neurosci 31:9789–9799
Eberhart J, Swartz ME, Koblar SA, Pasquale EB, Krull CE (2002) EphA4 constitutes a population-specific guidance cue for motor neurons. Dev Biol 247:89–101
Eccles JC, Eccles RM, Lundberg A (1957) The convergence of monosynaptic excitatory afferents on to many different species of alpha motoneurones. J Physiol 137:22–50
El Manira A (2014) Dynamics and plasticity of spinal locomotor circuits. Curr Opin Neurobiol 29:133–141
Ericson J, Rashbass P, Schedl A, Brenner-Morton S, Kawakami A, van Heyningen V, Jessell TM, Briscoe J (1997) Pax6 controls progenitor cell identity and neuronal fate in response to graded Shh signaling. Cell 90:169–180
Espinosa-Medina I, Saha O, Boismoreau F, Chettouh Z, Rossi F, Richardson WD, Brunet JF (2016) The sacral autonomic outflow is sympathetic. Science 354:893–897
Esposito MS, Capelli P, Arber S (2014) Brainstem nucleus MdV mediates skilled forelimb motor tasks. Nature 508:351–356
Falgairolle M, Puhl JG, Pujala A, Liu W, O'Donovan MJ (2017) Motoneurons regulate the central pattern generator during drug-induced locomotor-like activity in the neonatal mouse. elife 6:e26622
Fetcho JR (1987) A review of the organization and evolution of motoneurons innervating the axial musculature of vertebrates. Brain Res 434:243–280
Fetcho JR (1992) The spinal motor system in early vertebrates and some of its evolutionary changes. Brain Behav Evol 40:82–97
Fritzsch B, Elliott KL, Glover JC (2017) Gaskell revisited: new insights into spinal autonomics necessitate a revised motor neuron nomenclature. Cell Tissue Res 370:195–209
Fukuhara K, Imai F, Ladle DR, Katayama K, Leslie JR, Arber S, Jessell TM, Yoshida Y (2013) Specificity of monosynaptic sensory-motor connections imposed by repellent Sema3E-PlexinD1 signaling. Cell Rep 5:748–758
Fulton BP, Miledi R, Takahashi T (1980) Electrical synapses between motoneurons in the spinal cord of the newborn rat. Proc R Soc Lond B Biol Sci 208:115–120
Goetz C, Pivetta C, Arber S (2015) Distinct limb and trunk premotor circuits establish laterality in the spinal cord. Neuron 85:131–144

Golden MG, Dasen JS (2012) Polycomb repressive complex 1 activities determine the columnar organization of motor neurons. Genes Dev 26:2236–2250
Goulding M (2009) Circuits controlling vertebrate locomotion: moving in a new direction. Nat Rev Neurosci 10:507–518
Griener A, Zhang W, Kao H, Wagner C, Gosgnach S (2015) Probing diversity within subpopulations of locomotor-related V0 interneurons. Dev Neurobiol 75:1189–1203
Grillner S (2006) Biological pattern generation: the cellular and computational logic of networks in motion. Neuron 52:751–766
Grillner S, Jessell TM (2009) Measured motion: searching for simplicity in spinal locomotor networks. Curr Opin Neurobiol 19:572–586
Gutman CR, Ajmera MK, Hollyday M (1993) Organization of motor pools supplying axial muscles in the chicken. Brain Res 609:129–136
Haase G, Dessaud E, Garces A, de Bovis B, Birling M, Filippi P, Schmalbruch H, Arber S, deLapeyriere O (2002) GDNF acts through PEA3 to regulate cell body positioning and muscle innervation of specific motor neuron pools. Neuron 35:893–905
Hanley O, Zewdu R, Cohen LJ, Jung H, Lacombe J, Philippidou P, Lee DH, Selleri L, Dasen JS (2016) Parallel Pbx-dependent pathways govern the coalescence and fate of motor columns. Neuron 91:1005–1020
Hanson MG, Landmesser LT (2004) Normal patterns of spontaneous activity are required for correct motor axon guidance and the expression of specific guidance molecules. Neuron 43:687–701
Hanson MG, Landmesser LT (2006) Increasing the frequency of spontaneous rhythmic activity disrupts pool-specific axon fasciculation and pathfinding of embryonic spinal motoneurons. J Neurosci 26:12769–12780
Hayashi M, Hinckley CA, Driscoll SP, Moore NJ, Levine AJ, Hilde KL, Sharma K, Pfaff SL (2018) Graded arrays of spinal and supraspinal V2a interneuron subtypes underlie forelimb and hindlimb motor control. Neuron 97:869–884 e865
Hinckley CA, Alaynick WA, Gallarda BW, Hayashi M, Hilde KL, Driscoll SP, Dekker JD, Tucker HO, Sharpee TO, Pfaff SL (2015) Spinal locomotor circuits develop using hierarchical rules based on motorneuron position and identity. Neuron 87:1008–1021
Hollyday M (1980) Organization of motor pools in the chick lumbar lateral motor column. J Comp Neurol 194:143–170
Hollyday M, Jacobson RD (1990) Location of motor pools innervating chick wing. J Comp Neurol 302:575–588
Hollyday M, Hamburger V, Farris JM (1977) Localization of motor neuron pools supplying identified muscles in normal and supernumerary legs of chick embryo. Proc Natl Acad Sci U S A 74:3582–3586
Holst RJ, Bone Q (1993) On bipedalism in skates and rays. Philos Trans R Soc B 339:105–108
Honig MG, Frase PA, Camilli SJ (1998) The spatial relationships among cutaneous, muscle sensory and motoneuron axons during development of the chick hindlimb. Development 125:995–1004
Hutchinson SA, Cheesman SE, Hale LA, Boone JQ, Eisen JS (2007) Nkx6 proteins specify one zebrafish primary motoneuron subtype by regulating late islet1 expression. Development 134:1671–1677
Imai F, Yoshida Y (2018) Molecular mechanisms underlying monosynaptic sensory-motor circuit development in the spinal cord. Dev Dyn 247:581–587
Inoue K, Ozaki S, Shiga T, Ito K, Masuda T, Okado N, Iseda T, Kawaguchi S, Ogawa M, Bae SC et al (2002) Runx3 controls the axonal projection of proprioceptive dorsal root ganglion neurons. Nat Neurosci 5:946–954
Jessell TM (2000) Neuronal specification in the spinal cord: inductive signals and transcriptional codes. Nat Rev Genet 1:20–29
Jessell TM, Surmeli G, Kelly JS (2011) Motor neurons and the sense of place. Neuron 72:419–424

Ji SJ, Zhuang B, Falco C, Schneider A, Schuster-Gossler K, Gossler A, Sockanathan S (2006) Mesodermal and neuronal retinoids regulate the induction and maintenance of limb innervating spinal motor neurons. Dev Biol 297:249–261

Jung H, Dasen JS (2015) Evolution of patterning systems and circuit elements for locomotion. Dev Cell 32:408–422

Jung H, Lacombe J, Mazzoni EO, Liem KF Jr, Grinstein J, Mahony S, Mukhopadhyay D, Gifford DK, Young RA, Anderson KV et al (2010) Global control of motor neuron topography mediated by the repressive actions of a single hox gene. Neuron 67:781–796

Jung H, Mazzoni EO, Soshnikova N, Hanley O, Venkatesh B, Duboule D, Dasen JS (2014) Evolving Hox activity profiles govern diversity in locomotor systems. Dev Cell 29:171–187

Jung H, Baek M, D'Elia KP, Boisvert C, Currie PD, Tay BH, Venkatesh B, Brown SM, Heguy A, Schoppik D et al (2018) The ancient origins of neural substrates for land walking. Cell 172:667–682 e615

Kania A, Jessell TM (2003) Topographic motor projections in the limb imposed by LIM homeodomain protein regulation of ephrin-A:EphA interactions. Neuron 38:581–596

Kania A, Johnson RL, Jessell TM (2000) Coordinate roles for LIM homeobox genes in directing the dorsoventral trajectory of motor axons in the vertebrate limb. Cell 102:161–173

Kanning KC, Kaplan A, Henderson CE (2010) Motor neuron diversity in development and disease. Annu Rev Neurosci 33:409–440

Kiehn O (2016) Decoding the organization of spinal circuits that control locomotion. Nat Rev Neurosci 17:224–238

King BL, Gillis JA, Carlisle HR, Dahn RD (2011) A natural deletion of the HoxC cluster in elasmobranch fishes. Science 334:1517

Kmita M, Duboule D (2003) Organizing Axes in Time and Space; 25 Years of Colinear Tinkering. Science 301(5631):331–333. https://doi.org/10.1126/science.1085753

Koch SC, Del Barrio MG, Dalet A, Gatto G, Gunther T, Zhang JM, Seidler B, Saur D, Schule R, Goulding M (2017) ROR beta spinal interneurons gate sensory transmission during locomotion to secure a fluid walking gait. Neuron 96:1419

Koester DM, Spirito CP (2003) Punting: an unusual mode of locomotion in the little skate, Leucoraja erinacea (Chondrichthyes: Rajidae). Copeia 2003:553–561

Lacombe J, Hanley O, Jung H, Philippidou P, Surmeli G, Grinstein J, Dasen JS (2013) Genetic and functional modularity of Hox activities in the specification of limb-innervating motor neurons. PLoS Genet 9:e1003184

Lallemend F, Ernfors P (2012) Molecular interactions underlying the specification of sensory neurons. Trends Neurosci 35:373–381

Landmesser L (1978a) The development of motor projection patterns in the chick hind limb. J Physiol 284:391–414

Landmesser L (1978b) The distribution of motoneurones supplying chick hind limb muscles. J Physiol 284:371–389

Landmesser LT (2001) The acquisition of motoneuron subtype identity and motor circuit formation. Int J Dev Neurosci 19:175–182

Lanuza GM, Gosgnach S, Pierani A, Jessell TM, Goulding M (2004) Genetic identification of spinal interneurons that coordinate left-right locomotor activity necessary for walking movements. Neuron 42:375–386

Leal F, Cohn MJ (2018) Developmental, genetic, and genomic insights into the evolutionary loss of limbs in snakes. Genesis 56:e23077

Lee SK, Pfaff SL (2003) Synchronization of neurogenesis and motor neuron specification by direct coupling of bHLH and homeodomain transcription factors. Neuron 38:731–745

Lee SK, Lee B, Ruiz EC, Pfaff SL (2005) Olig2 and Ngn2 function in opposition to modulate gene expression in motor neuron progenitor cells. Genes Dev 19:282–294

Lee S, Lee B, Joshi K, Pfaff SL, Lee JW, Lee SK (2008) A regulatory network to segregate the identity of neuronal subtypes. Dev Cell 14:877–889

Lee S, Lee B, Lee JW, Lee SK (2009) Retinoid signaling and neurogenin2 function are coupled for the specification of spinal motor neurons through a chromatin modifier CBP. Neuron 62:641–654

Lee S, Cuvillier JM, Lee B, Shen R, Lee JW, Lee SK (2012) Fusion protein Isl1-Lhx3 specifies motor neuron fate by inducing motor neuron genes and concomitantly suppressing the interneuron programs. Proc Natl Acad Sci U S A 109:3383–3388

Liang X, Song MR, Xu Z, Lanuza GM, Liu Y, Zhuang T, Chen Y, Pfaff SL, Evans SM, Sun Y (2011) Isl1 is required for multiple aspects of motor neuron development. Mol Cell Neurosci 47:215–222

Lin JH, Saito T, Anderson DJ, Lance-Jones C, Jessell TM, Arber S (1998) Functionally related motor neuron pool and muscle sensory afferent subtypes defined by coordinate ETS gene expression. Cell 95:393–407

Liu DW, Westerfield M (1988) Function of identified motoneurones and coordination of primary and secondary motor systems during zebra fish swimming. J Physiol Lond 403:73–89

Liu JP, Laufer E, Jessell TM (2001) Assigning the positional identity of spinal motor neurons: rostrocaudal patterning of Hox-c expression by FGFs, Gdf11, and retinoids. Neuron 32:997–1012

Livet J, Sigrist M, Stroebel S, De Paola V, Price SR, Henderson CE, Jessell TM, Arber S (2002) ETS gene Pea3 controls the central position and terminal arborization of specific motor neuron pools. Neuron 35:877–892

Lucifora LO, Vassallo AI (2002) Walking in skates (Chondrichthyes, Rajidae): anatomy, behaviour and analogies to tetrapod locomotion. Biol J Linn Soc 77:35–41

Luria V, Krawchuk D, Jessell TM, Laufer E, Kania A (2008) Specification of motor axon trajectory by ephrin-B:EphB signaling: symmetrical control of axonal patterning in the developing limb. Neuron 60:1039–1053

Ma YC, Song MR, Park JP, Henry Ho HY, Hu L, Kurtev MV, Zieg J, Ma Q, Pfaff SL, Greenberg ME (2008) Regulation of motor neuron specification by phosphorylation of neurogenin 2. Neuron 58:65–77

Macesic LJ, Kajiura SM (2010) Comparative punting kinematics and pelvic fin musculature of benthic batoids. J Morphol 271:1219–1228

Machado CB, Kanning KC, Kreis P, Stevenson D, Crossley M, Nowak M, Iacovino M, Kyba M, Chambers D, Blanc E et al (2014) Reconstruction of phrenic neuron identity in embryonic stem cell-derived motor neurons. Development 141:784–794

Machado TA, Pnevmatikakis E, Paninski L, Jessell TM, Miri A (2015) Primacy of flexor locomotor pattern revealed by ancestral reversion of motor neuron identity. Cell 162:338–350

Mazzoni EO, Mahony S, Closser M, Morrison CA, Nedelec S, Williams DJ, An D, Gifford DK, Wichterle H (2013a) Synergistic binding of transcription factors to cell-specific enhancers programs motor neuron identity. Nat Neurosci 16:1219–1227

Mazzoni EO, Mahony S, Peljto M, Patel T, Thornton SR, McCuine S, Reeder C, Boyer LA, Young RA, Gifford DK et al (2013b) Saltatory remodeling of Hox chromatin in response to rostrocaudal patterning signals. Nat Neurosci 16:1191–1198

McLean DL, Fan J, Higashijima S, Hale ME, Fetcho JR (2007) A topographic map of recruitment in spinal cord. Nature 446:71–75

Mears SC, Frank E (1997) Formation of specific monosynaptic connections between muscle spindle afferents and motoneurons in the mouse. J Neurosci 17:3128–3135

Mendell LM, Henneman E (1968) Terminals of single Ia fibers: distribution within a pool of 300 homonymous motor neurons. Science 160:96–98

Mendelsohn AI, Simon CM, Abbott LF, Mentis GZ, Jessell TM (2015) Activity regulates the incidence of heteronymous sensory-motor connections. Neuron 87:111–123

Mendelsohn AI, Dasen JS, Jessell TM (2017) Divergent Hox coding and evasion of retinoid signaling specifies motor neurons innervating digit muscles. Neuron 93:792–805 e794

Mendelson B, Frank E (1991) Specific monosynaptic sensory-motor connections form in the absence of patterned neural activity and motoneuronal cell death. J Neurosci 11:1390–1403

Menelaou E, McLean DL (2012) A gradient in endogenous rhythmicity and oscillatory drive matches recruitment order in an axial motor pool. J Neurosci 32:10925–10939

Mentis GZ, Alvarez FJ, Bonnot A, Richards DS, Gonzalez-Forero D, Zerda R, O'Donovan MJ (2005) Noncholinergic excitatory actions of motoneurons in the neonatal mammalian spinal cord. Proc Natl Acad Sci U S A 102:7344–7349

Metzis V, Steinhauser S, Pakanavicius E, Gouti M, Stamataki D, Ivanovitch K, Watson T, Rayon T, Mousavy Gharavy SN, Lovell-Badge R et al (2018) Nervous system regionalization entails axial allocation before neural differentiation. Cell 175:1105–1118 e1117

Milner LD, Landmesser LT (1999) Cholinergic and GABAergic inputs drive patterned spontaneous motoneuron activity before target contact. J Neurosci 19:3007–3022

Mizuguchi R, Sugimori M, Takebayashi H, Kosako H, Nagao M, Yoshida S, Nabeshima Y, Shimamura K, Nakafuku M (2001) Combinatorial roles of olig2 and neurogenin2 in the coordinated induction of pan-neuronal and subtype-specific properties of motoneurons. Neuron 31:757–771

Muller D, Cherukuri P, Henningfeld K, Poh CH, Wittler L, Grote P, Schluter O, Schmidt J, Laborda J, Bauer SR et al (2014) Dlk1 promotes a fast motor neuron biophysical signature required for peak force execution. Science 343:1264–1266

Murakami Y, Tanaka M (2011) Evolution of motor innervation to vertebrate fins and limbs. Dev Biol 355:164–172

Myers PZ, Eisen JS, Westerfield M (1986) Development and axonal outgrowth of identified motoneurons in the zebrafish. J Neurosci 6:2278–2289

Narendra V, Rocha PP, An D, Raviram R, Skok JA, Mazzoni EO, Reinberg D (2015) CTCF establishes discrete functional chromatin domains at the Hox clusters during differentiation. Science 347:1017–1021

Nishimaru H, Restrepo CE, Ryge J, Yanagawa Y, Kiehn O (2005) Mammalian motor neurons corelease glutamate and acetylcholine at central synapses. Proc Natl Acad Sci U S A 102:5245–5249

Nordström U, Maier E, Jessell TM, Edlund T, Nusse R (2006) An Early Role for Wnt Signaling in Specifying Neural Patterns of Cdx and Hox Gene Expression and Motor Neuron Subtype Identity. PLoS Biology 4(8):e252. https://doi.org/10.1371/journal.pbio.0040252

Novitch BG, Chen AI, Jessell TM (2001) Coordinate regulation of motor neuron subtype identity and pan-neuronal properties by the bHLH repressor Olig2. Neuron 31:773–789

Novitch BG, Wichterle H, Jessell TM, Sockanathan S (2003) A requirement for retinoic acid-mediated transcriptional activation in ventral neural patterning and motor neuron specification. Neuron 40:81–95

Parker HJ, Krumlauf R (2020) A Hox gene regulatory network for hindbrain segmentation. Curr Top Dev Biol 139:169–203

Patel TD, Kramer I, Kucera J, Niederkofler V, Jessell TM, Arber S, Snider WD (2003) Peripheral NT3 signaling is required for ETS protein expression and central patterning of proprioceptive sensory afferents. Neuron 38:403–416

Pecho-Vrieseling E, Sigrist M, Yoshida Y, Jessell TM, Arber S (2009) Specificity of sensory-motor connections encoded by Sema3e-Plxnd1 recognition. Nature 459:842–846

Peljto M, Dasen JS, Mazzoni EO, Jessell TM, Wichterle H (2010) Functional Diversity of ESC-Derived Motor Neuron Subtypes Revealed through Intraspinal Transplantation. Cell Stem Cell 7(3):355–366. https://doi.org/10.1016/j.stem.2010.07.013

Philippidou P, Dasen JS (2013) Hox genes: choreographers in neural development, architects of circuit organization. Neuron 80:12–34

Philippidou P, Walsh CM, Aubin J, Jeannotte L, Dasen JS (2012) Sustained Hox5 gene activity is required for respiratory motor neuron development. Nat Neurosci 15:1636–1644

Poliak S, Norovich AL, Yamagata M, Sanes JR, Jessell TM (2016) Muscle-type identity of proprioceptors specified by spatially restricted signals from limb mesenchyme. Cell 164:512–525

Prasad A, Hollyday M (1991) Development and migration of avian sympathetic preganglionic neurons. J Comp Neurol 307:237–258

Price SR, De Marco Garcia NV, Ranscht B, Jessell TM (2002) Regulation of Motor Neuron Pool Sorting by Differential Expression of Type II Cadherins. Cell 109(2):205–216. https://doi.org/10.1016/S0092-8674(02)00695-5

Rhee HS, Closser M, Guo Y, Bashkirova EV, Tan GC, Gifford DK, Wichterle H (2016) Expression of terminal effector genes in mammalian neurons is maintained by a dynamic relay of transient enhancers. Neuron 92:1252–1265

Ribes V, Briscoe J (2009) Establishing and interpreting graded Sonic Hedgehog signaling during vertebrate neural tube patterning: the role of negative feedback. Cold Spring Harb Perspect Biol 1:a002014

Romanes GJ (1941) Cell columns in the spinal cord of a human foetus of fourteen weeks. J Anat 75(145–152):141

Romanes GJ (1942) The development and significance of the cell columns in the ventral horn of the cervical and upper thoracic spinal cord of the rabbit. J Anat Lond 76:112–130

Romer AS, Parsons TS (1977) The vertebrate body, 5th edn. Saunders, Philadelphia

Rosenberger LJ (2001) Pectoral fin locomotion in batoid fishes: undulation versus oscillation. J Exp Biol 204:379–394

Rousso DL, Gaber ZB, Wellik D, Morrisey EE, Novitch BG (2008) Coordinated actions of the forkhead protein Foxp1 and Hox proteins in the columnar organization of spinal motor neurons. Neuron 59:226–240

Sagner A, Briscoe J (2019) Establishing neuronal diversity in the spinal cord: a time and a place. Development 146:dev182154

Sagner A, Gaber ZB, Delile J, Kong JH, Rousso DL, Pearson CA, Weicksel SE, Melchionda M, Mousavy Gharavy SN, Briscoe J et al (2018) Olig2 and Hes regulatory dynamics during motor neuron differentiation revealed by single cell transcriptomics. PLoS Biol 16:e2003127

Seredick SD, Van Ryswyk L, Hutchinson SA, Eisen JS (2012) Zebrafish Mnx proteins specify one motoneuron subtype and suppress acquisition of interneuron characteristics. Neural Dev 7:35

Seredick S, Hutchinson SA, Van Ryswyk L, Talbot JC, Eisen JS (2014) Lhx3 and Lhx4 suppress Kolmer-Agduhr interneuron characteristics within zebrafish axial motoneurons. Development 141:3900–3909

Shah V, Drill E, Lance-Jones C (2004) Ectopic expression of Hoxd10 in thoracic spinal segments induces motoneurons with a lumbosacral molecular profile and axon projections to the limb. Dev Dyn 231:43–56

Sharma K, Sheng HZ, Lettieri K, Li H, Karavanov A, Potter S, Westphal H, Pfaff SL (1998) LIM homeodomain factors Lhx3 and Lhx4 assign subtype identities for motor neurons. Cell 95:817–828

Sharma K, Leonard AE, Lettieri K, Pfaff SL (2000) Genetic and epigenetic mechanisms contribute to motor neuron pathfinding. Nature 406:515–519

Sharma N, Flaherty K, Lezgiyeva K, Wagner DE, Klein AM, Ginty DD (2020) The emergence of transcriptional identity in somatosensory neurons. Nature 577:392–398

Shin MM, Catela C, Dasen J (2020) Intrinsic control of neuronal diversity and synaptic specificity in a proprioceptive circuit. elife 9:e56374

Shirasaki R, Pfaff SL (2002) Transcriptional codes and the control of neuronal identity. Annu Rev Neurosci 25:251–281

Shirasaki R, Lewcock JW, Lettieri K, Pfaff SL (2006) FGF as a target-derived chemoattractant for developing motor axons genetically programmed by the LIM code. Neuron 50:841–853

Smith CL, Hollyday M (1983) The development and postnatal organization of motor nuclei in the rat thoracic spinal cord. J Comp Neurol 220:16–28

Sockanathan S, Jessell TM (1998) Motor neuron-derived retinoid signaling specifies the subtype identity of spinal motor neurons. Cell 94:503–514

Song J, Ampatzis K, Bjornfors ER, El Manira A (2016) Motor neurons control locomotor circuit function retrogradely via gap junctions. Nature 529:399–402

Soshnikova N, Duboule D (2009) Science 324(5932):1320–1323. https://doi.org/10.1126/science.1171468

Stepien AE, Tripodi M, Arber S (2010) Monosynaptic rabies virus reveals premotor network organization and synaptic specificity of cholinergic partition cells. Neuron 68:456–472
Stifani N (2014) Motor neurons and the generation of spinal motor neuron diversity. Front Cell Neurosci 8:293
Surmeli G, Akay T, Ippolito GC, Tucker PW, Jessell TM (2011) Patterns of spinal sensory-motor connectivity prescribed by a dorsoventral positional template. Cell 147:653–665
Sweeney LB, Bikoff JB, Gabitto MI, Brenner-Morton S, Baek M, Yang JH, Tabak EG, Dasen JS, Kintner CR, Jessell TM (2018) Origin and segmental diversity of spinal inhibitory interneurons. Neuron 97:341–355 e343
Talpalar AE, Bouvier J, Borgius L, Fortin G, Pierani A, Kiehn O (2013) Dual-mode operation of neuronal networks involved in left-right alternation. Nature 500:85–88
Tan GC, Mazzoni EO, Wichterle H (2016) Iterative role of notch signaling in spinal motor neuron diversification. Cell Rep 16:907–916
Tanabe Y, William C, Jessell TM (1998) Specification of motor neuron identity by the MNR2 homeodomain protein. Cell 95:67–80
Thaler J, Harrison K, Sharma K, Lettieri K, Kehrl J, Pfaff SL (1999) Active suppression of interneuron programs within developing motor neurons revealed by analysis of homeodomain factor HB9. Neuron 23:675–687
Thaler JP, Lee SK, Jurata LW, Gill GN, Pfaff SL (2002) LIM factor Lhx3 contributes to the specification of motor neuron and interneuron identity through cell-type-specific protein-protein interactions. Cell 110:237–249
Thaler JP, Koo SJ, Kania A, Lettieri K, Andrews S, Cox C, Jessell TM, Pfaff SL (2004) A postmitotic role for Isl-class LIM homeodomain proteins in the assignment of visceral spinal motor neuron identity. Neuron 41:337–350
Tiret L, Le Mouellic H, Maury M, Brulet P (1998) Increased apoptosis of motoneurons and altered somatotopic maps in the brachial spinal cord of Hoxc-8-deficient mice. Development 125:279–291
Tosney KW (1987) Proximal tissues and patterned neurite outgrowth at the lumbosacral level of the chick-embryo – deletion of the dermamyotome. Dev Biol 122:540–558
Tosney KW (1988) Proximal tissues and patterned neurite outgrowth at the lumbosacral level of the chick-embryo – partial and complete deletion of the somite. Dev Biol 127:266–286
Tosney KW, Landmesser LT (1985a) Development of the major pathways for neurite outgrowth in the chick hindlimb. Dev Biol 109:193–214
Tosney KW, Landmesser LT (1985b) Growth cone morphology and trajectory in the lumbosacral region of the chick embryo. J Neurosci 5:2345–2358
Tripodi M, Stepien AE, Arber S (2011) Motor antagonism exposed by spatial segregation and timing of neurogenesis. Nature 479:61–U84
Tsuchida T, Ensini M, Morton SB, Baldassare M, Edlund T, Jessell TM, Pfaff SL (1994) Topographic organization of embryonic motor neurons defined by expression of LIM homeobox genes. Cell 79:957–970
Tuthill JC, Azim E (2018) Proprioception. Curr Biol 28:R194–R203
Vagnozzi AN, Garg K, Dewitz C, Moore MT, Cregg JM, Jeannotte L, Zampieri N, Landmesser LT, Philippidou P (2020) Phrenic-specific transcriptional programs shape respiratory motor output. elife 9:e52859
Vallstedt A, Muhr J, Pattyn A, Pierani A, Mendelsohn M, Sander M, Jessell TM, Ericson J (2001) Different levels of repressor activity assign redundant and specific roles to Nkx6 genes in motor neuron and interneuron specification. Neuron 31:743–755
Vanderhorst VG, Holstege G (1997) Organization of lumbosacral motoneuronal cell groups innervating hindlimb, pelvic floor, and axial muscles in the cat. J Comp Neurol 382:46–76
Velasco S, Ibrahim MM, Kakumanu A, Garipler G, Aydin B, Al-Sayegh MA, Hirsekorn A, Abdul-Rahman F, Satija R, Ohler U et al (2017) A multi-step transcriptional and chromatin state cascade underlies motor neuron programming from embryonic stem cells. Cell Stem Cell 20:205–217 e208

Vermot J, Schuhbaur B, Le Mouellic H, McCaffery P, Garnier JM, Hentsch D, Brulet P, Niederreither K, Chambon P, Dolle P et al (2005) Retinaldehyde dehydrogenase 2 and Hoxc8 are required in the murine brachial spinal cord for the specification of Lim1+ motoneurons and the correct distribution of Islet1+ motoneurons. Development 132:1611–1621

Vrieseling E, Arber S (2006) Target-induced transcriptional control of dendritic patterning and connectivity in motor neurons by the ETS gene Pea3. Cell 127:1439–1452

Wang Y, Wu H, Zelenin P, Fontanet P, Wanderoy S, Petitpre C, Comai G, Bellardita C, Xue-Franzen Y, Huettl RE et al (2019) Muscle-selective RUNX3 dependence of sensorimotor circuit development. Development 146:dev181750

Wichterle H, Lieberam I, Porter JA, Jessell TM (2002) Directed differentiation of embryonic stem cells into motor neurons. Cell 110:385–397

Wu Y, Wang G, Scott SA, Capecchi MR (2008) Hoxc10 and Hoxd10 regulate mouse columnar, divisional and motor pool identity of lumbar motoneurons. Development 135:171–182

Wu D, Schieren I, Qian Y, Zhang C, Jessell TM, de Nooij JC (2019) A role for sensory end organ-derived signals in regulating muscle spindle proprioceptor phenotype. J Neurosci 39:4252–4267

Yuengert R, Hori K, Kibodeaux EE, McClellan JX, Morales JE, Huang TWP, Neul JL, Lai HC (2015) Origin of a non-Clarke's column division of the dorsal spinocerebellar tract and the role of caudal proprioceptive neurons in motor function. Cell Rep 13:1258–1271

Zhang J, Lanuza GM, Britz O, Wang Z, Siembab VC, Zhang Y, Velasquez T, Alvarez FJ, Frank E, Goulding M (2014) V1 and v2b interneurons secure the alternating flexor-extensor motor activity mice require for limbed locomotion. Neuron 82:138–150

Zhou Q, Anderson DJ (2002) The bHLH transcription factors OLIG2 and OLIG1 couple neuronal and glial subtype specification. Cell 109:61–73

Chloride Homeostasis in Developing Motoneurons

Pascal Branchereau and Daniel Cattaert

Abstract Maturation of GABA/Glycine chloride-mediated synaptic inhibitions is crucial for the establishment of a balance between excitation and inhibition. GABA and glycine are excitatory neurotransmitters on immature neurons that exhibit elevated $[Cl^-]_i$. Later in development $[Cl^-]_i$ drops leading to the occurrence of inhibitory synaptic activity. This ontogenic change is closely correlated to a differential expression of two cation-chloride cotransporters that are the Cl^- channel K^+/Cl^- co-transporter type 2 (KCC2) that extrudes Cl^- ions and the Na^+-K^+-$2Cl^-$ cotransporter NKCC1 that accumulates Cl^- ions. The classical scheme built from studies performed on cortical and hippocampal networks proposes that immature neurons display high $[Cl^-]_i$ because NKCC1 is overexpressed compared to KCC2 and that the co-transporters ratio reverses in mature neurons, lowering $[Cl^-]_i$. In this chapter, we will see that this classical scheme is not true in motoneurons (MNs) and that an early alteration of the chloride homeostasis may be involved in pathological conditions.

Keywords Chloride co-transporters (CCCs) · KCC2 · NKCC1 · ALS disease · SOD1^{G93A} mouse; time course · Inhibitory synaptic events · GABA/glycine · Patch-clamp · Modelling · Spinal cord motoneuron

1 Introduction

In the adult mammalian central nervous system (CNS), the two main inhibitory transmitters are γ-aminobutyric acid (GABA) in the brain and glycine in the brainstem and spinal cord. GABA and glycine open ionotropic $GABA_A$ (and $GABA_C$ in

P. Branchereau (✉) · D. Cattaert
Institut de Neurosciences Cognitives et Intégratives d'Aquitaine (INCIA), Univ. Bordeaux, UMR 5287, CNRS, Bordeaux, France
e-mail: pascal.branchereau@u-bordeaux.fr

M. J. O'Donovan, M. Falgairolle (eds.), *Vertebrate Motoneurons*, Advances in Neurobiology 28, https://doi.org/10.1007/978-3-031-07167-6_2

diencephalic structures (Chebib 2004)) and Gly receptors, respectively. These so-called inhibitory amino-acids evoke a chloride current that depends on the intracellular chloride concentration $[Cl^-]_i$, both receptors being largely permeable to Cl^- but also, to a lesser extent to $HC0_3^-$ ions (Bormann et al. 1987). Because $[Cl^-]_i$ largely determines the direction and magnitude of current flow through $GABA_A$ and Gly channels, the stability of $[Cl^-]_i$ is important to maintain consistent synaptic inhibition in the mature CNS. $[Cl^-]_i$ is regulated by two cation-chloride cotransporters (CCCs) that are the Cl^- channel K^+/Cl^- co-transporter type 2 (KCC2) (Payne et al. 1996), preferentially in neurons, that extrudes Cl^- ions using the K^+ gradient (Rivera et al. 1999), and the Na^+-K^+-$2Cl^-$ cotransporter NKCC1, found in nearly all cell types, that accumulates Cl^- ions using the Na^+ gradient (Russell 2000) imposed by the Na^+-K^+ ATPase (Lees 1991). Both KCC2 and NKCC1 messenger RNAs are widely expressed in the CNS (Kanaka et al. 2001). KCC3, which belongs to the four outwardly directed K^+-Cl^- cotransporters KCC1–4, is also described as playing a major role in GABAergic/glycinergic transmission in adult sensory neurons (Lucas et al. 2012). Water is transported with cations and Cl^- by CCCs and water movement contributes to the free energy of Cl^- transport (Zeuthen 2010). $[Cl^-]_i$ is not solely determined by transport activity but also by local impermeant anions mainly present in the cytoplasmic nuclear nucleic acids and polysulfated proteoglycans located in the extracellular matrix (Glykys et al. 2014). Other mechanisms have been reported in the control of $[Cl^-]_i$ in CNS neurons such as Na^+-dependent and Na^+-independent anion exchangers (NDAE and AE, respectively). The anion exchanger AE3 that moves chloride into the cell as it expels bicarbonate is expressed in the brain and AE3-knockout mice are more sensitive to seizure-inducing agents (Hentschke et al. 2006). The $HCO3^-$ current across $GABA_A$ or Gly receptors is depolarizing in all neurons, the equilibrium potential of $HCO3^-$ ions being fixed by pH requirements. Interestingly, the cytosolic carbonic anhydrase (CA) isoform CAVII that converts CO_2 into HCO_3^- (and protons) is upregulated in the brain during development (Rivera et al. 2005), possibly rendering GABAergic transmission excitatory in mature neurons in case of massive activation of $GABA_A$ receptors (Ruusuvuori et al. 2004). Hence, many other various regulatory mechanisms, including those mediated by plasma membrane Cl^- channels and transporters, are involved in the regulation of $[Cl^-]_i$ (Rahmati et al. 2018).

In this chapter we will consider KCC2 and NKCC1 as being the primary cation-chloride cotransporters (CCCs) in CNS neurons that converts excitatory GABA/glycine responses of immature neurons to inhibitory responses in mature neurons. After giving a rapid overview of the molecular properties of these chloride co-transporters, we will present their involvement in the maturation of chloride homeostasis in rodent spinal motoneurons and the associated functional consequences. The involvement of these chloride co-transporters in the amyotrophic lateral sclerosis disease affecting the motoneuron will be presented in a third part.

2 NKCC1 and KCC2 Molecular Properties

NKCC1 is one of the two isoforms of the Na^+-K^+-Cl^- cotransporter (NKCC) protein found in all cell types, NKCC2 being specific to the kidney (Russell 2000). Three phosphoacceptor sites were identified in the N-terminal domain of the protein NKCC1 (at Thr184, Thr189, and Thr202), Thr189 being necessary for activation of the protein, whereas phosphorylation at Thr184 and Thr202 being modulatory (Darman and Forbush 2002).

Phosphorylation of S940 (Silayeva et al. 2015) and dephosphorylation of T906/T1007 (Moore et al. 2018; Watanabe et al. 2019) are essential for the potentiation of its activity. In addition, it has been established that the mammalian KCC2 (alias Slc12a5) gene generates a novel KCC2a isoform that differs from the only previously known KCC2 isoform (now termed KCC2b) by 40 unique N-terminal amino acid residues (Uvarov et al. 2007). Like KCC2b, the expression of KCC2a mRNA is restricted to CNS neurons. In the cortical cultures KCC2b is likely responsible for the extensively studied "developmental shift" from depolarizing to hyperpolarizing GABAergic responses because GABAergic responses remain depolarizing from KCC2 knockout mice (Zhu et al. 2005), which are now known to lack KCC2b only (Uvarov et al. 2007).

3 Chloride Homeostasis in Developing Motoneurons

Anatomical Maturation of CCCs

The classical scheme built from studies performed on cortical and hippocampal networks proposes that, in immature neurons, NKCC1 is highly expressed compared to KCC2 and therefore $[Cl^-]_i$ is high, leading to excitatory GABAergic effects, and that in mature neurons the co-transporter ratio is reversed leading to lower $[Cl^-]_i$ and inhibitory GABAergic effects (Ben-Ari 2014; Owens and Kriegstein 2002; Watanabe and Fukuda 2015; Come et al. 2019; Fukuda 2020). KCC2a protein level is important around birth and becomes reduced in adult brainstem and spinal cord, most of the KCC2 protein being of the KCC2b isoform (Uvarov et al. 2009). In brainstem region it has been shown that monomeric KCC2 is present in immature neurons whereas KCC2 oligomers with molecular masses of ~270, ~400, and ~500 kDa are identified in the mature neurons (Blaesse et al. 2006). The KCC2a isoform is expressed in brainstem regions containing neuronal populations involved in respiration and is important for establishing proper breathing behavior at the time of birth (Dubois et al. 2018). In spinal cord motoneurons (MNs) little data on the maturation of E_{Cl} during perinatal stages were available (Wu et al. 1992; Jean-Xavier et al. 2006; Stein et al. 2004).

In the spinal cord, the exploration of the maturation of KCC2 and NKCC1 between embryonic (E) stage 11.5 (E11.5) and birth, i.e. postnatal (P) stage 0 (P0),

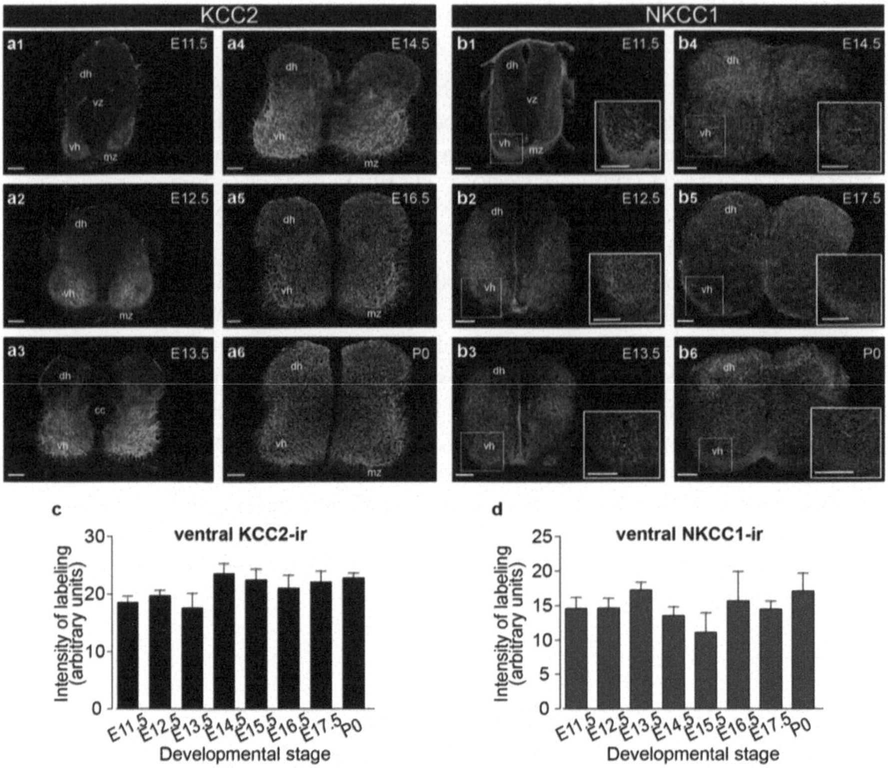

Fig. 1 Embryonic maturation of KCC2 and NKCC1 at lumbar level. (**a1**) KCC2-ir on a frontal section of lumbar spinal cord at E11.5. Note that the staining is restricted to the ventro-lateral grey matter in the ventral horn (vh). (**a2**) at E12.5, KCC2-ir has expanded to the whole ventral grey matter. (**a3**) at E13.5, the KCC2-ir may be detected in the dorsal zone and in dorso-medial areas. (**a4**) at E14.5, the whole grey matter shows KCC2-ir with a stronger staining in ventral areas compared to the dorsal one. (**a5–a6**) from E16.5 to P0, KCC2-ir spreads across the entire grey matter. (**b1**) at E11.5, NKCC1-ir is detected in the ventro-lateral and dorso-lateral grey matter and in the marginal zone. The inset, on the bottom right hand corner shows, as for all panels, a higher magnification of the NKCC1 staining within the motoneuronal area corresponding to boxed areas drawn on the global view. (**b2–b3**) from E12.5 to E13.5, the NKCC1 staining remains stable in the ventral area whereas immunoreactivity progressively invades the entire dorsal zone. (**b4–b6**) from E14.5 to P0, NKCC1-ir remains detectable in the ventral horn but appears stronger in the dorsal area. Staining was performed with the monoclonal T4 antibody. Scale bar = 100 µm. cc central canal, *dh* dorsal horn, *mz* marginal zone, *v* ventricle, *vh* ventral horn, *vz* ventricular zone. (**c**) and (**d**) quantitative analysis of KCC2- (**c**) and NKCC1-ir (**d**) in the ventral zone. Note that the intensity of both KCC2 and NKCC1 staining remains stable in the lumbar ventral area during ontogeny (no statistical difference between stages, One Way ANOVA). Each histogram corresponds to 4–12 preparations. (Data are means ± S.E.M. Modified from Delpy et al. 2008)

using immunohistochemistry and an anti-KCC2pan that does not differentiate between KCC2a and KCC2b (Markkanen et al. 2014), shows that the KCC2 protein is detected as early as E11.5, like the NKCC1 protein, in the ventral motor networks, and that both CCCs exhibit a constant staining over development in this area

(Delpy et al. 2008) (Fig. 1). The KCC2 immunoreactivity (ir) (KCC2-ir) is therefore quite comparable in immature and mature motor networks of the spinal cord. An early detection of KCC2-ir as well as NKCC1-ir is also found in the embryonic chicken spinal motor circuit with a differential expression in the medial and lateral division of the lateral motor column (LMC) (Law et al. 2014). Remarkably, the high expression of NKCC1 at early stages may be responsible for the elevated $[Na^+]$ (~ 60 mM) detected in mid-embryonic chicken MNs compared to late-embryonic MNs (~ 30 mM) (Lindsly et al. 2017).

Brainstem structures are also enriched with KCC2 at early stages. KCC2-ir is densely found in the mouse pre-Bötzinger complex at E18.5, a brainstem respiratory region (Chapuis et al. 2014). Interestingly, neonatal and adult neurons of the rat lateral superior olive (Blaesse et al. 2006) and cochlear nucleus (Vale et al. 2005) exhibit a similar KCC2-ir.

Modulation of the Chloride Homeostasis in Developing Motoneurons

The lifetime of the KCC2 transporter is >4 h (Puskarjov et al. 2012) whereas KCC2 turn-over rate measured on HEK-293 cell membranes (Lee et al. 2007) or hippocampal slices (Rivera et al. 2004) is in the range of 20–30 min. KCC2 surface stability is therefore far from being static and identifying signaling pathways and molecular partners that control this stability is central in order to better understand pathological changes in relation to changes in the excitation/inhibition balance. KCC2 expression and function can be rapidly modulated *via* large number of signaling pathways including Brain-Derived Neurotrophic Factor (BDNF), Insulin-Like Growth Factor1 (IGF-1), and Neurturin, the regulatory effects of BDNF on KCC2 being the most thoroughly characterized (Medina et al. 2014). In developing hippocampal neurons BDNF strongly increases KCC2 mRNA (Aguado et al. 2003; Ludwig et al. 2011) whereas in mature neurons BDNF decreased KCC2 mRNA and protein (Rivera et al. 2004), KCC2 phosphorylation (Wake et al. 2007) and KCC2 surface expression (Boulenguez et al. 2010; Wake et al. 2007). In the developing mouse spinal cord, KCC2 is expressed at early stages in the motoneuronal area (Fig. 1). Using organotypic cultures of E11.5 mouse spinal cords, we found that applying the BDNF scavenger (TrkB-IgG) does not prevent the early KCC2 expression, indicating that it does not rely on BDNF (Allain et al. 2015). Other potentials candidates are GABA and glycine that are expressed early in development (Allain et al. 2004, 2006). These classical neurotransmitters may interact with NKCC1 and/or KCC2 and play a trophic role during development. GABA appears essential for the development of KCC2-dependent inhibition in the cortex and retina. In fact, GABAergic activity modulates the mRNA levels or protein levels of KCC2, whose expression correlates with the switch between excitatory to inhibitory GABAergic effect (Ganguly et al. 2001; Leitch et al. 2005). In mouse spinal motoneuronal networks, GABA is not involved in the maturation of KCC2. Blocking $GABA_AR$ with

Gabazine does not affect the maturation of KCC2 in organotypic cultures of E11.5 spinal cord (Allain et al. 2015). Glycine, however, through non-synaptic release, is necessary for the maturation of the KCC2 protein but not the NKCC1 protein (Allain et al. 2015).

Different receptors are involved in KCC2 regulation including group I mGluR, 5-HT$_{2A}$R, A$_3$AR, OXTR, α1-Adrenergic R, mAchR (Mahadevan and Woodin 2016). In the postnatal P5-P7 rat lumbar spinal MNs, the 5-HT$_{2A}$ agonist TCB-2 rescues the Cl$^-$ homeostasis defects in SCI model by hyperpolarizing inhibitory postsynaptic potentials (IPSPs) (Bos et al. 2013). In the embryonic E17.5 mouse lumbar spinal MNs, 5-HT strongly hyperpolarizes the E_{GABAAR} by interacting with KCC2 (Martin et al. 2020).

Functional Consequences

Gramicidin perforated patch-clamp recording that preserves physiological [Cl$^-$]$_i$ indicates that E_{GABAAR} is very depolarizing in mouse lumbar spinal MNs at early developmental stages (E13.5) and drops after E15.5 (Delpy et al. 2008) (Fig. 2). E_{GABAAR} is equivalent to E_{Cl} because Cl$^-$ ions are mainly involved in GABAAR currents in fetal MNs (Bormann et al. 1987; Gao and Ziskind-Conhaim 1995). Therefore, [Cl$^-$]$_i$ that is calculated from the Nernst equation and the experimentally

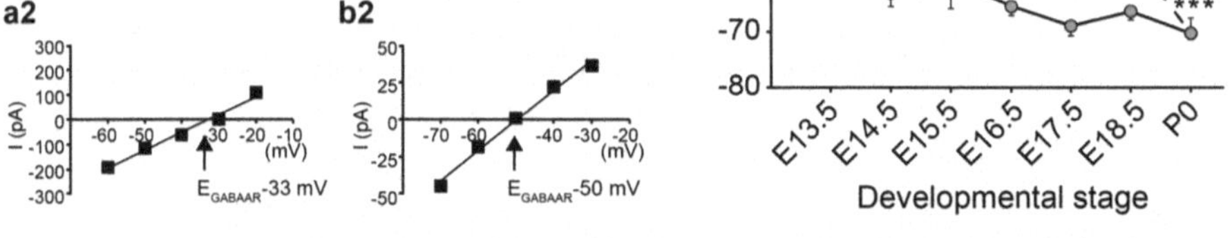

Fig. 2 Embryonic maturation of GABA$_A$R-related effects on membrane potential. (**a1**, **b1**) perforated-patch-clamp recordings showing currents evoked by brief application of isoguvacine, from different holding potentials, at E13.5 (**a1**) and E16.5 (**b1**). Current versus voltage plots (same neurons shown in **a1** and **b1**) allowing calculation of GABA$_A$R equilibrium potential (E_{GABAAR}): −33 mV at E13.5 (**a2**) and −50 mV at E16.5 (**b2**) (measured by linear regression analysis, $r^2 > 0.96$ in both cases). (**c**) evolution of E_{REST} (blue circles), spike threshold (yellow triangles) and E_{GABAAR} (black circles) in motoneurons during embryonic development. From E13.5 to E15.5, spike threshold and E_{GABAAR} are not significantly different (non parametric unpaired t test), whereas from E16.5 to birth (P0), E_{GABAAR} is significantly lower that spike threshold (**$P < 0.01$; ***$P < 0.0001$). Data are means ± S.E.M. (Modified from Delpy et al. 2008)

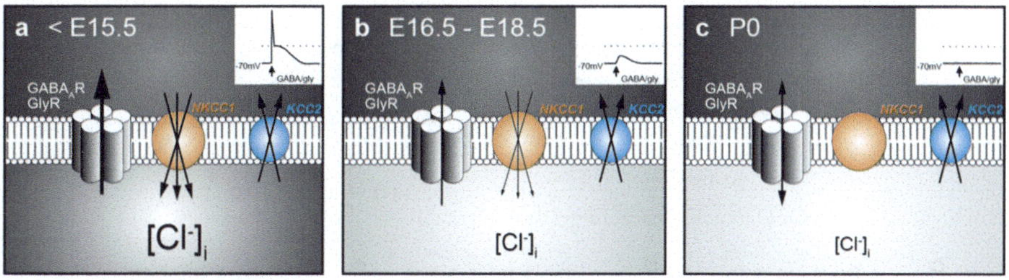

Fig. 3 Mechanisms underlying the developmental shift the GABA$_A$R/GlyR responses in the mouse spinal cord. (**a**) before E15.5, both KCC2 and NKCC1 are active and $[Cl^-]_i$ is high. Therefore, the activation of Cl^--gated receptors induces a large outward chloride current that depolarizes the membrane and triggers action potentials (see inset, right upper corner). (**b**) between E16.5 and E18.5, $[Cl^-]_i$ is reduced due to a decrease of the NKCC1 efficacy. At these stages, although the activation of GABA$_A$R and GlyR leads to depolarization, no action potential is elicited (inset). (**c**) at birth, NKCC1 has lost its efficiency and only the chloride extruder KCC2 remains functional; lowering $[Cl^-]_i$. Inset: around resting potential (E_{REST} −70 mV), the activation of GABA$_A$R/GlyR does not modify the membrane potential. (Modified from Delpy et al. 2008)

assessed E_{Cl} value, exhibits a dramatic reduction from 30 to 8 mM between E13.5 and P0 (Delpy et al. 2008).

Both co-transporters KCC2 and NKCC1 are involved in the control of $[Cl^-]_i$ in mouse spinal embryonic MNs and the relative efficacy of NKCC1 and KCC2 changes as the animal matures: the chloride extruder KCC2 is active in immature and in behaviorally mature spinal networks while NKCC1 becomes inefficient during maturation. In the mouse spinal cord, E_{Cl} maturation is related to a NKCC1 inactivation after E15.5 and KCC2 becomes mainly responsible for E_{Cl} values after this stage (Delpy et al. 2008) (Fig. 3).

Stil and collaborators show that the maturation of chloride homeostasis is not completed at birth in rat lumbar MNs and that the upregulation of KCC2 plays a key role in the shift from depolarizing to hyperpolarizing IPSPs (Stil et al. 2009). Between P0 and P7 E_{Cl}, obtained by assessing the reversal of inhibitory postsynaptic potentials (E_{IPSP}), becomes ~8 mV more hyperpolarized (Fig. 4). Interestingly, because the resting membrane potential (E_{REST}) remains stable between P0 and P7, E_{IPSP} is described as being depolarizing in the P0-P3 developmental time windows before becoming hyperpolarizing at P7 (Fig. 4). Depolarizing E_{GABA} is also described in mouse spinal MNs at P1-P3, before dropping to E_{REST} level at P8-P11 (Stein et al. 2004).

KCC2 Knockout mice die at birth due to respiratory insufficiency highlighting the importance of this neuronal CCC in the maintenance of inhibition in postnatal neuronal networks (Hubner et al. 2001). As we have seen KCC2 expression is developmentally regulated. KCC2 overexpression reverses the neuronal Cl^- gradient and results in fewer mature spontaneously active spinal neurons, more immature silent neurons, and disrupted motor activity as demonstrated in developing zebrafish (Reynolds et al. 2008). This indicates that Cl^--mediated excitation plays a role in promoting neurogenesis.

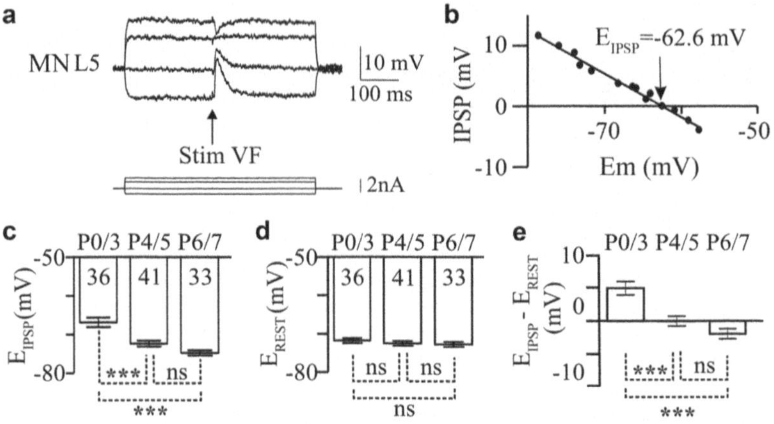

Fig. 4 Inhibitory postsynaptic potentials (IPSPs) become hyperpolarizing during the first postnatal week. (**a**) IPSPs evoked by ventral funiculus (VF) stimulation (arrow) at different holding potentials in a lumbar MN at P0. Conventional intracellular recordings. At the resting membrane potential (E_{REST}; −71.6 mV), the evoked IPSP is depolarizing. (**b**) Amplitude of IPSPs plotted against holding potential giving a reversal potential of the IPSP (E_{IPSP}) at −62.6 mV. (**c**) The E_{IPSP} is significantly more hyperpolarized at P4–5 than at P0–3 ($P < 0.001$, one-way ANOVA, Tukey post-test). (**d**) The resting membrane potential (E_{REST}) is not statistically different (~ −72 mV) in the three age groups. (**e**) Amplitude of depolarizing or hyperpolarizing IPSPs recorded at E_{REST} ($P < 0.001$ between P0–3 and the other two groups, one-way ANOVA, Tukey post-test). (Modified from Stil et al. 2009)

Whereas NKCC1 plays a role in maintaining a depolarizing E_{Cl} in immature mouse lumbar spinal MNs (Delpy et al. 2008), in the auditory brainstem MNs NKCC1 mRNA is not detected during the depolarizing phase, implying that this transporter does not contribute to the high $[Cl^-]_i$ in this brain area (Balakrishnan et al. 2003). Similarly, NKCC1 does not serve to accumulate chloride in immature retinal neurons (Zhang et al. 2007). What other transporter may be involved in creating a depolarized E_{Cl} in immature brainstem MNs? One candidate is the $Cl^-/HCO3^-$ exchanger AE3. Physiological data collected by P. Wenner's lab indicate that, in addition to NKCC1, the anion exchanger AE3 is likely to contribute to chloride accumulation in embryonic MNs (Gonzalez-Islas et al. 2009).

E_{Cl} values drop in mouse lumbar MNs after E15.5 but the net effect of GABA/glycine remains largely depolarizing at E17.5 (Delpy et al. 2008), leading to depolarizing GABAergic/glycinergic postsynaptic potentials (dGPSPs) composed of an initial and inhibitory (shunting) phase, carried by g_{Cl}, followed by an excitatory phase due to depolarization. While dGPSPs are purely excitatory at E13.5 because E_{Cl} is highly depolarized at this stage, they become exclusively inhibitory at E17.5 because E_{Cl} has dropped (Branchereau et al. 2016). E_{Cl} plays a determinant role in the inhibitory/excitatory components of dGPSP because (1) the initial excitatory action of dGPSP observed at E13.5 can be experimentally reversed to inhibition if E_{Cl} is hyperpolarized and (2) the inhibitory $GABA_AR$ response observed at E17.5 can be reversed to excitatory when E_{Cl} is experimentally increased (Branchereau et al. 2016) (Fig. 5). As a functional consequence, the E_{Cl} values drop in mouse

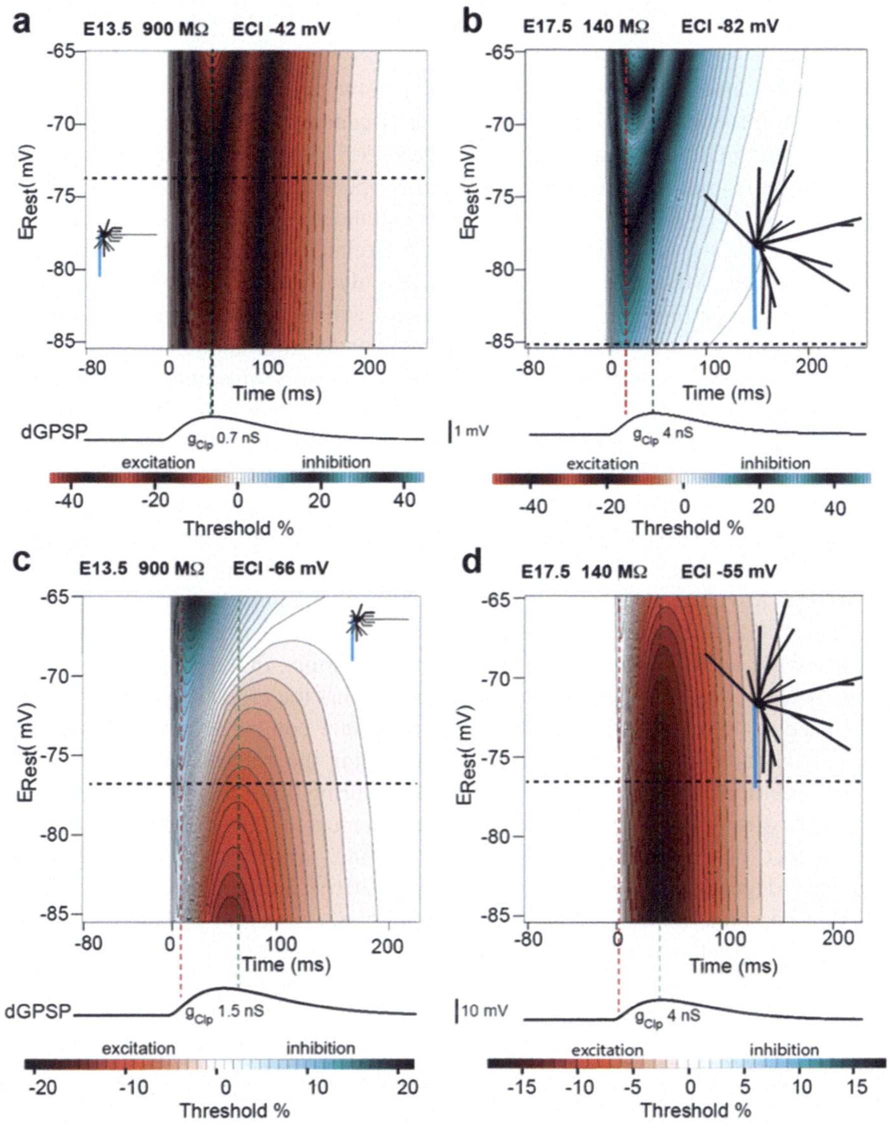

Fig. 5 Time course of the excitatory and inhibitory components of dGPSPs in E13.5 and E17.5 neurons according to physiological E_{Cl} values. Inhibitory excitatory time course (IETC) maps representing the time course (abscissa) of the threshold current (duration: 10 ms) during the simulation of a dGPSP in E13.5 (**a**, E_{Cl} −42 mV; **c**, E_{Cl} −66 mV) and E17.5 (**b**, E_{Cl} −66 mV; **d**, E_{Cl} −55 mV) neuron models as measured for a series of E_{Rest} (ordinate). Horizontal dashed lines represent the average of physiological E_{Rest}. Vertical black dashed lines represent the simulated dGPSP peak. The magnitude of the excitatory (red colors) and inhibitory (cyan colors) effects is indicated by the color scale (% of change of the threshold current, see bottom bars for values). Contours are disposed at every 1% change. Identical rules hold true in all of the subsequent IETC maps. The time course of the simulated dGPSP is presented below each IETC map (same time scale). Blue and red dashed vertical lines indicate the maximum excitatory and inhibitory effects, respectively. The shape and size of the simulated neurons are presented in the insets on the IETC maps (proximal portion of the axon in blue). Note that $[Cl^-]_i$ varies between subcellular regions such as dendrites, soma or axon (Szabadics et al. 2006; Glykys et al. 2014). Here E_{Cl} values are simulated based on $[Cl^-]_i$ assessed in the somatic compartment. (Modified from Branchereau et al. 2016)

lumbar MNs after E15.5 when spinal motor networks start exhibiting left-right alternation in locomotor-like activity due to half-centers organization and commissural inhibition (Branchereau et al. 2000). At E17.5 a clear left-right alternation is recorded, indicating that shunting inhibition is operating at this prenatal developmental stage. E_{Cl} values also drops when spontaneous network activity vanishes (Yvert et al. 2004), indicating that the NKCC1-related elevated $[Cl^-]_i$ detected in chick embryonic MNs using a chloride sensitive dye (Chub et al. 2006) is closely related to the occurrence of spontaneous network activity is immature networks.

During embryonic development, spinal MNs undergo dramatic changes in size and passive properties between E13.5 and P0, while spike threshold remains stable around −40 mV, resting membrane potential (Em) becomes more hyperpolarized from −55 to −70 mV, input resistance (Rin) decreases from ~900 MΩ at E13.5 to <200 MΩ after E16.5, membrane capacitance (Cm) increases continuously during the course of embryonic development from <30 pF at E13.5 to ~200 pF at P0 suggesting that MNs become gradually larger (Delpy et al. 2008). Of course the size and passive properties of spinal MNs continue to evolve during postnatal development (Jean-Xavier et al. 2006; Vinay et al. 2000). Computer simulations show that inhibitory dGPSP effects are moderately favored in large neurons with low Rin compared to small neurons with high Rin for similar dGPSP and similar E_{Cl} (Branchereau et al. 2016). Simulations also show that the effect of passive properties linked to the morphology of MNs remains minor compared with the effect of E_{Cl}. Finally, simulations demonstrate that increasing the chloride conductance g_{Cl} favors inhibition either during a single dGPSP or during trains in which g_{Cl} can summate increasingly with train frequency (Branchereau et al. 2016). Indeed, the excitatory effect of EPSPs is overcome by shunting inhibition in dGPSPs in a frequency-dependent manner (Branchereau et al. 2016). As a conclusion, among the different parameters that are Em, E_{Cl}, g_{Cl}, neuron size and Rin, E_{Cl} and g_{Cl} are the two main parameters controlling the magnitude and time course of inhibition and excitation during single dGPSPs. Interestingly, Jean-Xavier and collaborators used computer simulations to mimic the effect of physiological IPSPs on postnatal rat spinal MNs but because their simulations were performed with large gClp (120 nS), these authors primarily observed inhibitory effects (Jean-Xavier et al. 2007).

4 Pathological Considerations

Growing evidence suggest that a KCC2-dependent alteration of chloride homeostasis is often related to CNS disorders such as stress (Hewitt et al. 2009; Ostroumov et al. 2016), neuropathic pain (Coull et al. 2003; see Kaila et al. 2014 for review). Amyotrophic Lateral Sclerosis (ALS), also known as Lou Gehrig's disease, is a rapidly progressive neurodegenerative disease that affects pyramidal cells in the motor cortex, brainstem and spinal cord (Van Damme et al. 2017). ALS is usually clinically diagnosed at late age, peak age at onset being 58–63 years for sporadic disease and 47–52 years for familial disease (Logroscino et al. 2010). However, mechanisms leading to the disease progression remain unknown. ALS is often associated with spasticity leading to important disabling complications that compromise manual dexterity and gait (Kiernan et al. 2011). Spasticity is described as implying a hyperexcitability in spinal networks linked to a KCC2-related reduction of chloride-related synaptic inhibition (Boulenguez et al. 2010). Understanding how the dysregulation of the chloride homeostasis evolves in ALS MNs is therefore crucial. Mouse models based on genetic defects described in familial ALS have been developed. The transgenic mouse model SOD1G93A (SOD, Gly93→Ala substitution), which expresses large amounts of human mutant SOD, faithfully recapitulates a vast majority of the pathology's abnormalities seen in ALS patients (Fogarty 2018). This ALS model reveals early deficit in chloride homeostasis that evolves in lumbar spinal MNs. In fact, a high $[Cl^-]_i$ is found in prenatal $SOD1^{G93A}$ E17.5 lumbar spinal MNs compared to WT MNs from the same littermate, associated with a KCC2 reduction (Branchereau et al. 2019). High $[Cl^-]_i$ slows down the decay of glycine- and GABA-mediated inhibitory synaptic events (Houston et al. 2009; Pitt et al. 2008), the structural basis of this effect being a direct effect of chloride ions acting in the pore of glycine and GABA channels (Moroni et al. 2011). In line with this direct modulation of the GABA/glycine synaptic transmission, $SOD1^{G93A}$ E17.5 lumbar spinal MNs exhibit a slower relaxation (higher tau_{decay}) of synaptic inhibitory events, strengthening inhibition and compensating their more depolarized E_{Cl} (Branchereau et al. 2019) (Fig. 6).

Prenatal lumbar $SOD1^{G93A}$ MNs exhibit a reduced level of KCC2 compared to WT MNs from the same littermate. How these early changes influence the later pathology remains an open question but we may hypothesize that low KCC2 may lead to lack of inhibition at birth time and long-lasting motoneuron burden. It remains to verify how this change evolves with the disease. No changes in the KCC2 oligomer/monomer ratio is described in the lumbar spinal motoneuronal network of $SOD1^{G93A}$ mice at 8, 12 and 16 weeks of age compared to WT mice (Modol et al. 2014). whereas a downregulation of the potassium chloride cotransporter KCC2 is present in vulnerable (large-size) MNs of the $SOD1^{G93A}$ mouse model of ALS at late stages (120 days) (Fuchs et al. 2010).

5 Concluding Remarks

The balance between excitation and Cl^--dependent inhibition is a key factor regulating the output of motoneurons controlling movement and vital functions such as respiration. We have seen that Cl^- homeostasis is closely regulated by KCC2 and NKCC1 but other mechanisms likely operate. Also, it must be kept in mind that $[Cl^-]_i$ is likely not constantly distributed in MNs as demonstrated in pyramidal cortical and hippocampal neurons (Szabadics et al. 2006; Glykys et al. 2014). Spinal MNs are large neurons lacking the calcium binding proteins calbindin and calretinin (Berg et al. 2018) and exhibiting a poor capacity for regulating $[Ca^{2+}]_i$. These properties put spinal MNs at risk of excessive glutamatergic stimulation, which in turn

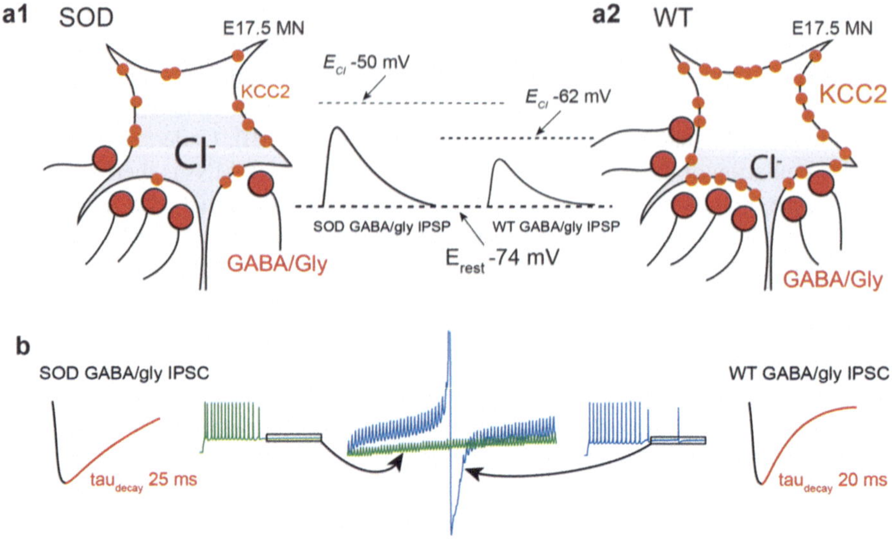

Fig. 6 (**a**) Altered inhibitory inputs to fetal SOD1^{G93A} MNs. $[Cl^-]_i$ is higher in SOD MNs (**a1**) than in WT MNs (**a2**) because of a KCC2 down-regulation, leading to an increased GABA/Gly-induced depolarizing effect (see insets). (**b**) Consequence of increasing tau_{decay} on the GABA/gly inhibitory effect in SOD-like MNs. Due to an accumulation in the intracellular compartment, E_{GABAAR} exerts a strong depolarizing effect. A burst of spikes generated by MNs is hardly blocked by a barrage of GABA/gly events (see blue traces) when tau_{decay} is set to 20 ms. Increasing tau_{decay} to 25 ms allows a better summation of the shunting component of the depolarizing GABA/gly post-synaptic event leading to a better clamp of Em towards E_{GABAAR} and to the blockade of MN discharge (see green traces). (Modified from Branchereau et al. 2019)

can lead to increased $[Ca^{2+}]_i$ and subsequent high demands on mitochondrial function, as they attempt to regulate excess $[Ca^{2+}]_i$. Therefore, Cl^--related inhibitory inputs from Renshaw cells and other spinal interneurons on MNs play a crucial role to regulate the output of MNs. This is true in immature spinal networks generating spontaneous network activity because of GABA and glycine release (Czarnecki et al. 2014; Chub and O'Donovan 2001; Hanson and Landmesser 2003) but also in mature networks that may easily exhibit depolarizing GABA/glycine synaptic activities in pathological conditions (Allain et al. 2011). Further research is necessary to better understand the complex mechanisms controlling chloride homeostasis in MNs and to identify disfunctions likely occurring during development and leading to pathologies.

References

Aguado F, Carmona MA, Pozas E, Aguilo A, Martinez-Guijarro FJ, Alcantara S, Borrell V, Yuste R, Ibanez CF, Soriano E (2003) BDNF regulates spontaneous correlated activity at early developmental stages by increasing synaptogenesis and expression of the K+/Cl- co-transporter KCC2. Development 130:1267–1280

Allain AE, Bairi A, Meyrand P, Branchereau P (2004) Ontogenic changes of the GABAergic system in the embryonic mouse spinal cord. Brain Res 1000:134–147

Allain AE, Bairi A, Meyrand P, Branchereau P (2006) Expression of the glycinergic system during the course of embryonic development in the mouse spinal cord and its co-localization with GABA immunoreactivity. J Comp Neurol 496:832–846

Allain AE, Le Corronc H, Delpy A, Cazenave W, Meyrand P, Legendre P, Branchereau P (2011) Maturation of the GABAergic transmission in normal and pathologic motoneurons. Neural Plast 2011:905624

Allain AE, Cazenave W, Delpy A, Exertier P, Barthe C, Meyrand P, Cattaert D, Branchereau P (2015) Nonsynaptic glycine release is involved in the early KCC2 expression. Dev Neurobiol 2015. https://doi.org/10.1002/dneu.22358

Balakrishnan V, Becker M, Lohrke S, Nothwang HG, Guresir E, Friauf E (2003) Expression and function of chloride transporters during development of inhibitory neurotransmission in the auditory brainstem. J Neurosci 23:4134–4145

Ben-Ari Y (2014) The GABA excitatory/inhibitory developmental sequence: a personal journey. Neuroscience 279:187–219

Berg EM, Bertuzzi M, Ampatzis K (2018) Complementary expression of calcium binding proteins delineates the functional organization of the locomotor network. Brain Struct Funct 223:2181–2196

Blaesse P, Guillemin I, Schindler J, Schweizer M, Delpire E, Khiroug L, Friauf E, Nothwang HG (2006) Oligomerization of KCC2 correlates with development of inhibitory neurotransmission. J Neurosci 26:10407–10419

Bormann J, Hamill OP, Sakmann B (1987) Mechanism of anion permeation through channels gated by glycine and gamma-aminobutyric acid in mouse cultured spinal neurones. J Physiol 385:243–286

Bos R, Sadlaoud K, Boulenguez P, Buttigieg D, Liabeuf S, Brocard C, Haase G, Bras H, Vinay L (2013) Activation of 5-HT2A receptors upregulates the function of the neuronal K-Cl cotransporter KCC2. Proc Natl Acad Sci U S A 110:348–353

Boulenguez P, Liabeuf S, Bos R, Bras H, Jean-Xavier C, Brocard C, Stil A, Darbon P, Cattaert D, Delpire E, Marsala M, Vinay L (2010) Down-regulation of the potassium-chloride cotransporter KCC2 contributes to spasticity after spinal cord injury. Nat Med 16:302–307

Branchereau P, Morin D, Bonnot A, Ballion B, Chapron J, Viala D (2000) Development of lumbar rhythmic networks: from embryonic to neonate locomotor-like patterns in the mouse. Brain Res Bull 53:711–718

Branchereau P, Cattaert D, Delpy A, Allain AE, Martin E, Meyrand P (2016) Depolarizing GABA/glycine synaptic events switch from excitation to inhibition during frequency increases. Sci Rep 6:21753

Branchereau P, Martin E, Allain AE, Cazenave W, Supiot L, Hodeib F, Laupenie A, Dalvi U, Zhu H, Cattaert D (2019) Relaxation of synaptic inhibitory events as a compensatory mechanism in fetal SOD spinal motor networks. elife 8:e51402

Chapuis C, Autran S, Fortin G, Simmers J, Thoby-Brisson M (2014) Emergence of sigh rhythmogenesis in the embryonic mouse. J Physiol 592:2169–2181

Chebib M (2004) GABAC receptor ion channels. Clin Exp Pharmacol Physiol 31:800–804

Chub N, O'Donovan MJ (2001) Post-episode depression of GABAergic transmission in spinal neurons of the chick embryo. J Neurophysiol 85:2166–2176

Chub N, Mentis GZ, O'Donovan MJ (2006) Chloride-sensitive MEQ fluorescence in chick embryo motoneurons following manipulations of chloride and during spontaneous network activity. J Neurophysiol 95:323–330
Come E, Heubl M, Schwartz EJ, Poncer JC, Levi S (2019) Reciprocal regulation of KCC2 trafficking and synaptic activity. Front Cell Neurosci 13:48
Coull JA, Boudreau D, Bachand K, Prescott SA, Nault F, Sik A, De Koninck P, De Koninck Y (2003) Trans-synaptic shift in anion gradient in spinal lamina I neurons as a mechanism of neuropathic pain. Nature 424:938–942
Czarnecki A, Le Corronc H, Rigato C, Le Bras B, Couraud F, Scain AL, Allain AE, Mouffle C, Bullier E, Mangin JM, Branchereau P, Legendre P (2014) Acetylcholine controls GABA-, glutamate-, and glycine-dependent giant depolarizing potentials that govern spontaneous motoneuron activity at the onset of synaptogenesis in the mouse embryonic spinal cord. J Neurosci 34:6389–6404
Darman RB, Forbush B (2002) A regulatory locus of phosphorylation in the N terminus of the Na-K-Cl cotransporter, NKCC1. J Biol Chem 277:37542–37550
Delpy A, Allain AE, Meyrand P, Branchereau P (2008) NKCC1 cotransporter inactivation underlies embryonic development of chloride-mediated inhibition in mouse spinal motoneuron. J Physiol 586:1059–1075
Dubois CJ, Cardoit L, Schwarz V, Markkanen M, Airaksinen MS, Uvarov P, Simmers J, Thoby-Brisson M (2018) Role of the K(+)-Cl(-) cotransporter KCC2a isoform in mammalian respiration at birth. eNeuro 5:ENEURO.0264-18.2018
Fogarty MJ (2018) Driven to decay: excitability and synaptic abnormalities in amyotrophic lateral sclerosis. Brain Res Bull 140:318–333
Fuchs A, Ringer C, Bilkei-Gorzo A, Weihe E, Roeper J, Schutz B (2010) Downregulation of the potassium chloride cotransporter KCC2 in vulnerable motoneurons in the SOD1-G93A mouse model of amyotrophic lateral sclerosis. J Neuropathol Exp Neurol 69:1057–1070
Fukuda A (2020) Chloride homeodynamics underlying modal shifts in cellular and network oscillations. Neurosci Res 156:14–23. https://doi.org/10.1016/j.neures.2020.02.010
Ganguly K, Schinder AF, Wong ST, Poo M (2001) GABA itself promotes the developmental switch of neuronal GABAergic responses from excitation to inhibition. Cell 105:521–532
Gao BX, Ziskind-Conhaim L (1995) Development of glycine- and GABA-gated currents in rat spinal motoneurons. J Neurophysiol 74:113–121
Glykys J, Dzhala V, Egawa K, Balena T, Saponjian Y, Kuchibhotla KV, Bacskai BJ, Kahle KT, Zeuthen T, Staley KJ (2014) Local impermeant anions establish the neuronal chloride concentration. Science 343:670–675
Gonzalez-Islas C, Chub N, Wenner P (2009) NKCC1 and AE3 appear to accumulate chloride in embryonic motoneurons. J Neurophysiol 101:507–518
Hanson MG, Landmesser LT (2003) Characterization of the circuits that generate spontaneous episodes of activity in the early embryonic mouse spinal cord. J Neurosci 23:587–600
Hentschke M, Wiemann M, Hentschke S, Kurth I, Hermans-Borgmeyer I, Seidenbecher T, Jentsch TJ, Gal A, Hubner CA (2006) Mice with a targeted disruption of the Cl-/HCO3- exchanger AE3 display a reduced seizure threshold. Mol Cell Biol 26:182–191
Hewitt SA, Wamsteeker JI, Kurz EU, Bains JS (2009) Altered chloride homeostasis removes synaptic inhibitory constraint of the stress axis. Nat Neurosci 12:438–443
Houston CM, Bright DP, Sivilotti LG, Beato M, Smart TG (2009) Intracellular chloride ions regulate the time course of GABA-mediated inhibitory synaptic transmission. J Neurosci 29:10416–10423
Hubner CA, Stein V, Hermans-Borgmeyer I, Meyer T, Ballanyi K, Jentsch TJ (2001) Disruption of KCC2 reveals an essential role of K-Cl cotransport already in early synaptic inhibition. Neuron 30:515–524
Jean-Xavier C, Pflieger JF, Liabeuf S, Vinay L (2006) Inhibitory postsynaptic potentials in lumbar motoneurons remain depolarizing after neonatal spinal cord transection in the rat. J Neurophysiol 96:2274–2281

Jean-Xavier C, Mentis GZ, O'Donovan MJ, Cattaert D, Vinay L (2007) Dual personality of GABA/glycine-mediated depolarizations in immature spinal cord. Proc Natl Acad Sci U S A 104:11477–11482
Kaila K, Price TJ, Payne JA, Puskarjov M, Voipio J (2014) Cation-chloride cotransporters in neuronal development, plasticity and disease. Nat Rev Neurosci 15:637–654
Kanaka C, Ohno K, Okabe A, Kuriyama K, Itoh T, Fukuda A, Sato K (2001) The differential expression patterns of messenger RNAs encoding K-Cl cotransporters (KCC1,2) and Na-K-2Cl cotransporter (NKCC1) in the rat nervous system. Neuroscience 104:933–946
Kiernan MC, Vucic S, Cheah BC, Turner MR, Eisen A, Hardiman O, Burrell JR, Zoing MC (2011) Amyotrophic lateral sclerosis. Lancet 377:942–955
Law C, Paquet M, Kania A (2014) Emergence of motor circuit activity. PLoS One 9:e93836
Lee HH, Walker JA, Williams JR, Goodier RJ, Payne JA, Moss SJ (2007) Direct protein kinase C-dependent phosphorylation regulates the cell surface stability and activity of the potassium chloride cotransporter KCC2. J Biol Chem 282:29777–29784
Lees GJ (1991) Inhibition of sodium-potassium-ATPase: a potentially ubiquitous mechanism contributing to central nervous system neuropathology. Brain Res Brain Res Rev 16:283–300
Leitch E, Coaker J, Young C, Mehta V, Sernagor E (2005) GABA type-A activity controls its own developmental polarity switch in the maturing retina. J Neurosci 25:4801–4805
Lindsly C, Gonzalez-Islas C, Wenner P (2017) Elevated intracellular Na(+) concentrations in developing spinal neurons. J Neurochem 140:755–765
Logroscino G, Traynor BJ, Hardiman O, Chio A, Mitchell D, Swingler RJ, Millul A, Benn E, Beghi E, Eurals (2010) Incidence of amyotrophic lateral sclerosis in Europe. J Neurol Neurosurg Psychiatry 81:385–390
Lucas O, Hilaire C, Delpire E, Scamps F (2012) KCC3-dependent chloride extrusion in adult sensory neurons. Mol Cell Neurosci 50:211–220
Ludwig A, Uvarov P, Soni S, Thomas-Crusells J, Airaksinen MS, Rivera C (2011) Early growth response 4 mediates BDNF induction of potassium chloride cotransporter 2 transcription. J Neurosci 31:644–649
Mahadevan V, Woodin MA (2016) Regulation of neuronal chloride homeostasis by neuromodulators. J Physiol 594:2593–2605
Markkanen M, Karhunen T, Llano O, Ludwig A, Rivera C, Uvarov P, Airaksinen MS (2014) Distribution of neuronal KCC2a and KCC2b isoforms in mouse CNS. J Comp Neurol 522:1897–1914
Martin E, Cazenave W, Allain AE, Cattaert D, Branchereau P (2020) Implication of 5-HT in the dysregulation of chloride homeostasis in prenatal spinal motoneurons from the G93A mouse model of amyotrophic lateral sclerosis. Int J Mol Sci 21(3):1107
Medina I, Friedel P, Rivera C, Kahle KT, Kourdougli N, Uvarov P, Pellegrino C (2014) Current view on the functional regulation of the neuronal K(+)-Cl(-) cotransporter KCC2. Front Cell Neurosci 8:27
Modol L, Mancuso R, Ale A, Francos-Quijorna I, Navarro X (2014) Differential effects on KCC2 expression and spasticity of ALS and traumatic injuries to motoneurons. Front Cell Neurosci 8:7
Moore YE, Deeb TZ, Chadchankar H, Brandon NJ, Moss SJ (2018) Potentiating KCC2 activity is sufficient to limit the onset and severity of seizures. Proc Natl Acad Sci U S A 115:10166–10171
Moroni M, Biro I, Giugliano M, Vijayan R, Biggin PC, Beato M, Sivilotti LG (2011) Chloride ions in the pore of glycine and GABA channels shape the time course and voltage dependence of agonist currents. J Neurosci 31:14095–14106
Ostroumov A, Thomas AM, Kimmey BA, Karsch JS, Doyon WM, Dani JA (2016) Stress increases ethanol self-administration via a shift toward excitatory GABA signaling in the ventral tegmental area. Neuron 92:493–504
Owens DF, Kriegstein AR (2002) Is there more to GABA than synaptic inhibition? Nat Rev Neurosci 3:715–727
Payne JA, Stevenson TJ, Donaldson LF (1996) Molecular characterization of a putative K-Cl cotransporter in rat brain. A neuronal-specific isoform. J Biol Chem 271:16245–16252

Pitt SJ, Sivilotti LG, Beato M (2008) High intracellular chloride slows the decay of glycinergic currents. J Neurosci 28:11454–11467

Puskarjov M, Ahmad F, Kaila K, Blaesse P (2012) Activity-dependent cleavage of the K-Cl cotransporter KCC2 mediated by calcium-activated protease calpain. J Neurosci 32:11356–11364

Rahmati N, Hoebeek FE, Peter S, De Zeeuw CI (2018) Chloride homeostasis in neurons with special emphasis on the olivocerebellar system: differential roles for transporters and channels. Front Cell Neurosci 12:101

Reynolds A, Brustein E, Liao M, Mercado A, Babilonia E, Mount DB, Drapeau P (2008) Neurogenic role of the depolarizing chloride gradient revealed by global overexpression of KCC2 from the onset of development. J Neurosci 28:1588–1597

Rivera C, Voipio J, Payne JA, Ruusuvuori E, Lahtinen H, Lamsa K, Pirvola U, Saarma M, Kaila K (1999) The K+/Cl- co-transporter KCC2 renders GABA hyperpolarizing during neuronal maturation. Nature 397:251–255

Rivera C, Voipio J, Thomas-Crusells J, Li H, Emri Z, Sipila S, Payne JA, Minichiello L, Saarma M, Kaila K (2004) Mechanism of activity-dependent downregulation of the neuron-specific K-Cl cotransporter KCC2. J Neurosci 24:4683–4691

Rivera C, Voipio J, Kaila K (2005) Two developmental switches in GABAergic signalling: the K+-Cl- cotransporter KCC2 and carbonic anhydrase CAVII. J Physiol 562:27–36

Russell JM (2000) Sodium-potassium-chloride cotransport. Physiol Rev 80:211–276

Ruusuvuori E, Li H, Huttu K, Palva JM, Smirnov S, Rivera C, Kaila K, Voipio J (2004) Carbonic anhydrase isoform VII acts as a molecular switch in the development of synchronous gamma-frequency firing of hippocampal CA1 pyramidal cells. J Neurosci 24:2699–2707

Silayeva L, Deeb TZ, Hines RM, Kelley MR, Munoz MB, Lee HH, Brandon NJ, Dunlop J, Maguire J, Davies PA, Moss SJ (2015) KCC2 activity is critical in limiting the onset and severity of status epilepticus. Proc Natl Acad Sci U S A. 112(11):3523–3528. https://doi.org/10.1073/pnas.1415126112. Epub 2015 Mar 2.

Stein V, Hermans-Borgmeyer I, Jentsch TJ, Hubner CA (2004) Expression of the KCl cotransporter KCC2 parallels neuronal maturation and the emergence of low intracellular chloride. J Comp Neurol 468:57–64

Stil A, Liabeuf S, Jean-Xavier C, Brocard C, Viemari JC, Vinay L (2009) Developmental up-regulation of the potassium-chloride cotransporter type 2 in the rat lumbar spinal cord. Neuroscience 164:809–821

Szabadics J, Varga C, Molnar G, Olah S, Barzo P, Tamas G (2006) Excitatory effect of GABAergic axo-axonic cells in cortical microcircuits. Science 311:233–235

Uvarov P, Ludwig A, Markkanen M, Pruunsild P, Kaila K, Delpire E, Timmusk T, Rivera C, Airaksinen MS (2007) A novel N-terminal isoform of the neuron-specific K-Cl cotransporter KCC2. J Biol Chem 282:30570–30576

Uvarov P, Ludwig A, Markkanen M, Soni S, Hubner CA, Rivera C, Airaksinen MS (2009) Coexpression and heteromerization of two neuronal K-Cl cotransporter isoforms in neonatal brain. J Biol Chem 284:13696–13704

Vale C, Caminos E, Martinez-Galan JR, Juiz JM (2005) Expression and developmental regulation of the K+-Cl- cotransporter KCC2 in the cochlear nucleus. Hear Res 206:107–115

Van Damme P, Robberecht W, Van Den Bosch L (2017) Modelling amyotrophic lateral sclerosis: progress and possibilities. Dis Model Mech 10:537–549

Vinay L, Brocard F, Pflieger JF, Simeoni-Alias J, Clarac F (2000) Perinatal development of lumbar motoneurons and their inputs in the rat. Brain Res Bull 53:635–647

Wake H, Watanabe M, Moorhouse AJ, Kanematsu T, Horibe S, Matsukawa N, Asai K, Ojika K, Hirata M, Nabekura J (2007) Early changes in KCC2 phosphorylation in response to neuronal stress result in functional downregulation. J Neurosci 27:1642–1650

Watanabe M, Fukuda A (2015) Development and regulation of chloride homeostasis in the central nervous system. Front Cell Neurosci 9:371

Watanabe M, Zhang J, Mansuri MS, Duan J, Karimy JK, Delpire E, Alper SL, Lifton RP, Fukuda A, Kahle KT (2019) Developmentally regulated KCC2 phosphorylation is essential for dynamic GABA-mediated inhibition and survival. Sci Signal 12:eaaw9315
Wu WL, Ziskind-Conhaim L, Sweet MA (1992) Early development of glycine- and GABA-mediated synapses in rat spinal cord. J Neurosci 12:3935–3945
Yvert B, Branchereau P, Meyrand P (2004) Multiple spontaneous rhythmic activity patterns generated by the embryonic mouse spinal cord occur within a specific developmental time window. J Neurophysiol 91:2101–2109
Zeuthen T (2010) Water-transporting proteins. J Membr Biol 234:57–73
Zhang LL, Delpire E, Vardi N (2007) NKCC1 does not accumulate chloride in developing retinal neurons. J Neurophysiol 98:266–277
Zhu L, Lovinger D, Delpire E (2005) Cortical neurons lacking KCC2 expression show impaired regulation of intracellular chloride. J Neurophysiol 93:1557–1568

Normal Development and Pathology of Motoneurons: Anatomy, Electrophysiological Properties, Firing Patterns and Circuit Connectivity

Joshua I. Chalif and George Z. Mentis

Abstract This chapter will provide an introduction into motoneuron anatomy, electrophysiological properties, firing patterns focusing on development and also describing several pathological conditions that affect mononeurons. It starts with a historical retrospective describing the early landmark work into motoneurons. The next section lays out the various types of motoneurons (alpha, beta, and gamma) and their subclasses (fast-twitch fatigable, fast-twitch fatigue-resistant, and slow-twitch fatigue resistant), highlighting the functional relevance of this classification scheme. The third section describes the development of motoneurons' passive and active electrophysiological properties. This section also defines the major terms one uses in describing how a neuron functions electrophysiologically. The electrophysiological aspects of a neuron is critical to understanding how it behaves within a circuit and contributes to behavior since the firing of an action potential is how neurons communicate with each other and with muscles. The electrophysiological changes of motoneurons over development underlies how their function changes over the lifetime of an organism. After describing the properties of individual motoneurons, the chapter then turns to revealing how motoneurons interact within complex neural circuits, with other motoneurons as well as sensory neurons, and how these circuits change over development. Finally, this chapter ends with highlighting some recent

J. I. Chalif
Departments of Neurology and Pathology & Cell Biology, Center for Motor Neuron Biology and Disease, Columbia University, New York, NY, USA

Department of Neurosurgery, Brigham and Women's Hospital, Harvard University, Boston, MA, USA
e-mail: jchalif@bwh.harvard.edu

G. Z. Mentis (✉)
Departments of Neurology and Pathology & Cell Biology, Center for Motor Neuron Biology and Disease, Columbia University, New York, NY, USA
e-mail: gzmentis@columbia.edu

M. J. O'Donovan, M. Falgairolle (eds.), *Vertebrate Motoneurons*, Advances in Neurobiology 28, https://doi.org/10.1007/978-3-031-07167-6_3

advances made in motoneuron pathology, focusing on spinal muscular atrophy, amyotrophic lateral sclerosis, and axotomy.

Keywords Motoneuron · Development · Spinal cord · Firing pattern · Motor unit · Excitability · Intrinsic properties · SMA · ALS

1 Introduction and Historical Retrospective

The human body's control of muscle activity has been recognized to be critical since antiquity when Hippocrates, Aristotle and Galen all wrote about how muscles permit movements such as breathing, swallowing, and navigating one's environment. Work continued in the Renaissance with Leonardo da Vinci, Andreas Vesalius, Rene Descartes, and Giovanni Borelli making significant insights into our understanding of the anatomy and physiology of the human body. Motor neurons came to be recognized as arguably the most important neurons in the animal body, since they are responsible for movements essential for life. Sir Charles Sherrington, the preeminent English neurophysiologist recognized the importance of motoneurons over a century ago, stating "to move is all that mankind can do, and that for such the sole executant is muscle, whether in whispering a syllable or in felling a forest" (1924).

The motoneuron cell body or soma resides within the spinal cord and their main axon projects through peripheral nerves to contact muscles in the periphery at a specialized synapse known as the neuromuscular junction. Due to their accessibility, they have long-served as a model for neuronal and circuit mechanisms operating during development. It has been known since the late 1700s that electricity is involved in neuronal activity (Galvani 1791; Volta 1800), and motoneurons have served as the cell type providing key information about the function of neurons more generally. However, the explosion in our understanding of motoneuron biology started with the first intracellular recordings from cat motoneurons. In the mid-1900s, John Eccles and colleagues in New Zealand (Brock et al. 1952, 1953), and J. Walter Woodbury and Harry Patton in Seattle, USA (Woodbury and Patton 1952) acquired the first intracellular recordings from spinal motoneurons. They used sharp microelectrodes to penetrate central nervous system tissue and a single cell membrane, high impedance input stages, and controlled current delivery systems, combined with a high-speed display system to achieve this major advance. Since then, our techniques to interrogate neurons have greatly expanded to include more complex intracellular recording techniques, such as whole-cell patch clamp as well as perforated patch, to more sophisticated approaches and imaging techniques (such as optogenetics, chemogenetics, and calcium and voltage sensitive dye imaging), all of which have been utilized to examine the development, electrophysiological properties, and function of motoneurons, both during normal behavior and in disease state.

2 Classification of Motoneurons

There are three main types of spinal motoneurons (Manuel and Zytnicki 2011). Alpha motoneurons (α-MN) innervate the large extrafusal striated muscle fibers that produce the force underlying movement. Gamma motoneurons (γ-MN) innervate the smaller intrafusal muscle fibers which reside within the muscle spindle, the specialized sensory endings of proprioceptive neurons. The main function of γ-MNs is to modulate the sensitivity of muscle spindles to changes in stretch. Beta motoneurons (β-MN) are an intermediary between α-MN and γ-MN, which innervate both intrafusal and extrafusal muscle fibers (Bessou et al. 1965). β-MN however, is the class of motoneurons that have been the least studied, mostly due to technical difficulties in identification and intracellular recordings. Due to these reasons, β-MNs are often grouped together with α-MNs, and this review will focus on α-MNs and γ-MNs.

A motor pool designates the collection of motoneurons that innervate a single muscle. A given motor pool contains a specific ratio of α-MNs to γ-MNs. Motoneurons within any given pool display certain molecular, morphological, and connectivity properties that are specific to that motor pool.

α-motoneurons are larger in soma size than γ-motoneurons and their axon conduction velocity is greater than that of γ-MNs, reflecting their larger axonal diameter (Simon et al. 1996). The dendritic architecture of γ-motoneurons tends to be less branched and simpler than those observed in α-MNs (Burke et al. 1994; Fleshman et al. 1988). Another important difference between α-and γ-MNs is the lack of proprioceptive synaptic inputs onto γ-MNs, whereas all α-motoneurons receive direct proprioceptive inputs (Eccles et al. 1960; Eccles et al. 1957). Furthermore, α-MNs possess recurrent axon collaterals which project locally within the ventral spinal cord, whereas γ-MNs have much fewer axon collaterals (Cullheim and Ulfhake 1979a, b; Westbury 1982). During development, it is difficult to distinguish between α-MNs and γ-MNs based on their soma size (Shneider et al. 2009). It has been demonstrated that confidence in distinguishing these two types of motoneurons based on soma size can be achieved around the second postnatal week (Shneider et al. 2009). Instead, investigators have focused their efforts on uncovering different genetic markers that can label these two motoneuron types. The development of γ-motoneurons appears to involve the transcription factor *Err3* (Friese et al. 2009), *Gfra1* and they also express higher levels of GDNF (Shneider et al. 2009). α-motoneurons, on the other hand, express higher levels of the *NeuN* antigen (Friese et al. 2009; Shneider et al. 2009).

Through the pioneering work by Bob Burke, α-motoneurons can be subdivided based upon the contractile properties of the motor units that they form with muscles into three subclasses: fast-twitch fatigable (FF), fast-twitch fatigue-resistant (FR), and slow-twitch fatigue resistant (S) (Burke et al. 1973, 1982). These subclasses of motoneurons have corresponding morphological and functional characteristics which suit them to their particular function. S motoneurons have smaller cell bodies and axons, whereas FF motoneurons are larger, with large-diameter axons. The average membrane area for FF motoneurons is ~20% larger than that of S motoneurons, reflecting their increased dendritic tree and increased axonal branching (Cullheim et al. 1987).

This simple difference in size has profound implications for the neurons' electrical properties and thereby their function. These ideas are embodied in the 'Size Principle' which posits that the smaller motoneurons are activated first during muscle activation (Henneman et al. 1965). Input resistance, or the voltage change over a cell membrane to a given steady current injection, is governed primarily by two factors: cell size (soma and dendritic arborization) and the properties of the cell's expressed channels. Due to their small size, the S motoneurons have a greater input resistance and are therefore more excitable, exhibiting greater voltage changes and reaching the threshold to initiate action potentials with less synaptic input. Conversely, the large FF motoneurons have a smaller input resistance and require greater synaptic input to reach the threshold to initiate action potentials. FR motoneurons, by contrast, have intermediate properties. During long bursts of sustained activation, such as during standing or walking, the S motoneurons are recruited in order to activate the slow motor units. During short bursts of forceful contraction, such as jumping or sprinting, the FF motoneurons are additionally recruited.

The firing rate is another important differentiating feature between FF, FR, and S motoneurons. S motoneurons exhibit continuous repetitive firing, which can persist even in the absence of synaptic excitatory input, partially due to long-lasting persistent inward currents (Heckman et al. 2008; Lee and Heckman 1998). FF motoneurons are fatigable, exhibiting high frequency action potentials for a brief period of time, which enables a time-limited forceful muscular contraction. Another important electrophysiological parameter that varies between S and FF motoneurons is the duration of the after-hyperpolarization (AHP), which is shaped by calcium-dependent potassium currents (Gardiner 1993). This current occurs after an action potential and mostly determines the maximal firing frequency of the neuron. FF motoneurons, which as previously stated have an increased maximal firing frequency, therefore have shorter AHPs than S motoneurons, which matches the contractile properties of their target muscle fibers (Manuel and Zytnicki 2011).

In general, neurons respond in a linear fashion to synaptic drive. With increasing excitatory synaptic input, neurons exhibit a higher frequency of action potentials. This can be easily visualized on F/I (frequency-current) plots when recording intracellularly from neurons. Upon increments of increasing current injection, the firing frequency increases. Motoneurons behave in this fashion for most currents. However, intriguingly, motoneurons have been shown to display non-linear behavior, known as "bistability" or "plateau potentials," where they can sustain repetitive firing even in the absence of a depolarizing drive (excitatory synaptic input physiologically or intracellular current injection experimentally) (Conway et al. 1988; Hounsgaard et al. 1988; Lee and Heckman 1999). This behavior increases in the presence of monoamine neuromodulators (Lee and Heckman 1998, 1999), which implies that motoneuron behavior and firing can be greatly regulated by descending inputs from the brain, which is the source of monoamine neurotransmission in the spinal cord. Plateau potentials are also present in motoneurons during development (Fig. 1).

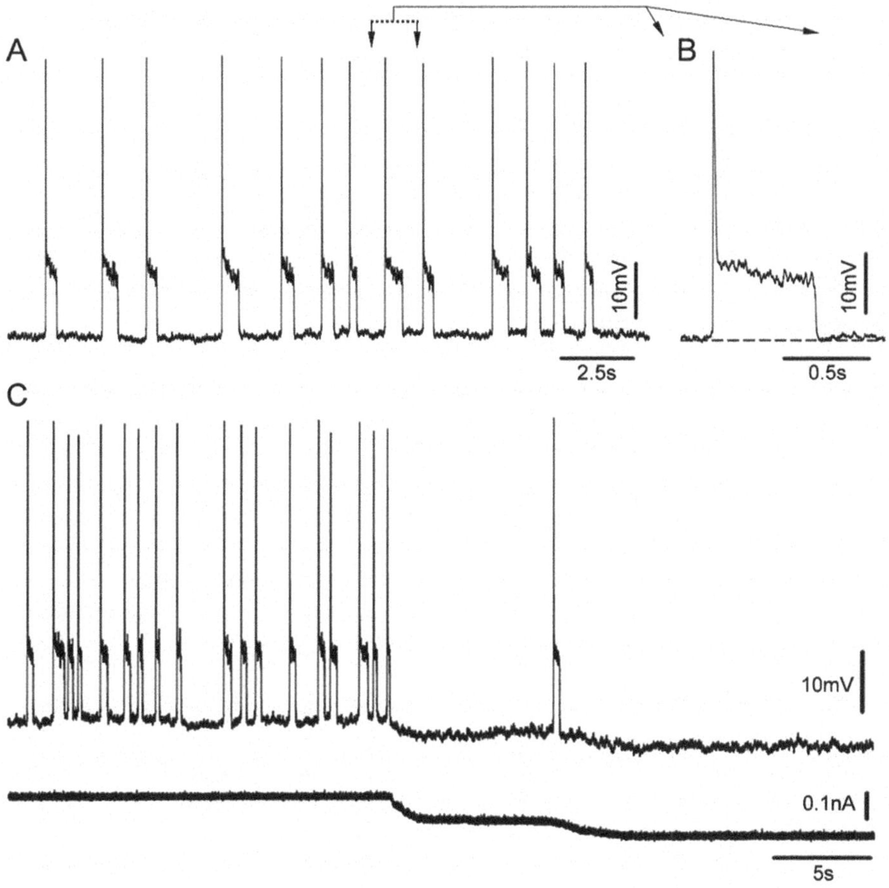

Fig. 1 Bistability in neonatal motoneurons. (**a**) Spontaneous activity recorded from a P3, L1 motoneuron with whole-cell patch clamp. The membrane potential exhibited a bistable behavior at its own resting membrane potential (Vm = -54 mV). An action potential was elicited at the onset of the bistable depolarization of approximately 10 mV which varied in duration. (**b**) Enlarged magnification of an event. (**c**) The bistable behavior was abolished with injection of hyperpolarizing current (bottom trace). (Unpublished observations by G.Z. Mentis and M.J. O'Donovan)

The plateau potential is thought to be driven by activation of a persistent inward current mediated by low-threshold L-type Ca^{2+} channels (L-type I_{Ca}) or persistent sodium channels (I_{NaP}). More recent work in *ex vivo* spinal cord preparations has shown that these potentials are temperature dependent as well, likely through a sodium-mediated current (I_{CaN}) flowing through putative TRPV2 channels (Bouhadfane et al. 2013).

In summary, the classes and subclasses of motoneurons demonstrate how morphological and molecular properties influence electrophysiological properties which determine the cell's ultimate function.

3 The Development of the Electrophysiological Properties and Firing Patterns of Motoneurons

Motoneurons are one of the most well-studied classes of neurons in the central nervous system. Although there are still numerous unanswered questions, we now know many of the processes involved in the functional and anatomical development of motoneurons. In addition to spinal motoneurons, the most studied types of motoneurons are oculomotor and hypoglossal, which reside in the brainstem. Overall, it is generally agreed that motoneuron excitability decreases while motoneuron firing rate increases over development.

A physiological property of motoneurons corresponding to their excitability, the input resistance (Fig. 2), decreases during postnatal development for both brainstem and spinal motoneurons. For oculomotor motoneurons, input resistance decreases ~25% over development (Carrascal et al. 2005). For spinal and hypoglossal motoneurons, input resistance decreases ~50% (Fulton and Walton 1986; Mentis et al. 2007; Núñez-Abades et al. 1993; Robinson and Cameron 2000). The developmental decrease in input resistance is likely due to two factors: (i) by an increase in leak-potassium currents and/or an increase in tonic inhibitory synaptic input (Cameron et al. 2000; Núñez-Abades et al. 2000) and (ii) by an increase in the somato-dendritic area of the motoneuron (Durand et al. 2006; Elbasiouny et al. 2010; Mentis et al. 2007).

The increase in channel function in the motoneuron membrane could be attributed to an increase in the number of channel(s) or due to an increase in existing channel function from modifications of their kinetics which are partially controlled by neuromodulators (Perrier and Hounsgaard 2000; Rekling et al. 2000). As discussed above, as neuronal size increases, the input resistance decreases. The soma size of motoneurons exhibits small changes during early postnatal development (Cameron et al. 1991; Carrascal et al. 2005; Núñez-Abades and Cameron 1995). However, the dendritic tree of hypoglossal, oculomotor and spinal motoneurons increases significantly during postnatal development (Cameron et al. 1991; Núñez-Abades and Cameron 1995; Ramírez and Ulfhake 1991; Ulfhake et al. 1988). At birth the dendritic tree represents about 90% of the total surface area of the motoneuron (Fig. 2), whereas in adulthood it represents 97–99%. In addition, the average branch order in oculomotor motoneurons increases, providing evidence that the dendritic architecture is also becoming more complex (Carrascal et al. 2005). This anatomical change causes an increase in the surface area of motoneurons which is believed to support the proliferation and development of synaptic inputs over the same time period (Núñez-Abades et al. 2000; Rekling et al. 2000; Vinay et al. 2000). This increase in overall neuronal size likely contributes to the decrease in input resistance during development. Regardless of the precise mechanism, the decrease in input resistance is functionally important since it determines the voltage change in a neuron to a given synaptic input, thereby making motoneurons less excitable.

The neuronal time constant is another passive membrane property used to describe the electrophysiological state of neurons. The time constant is calculated as

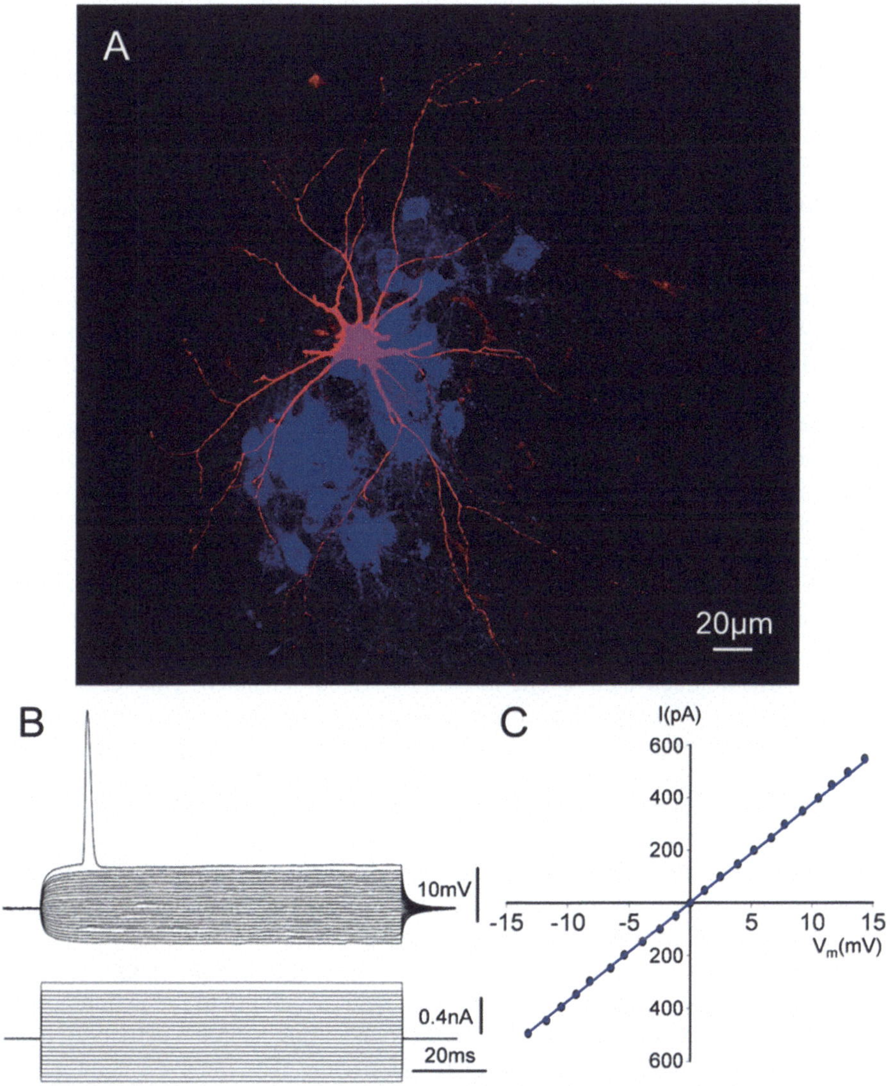

Fig. 2 Morphological and physiological characteristics of neonatal motoneurons. (**a**) Confocal Z-stack projection of a P5, L5 mouse motoneuron, recorded and subsequently filled with Neurobiotin and revealed with Avidin-Biotin Complex (in red). ChAT immunoreactivity is shown in blue. (**b**) Current-to-voltage relationship in the L5 motoneuron shown in (**a**), using the ex vivo spinal cord preparation. Superimposed traces of the voltage response (top traces) following steps of negative and positive current (bottom trace) injection through whole-cell patch clamp electrode. (**c**) The I-V plot revealed a linear response, indicating the passive membrane nature. The input resistance is calculated from the slope of the linear response (Rn = 26.3 MΩ). (Unpublished observations by G.Z. Mentis and M.J. O'Donovan)

the time taken for the transmembrane voltage to reach 63% of the plateau voltage in response to a positive or negative current injection during intracellular recordings. The passive membrane time constant is the product of the membrane resistance and the membrane capacitance. The time constant decreases in motoneurons during postnatal development, reflecting the reduction in input resistance (Carrascal et al. 2005; Mentis et al. 2007). The resting membrane potential, on the other hand, does not change between neonatal and adult motoneurons for hypoglossal, oculomotor, or spinal motoneurons (Carrascal et al. 2005; Fulton and Walton 1986; Núñez-Abades et al. 1993).

For newly born motoneurons, the voltage response to a negative current injection is nearly linear (Di Pasquale et al. 2001). However, for several decades it has been reported that adult hypoglossal motoneurons exhibit a membrane potential rectification in response to a negative current injection (Mosfeldt Laursen and Rekling 1989; Núñez-Abades and Cameron 1995; Viana et al. 1995). This membrane potential rectification can be viewed on voltage responses as a "sag" that is voltage-dependent and is most frequently seen with hyperpolarization of more than 20 mV below the neuron's resting membrane potential. In addition to hypoglossal motoneurons, this property has also been reported in spinal motoneurons and other brainstem motoneurons (Bertrand and Cazalets 1998; Magariños-Ascone et al. 1999; Russier et al. 2003), and occurs in about 40% of motoneurons (Carrascal et al. 2005). This "sag" is due to the h-current, mediated through the inward rectifying HCN (hyperpolarization-activated cyclic-nucleotide gated) channel. This channel is responsible for the pacemaking ability of the heart and has been proposed to be involved in rhythmogenesis in neurons (Harris-Warrick 2010). Correspondingly, in oculomotor, hypoglossal, and spinal motoneurons, a "post-inhibitory rebound" is sometimes seen, which is a rebound depolarization following a hyperpolarizing current which may be strong enough to elicit an action potential (Núñez-Abades et al. 1993; Viana et al. 1994). Although the h-current may also be involved, this characteristic is thought to be additionally mediated by a low-threshold calcium current (Umemiya and Berger 1994). Like the sag, this post-inhibitory rebound is voltage-dependent and increases over the postnatal development of motoneurons.

Active membrane properties determine the neuron's ability to fire action potentials, which are subsequently responsible for transmitting neuronal activity to their postsynaptic neuronal targets within neuronal circuits. The rheobase is an active membrane property defined as the minimum current required to generate an action potential. Rheobase is usually inversely correlated with input resistance such that a decrease in input resistance results in an increase in the rheobase. As expected therefore, the rheobase increases about two-fold during postnatal development in hypoglossal motoneurons (Carrascal et al. 2005). However, despite the decrease in input resistance, surprisingly the rheobase also decreases in oculomotor motoneurons during postnatal development. The voltage threshold ($V_{threshold}$) required to produce an action potential remains largely unchanged in hypoglossal motoneurons over development, but is actually decreased in oculomotor motoneurons in adulthood. Together, these findings suggest that despite the decrease in input resistance in oculomotor motoneurons, paradoxically it may become easier for them to fire action potentials in adulthood. There are two currents proposed to underlie this difference, a persistent Na^+ current and a long-lasting Ca^{2+} current, which are thought to undergo a differential modulation during postnatal maturation (Carrascal et al. 2005; Hornby et al. 2002; Russo and Hounsgaard 1999).

The action potential itself exhibits differences between postnatal and adult motoneurons as well. The spike duration decreases by ~50% in oculomotor, hypoglossal, and spinal motoneurons over development (Carrascal et al. 2005; Fulton and Walton 1986; Núñez-Abades et al. 1993; O'Dowd et al. 1988; Viana et al. 1994). This decrease in spike duration is driven by an increase in the slope of the both the rising and falling phase of the action potential, which is due to increases in the sodium and potassium currents respectively. Indeed, in *Xenopus* embryos, the Na^+ current doubles and K^+ current triples over development resulting in a narrowing of the action potential in spinal neurons (O'Dowd et al. 1988).

Motoneurons, like most other neurons, also exhibit an after-hyperpolarization, a reduction in the membrane voltage immediately after the action potential which is lower than the voltage membrane prior to the initiation of the action potential. The duration of the after-hyperpolarization decreases over postnatal development, while its amplitude remains unchanged in oculomotor, hypoglossal, and spinal motoneurons (Carrascal et al. 2005; Fulton and Walton 1986; Núñez-Abades et al. 1993; Russier et al. 2003; Viana et al. 1994). The after-hyperpolarization is mediated by a calcium-dependent potassium current, which can be modified by neuromodulators. The duration of the after-hyperpolarization is directly correlated with the spike duration (Núñez-Abades et al. 1993; Viana et al. 1994). During earlier development, the longer duration of the action potential in motoneurons leads to a greater influx of Ca^{++}, which causes an increase in the calcium-dependent potassium current and therefore the after-hyperpolarization (Viana et al. 1994, 1995).

The after-hyperpolarization is an important active property of neurons since its duration directly affects the firing frequency and repetitive firing pattern of the neuron. Thus, with a shorter after-hyperpolarization later in development in motoneurons, it is expected that the firing frequency would be increased. Motoneurons in the oculomotor and hypoglossal nuclei, as well as in the spinal cord are able to fire repetitive action potentials either in tonic (constant firing frequency) or burst-tonic (initially high frequency followed by a slower steady state frequency) modes at birth (Carrascal et al. 2005; Núñez-Abades et al. 1993; Viana et al. 1995). As expected from the after-hyperpolarization decrease, the maximum firing frequency is increased over development in all classes of motoneurons (Carrascal et al. 2005; Núñez-Abades et al. 1993; Viana et al. 1995). In particular, the maximal frequency of oculomotor motoneurons is increased about three-fold in the adult compared to the neonate, while hypoglossal motoneuron firing frequency increases about two-fold (Núñez-Abades et al. 1993). Of note, oculomotor motoneurons exhibit burst-tonic firing activity likely responsible for the saccades and subsequent fixation that the extraocular muscles control (De La Cruz et al. 1989; Fuchs et al. 1988). In addition to the calcium-dependent potassium channels responsible for the after-hyperpolarization, several other channels likely contribute to the increased firing frequency in adult motoneurons. A hyperpolarization-activated inward current

(Bayliss et al. 1994), a long-lasting calcium current and persistent sodium current (Li et al. 2004; Russo and Hounsgaard 1999), and an A-type potassium current (Russier et al. 2003) are all increased over development, and may also contribute towards the increase in firing frequency.

4 Development of Motoneuron Spinal Cord Circuitry

During normal development motoneurons are tasked to participate in the formation of sensory-motor circuits. This process involves the correct connections between motoneurons and their appropriate peripheral targets, the skeletal muscles, a process that is achieved during mid-embryonic development in mice. However, a critical event is the connection between sensory neurons and motoneurons in order to convey peripheral information directly to motoneurons. This is achieved by proprioceptive sensory neurons from a single muscle contacting homonymous motoneurons (Chen et al. 2003). In addition, motoneurons receive proprioceptive synaptic inputs not only from homonymous muscles but also from heteronymous muscles which are synergistic in nature, while avoiding synaptic inputs from proprioceptive neurons originating from antagonist muscles (Fig. 3) (Chen et al. 2003; Mendelsohn et al. 2015).

Although by adulthood motoneurons receive proprioceptive synapses only from homonymous and heteronymous (Eccles et al. 1957) but not from antagonistic muscles, during late embryonic development inappropriate proprioceptive synapses are formed between motoneurons and sensory synapses from antagonistic muscles (Poliak et al. 2016; Seebach and Ziskind-Conhaim 1994). Refinement of sensory-motor circuits occurs late during embryonic development and early in postnatal development, such that the homonymous and heteronymous inputs are reinforced, while the antagonist input is pruned. The molecular pathways responsible for this synaptic refinement, through the activation of the classical complement pathway via C1q, its originating protein (Vukojicic et al. 2019), are beginning to emerge. The classical complement pathway is one of the three pathways for activating the complement system which is an important component of the innate immune system. Through the classical complement pathway, multiple complement proteins are recruited ultimately leading to assembly of the membrane attack complex which creates pores on target cell's membranes leading to cell lysis and death. Interestingly, emerging evidence indicates that this pathway is also used in synaptic refinement during neurodevelopment (Stevens et al. 2007; Vukojicic et al. 2019).

Interestingly, during early development, motoneurons are electrically connected through gap junctions (Mentis et al. 2002; Personius et al. 2007; Walton and Navarrete 1991). This electrical coupling has been confirmed both electrophysiologically as well as morphologically through dye-coupling (Fig. 4). The electrical coupling is mediated by at least five different connexins (Cx), Cx36, Cx37, Cx40, Cx43 and Cx45 (Chang et al. 1999). The electrical communication amongst spinal motoneurons diminishes by the end of the second postnatal week (Walton and Navarrete 1991) and by the end of the first postnatal week in brainstem

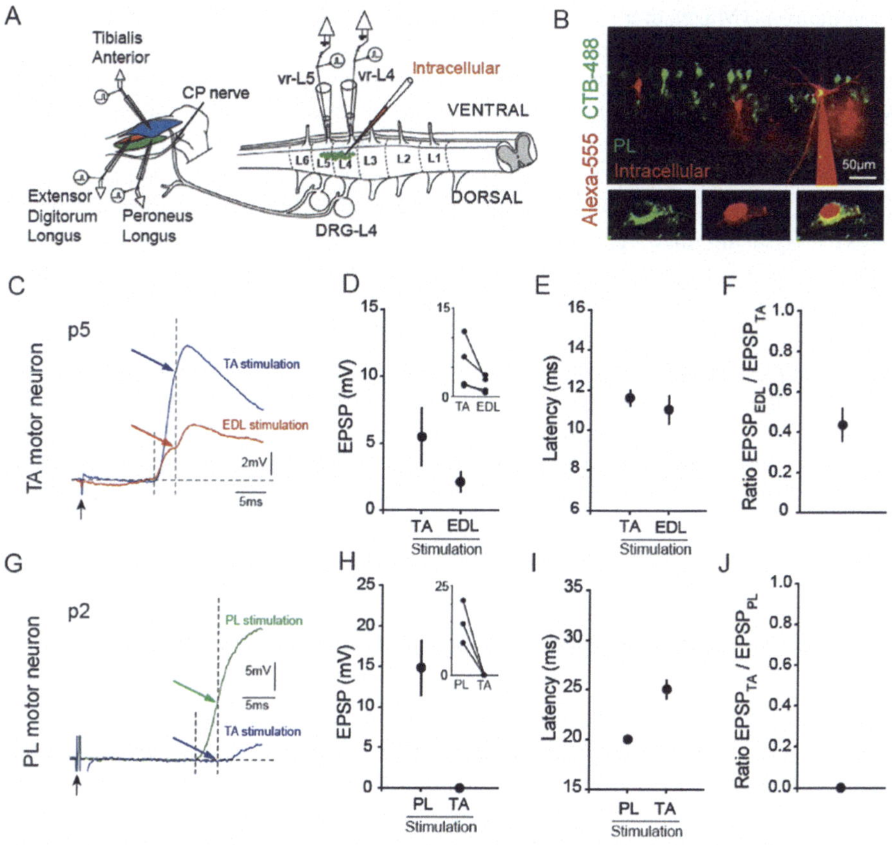

Fig. 3 Functional validation of heteronymous connections in neonatal motoneurons. (**a**) Schematic of the ex vivo lumbar spinal cord-hindlimb preparation. Stimulating electrodes were placed in tibialis anterior (TA), extensor digitorum longus (EDL), and peroneus longus (PL) muscles to activate proprioceptive fibers. Ventral roots were cut and placed into suction electrodes for either stimulation or recording. Motoneurons (in green) were visually identified following muscle-specific labeling by CTb-488 and recorded intracellularly using whole-cell patch clamp. (**b**) Image of PL motoneurons retrogradely labeled at birth with CTb-488 and showing three cells filled with intracellular dye after whole-cell recording (in red). (**c**) Intracellularly recorded EPSPs in a single retrogradely labeled TA motoneuron upon stimulation of TA or EDL muscle. Black arrow indicates stimulus artifact. First dashed line indicates onset of EPSP response. Second dashed line indicates the maximum amplitude of the monosynaptic response, as determined at 3 ms after EPSP onset. (**d**) Average EPSP amplitudes induced in TA motoneurons upon TA or EDL muscle stimulation. Inset represents corresponding relationship within each recorded motoneuron. (**e**) Average latency of EPSP onset upon TA or EDL stimulation, as defined in relation to stimulus artifact. (**f**) Average ratio of the EPSP amplitude induced in each TA motoneuron by EDL stimulation to the EPSP amplitude induced by TA stimulation. (**g**) Intracellularly recorded EPSPs in a single retrogradely labeled PL motoneuron upon stimulation of PL or TA muscle. The longer latency of response is due to the younger age (P2) at the time of recording. (**h**) Average EPSP amplitudes induced in PL motoneurons upon PL or TA muscle stimulation. (**i**) Average latency of EPSP onset upon PL or TA stimulation. (**j**) Average ratio of the EPSP amplitude induced in each PL motoneuron by TA stimulation to the EPSP amplitude induced by PL stimulation. (Adapted from Mendelsohn et al. 2015; Neuron 87:111–23)

motoneurons (Mazza et al. 1992). The precise function of electrical communication amongst motoneurons has not been clarified, but evidence suggests that it may contribute towards the development of adult motor units, since the polyneuronal innervation of muscle fibers is eliminated along the same time course of spinal motoneuron electrical communication during the first two postnatal weeks (Walsh and Lichtman 2003). To this end, during late embryonic and early postnatal life, neuromuscular junctions undergo synapse elimination that is modulated by patterns of motoneuron activity. By using mice that lack Cx40, electrical coupling among lumbar motoneurons, measured by whole-cell recordings, was reduced, and single motor unit recordings in awake, behaving neonates showed that temporally correlated motoneuron activity was also reduced (Personius et al. 2007).

This study suggests that gap junctional coupling modulates neuronal activity patterns that, in turn, mediate synaptic competition. Furthermore, the emergence of the adult motor unit along with the disappearance of gap junction coupling appear to depend upon muscle activation and its afferent feedback since the process can be disrupted by muscle paralysis (Fig. 4), thereby prolonging polyneuronal innervation (Pastor et al. 2003).

5 Motoneuron Function Is Altered in Disease States

Pathological conditions that impact motoneuron function are readily apparent in clinical practice due to the importance of motoneurons in controlling movement. Neurodegenerative conditions such as spinal muscular atrophy and amyotrophic lateral sclerosis, lesions of peripheral nerves such as axotomy, and compressive lesions of the spinal cord caused by trauma, tumor, or degenerative spine disease are among the more commonly encountered pathologies of motoneurons.

During normal aging, alpha motoneurons are thought to survive until middle age and steadily decline in numbers at older ages (Hashizume et al. 1988). However, in

Fig.4 (continued) between the SLD amplitude and the intensity of the injected current, confirming the electrotonic nature of the SLD potential. (**d**) Antidromic spike evoked in a P6 Botulinum neurotoxin-treated TA motoneuron by stimulation of the VR L4 at a shorter latency than its SLD (SLD, arrow). Inset: a higher magnification of the SLD at the base of the spike. Arrowhead denotes the onset of the stimulus artifact. (**e**) Insensitivity of SLD to collision test. Three traces are superimposed, corresponding to the antidromic action potential evoked after stimulating the VR L4 (arrowhead), the underlying SLD evoked by subthreshold VR stimulation (arrow) and the collision of the antidromic spike with a preceding orthodromic spike elicited by intracellular current injection which shows the preservation of the SLD (circle). (**f, g**) Photomicrographs showing a cluster of coupled motoneurons labeled with biocytin in a P7 Botulinum neurotoxin-treated spinal cord. (**f**) Several motoneurons appeared in close apposition to the dorsally directed dendritic trunk (thick arrows in both **f** and **g**) of the intracellularly stained motoneuron shown in **g** (located in an adjacent section). a: axon (small arrow in **f**). (**h**) Camera lucida reconstruction of the injected cell and its cluster of coupled motoneurons (stippled somata) shown in **f** and **g**. The reconstruction was performed using 7 parasagittal 100-μm-thick sections. Note two axon collateral branches (arrows) arising from the main axon and dorsally directed toward the motoneuron pool. (Adapted from Pastor et al., 2003; J. Neurophysiol. 89:793–805)

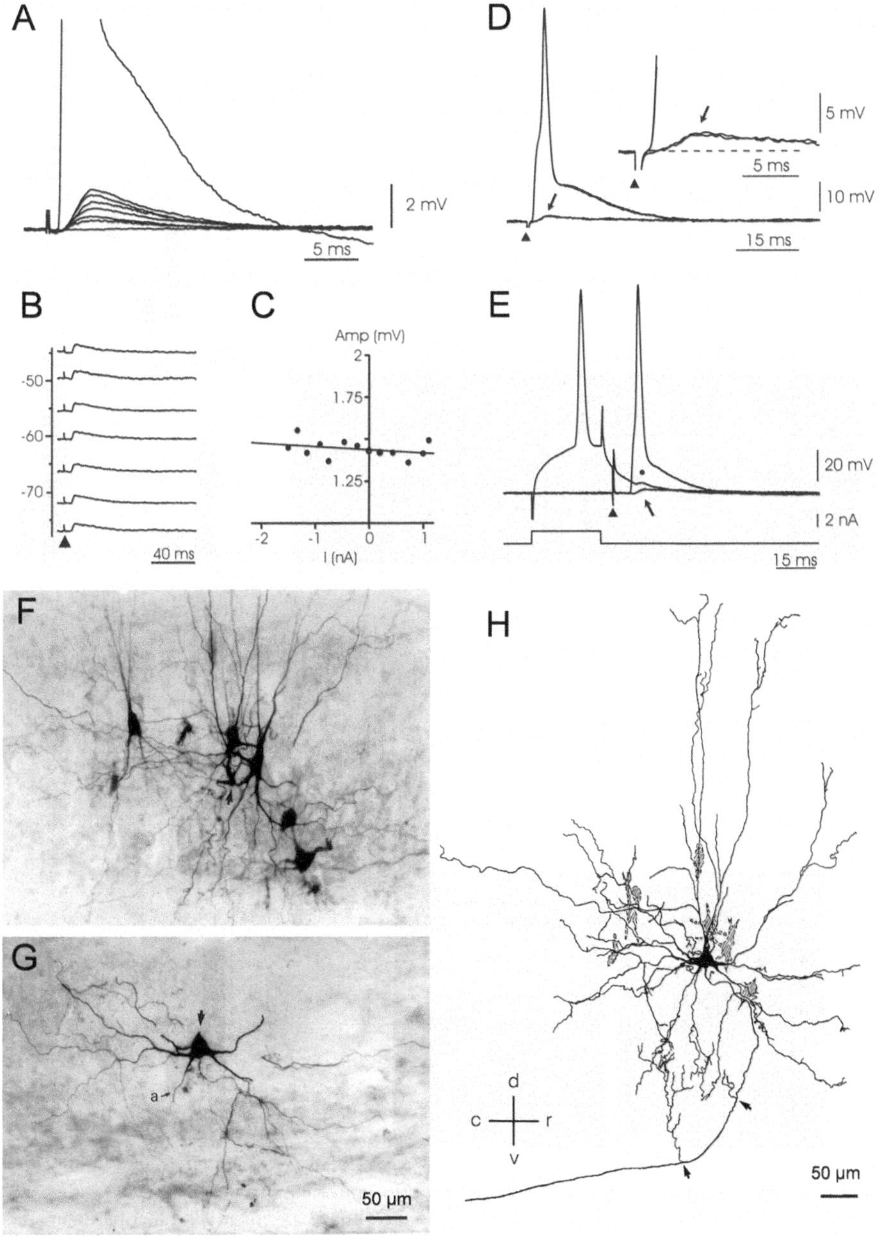

Fig. 4 Electrical coupling in neonatal motoneurons. (**a**) Subthreshold electrical stimulation of the ventral root L4 at increasing strength demonstrated the gradation of the short latency depolarization (SLD) amplitude into several discrete components until an all-or-none antidromic spike was elicited in a P6 Botulinum neurotoxin-treated TA motoneuron. The Botulinum toxin blocks chemical synaptic input, thereby isolating electrical synpases. (**b**) Insensitivity to injection of transmembrane currents. After injection of depolarizing and hyperpolarizing current, the amplitude of the SLD remained largely unaltered. (**c**) Plot showing the absence of linear correlation ($P > 0.2$)

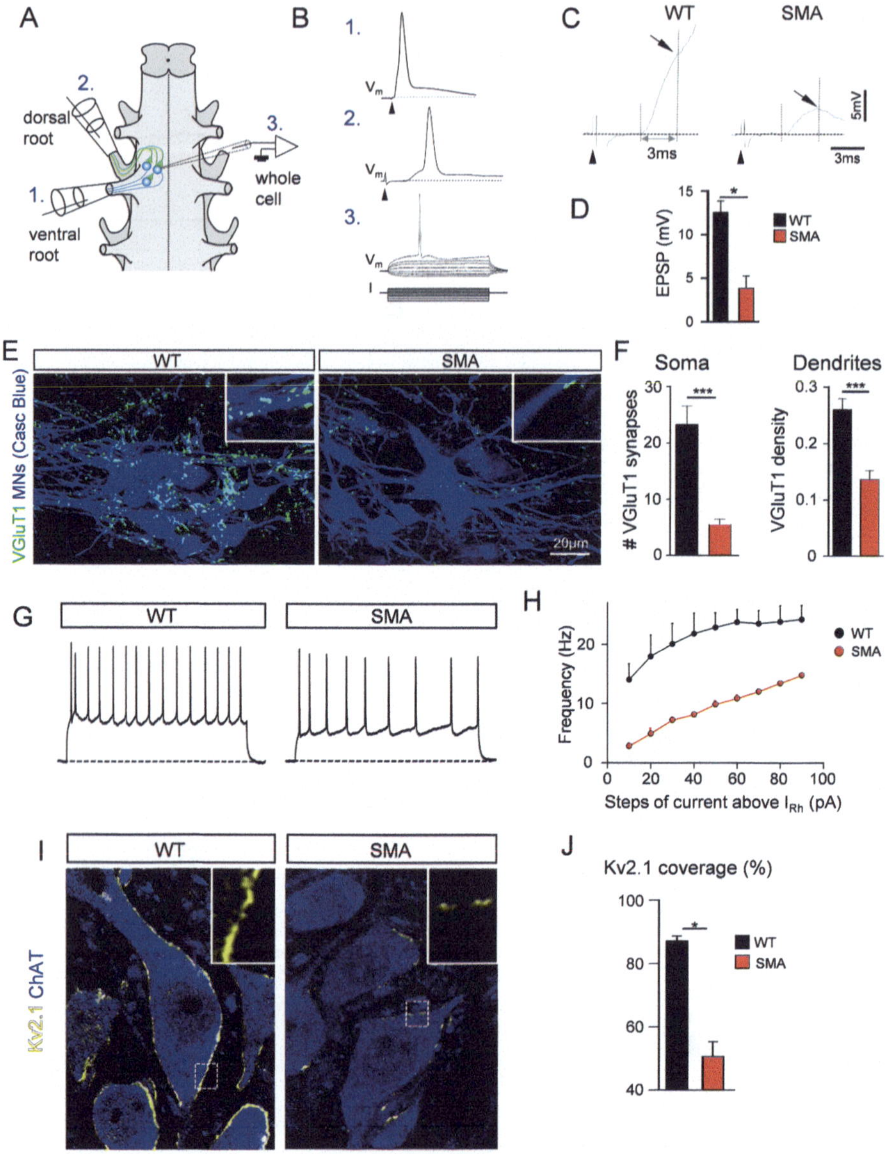

Fig. 5 SMA motoneurons exhibit dysfunctional proprioceptive input and fire at lower frequencies due to the reduction of the Kv2.1 channel. (**a**) Schematic of the experimental protocol utilizing the intact ex vivo spinal cord preparation. Extracellular electrodes in the L2 ventral root (1) and L2 dorsal root (2) (sensory fibers in green) were used for stimulation or recording. Responses from individual motoneurons (in blue) were recorded intracellularly (3) (whole-cell patch clamp). (**b**) Traces show: (1) an antidromically evoked action potential following ventral root stimulation, (2) a synaptic response from a motoneuron following dorsal root stimulation and (3) current-to-voltage relationship in whole-cell configuration from an L2 motoneuron. (**c**) Intracellular responses of monosynaptic EPSPs following supramaximal stimulation of the L2 dorsal root in homonymous

the devastating neurodegenerative diseases spinal muscular atrophy (SMA) and amyotrophic lateral sclerosis (ALS), motoneurons die prematurely. SMA is a hereditary condition, due to deficiency of the SMN protein, in which motoneuron death, typically early in development, is a hallmark of the disease (Burghes and Beattie 2009; Tisdale and Pellizzoni 2015). In ALS, on the other hand, most of the human cases are of sporadic origin (>90%) and only ~10% are familial cases (Al-Chalabi et al. 2017; Srinivasan and Rajasekaran 2020). In both diseases, specific motoneuron pools and subtypes of motoneurons are selectively affected in a spatiotemporal manner.

In SMA, motoneurons degenerate through cell-autonomous mechanisms imposed by SMN deficiency, through convergent mechanisms of p53 pathway activation (Simon et al. 2017; Van Alstyne et al. 2018). The p53 pathway uses a complex array of genes which are usually activated by intrinsic or extrinsic stress to ultimately halt DNA replication, chromosome segregation, and cell division which can lead to apoptosis (Vogelstein et al. 2000). Interestingly, SMA motoneurons that are destined to die exhibit signs of dysfunction (electrophysiological and synaptic) before their p53-dependent death (Mentis et al. 2011). Motoneurons innervating proximal or axial musculature are more vulnerable to dysfunction and degeneration in mouse models of the disease (Fletcher et al. 2017; Mentis et al. 2011), consistent with the proximo-distal progression of the disease in SMA patients. This differential effect between motoneurons has been recapitulated in SMA mouse models of disease, in which rostral lumbar motoneurons (L1/L2) – which preferentially innervate axial muscles such as iliopsoas and quadratus lumborum - are more affected than caudal lumbar motoneurons (L4/L5) – which preferentially innervate distal hindlimb muscles such as tibialis anterior and gastrocnemius (Mentis et al. 2011). Although SMA has long been considered to be solely a motoneuron disease, evidence provided by many laboratories points to SMA being a disease of motor circuits (Imlach et al. 2012; Ling et al. 2010; Lotti et al. 2012; Mentis et al. 2011). One of the earliest pathological events is the dysfunction and eventual elimination of proprioceptive synapses contacting vulnerable motoneurons (Fig. 5). The molecular

Fig. 5 (continued) motoneurons for a wild type (WT) and an SMA motoneuron. Arrows indicate the peak EPSP amplitude measured at 3 ms after the onset of response and arrowheads the stimulus artifact. (**d**) The average peak EPSP amplitude in motoneurons in WT and SMA motoneurons, indicating the significant reduction. * $P < 0.05$, t-test. (**e**) Z-stack projection of confocal images from retrogradely labeled L2 motoneurons (blue; with Cascade Blue) and VGluT1 synaptic boutons (green) in a WT and an SMA mouse at P4. Insets show VGluT1 synaptic appositions on proximal dendrites at higher magnification. (**f**) The average number of VGluT1 boutons on somata and proximal dendrites (0–50 μm from the soma) of L2 WT and SMA motoneurons. *** $P < 0.001$, *t*-test. (**g**) Intracellular responses of repetitive firing following 50 pA current injection above the minimum current required for continuous spiking, in a WT and an SMA motoneuron at P4. (**h**) Frequency-to-current relationship for WT and SMA motoneurons at P4. There was a significant reduction in firing frequency at all current steps. (**i**) Single optical plane confocal images of L2 motoneurons (ChAT; in blue) expressing Kv2.1 channels (in yellow) in WT and SMA mice at P4. Insets show Kv2.1 immunoreactivity at higher magnification of the boxed dotted area. (**j**) Percentage somatic coverage of Kv2.1 in motoneurons in WT and SMA motoneurons. There is a significant reduction in the expression of Kv2.1 channels in SMA mice. * $P < 0.05$, *t*-test. (Adapted from Fletcher et al. 2017; Nature Neurosci. 20:905–916) mechanisms for the selective synaptic dysfunction and subsequent elimination in SMA has been shown to involve the classical complement pathway through the initiating protein C1q (Vukojicic et al. 2019). The synaptic dysfunction (Fletcher et al. 2017) and its subsequent reduction (Fletcher et al. 2017; Mentis et al. 2011) occur prior to any motoneuron death (Fletcher et al. 2017; Mentis et al. 2011).

As a consequence of synaptic dysfunction, motoneurons respond by increasing their excitability. Motoneuron dysfunction in SMA is characterized by an initial hyperexcitability in terms of input resistance, time constant and voltage threshold for the initiation of an action potential (Mentis et al. 2011). Paradoxically, although SMA motoneurons during the onset of the disease exhibit these characteristics of hyperexcitability, at the same time, they are unable to fire repetitively at frequencies similar to their healthy age-matched counterparts (Fletcher et al. 2017). The inability to fire repetitively at relatively high frequencies has been associated with reduction of the expression of the delayed rectifier potassium channel Kv2.1 (Fig. 5) (Fletcher et al. 2017), which has been shown to regulate repetitive firing (Liu and Bean 2014; Misonou 2010). The inability of vulnerable SMA motoneurons to fire repetitively has been proposed to be one of the main mechanisms responsible for muscle paralysis (Fletcher et al. 2017).

In ALS, the muscular progression of disease is opposite to that occurring in SMA. Motoneurons innervating distal muscles are more vulnerable than motoneurons innervating proximal or axial muscles, resulting in disto-proximal progression of the disease. In addition, since ALS is a disease that tends to affect mature motoneurons, studies have identified vulnerable motoneurons in the context of their motor unit identity. FF motoneurons are the most vulnerable and the first to die, while S motoneurons can survive late into the disease (Dengler et al. 1990; Frey et al. 2000; Kanning et al. 2010; Pun et al. 2006). Furthermore, specific motor pools, namely the oculomotor, trochlear, and abducens nuclei in the midbrain/hindbrain responsible for extraocular eye movements, and Onuf's nucleus in the lumbosacral spinal cord responsible for urinary and anal sphincter contraction appear resistant to the neurodegeneration seen in ALS (Ferrucci et al. 2010; Kaminski et al. 2002). Dysfunction of ALS motoneurons has been proposed to be manifested as the inability to induce repetitive firing in lumbar motoneurons in SOD1 and FUS mutants of ALS mouse models of disease (Delestrée et al. 2014; Martínez-Silva et al. 2018). The molecular mechanisms responsible for motoneuron dysfunction in ALS are currently under investigation.

In traumatic situations such a peripheral nerve injury, motoneurons are temporarily or permanently – depending on the severity of the trauma – disconnected from their peripheral muscle targets, an event known as axotomy. The effect of axotomy on the motoneuron function varies and depends on the timing of the insult. During early development, axotomy results in significant motoneuron degeneration through Wallerian degeneration (Lowrie and Vrbová 1992, 2001; Mentis et al. 2007), whereas, in contrast, axotomy in adulthood results in little or no motoneuron death. Disconnecting motoneurons from their peripheral target muscles shortly after birth experimentally results in profound changes in their function. Axotomized motoneurons become hyperexcitable, a state that is reflected by an increase in input resistance and a reduction in the voltage threshold for action potential induction (Mentis et al. 2007). In addition, following neonatal axotomy, those motoneurons that do survive into adulthood through reinnervation of muscles exhibit a shift in their firing pattern from phasic to tonic (Navarrete and Vrbová 1984; Vejsada et al. 1991). Chronic electromyogram recordings from the tibialis anterior muscle in freely moving rats revealed that the discharge pattern of the muscle was altered permanently after a neonatal sciatic nerve crush. In contrast, axotomy in the adult animal does not result in any changes in the firing pattern of motoneurons reinnervating peripheral muscles (Navarrete and Vrbová 1984).

Peripheral nerve injury damages both motoneuron axons as well as sensory axons. After axotomy (experimentally often by cutting or crushing the nerve, but presenting in patients usually through crush injuries), motoneurons experience drastic changes in their synaptic input which is associated with a dramatic neuroinflammatory response involving microglia and astrocytes. It has been recently postulated that this process is subdivided into two distinct mechanisms (Alvarez et al. 2020).

First, a rapid cell-autonomous shedding of synapses from motoneurons occurs independent of microglia, which is reversible after muscle reinnervation. In this phase, the rapid shedding of synapses after motoneuron axonal injury leads to complex changes in glutamatergic synaptic function resulting in consistent depression (Ikeda and Kato 2005; Yamada et al. 2011). When the regenerating motoneurons reinnervate their skeletal muscle, the functional depression reverts. Intriguingly, GABA/glycine synapses remain on motoneurons during this period and may in fact be a source of excitatory drive. Although traditionally thought of as "inhibitory synapses" the actual impact of GABA/glycine synapses on post-synaptic membrane potential is dependent on many factors including the bouton number, the proportion of GABA vs. glycine synapses, the receptor subunit composition, and the driving forces through the receptors as dictated by the internal chloride concentration (Alvarez 2017). The potassium chloride transporter 2 (KCC2) is downregulated in axotomized motoneurons (Akhter et al. 2019; Nabekura et al. 2002; Toyoda et al. 2003), causing alterations in the chloride concentration which results in the reversal of GABA/glycine synaptic actions such that activation of these receptors results in depolarizing post-synaptic events. This may help explain previous findings of suppression of inhibitory post-synaptic potentials and their replacement by excitatory post-synaptic potentials after axotomy in early experiments (Takata and Nagahama 1983). These depolarizing post-synaptic events elicited by GABA/glycine result in activation of NMDA receptors and voltage-gated Ca^{2+} channels (Toyoda et al. 2003), which may be part of the regenerative process, similar to the rhythmic depolarizations which are characteristic of early developing motoneurons (Hanson et al. 2008; O'Donovan et al. 1998). In summary, axotomized motoneurons may replace the high frequency glutamatergic excitatory input with low-frequency voltage oscillations driven by GABA/glycine to allow for calcium entry into the cell which is adapted for promoting regeneration.

Second, a slower process eliminates the axon collaterals from injured proprioceptive afferents, which permanently alters spinal cord circuitry by removing synaptic inputs not just to the axotomized motoneurons but also to many other spinal neurons including non-injured heteronymous motoneurons and spinal interneurons. This process has a slower time course than the first phase and is not reversible by muscle reinnervation. Genetic prevention of microglia activation after nerve injury has shown that this second process depends upon microglia (Rotterman et al. 2019). The most obvious consequence of the loss of proprioceptor afferents is the loss of the stretch reflex, which was first analyzed in soldiers with nerve injuries during World War II (Barker and Young 1947). Sensory fiber die-back is dependent on the type of injury and does not occur after crush injuries, likely due to preservation of the endoneurium in crush injuries, which helps guide the regenerating axons towards their original targets. In contrast, after transection injuries, peripheral targeting errors scramble the motor pool organization in the spinal cord where proprioceptive synaptic inputs normally avoid antagonist muscles, which likely explains the presence of reciprocal excitation between antagonist muscles and higher rates of co-contraction after nerve transection (Horstman et al. 2019; Sabatier et al. 2011).

References

Akhter ET, Griffith RW, English AW, Alvarez FJ (2019) Removal of the potassium chloride co-transporter from the somatodendritic membrane of axotomized motoneurons is independent of BDNF/TrkB signaling but is controlled by neuromuscular innervation. eNeuro 6

Al-Chalabi A, van den Berg LH, Veldink J (2017) Gene discovery in amyotrophic lateral sclerosis: implications for clinical management. Nat Rev Neurol 13:96–104

Alvarez FJ (2017) Gephyrin and the regulation of synaptic strength and dynamics at glycinergic inhibitory synapses. Brain Res Bull 129:50–65

Alvarez FJ, Rotterman TM, Akhter ET, Lane AR, English AW, Cope TC (2020) Synaptic plasticity on motoneurons after axotomy: a necessary change in paradigm. Front Mol Neurosci 13:68

Barker D, Young JZ (1947) Recovery of stretch reflexes after nerve injury, vol 1. Lancet, London, pp 704–707

Bayliss DA, Viana F, Bellingham MC, Berger AJ (1994) Characteristics and postnatal development of a hyperpolarization-activated inward current in rat hypoglossal motoneurons in vitro. J Neurophysiol 71:119–128

Bertrand S, Cazalets JR (1998) Postinhibitory rebound during locomotor-like activity in neonatal rat motoneurons in vitro. J Neurophysiol 79:342–351

Bessou P, Emonet-Dénand F, Laporte Y (1965) Motor fibres innervating extrafusal and intrafusal muscle fibres in the cat. J Physiol 180:649–672

Bouhadfane M, Tazerart S, Moqrich A, Vinay L, Brocard F (2013) Sodium-mediated plateau potentials in lumbar motoneurons of neonatal rats. J Neurosci Off J Soc Neurosci 33:15626–15641

Brock LG, Coombs JS, Eccles JC (1952) The recording of potentials from motoneurones with an intracellular electrode. J Physiol 117:431–460

Brock LG, Coombs JS, Eccles JC (1953) Intracellular recording from antidromically activated motoneurones. J Physiol 122:429–461

Burghes AH, Beattie CE (2009) Spinal muscular atrophy: why do low levels of survival motor neuron protein make motor neurons sick? Nat Rev Neurosci 10:597–609

Burke RE, Levine DN, Tsairis P, Zajac FE 3rd (1973) Physiological types and histochemical profiles in motor units of the cat gastrocnemius. J Physiol 234:723–748

Burke RE, Dum RP, Fleshman JW, Glenn LL, Lev-Tov A, O'Donovan MJ, Pinter MJ (1982) A HRP study of the relation between cell size and motor unit type in cat ankle extensor motoneurons. J Comp Neurol 209:17–28

Burke RE, Fyffe RE, Moschovakis AK (1994) Electrotonic architecture of cat gamma motoneurons. J Neurophysiol 72:2302–2316

Cameron WE, He F, Kalipatnapu P, Jodkowski JS, Guthrie RD (1991) Morphometric analysis of phrenic motoneurons in the cat during postnatal development. J Comp Neurol 314:763–776

Cameron WE, Núñez-Abades PA, Kerman IA, Hodgson TM (2000) Role of potassium conductances in determining input resistance of developing brain stem motoneurons. J Neurophysiol 84:2330–2339

Carrascal L, Nieto-Gonzalez JL, Cameron WE, Torres B, Nunez-Abades PA (2005) Changes during the postnatal development in physiological and anatomical characteristics of rat motoneurons studied in vitro. Brain Res Brain Res Rev 49:377–387

Chang Q, Gonzalez M, Pinter MJ, Balice-Gordon RJ (1999) Gap junctional coupling and patterns of connexin expression among neonatal rat lumbar spinal motor neurons. J Neurosci Off J Soc Neurosci 19:10813–10828

Chen HH, Hippenmeyer S, Arber S, Frank E (2003) Development of the monosynaptic stretch reflex circuit. Curr Opin Neurobiol 13:96–102

Conway BA, Hultborn H, Kiehn O, Mintz I (1988) Plateau potentials in alpha-motoneurones induced by intravenous injection of L-dopa and clonidine in the spinal cat. J Physiol 405:369–384

Cullheim S, Ulfhake B (1979a) Observations on the morphology of intracellularly stained gamma-motoneurons in relation to their axon conduction velocity. Neurosci Lett 13:47–50

Cullheim S, Ulfhake B (1979b) Relations between cell body size, axon diameter and axon conduction velocity of triceps surae alpha montoneurons during the postnatal development in the cat. J Comp Neurol 188:679–686

Cullheim S, Fleshman JW, Glenn LL, Burke RE (1987) Three-dimensional architecture of dendritic trees in type-identified alpha-motoneurons. J Comp Neurol 255:82–96

De La Cruz RR, Escudero M, Delgado-García JM (1989) Behaviour of medial rectus motoneurons in the alert cat. Eur J Neurosci 1:288–295

Delestrée N, Manuel M, Iglesias C, Elbasiouny SM, Heckman CJ, Zytnicki D (2014) Adult spinal motoneurones are not hyperexcitable in a mouse model of inherited amyotrophic lateral sclerosis. J Physiol 592:1687–1703

Dengler R, Konstanzer A, Küther G, Hesse S, Wolf W, Struppler A (1990) Amyotrophic lateral sclerosis: macro-EMG and twitch forces of single motor units. Muscle Nerve 13:545–550

Di Pasquale E, Tell F, Ptak K, Monteau R, Hilaire G (2001) Perinatal changes of I(h) in phrenic motoneurons. Eur J Neurosci 13:1403–1410

Durand J, Amendola J, Bories C, Lamotte d'Incamps B (2006) Early abnormalities in transgenic mouse models of amyotrophic lateral sclerosis. J Physiol Paris 99:211–220

Eccles JC, Eccles RM, Lundberg A (1957) The convergence of monosynaptic excitatory afferents on to many different species of alpha motoneurones. J Physiol 137:22–50

Eccles JC, Eccles RM, Iggo A, Lundberg A (1960) Electrophysiological studies on gamma motoneurones. Acta Physiol Scand 50:32–40

Elbasiouny SM, Amendola J, Durand J, Heckman CJ (2010) Evidence from computer simulations for alterations in the membrane biophysical properties and dendritic processing of synaptic inputs in mutant superoxide dismutase-1 motoneurons. J Neurosci Off J Soc Neurosci 30:5544–5558

Ferrucci M, Spalloni A, Bartalucci A, Cantafora E, Fulceri F, Nutini M, Longone P, Paparelli A, Fornai F (2010) A systematic study of brainstem motor nuclei in a mouse model of ALS, the effects of lithium. Neurobiol Dis 37:370–383

Fleshman JW, Segev I, Burke RB (1988) Electrotonic architecture of type-identified alpha-motoneurons in the cat spinal cord. J Neurophysiol 60:60–85

Fletcher EV, Simon CM, Pagiazitis JG, Chalif JI, Vukojicic A, Drobac E, Wang X, Mentis GZ (2017) Reduced sensory synaptic excitation impairs motor neuron function via Kv2.1 in spinal muscular atrophy. Nat Neurosci 20:905–916

Frey D, Schneider C, Xu L, Borg J, Spooren W, Caroni P (2000) Early and selective loss of neuromuscular synapse subtypes with low sprouting competence in motoneuron diseases. J Neurosci Off J Soc Neurosci 20:2534–2542

Friese A, Kaltschmidt JA, Ladle DR, Sigrist M, Jessell TM, Arber S (2009) Gamma and alpha motor neurons distinguished by expression of transcription factor Err3. Proc Natl Acad Sci U S A 106:13588–13593

Fuchs AF, Scudder CA, Kaneko CR (1988) Discharge patterns and recruitment order of identified motoneurons and internuclear neurons in the monkey abducens nucleus. J Neurophysiol 60:1874–1895

Fulton BP, Walton K (1986) Electrophysiological properties of neonatal rat motoneurones studied in vitro. J Physiol 370:651–678

Galvani A (1791) De Viribus Electricitatis in Motu Musculari Commentarius (note on the effect of electricity on muscular motion) Typographia Instuti Scientarium. Bologna 7

Gardiner PF (1993) Physiological properties of motoneurons innervating different muscle unit types in rat gastrocnemius. J Neurophysiol 69:1160–1170

Hanson MG, Milner LD, Landmesser LT (2008) Spontaneous rhythmic activity in early chick spinal cord influences distinct motor axon pathfinding decisions. Brain Res Rev 57:77–85

Harris-Warrick RM (2010) General principles of rhythmogenesis in central pattern generator networks. Prog Brain Res 187:213–222

Hashizume K, Kanda K, Burke RE (1988) Medial gastrocnemius motor nucleus in the rat: age-related changes in the number and size of motoneurons. J Comp Neurol 269:425–430

Heckman CJ, Johnson M, Mottram C, Schuster J (2008) Persistent inward currents in spinal motoneurons and their influence on human motoneuron firing patterns. The Neuroscientist 14:264–275

Henneman E, Somjen G, Carpenter DO (1965) Functional significance of cell size in spinal motoneurons. J Neurophysiol 28:560–580

Hornby TG, McDonagh JC, Reinking RM, Stuart DG (2002) Effects of excitatory modulation on intrinsic properties of turtle motoneurons. J Neurophysiol 88:86–97

Horstman GM, Housley SN, Cope TC (2019) Dysregulation of mechanosensory circuits coordinating the actions of antagonist motor pools following peripheral nerve injury and muscle reinnervation. Exp Neurol 318:124–134

Hounsgaard J, Hultborn H, Jespersen B, Kiehn O (1988) Bistability of alpha-motoneurones in the decerebrate cat and in the acute spinal cat after intravenous 5-hydroxytryptophan. J Physiol 405:345–367

Ikeda R, Kato F (2005) Early and transient increase in spontaneous synaptic inputs to the rat facial motoneurons after axotomy in isolated brainstem slices of rats. Neuroscience 134:889–899

Imlach WL, Beck ES, Choi BJ, Lotti F, Pellizzoni L, McCabe BD (2012) SMN is required for sensory-motor circuit function in drosophila. Cell 151:427–439

Kaminski HJ, Richmonds CR, Kusner LL, Mitsumoto H (2002) Differential susceptibility of the ocular motor system to disease. Ann N Y Acad Sci 956:42–54

Kanning KC, Kaplan A, Henderson CE (2010) Motor neuron diversity in development and disease. Annu Rev Neurosci 33:409–440

Lee RH, Heckman CJ (1998) Bistability in spinal motoneurons in vivo: systematic variations in persistent inward currents. J Neurophysiol 80:583–593

Lee RH, Heckman CJ (1999) Enhancement of bistability in spinal motoneurons in vivo by the noradrenergic alpha1 agonist methoxamine. J Neurophysiol 81:2164–2174

Li Y, Gorassini MA, Bennett DJ (2004) Role of persistent sodium and calcium currents in motoneuron firing and spasticity in chronic spinal rats. J Neurophysiol 91:767–783

Ling KK, Lin MY, Zingg B, Feng Z, Ko CP (2010) Synaptic defects in the spinal and neuromuscular circuitry in a mouse model of spinal muscular atrophy. PLoS One 5:e15457

Liu PW, Bean BP (2014) Kv2 channel regulation of action potential repolarization and firing patterns in superior cervical ganglion neurons and hippocampal CA1 pyramidal neurons. J Neurosci Off J Soc Neurosci 34:4991–5002

Lotti F, Imlach WL, Saieva L, Beck ES, Hao le T, Li DK, Jiao W, Mentis GZ, Beattie CE, McCabe BD et al (2012) An SMN-dependent U12 splicing event essential for motor circuit function. Cell 151:440–454

Lowrie MB, Vrbová G (1992) Dependence of postnatal motoneurones on their targets: review and hypothesis. Trends Neurosci 15:80–84

Lowrie MB, Vrbová G (2001) Repeated injury to the sciatic nerve in immature rats causes motoneuron death and impairs muscle recovery. Exp Neurol 171:170–175

Magariños-Ascone C, Núñez A, Delgado-García JM (1999) Different discharge properties of rat facial nucleus motoneurons. Neuroscience 94:879–886

Manuel M, Zytnicki D (2011) Alpha, beta and gamma motoneurons: functional diversity in the motor system's final pathway. J Integr Neurosci 10:243–276

Martínez-Silva ML, Imhoff-Manuel RD, Sharma A, Heckman CJ, Shneider NA, Roselli F, Zytnicki D, Manuel M (2018) Hypoexcitability precedes denervation in the large fast-contracting motor units in two unrelated mouse models of ALS. eLife:7

Mazza E, Núñez-Abades PA, Spielmann JM, Cameron WE (1992) Anatomical and electrotonic coupling in developing genioglossal motoneurons of the rat. Brain Res 598:127–137

Mendelsohn AI, Simon CM, Abbott LF, Mentis GZ, Jessell TM (2015) Activity regulates the incidence of heteronymous sensory-motor connections. Neuron 87:111–123

Mentis GZ, Díaz E, Moran LB, Navarrete R (2002) Increased incidence of gap junctional coupling between spinal motoneurones following transient blockade of NMDA receptors in neonatal rats. J Physiol 544:757–764

Mentis GZ, Díaz E, Moran LB, Navarrete R (2007) Early alterations in the electrophysiological properties of rat spinal motoneurones following neonatal axotomy. J Physiol 582:1141–1161

Mentis GZ, Blivis D, Liu W, Drobac E, Crowder ME, Kong L, Alvarez FJ, Sumner CJ, O'Donovan MJ (2011) Early functional impairment of sensory-motor connectivity in a mouse model of spinal muscular atrophy. Neuron 69:453–467

Misonou H (2010) Homeostatic regulation of neuronal excitability by K(+) channels in normal and diseased brains. The Neuroscientist 16:51–64

Mosfeldt Laursen A, Rekling JC (1989) Electrophysiological properties of hypoglossal motoneurons of Guinea-pigs studied in vitro. Neuroscience 30:619–637

Nabekura J, Ueno T, Okabe A, Furuta A, Iwaki T, Shimizu-Okabe C, Fukuda A, Akaike N (2002) Reduction of KCC2 expression and GABAA receptor-mediated excitation after in vivo axonal injury. J Neurosci Off J Soc Neurosci 22:4412–4417

Navarrete R, Vrbová G (1984) Differential effect of nerve injury at birth on the activity pattern of reinnervated slow and fast muscles of the rat. J Physiol 351:675–685

Núñez-Abades PA, Cameron WE (1995) Morphology of developing rat genioglossal motoneurons studied in vitro: relative changes in diameter and surface area of somata and dendrites. J Comp Neurol 353:129–142

Núñez-Abades PA, Spielmann JM, Barrionuevo G, Cameron WE (1993) In vitro electrophysiology of developing genioglossal motoneurons in the rat. J Neurophysiol 70:1401–1411

Núñez-Abades PA, Pattillo JM, Hodgson TM, Cameron WE (2000) Role of synaptic inputs in determining input resistance of developing brain stem motoneurons. J Neurophysiol 84:2317–2329

O'Donovan MJ, Chub N, Wenner P (1998) Mechanisms of spontaneous activity in developing spinal networks. J Neurobiol 37:131–145

O'Dowd DK, Ribera AB, Spitzer NC (1988) Development of voltage-dependent calcium, sodium, and potassium currents in Xenopus spinal neurons. J Neurosci Off J Soc Neurosci 8:792–805

Pastor AM, Mentis GZ, De La Cruz RR, Díaz E, Navarrete R (2003) Increased electrotonic coupling in spinal motoneurons after transient botulinum neurotoxin paralysis in the neonatal rat. J Neurophysiol 89(2):793–805. https://doi.org/10.1152/jn.00498.2002. PMID: 12574457

Perrier JF, Hounsgaard J (2000) Development and regulation of response properties in spinal cord motoneurons. Brain Res Bull 53:529–535

Personius KE, Chang Q, Mentis GZ, O'Donovan MJ, Balice-Gordon RJ (2007) Reduced gap junctional coupling leads to uncorrelated motor neuron firing and precocious neuromuscular synapse elimination. Proc Natl Acad Sci U S A 104:11808–11813

Poliak S, Norovich AL, Yamagata M, Sanes JR, Jessell TM (2016) Muscle-type identity of proprioceptors specified by spatially restricted signals from limb mesenchyme. Cell 164:512–525

Pun S, Santos AF, Saxena S, Xu L, Caroni P (2006) Selective vulnerability and pruning of phasic motoneuron axons in motoneuron disease alleviated by CNTF. Nat Neurosci 9:408–419

Ramírez V, Ulfhake B (1991) Postnatal development of cat hind limb motoneurons supplying the intrinsic muscles of the foot sole. Brain Res Dev Brain Res 62:189–202

Rekling JC, Funk GD, Bayliss DA, Dong XW, Feldman JL (2000) Synaptic control of motoneuronal excitability. Physiol Rev 80:767–852

Robinson DW, Cameron WE (2000) Time-dependent changes in input resistance of rat hypoglossal motoneurons associated with whole-cell recording. J Neurophysiol 83:3160–3164

Rotterman TM, Akhter ET, Lane AR, Mac Pherson KP, García VV, Tansey MG, Alvarez FJ (2019) Spinal motor circuit synaptic plasticity after peripheral nerve injury depends on microglia activation and a CCR2 mechanism. J Neurosci Off J Soc Neurosci 39:3412–3433

Russier M, Carlier E, Ankri N, Fronzaroli L, Debanne D (2003) A-, T-, and H-type currents shape intrinsic firing of developing rat abducens motoneurons. J Physiol 549:21–36

Russo RE, Hounsgaard J (1999) Dynamics of intrinsic electrophysiological properties in spinal cord neurones. Prog Biophys Mol Biol 72:329–365

Sabatier MJ, To BN, Nicolini J, English AW (2011) Effect of axon misdirection on recovery of electromyographic activity and kinematics after peripheral nerve injury. Cells Tissues Organs 193:298–309

Seebach BS, Ziskind-Conhaim L (1994) Formation of transient inappropriate sensorimotor synapses in developing rat spinal cords. J Neurosci Off J Soc Neurosci 14:4520–4528

Sherrington CS (1924) Problems of muscular receptivity. Nature 113:732–732

Shneider NA, Brown MN, Smith CA, Pickel J, Alvarez FJ (2009) Gamma motor neurons express distinct genetic markers at birth and require muscle spindle-derived GDNF for postnatal survival. Neural Dev 4:42

Simon M, Destombes J, Horcholle-Bossavit G, Thiesson D (1996) Postnatal development of alpha- and gamma-peroneal motoneurons in kittens: an ultrastructural study. Neurosci Res 25:77–89

Simon CM, Dai Y, Van Alstyne M, Koutsioumpa C, Pagiazitis JG, Chalif JI, Wang X, Rabinowitz JE, Henderson CE, Pellizzoni L et al (2017) Converging mechanisms of p53 activation drive motor neuron degeneration in spinal muscular atrophy. Cell Rep 21:3767–3780

Srinivasan E, Rajasekaran R (2020) A systematic and comprehensive review on disease-causing genes in amyotrophic lateral sclerosis. J Molecul Neurosci: MN 70:1742–1770

Stevens B, Allen NJ, Vazquez LE, Howell GR, Christopherson KS, Nouri N, Micheva KD, Mehalow AK, Huberman AD, Stafford B et al (2007) The classical complement cascade mediates CNS synapse elimination. Cell 131:1164–1178

Takata M, Nagahama T (1983) Synaptic efficacy of inhibitory synapses in hypoglossal motoneurons after transection of the hypoglossal nerves. Neuroscience 10:23–29

Tisdale S, Pellizzoni L (2015) Disease mechanisms and therapeutic approaches in spinal muscular atrophy. J Neurosci Off J Soc Neurosci 35:8691–8700

Toyoda H, Ohno K, Yamada J, Ikeda M, Okabe A, Sato K, Hashimoto K, Fukuda A (2003) Induction of NMDA and GABAA receptor-mediated Ca2+ oscillations with KCC2 mRNA downregulation in injured facial motoneurons. J Neurophysiol 89:1353–1362

Ulfhake B, Cullheim S, Franson P (1988) Postnatal development of cat hind limb motoneurons. I: changes in length, branching structure, and spatial distribution of dendrites of cat triceps surae motoneurons. J Comp Neurol 278:69–87

Umemiya M, Berger AJ (1994) Properties and function of low- and high-voltage-activated Ca2+ channels in hypoglossal motoneurons. J Neurosci Off J Soc Neurosci 14:5652–5660

Van Alstyne M, Simon CM, Sardi SP, Shihabuddin LS, Mentis GZ, Pellizzoni L (2018) Dysregulation of Mdm2 and Mdm4 alternative splicing underlies motor neuron death in spinal muscular atrophy. Genes Dev 32:1045–1059

Vejsada R, Hník P, Navarrete R, Palecek J, Soukup T, Borecka U, Payne R (1991) Motor functions in rat hindlimb muscles following neonatal sciatic nerve crush. Neuroscience 40:267–275

Viana F, Bayliss DA, Berger AJ (1994) Postnatal changes in rat hypoglossal motoneuron membrane properties. Neuroscience 59:131–148

Viana F, Bayliss DA, Berger AJ (1995) Repetitive firing properties of developing rat brainstem motoneurones. J Physiol 486(Pt 3):745–761

Vinay L, Brocard F, Pflieger JF, Simeoni-Alias J, Clarac F (2000) Perinatal development of lumbar motoneurons and their inputs in the rat. Brain Res Bull 53:635–647

Vogelstein B, Lane D, Levine AJ (2000) Surfing the p53 network. Nature 408:307–310

Volta A (1800) On the electricity excited by the mere contact of conductors of different kinds. Philos Trans R Soc Lond A 90:403–431

Vukojicic A, Delestrée N, Fletcher EV, Pagiazitis JG, Sankaranarayanan S, Yednock TA, Barres BA, Mentis GZ (2019) The classical complement pathway mediates microglia-dependent remodeling of spinal motor circuits during development and in SMA. Cell Rep 29:3087–3100.e3087

Walsh MK, Lichtman JW (2003) In vivo time-lapse imaging of synaptic takeover associated with naturally occurring synapse elimination. Neuron 37:67–73

Walton KD, Navarrete R (1991) Postnatal changes in motoneurone electrotonic coupling studied in the in vitro rat lumbar spinal cord. J Physiol 433:283–305

Westbury DR (1982) A comparison of the structures of alpha and gamma-spinal motoneurones of the cat. J Physiol 325:79–91

Woodbury JW, Patton HD (1952) Electrical activity of single spinal cord elements. Cold Spring Harb Symp Quant Biol 17:185–188

Yamada J, Nakanishi H, Jinno S (2011) Differential involvement of perineuronal astrocytes and microglia in synaptic stripping after hypoglossal axotomy. Neuroscience 182:1–10

Homeostatic Regulation of Motoneuron Properties in Development

Peter A. Wenner and Dobromila Pekala

Abstract Homeostatic plasticity represents a set of compensatory mechanisms that are engaged following a perturbation to some feature of neuronal or network function. Homeostatic mechanisms are most robustly expressed during development, a period that is replete with various perturbations such as increased cell size and the addition/removal of synaptic connections. In this review we look at numerous studies that have advanced our understanding of homeostatic plasticity by taking advantage of the accessibility of developing motoneurons. We discuss the homeostatic regulation of embryonic movements in the living chick embryo and describe the spinal compensatory mechanisms that act to recover these movements (homeostatic intrinsic plasticity) or stabilize synaptic strength (synaptic scaling). We describe the expression and triggering mechanisms of these forms of homeostatic plasticity and thereby gain an understanding of their roles in the motor system. We then illustrate how these findings can be extended to studies of developing motoneurons in other systems including the rodents, zebrafish, and fly. Furthermore, studies in developing drosophila have been critical in identifying some of the molecular signaling cascades and expression mechanisms that underlie homeostatic intrinsic membrane excitability. This powerful model organism has also been used to study a presynaptic form of homeostatic plasticity where increases or decreases in synaptic transmission are associated with compensatory changes in probability of release at the neuromuscular junction. Further, we describe studies that demonstrate homeostatic adjustments of ion channel expression following perturbations to other kinds of ion channels. Finally, we discuss work in xenopus that shows a homeostatic regulation of neurotransmitter phenotype in developing motoneurons following activity perturbations. Together, this work illustrates the importance of developing motoneurons in elucidating the mechanisms and roles of homeostatic plasticity.

Keywords Motoneuron · Muscle · Plasticity · Homeostatic · Development

---

P. A. Wenner (✉) · D. Pekala
Department of Cell Biology, Whitehead Biomedical Research Building, Emory University School of Medicine, Atlanta, GA, USA
e-mail: pwenner@emory.edu

M. J. O'Donovan, M. Falgairolle (eds.), *Vertebrate Motoneurons*, Advances in Neurobiology 28, https://doi.org/10.1007/978-3-031-07167-6_4

1 Introduction

Homeostatic plasticity represents a set of mechanisms that are triggered in response to a perturbation of neural activity that acts to homeostatically restore neural function. This was first appreciated by Eve Marder's lab studying the activity patterns of neurons from the crustacean stomatogastric ganglion. These cells normally fire rhythmic bursts of action potentials *in vivo*, but this activity was lost when stomatogastric ganglion cells were individually plated in culture. However, within few days the original firing pattern was restored due to homeostatic plasticity (Turrigiano et al. 1994). Later studies identified several different classes of homeostatic mechanisms following activity or neurotransmitter receptor blockade that compensated by altering synaptic strength (homeostatic synaptic plasticity) or membrane excitability (homeostatic intrinsic plasticity). For instance, following 2 days of spike blockade cortical neurons in culture strengthened excitatory synapses, weakened inhibitory synapses, and increased their intrinsic membrane excitability (Turrigiano et al. 1998; Desai et al. 1999; Kilman et al. 2002). This kind of work has been carried out in several different classes of cell, including motoneurons from many different species. These studies of homeostatic plasticity in motoneurons have dramatically advanced the field due to the accessible nature of the motoneuron. In this chapter we look at the advances in the field of homeostatic plasticity that have been made by studying motoneurons. We focus on developing systems as these homeostatic mechanisms will shape the developmental trajectory of synaptic strength and cellular excitability in the more mature system. We begin by discussing homeostatic recovery of embryonic movements in the living chick embryo and follow up by describing both homeostatic synaptic plasticity and homeostatic intrinsic plasticity mechanisms. We consider the molecular mechanisms underlying these forms of plasticity, along with their triggers, and functions. We then discuss how these findings translate to motoneurons in other systems by examining studies that carry out *in vivo* perturbations in rat, zebrafish, and fly motoneurons. Finally, we consider the many novel advances in homeostatic plasticity that have been made in the developing fly and frog. These include: the signaling pathway for intrinsic plasticity in the fly, presynaptic plasticity at the neuromuscular junction of the the fly, voltage-gated channel co-regulation in the fly, and regulation of neurotransmitter phenotype in the frog.

2 Chick Embryo Homeostatic Plasticity

One of the great advantages that motoneurons provide is that they drive muscle contractions, which can be easily observed. Nowhere is this more clear than when watching embryonic movements in the chick *in ovo* (Preyer 1937). People have studied these movements for hundreds of years. The movements are driven by the spinal activation of motoneurons, and thus can serve as a measure of motoneuron activity. We have taken advantage of this feature to study homeostatic plasticity in a living system at a developmental stage where there is a robust expression of homeostatic mechanisms. Midway through embryonic development, from embryonic day 8 to 10 (E8 to E10), we have injected antagonists to either glutamatergic (CNQX/APV) or GABAergic (gabazine or bicuculline) receptors (Fig. 1). GABA is depolarizing at this stage of development because neurons have higher intracellular chloride (Cl^{-}_{in}) than in mature neurons (Rivera et al. 1999; Farrant and Kaila 2007; Gonzalez-Islas et al. 2009; Lindsly et al. 2014). This elevated Cl^{-}_{in} depolarizes the reversal potential for $GABA_A$ receptors to levels above the resting membrane potential. Because both of these transmitter systems are depolarizing and excitatory at this stage in development, injection of either antagonist transiently abolishes bouts of kicking activity. However, within an hour, the kicks begin to return, and by 12 h after addition of the antagonists the embryonic movements have homeostatically recovered to control levels. The results of these studies are of great importance for a couple of reasons. First, this is one of the very few homeostatic studies that actually show a homeostatic recovery of a network behavior, and do so *in vivo*. Second, the recovery following blockade of GABAergic or glutamatergic neurotransmission is highly similar, even though these movements are now driven predominantly by either GABAergic or glutamatergic transmission, rather than both. This study sets up the first part of this chapter, by establishing the advantages of studying motoneurons in this model system to identify the homeostatic mechanisms that underlie the recovery of spontaneous network-driven movements.

Similar studies had shown previously that spontaneous network activity (SNA) observed in the isolated spinal cord preparation *in vitro* could recover its expression after GABAergic or glutamatergic receptor blockade, although at a slower frequency (Chub and O'Donovan 1998). On the other hand, recovery was not observed following both glutamatergic and GABA/glycinergic blockade. SNA in the intact system drives embryonic movements and this kind of activity has been described in most developing networks shortly after synaptic connections first form, including

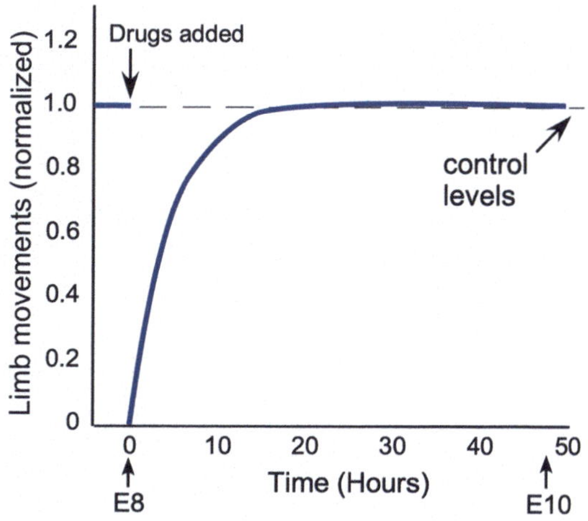

Fig. 1 Heomeostatic recovery of embryonic movements 12 h after GABAergic or glutamatergic antagonists added *in ovo*

the spinal cord, retina, hippocampus, among others (Ben-Ari et al. 1989; Gummer and Mark 1994; Fortin et al. 1995; Itaya et al. 1995; Lippe 1995; O'Donovan et al. 1998; Fedirchuk et al. 1999; Feller 1999; Ho and Waite 1999; Wong 1999; Ben-Ari 2001). This activity comprises episodic bursts of network spiking activity that last for seconds and are followed by longer lasting quiescent periods, as shown in the isolated chick embryo spinal preparation in Fig. 2. These bursts, or episodes, are a consequence of the highly excitable nature of a synaptic circuit, in which GABAergic neurotransmission is excitatory in early development due to an elevated Cl^{-}_{in} (Gonzalez-Islas et al. 2009; Lindsly et al. 2014). During these episodes the majority of cells fire action potentials and experience increases in cytoplasmic calcium, important in multiple aspects of development. SNA itself is important for several aspects of development including synaptic strength, cellular excitability, motoneuron pathfinding, bone and muscle maturation (Ruano-Gil et al. 1978; Toutant et al. 1979; Hall and Herring 1990; Hanson and Landmesser 2004; Gonzalez-Islas and Wenner 2006). Despite its widespread nature, we have an incomplete understanding of how SNA influences the maturation of the networks that generate it. The importance of SNA is highlighted by its ubiquitous expression throughout the developing nervous system and the finding that it is under strong homeostatic regulation.

The fact that SNA-driven embryonic movements are restored after neurotransmission blockade led to questions about the compensatory mechanisms that underlie this homeostatic process. We break this down into 2 basic components – the effector mechanisms that actually alter the system to compensate for the loss of excitatory synaptic input and the sensors that detect the altered activity which then trigger the downstream effector mechanisms.

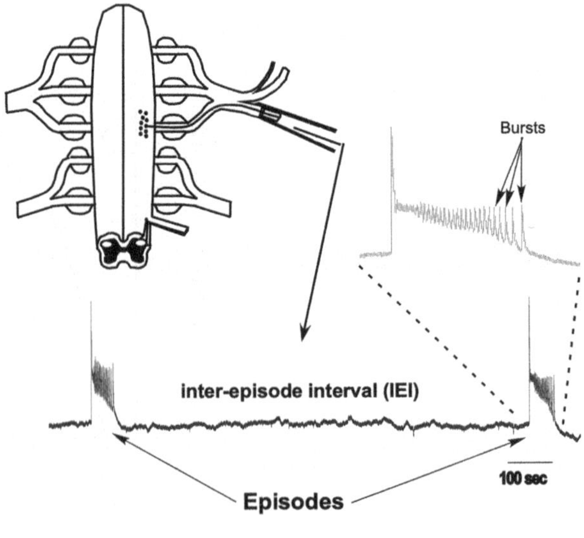

Fig. 2 Schematic of the isolated spinal cord and extracellular suction electrode recording (DC-coupled) showing 2 episodes of SNA. These recordings represent population motoneuron (black dots) recordings

Synaptic Scaling

While changes in cellular excitability clearly play a role in homeostatic plasticity, far more attention has focused on compensatory changes in synaptic strength. Synaptic strength for these studies is often measured by the amplitude of miniature postsynaptic currents (mPSC or quantal amplitude), which represent the postsynaptic responses to the spontaneous release of a single vesicle of transmitter. Many labs have demonstrated that altering activity levels of synaptically connected neuronal networks in culture for 1–3 days led to compensatory changes in synaptic strength that could contribute to the homeostatic recovery of normal activity levels (Turrigiano et al. 1998; Lissin et al. 1998; O'Brien et al. 1998; Davis 2006; Rich and Wenner 2007; Turrigiano 2008). When activity was blocked for 2 days in cultured neuronal networks the entire distribution of glutamatergic mPSC amplitudes increased by a scaling factor, and this was termed synaptic scaling (Turrigiano et al. 1998). In addition, glutamatergic AMPA mPSCs scale down after increased network activity. Typically, scaling is mediated by changes in the number or subunit composition of synaptic receptors (Turrigiano 2012). In both cases the strength of the synapses changed in a direction that would be expected to compensate for the perturbation.

In order to determine whether scaling is present and could underlie the homeostatic recovery of embryonic movements we reduced SNA *in ovo* in the chick from embryonic day 8–10 (E8–E10) infusing the Na^+ channel blocker lidocaine (Gonzalez-Islas and Wenner 2006). We confirmed the activity reduction by observing a cessation of embryonic movements through a window in the eggshell. This suggested that there was no motoneuron output driven by spontaneous network activity that produced visibly detectable limb movements. Embryonic movements were dramatically reduced when lidocaine was infused. Whole cell recordings from activity-blocked motoneurons 2 days later (E10) in the isolated cord *in vitro* demonstrated compensatory increases in the amplitude of AMPAergic and GABAergic mPSCs recorded from motoneurons (~40%). The strengthening was compensatory, because GABA is depolarizing and excitatory in the embryo.

The next question was directed at the mechanisms that underly the increase in mPSC amplitude. As mentioned above, scaling is typically mediated by changes in neurotransmitter receptor expression at the synapse. Indeed, we found that AMPAergic upscaling was mediated by an insertion of GluA2-lacking AMPA receptors (Garcia-Bereguiain et al. 2013). These receptors are calcium permeable and have been described in several previous scaling studies where neurotransmission is blocked (Thiagarajan et al. 2005; Sutton et al. 2006; Aoto et al. 2008). On the other hand, GABAergic scaling was a very different mechanism. Rather than altering conductance by regulating neurotransmitter receptors we found that GABAergic upscaling was mediated by increasing the driving force though intracellular chloride accumulation (Gonzalez-Islas et al. 2010). In control E10 motoneurons chloride levels are ~ 50 mM corresponding to a reversal potential of ~ −30 mV, while activity blockade resulted in an ~ 80 mM intracellular chloride concentration corresponding to a −10 mV reversal potential. Due to the fact that embryonic chick motoneurons exist in a characteristic ventro-lateral location in the developing cord, we were also able to easily identify motoneurons and confirmed an increased Cl^{-}_{in} using the chloride indicator Clomeleon (Lindsly et al. 2014; Kuner and Augustine 2000). This represents an elegant means of scaling up as it could increase all synapses by a multiplicative manner, as predicted by the scaling hypothesis. Both the high baseline and activity dependent increase in intracellular chloride is mediated by at least 2 transporters (NKCC1 and AE3) ((Gonzalez-Islas et al. 2009, 2010) and unreported observations).

We first thought that scaling was a homeostatic mechanism to recover the network spiking activity by increasing excitatory synaptic strength. If true then we would expect the recovery following either gabazine or CNQX/APV would trigger synaptic scaling by the time the movements recovered. We found that gabazine did trigger scaling, and increased mPSC amplitude even more so than dramatic reductions of spiking activity, but interestingly glutamatergic blockade did not trigger any scaling of either GABA or AMPA mPSCs (Wilhelm and Wenner 2008). This provided the first indication that scaling was not simply triggered by a reduction in spiking activity. Next, we determined that although scaling was triggered by 48 h of GABAergic block, it was not expressed at 12 h of blockade, even though this was when the embryonic movements had recovered. Yet another discrepancy was that while episodes of SNA expressed in the isolated cord were more frequent after 48 h lidocaine treatment, they were not more frequent after 48 h of gabazine treatment despite the fact that scaling was robustly expressed at this point. Together, the findings suggested that homeostatic recovery of spiking activity in the network was not mediated by scaling. This idea was largely confirmed in a study in which we manipulated the frequency of action potential-independent GABAergic mPSCs (Garcia-Bereguiain et al. 2016). Since it appeared that blocking $GABA_A$ receptor activation triggered scaling, we tested whether GABAergic transmission associated with spontaneous release of GABA vesicles could influence scaling. In an earlier paper we were able to show that blocking nicotinic transmission decreased, and nicotine increased, GABAergic mPSC frequency, which gave us a tool to manipulate spontaneous GABAergic transmission (Gonzalez-Islas et al. 2016). Although not demonstrated in chick embryo, previous work suggests that presynaptic nicotinic receptors act to gate vesicle release in several systems (Barik and Wonnacott 2009). We found that reducing GABA mPSC frequency for 2 days (E8–10) induced upward scaling of both GABA and AMPA mPSCs, and increasing GABA mPSC frequency caused a downscaling of both GABA and AMPA mPSCs. Further, nicotinic induction of scaling was occluded by $GABA_A$ receptor blockade, demonstrating the importance of GABAergic transmission in this process. Importantly, when we combined activity blockade (lidocaine) with nicotine treatment we found a downscaling. This result was critical because lidocaine blocks spiking which by itself triggers upscaling, but this is converted to a downscaling by merely increasing spontaneous GABA mPSCs. The result also suggests that spiking activity plays little role in scaling because both nicotine treatment and nicotine plus lidocaine treatment produces the same downscaled mPSC amplitude but one increases embryonic movements and the other blocks them. These results combined with other parts of this study made a strong case for the idea that scaling was tied to $GABA_A$ receptor activation due to spontaneous quantal release of a single vesicle rather than to spiking activity. How could this impact signaling cascades important for triggering scaling? While this is currently unclear, we hypothesize that GABAergic mPSCs may produce small depolarizations that lead to small calcium transients that are important for the scaling process. Alternatively, the large calcium transients associated with strong GABAergic transmission and network activation may not be in the appropriate range to impact scaling. Our work in embryonic motoneurons in the chick was consistent with the idea that scaling was a process that homeostatically controlled synaptic strength at individual synapses. Similar results demonstrating the importance of glutamatergic mPSCs in triggering scaling have been observed in in cortical or hippocampal cultures at a stage when GABA is no longer depolarizing (Sutton et al. 2006; Hou et al. 2008; Maghsoodi et al. 2008; Fong et al. 2015). However, alterations of spiking activity can trigger scaling in certain circumstances, yet it is not clear if the signaling pathways underlying these two forms of scaling are similar or distinct. On the other hand, it was clear in the chick that the homeostatic recovery following transmitter blockade *in ovo* was not mediated by scaling, but rather must be through distinct homeostatic mechanisms. If scaling does not mediate homeostatic regulation of spiking activity, then what is its function? One possibility is that it ensures that a given synapse or dendritic compartment remains sensitive to its inputs during developmental challenges where cells get bigger and input resistance is reduced.

Homeostatic Intrinsic Plasticity

In addition to homeostatic synaptic plasticity, cells have been shown to make compensations in membrane excitability, thus changing the likelihood that a synaptic input will bring the cell to threshold. For instance, following spike or neurotransmission blockade cortical and hippocampal neurons reduce their threshold current through changes in voltage gated conductances (e.g. Na^+ and K^+ channels)(Desai et al. 1999; Lee et al. 2015). We thought it was likely that homeostatic intrinsic plasticity (HIP) had contributed to the recovery of embryonic movements following neurotransmitter blockade since scaling had not occurred by the time recovery was achieved. Indeed, we did find evidence that HIP was triggered following GABAergic block, and was observed by 12 h of gabazine treatment (Wilhelm et al. 2009). We found that the threshold current was reduced and the frequency-current (FI) curve shifted to the left by 12 h of gabazine treatment in the isolated cord in the absence of any drugs. This suggested that these changes could have contributed to the recovery of embryonic movements following GABAergic blockade. Underlying these changes were increases in voltage-gated Na^+ channel currents and decreases in two different K^+ channel currents (I_A and I_{Kca}) (Wilhelm et al. 2009). Surprisingly, no changes in threshold current or FI curve was observed following 12 h glutamatergic blockade. Therefore, while we did see some changes in classical HIP mechanisms triggered by gabazine that could help recover embryonic movements, we were still in the dark in terms of the mechanisms that underlie the recovery in glutamatergic blockade.

Recently we have described a new form of homeostatic intrinsic plasticity that is likely to significantly contribute to the recovery of embryonic movements in the first hours of either glutamatergic or GABAergic block. Within the first hour of applying a neurotransmitter antagonist (either gabazine or CNQX/APV) we see a 10–15 mV depolarization of the resting membrane potential (Gonzalez-Islas et al. 2020). This brings both motoneurons and interneurons closer to threshold. It is particularly important that motoneurons display this form of increased excitability as they are responsible for triggering the initiation of an episode of SNA (Wenner and O'Donovan 1999). We had not noticed these changes in previous studies because they are only observed in the presence of the neurotransmitter receptor antagonists. In previous studies the antagonists were injected in the egg, but the drugs were washed out when we isolated the spinal cord to look for changes in excitability of these cells. The mechanism of this compensatory depolarization of resting membrane potential remains unclear, but one possibility is that it is mediated by an inactivation of the alpha 3 isoform of the Na/K ATPase as shown in *Xenopus, Drosophila,* and in the mouse (Zhang and Sillar 2012; Picton et al. 2017, 2018). This form of the Na^+ pump has a low affinity for Na^+ but generates a hyperpolarizing current when Na^+ levels are high as occurs in embryonic motoneurons in the chick (Lindsly et al. 2017).

In summary, we have been able to significantly advance the field of homeostatic plasticity because we could follow the output of the spinal cord, motoneurons, as they drive embryonic movements. In this way we could observe the actual homeostatic recovery of network behavior and compare it to the timing of the expression of homeostatic mechanisms. Using this methodology we have determined that blocking episodes of SNA with either glutamatergic or GABAergic neurotransmitter antagonists triggers a depolarization of spinal motoneurons and interneurons that are likely responsible for the initial recovery (first hours) of embryonic movements following this perturbation *in ovo*. We further show that other mechanisms of HIP likely contribute to subsequent aspects of the recovery in the first 12 h following GABAergic blockade. Finally, due to the accessibility of motoneurons we were able to recognize that while $GABA_A$ receptor block triggers scaling, this form of homeostatic synaptic plasticity does not contribute to the homeostatic recovery of spinal network behavior in the embryo. It will be important to see if this principle extends to other systems where scaling is induced.

3 Translation of Findings in Chick Embryo to Other Systems

These studies in the chick embryo have provided a relatively full picture of homeostatic plasticity in motoneurons. However, several studies in other systems have provided important contributions that deliver more context to these results and in many cases they demonstrate that work in the chick translates to other systems. Here we discuss these findings.

Nicotine Exposure to Developing Rat Hypoglossal Motoneurons

Similar to our work in the chick embryo (Garcia-Bereguiain et al. 2016; Gonzalez-Islas et al. 2016), the Fregosi lab has published a set of papers where vesicle release was manipulated during embryonic development by *in vivo* introduction of nicotine. The studies demonstrate what may be homeostatic responses that were observed at the neonatal stage in hypoglossal rat motoneurons. Embryos from E5 to the first postnatal week were exposed to nicotine *in vivo* through minipumps implanted into the pregnant dam. While it was not completely clear what nicotine did in the rat embryo, the authors do show that acute application of nicotine led to an increase in spontaneous glutamatergic and GABAergic vesicle release onto hypoglossal motoneurons in the P 0–5 neonatal slice (Wollman et al. 2018a; Buls Wollman et al. 2019). This chronic nicotine exposure *in vivo* during embryogenesis through the first neonatal week led to increased intrinsic membrane excitability (Pilarski et al. 2011), decreased GABAergic vesicle release (Wollman et al. 2018b), and either decreased (Pilarski et al. 2011) or no change in glutamatergic vesicle release (Buls Wollman et al. 2019). This could be homeostatic if vesicle release was increased by nicotine through embryonic development and caused the compensatory decrease in release observed in the first neonatal week. In the chick embryo nicotine, when injected into the egg from E8–10 (thus likely increasing GABAergic vesicle release) led to a downscaling of AMPA and GABA mPSC amplitudes, and therefore we might have expected similar results in rat hypoglossal motoneurons. The Fregosi group did find that GABAergic mPSCs were of smaller amplitude (Wollman et al. 2018b), however in 2 different studies glutamatergic mPSC amplitude was either not changed (Buls Wollman et al. 2019) or was increased (Cholanian et al. 2017). In addition, while the Fregosi lab saw changes in glutamatergic and GABAergic release after chronic nicotine treatment, we did not observe changes in glutamatergic or GABAergic mPSC frequency. Therefore, some of the results are consistent with observations in the chick embryo, and some could be seen as homeostatic, but there were several examples of distinct responses in the two systems and there were cases where there was no simple homeostatic rule. For instance, if GABA and AMPA vesicle release was increased throughout embryonic development, when both are likely depolarizing, then we might expect this to increase activity levels, and trigger a homeostatic reduction in intrinsic membrane excitability. However, chronic embryonic nicotine exposure led to hyperexcitable membranes in rat neonatal hypoglossal motoneurons. The possible reasons behind the differences in the chick and rat are manyfold (e.g. if the critical effect of nicotine exposure in the rat was to enhance inhibition). The absolute effect of nicotine on vesicle release or other aspects of network behavior in the rat embryo are unknown and may change at different stages of embryonic development. In the chick embryo nicotine only enhanced GABAergic vesicle release, not both GABAergic and glutamatergic release as described in the neonatal rat. To gain a better understanding it will be important to identify the effect of nicotine in the rat embryo throughout this period and determine the timing or sequence of events that are altered.

Scaling Induced in Zebrafish Motoneurons Did Not Alter Motor Activity

In the isolated cord of the chick embryo, we found that synaptic scaling was not accompanied by a change in the frequency or expression of motor activity. A similar result was demonstrated in the zebrafish embryo, where scaling could be produced but did not influence motor output (Knogler et al. 2010). Drapeau and collaborators found that treating embryos chronically with drugs that could alter activity/neurotransmission (e.g. TTX, CNQX, TNFα) triggered AMPAergic synaptic scaling in an apparent compensatory manner (Knogler et al. 2010). However, this had no influence on motoneuron activity patterns. The study also acutely increased or decreased AMPAergic synaptic strength. This was similar to imposing scaling acutely, but again there was no change in motoneuron activity patterns or swimming behavior. These results support our findings that scaling does not significantly alter activity levels or motor behavior. Instead, it is possible that some other form of homeostatic plasticity compensates for scaling (e.g. homeostatic intrinsic plasticity), and that scaling may have a distinct functional role.

Altered Levels of Neurotransmission Can Trigger HIP in Fly Motoneurons

Work in the fly *Drosophila* has provided significant advancements in our understanding of homeostatic plasticity in motoneurons due to the flies genetic accessibility, fast generation times, and ease to screen for motor deficits. Genetically increasing or decreasing synaptic input throughout the nervous system in embryonic and larval development produced compensatory changes in intrinsic excitability at later larval stages (Baines et al. 2001; Baines 2003). For instance, genetically blocking synaptic input triggered increases in membrane excitability (leftward shift in FI curve). These changes in intrinsic excitability occurred through changes in voltage-gated conductances, including alterations in Na^+ channel currents. Further, hyperpolarizing embryonic motoneurons while leaving synaptic inputs intact did not produce the changes in voltage-gated conductances that were seen in larval motoneurons from synaptically blocked mutants; this suggested that HIP was triggered through changes in synaptic activation rather than the associated reductions in spiking activity. These results fit nicely with our findings that reductions in GABAergic transmission, rather than reduced spiking, triggered increases in intrinsic excitability, and that these changes were mediated in part by increased Na^+ channel currents (Wilhelm et al. 2009). HIP was triggered by blocking GABAergic transmission in the chick embryo, but blocking nicotinic transmission in the fly, suggesting the importance of different kinds of neurotransmission in different systems.

4 Novel Homeostatic Principles in Motoneurons from Studies in the Developing Fly and Frog

Identification of Signaling Pathways in Fly Embryo Motoneurons

As described above, studies carried out in fly motoneurons show similarities with our work, however different studies in this system have dramatically expanded our understanding of homeostatic plasticity in motoneurons. Baines and collaborators have identified one of the key signaling molecules that mediates the compensatory changes in HIP that are triggered by reducing or increasing their cholinergic synaptic input (Mee et al. 2004). This group found that altering synaptic input in the embryo led to changes in the expression of a transcriptional repressor called Pumilio, which repressed the fly Na^+ channel mRNA *paralytic* or *para*. For instance, when synaptic activation of motoneurons was blocked during the embryonic period, Pumilio expression was reduced, which resulted in an increase in *para* and the persistent Na^+ current; this led to an increase in membrane excitability (HIP) at the larval stage (Mee et al. 2004; Muraro et al. 2008). This shift in persistent Na^+ current can be regulated by altering the splice variant of the *para* gene (Muraro et al. 2008).

In a series of related but distinct studies, this group went on to show that a different set of fly mutants had larger synaptic inputs at the larval stage and experienced a seizure like behavior (Marley and Baines 2011; Lin et al. 2012). Interestingly, these mutants had a reduced membrane excitability (right shifted FI curve), as if the motoneurons had homeostatically adjusted for the increased synaptic input. Despite this shift in membrane excitability the larger synaptic inputs triggered more spikes and these flies experienced seizure-like phenotypes, which were associated with an increase in persistent Na^+ currents. The Baines lab has also shown that different mutants that experience seizure-like activity can be rescued by the introduction of antiepileptic drugs during the embryonic period (Lin et al. 2012; Giachello and Baines 2015). Further, they determined that optogenetically altering activity in the

WT motor system during a tightly defined critical period (17–19 h after egg laying) could trigger increases in motoneuron synaptic currents as well as inducing seizure-like activity at the larval stage (Giachello and Baines 2015). This period of embryonic development corresponds to the stage at which the motor circuit is first functional. These findings suggest an embryonic critical period for triggering hyperexcitability in the later larval stage that could arise by altering the set point for homeostasis (Giachello and Baines 2017).

Presynaptic Homeostatic Plasticity at the Fly Neuromuscular Junction

So far, we have mostly focused on homeostatic mechanisms that are expressed in the postsynaptic motoneuron cell body and dendrites. However, a prominent form of homeostatic plasticity in the motor system exists in the presynaptic motoneuron terminals at the neuromuscular junction (NMJ). This form of presynaptic homeostatic plasticity was first described at the adult mammalian NMJ and is discussed extensively in a separate chapter in this book (see Chap. 5) by Mark Rich, (Plomp et al. 1992; Plomp 2017; Wang and Rich 2018)). Instead, we will focus on this presynaptic plasticity as it has been described in development at the *Drosophila* larval NMJ. One of the early observations of presynaptic homeostatic regulation was at the NMJ of a *Drosophila* mutant that exhibited hyperinnervation of a muscle by its motoneuron terminal (Davis and Goodman 1998a). Despite a dramatic increase in bouton number innervating the muscle, the authors observed a normal evoked response upon motoneuron stimulation. This result was shown to arise due to a compensatory reduction in the probability of transmitter release from the innervating motoneuron and was referred to as presynaptic homeostatic depression (PHD).

The pioneering physiological evidence for the mechanism that senses the perturbation and triggers the compensatory change in probability of release (*Pr*) came from experiments in *Drosophila* with genetic manipulations of muscle-specific glutamate receptors (Petersen et al. 1997; DiAntonio et al. 1999). These studies confirmed that loss of *Drosophila* glutamate receptor subunit (GluRIIA) diminished quantal size but left the evoked release intact, due to compensatory increase in quantal content (the number of vesicles released by an action potential in the innervating motoneuron). This compensatory increase in release was referred to as presynaptic homeostatic potentiation (PHP). Results of these studies indicated a postsynaptic component was involved in triggering the plasticity following these chronic perturbations. In another study Frank et al. (Frank et al. 2006) showed a similar PHP could be observed following an acute pharmacological blockade of postsynaptic receptors with a sub-saturating concentration of philanthotoxin (PhTx), an GluRIIA antagonist. Following bath application of PhTx to a nerve muscle preparation, it was demonstrated that miniature postsynaptic potentials (mEPSPs) in the muscle were quickly reduced in amplitude but within 10 min after drug application, quantal content had increased and thus homeostatically maintained the evoked response. We illustrate a schematic to summarize the basic principles of acutely induced PHP (Fig. 3). This study showed that PHP could occur within minutes, and did not require the chronic perturbations throughout development, characteristic of genetic modifications (Petersen et al. 1997; DiAntonio et al. 1999). This fast PHP was dependent on a reduction of postsynaptic glutamate receptor activation that depended on action potential-independent neurotransmission (Frank et al. 2006). This was surprising as it suggested the action potential-evoked depolarization that was homeostatically maintained was not necessary as a feedback signal, rather it was the reduction in receptor activation associated with spontaneous vesicle release that was used as a feedback mechanism to calculate the necessary increase in vesicles released during an evoked response. However, one early genetic study of PHP suggests that the impaired muscle excitability alone, without altering the function of glutamatergic receptors, could initiate a compensatory increase in quantal content (Paradis et al. 2001).

While it appears relatively clear that inhibition or loss of postsynaptic glutamate receptor function triggered PHP, the retrograde transsynaptic signal back to the presynaptic terminal has been quite elusive. However, several different molecules and signaling cascades have been implicated and they are discussed in detail in two excellent recent reviews of homeostatic plasticity at the fly NMJ (Frank et al. 2020; Goel and Dickman 2021). The postsynaptic trigger for chronic and acute PHP is

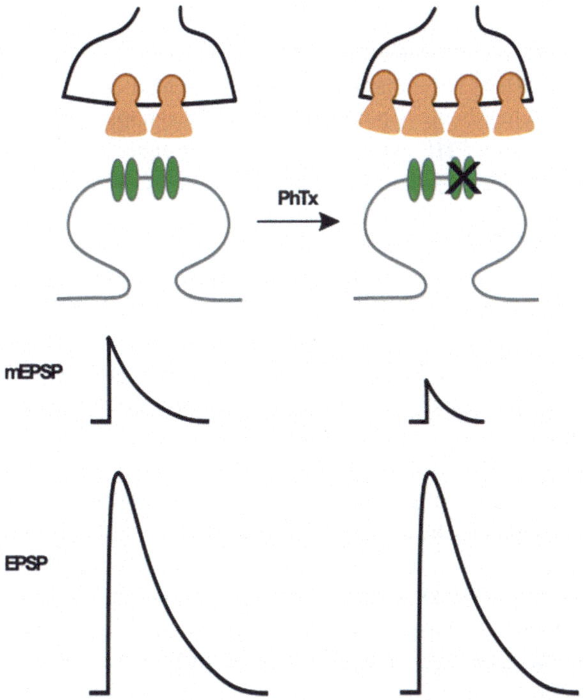

Fig. 3 PHP is triggered by partial blockade of postsynaptic glutamate receptors (green). PHTx application reduces the miniature excitatory postsynaptic potential (mEPSP) while the evoked excitatory postsynaptic potential (EPSP) is maintained due to an increase in the number of vesicles released during an action potential

associated with a reduction of Ca^{2+}/calmodulin dependent protein kinase II (CaMKII) (Haghighi et al. 2003; Goel et al. 2017; Li et al. 2018a). While this could be the result of a reduction of calcium influx due to blockade or loss of GluRIIA, another study showed that the reduction in CaMKII still occurred in the absence of extracellular calcium (Goel et al. 2017). Hauswirth et al. (Hauswirth et al. 2018) proposed another kinase, phosphoinositide-3-kinase (PI3K), was involved in rapid and sustained PHP expression. Finally, ubiquitin ligase Cullin-3 and its adaptor insomniac, discovered through a genetic screen of postsynaptic factors in PHP signaling, interacts with multiplexin, an extracellular matrix component which could provide a transynaptic bridge for signaling to the presynaptic terminal (Kikuma et al. 2019).

The mechanisms of expression of increases in quantal content or vesicle release associated with PHP have been better identified than the sensor and retrograde signals. Reliable neurotransmission during presynaptic plasticity depends on the expression of Ca_V2-type calcium channel Cacophony (Cac), which increases during chronic and acute PHP (Frank et al. 2006, 2009; Gratz et al. 2019). In addition to increased Cac and increased level of presynaptic Ca^{2+} during PHP (Gratz et al. 2019; Muller and Davis 2012), presynaptic kainite-type ionotropic glutamate receptor (DKaiR1D) (Kiragasi et al. 2017, 2020) and epithelial sodium channel ENaC (Younger et al. 2013) have been proposed to contribute to the enhanced release during PHP. Interestingly, changes in the size and organization of active zone components have also been reported during PHP (Goel et al. 2017; Frank et al. 2009; Gratz et al. 2019; Muller and Davis 2012; Weyhersmuller et al. 2011; Muller et al. 2012) and these changes can lead to the observed increase in *Pr*, but also an increase in the size of readily releasable synaptic vesicle pool (RRP) (Li et al. 2018a, b; Kiragasi et al. 2017; Weyhersmuller et al. 2011; Muller et al. 2012; Gavino et al. 2015). A set of studies have concluded that some synapses might be more or less efficient at expressing PHP, which could have implications for presynaptic plasticity in mammalian CNS, where there is a large range of synapses with different *Pr (*Newman et al. 2017*;* Genc and Davis 2019*)*.

PHP is far more studied, but there are several examples of PHD, where genetic mutations led to a compensatory reduction in quantal content as described in the initial Davis and Goodman study, mentioned above, where hyperinnervation produced the same evoked response as in controls (Davis and Goodman 1998b). In a separate study PHD was observed in mutants that expressed higher levels of the vesicular glutamate transporter, where vesicles were bigger and contained more glutamate per vesicle. This produced miniature postsynaptic potentials with larger amplitudes, but the evoked response was of normal size due to a reduction in the quantal content (Daniels et al. 2004). While this seems very similar to PHP, but in the opposite direction, PHD appears to be a fairly distinct process. First, the core genes involved in PHP expression do not seem to be necessary for PHD expression (Goel and Dickman 2021). Further, while PHP appears to be dependent on the postsynaptic response and is triggered by inhibiting postsynaptic receptors, PHD was not triggered by the opposite perturbation – increasing postsynaptic glutamate receptors on the muscle (Petersen et al. 1997; Li et al. 2018a, b). In fact, it has been proposed that induction of PHD depends on excess glutamate in the synaptic cleft (Daniels et al. 2004). Another study also reinforced the lack of sensitivity of PHD in the postsynaptic compartment and proposed that this form of plasticity may be autonomous, where the presynaptic neuron uses a negative feedback system, assessing glutamate concentration in the synaptic cleft as a way to homeostatically control glutamate release (Li et al. 2018b). In addition, unlike PHP, in PHD observed in mutants overexpressing the vesicular glutamate transporter, there did not appear to be obvious changes in active zone components, changes to the readily releasable pool, or CaMKII activity (Gratz et al. 2019; Gavino et al. 2015; Li et al. 2018b; Goel et al. 2019). On the other hand both PHP and PHD share inverse changes in *Pr* (Frank et al. 2009; Muller et al. 2012; Gavino et al. 2015). Observed changes in neurotransmitter release in both PHP and PHD are mediated by altered calcium influx in the presynaptic terminal, which increases in PHP and decreases in PHD (Gratz et al. 2019; Gavino et al. 2015).

Ultimately presynaptic homeostatic plasticity is a strong adaptive tool that provides differential control of synaptic efficacy in response to various challenges. PHP acts as a homeostat that stabilizes synaptic strength and safeguards NMJ function preventing muscle contraction failure (Petersen et al. 1997). PHD could act as glutamate homeostat with possible neuroprotective role from glutamate toxicity (Li et al. 2018b; Daniels et al. 2004).

Channel Coregulation Mediates HIP

In three separate studies in fly larval motoneurons, it was shown that there were compensations in ion channel expression that occurred when other ion channels were perturbed. In one study the authors showed that two different genes that encode the transient A-type potassium channel compensate for each other when one of the two is removed (Bergquist et al. 2010). The authors were able to show that loss of *Shal* led to an upregulation of *Shaker* and vice versa, and this was demonstrated to be a transcriptionally coupled process. A separate study demonstrated a coupling between *para* and *Shal,* such that a Pum-mediated reduction in para resulted in a compensatory change in *Shal* (Muraro et al. 2008). In a final study, a non-reciprocal relationship was observed between a delayed rectifier K^+ channel encoded by *shab* and a calcium-dependent K^+ conductance encoded by *slo*, a delayed rectifier K^+ channel, which again was transcription-dependent (Kim et al. 2017). In this study, the loss of *shal* triggered a compensatory upregulation of *slo*, but loss of *slo* did not result in an upregulation of *shal*. Therefore, in this case the homeostatic intrinsic plasticity was dependent on the nature of the specific current that was compromised. The first description of this sort of ion channel coregulation was demonstrated in the adult lobster stomatogastric ganglion (MacLean et al. 2003).

Homeostatic Regulation of Neurotransmitter Specification in Embryonic Xenopus Motoneurons

We have discussed homeostatic synaptic plasticity and homeostatic intrinsic plasticity, but another kind of homeostatic plasticity has been demonstrated in motoneurons that involves neurotransmitter specification. In the *Xenopus* embryonic neural tube it was shown that increasing the frequency of calcium spikes *in vivo*, triggered an increase in spinal neurons expressing inhibitory transmitters (GABA or glycine) and a decrease in cells expressing excitatory neurotransmitters (glutamate or acetylcholine) (Borodinsky et al. 2004). Motoneurons normally only release acetylcholine which acts on nicotinic receptors to produce depolarizing currents in the muscle. In fact this group went on to show that when calcium spikes were reduced in frequency in embryonic motoneurons these cells now also produced functional glutamatergic depolarizations in muscle (Borodinsky and Spitzer 2007). Similarly, when calcium spikes were increased, motoneurons then produced functional GABAergic and glycinergic currents in the muscles they innervated. In addition, further studies demonstrated that there was a functional gradient of calcium spike frequency from high in ventral neural tube to lower in dorsal neural tube, and this was set up by morphogenic proteins Sonic hedgehog (Belgacem and Borodinsky 2011) and bone morphogenic protein (Swapna and Borodinsky 2012). This then sets up a transcription factor and activity dependent interplay to determine neurotransmitter phenotype in a manner that appears to pay attention to activity levels, favoring inhibition when activity is increased and excitation when activity is reduced (Rosenberg and Spitzer 2011; Borodinsky and Belgacem 2016).

5 Summary

We have gained a great deal of information on the process and mechanisms of homeostatic plasticity during development thanks to the access that motoneurons provide due to their well characterized locations in the spinal cord, the access one has to their axons traveling in separate muscle nerves, and because they drive movements that can be easily observed. Across multiple developing preparations we have determined the existence of homeostatic synaptic plasticity (synaptic scaling and presynaptic plasticity), HIP, and neurotransmitter switching. Because we can actually observe a homeostatic recovery of SNA-driven embryonic movements in the chick embryo we have determined that different forms of HIP can recover these movements. We have also come to the realization that synaptic scaling does not contribute to this recovery.

References

Aoto J et al (2008) Synaptic signaling by all-trans retinoic acid in homeostatic synaptic plasticity. Neuron 60(2):308–320

Baines RA (2003) Postsynaptic protein kinase A reduces neuronal excitability in response to increased synaptic excitation in the Drosophila CNS. J Neurosci 23(25):8664–8672

Baines RA et al (2001) Altered electrical properties in Drosophila neurons developing without synaptic transmission. J Neurosci 21(5):1523–1531

Barik J, Wonnacott S (2009) Molecular and cellular mechanisms of action of nicotine in the CNS. Handb Exp Pharmacol 192:173–207

Belgacem YH, Borodinsky LN (2011) Sonic hedgehog signaling is decoded by calcium spike activity in the developing spinal cord. Proc Natl Acad Sci U S A 108(11):4482–4487

Ben-Ari Y (2001) Developing networks play a similar melody. Trends Neurosci 24(6):353–360

Ben-Ari Y et al (1989) Giant synaptic potentials in immature rat CA3 hippocampal neurones. J Physiol Lond 416:303–325

Bergquist S, Dickman DK, Davis GW (2010) A hierarchy of cell intrinsic and target-derived homeostatic signaling. Neuron 66(2):220–234

Borodinsky LN, Belgacem YH (2016) Crosstalk among electrical activity, trophic factors and morphogenetic proteins in the regulation of neurotransmitter phenotype specification. J Chem Neuroanat 73:3–8

Borodinsky LN, Spitzer NC (2007) Activity-dependent neurotransmitter-receptor matching at the neuromuscular junction. Proc Natl Acad Sci U S A 104(1):335–340

Borodinsky LN et al (2004) Activity-dependent homeostatic specification of transmitter expression in embryonic neurons. Nature 429(6991):523–530

Buls Wollman L et al (2019) Developmental nicotine exposure alters synaptic input to hypoglossal motoneurons and is associated with altered function of upper airway muscles. eNeuro 6(6)

Cholanian M et al (2017) Influence of developmental nicotine exposure on glutamatergic neurotransmission in rhythmically active hypoglossal motoneurons. Exp Neurol 287(Pt 2):254–260

Chub N, O'Donovan MJ (1998) Blockade and recovery of spontaneous rhythmic activity after application of neurotransmitter antagonists to spinal networks of the chick embryo. J Neurosci 18(1):294–306

Daniels RW et al (2004) Increased expression of the Drosophila vesicular glutamate transporter leads to excess glutamate release and a compensatory decrease in quantal content. J Neurosci 24(46):10466–10474

Davis GW (2006) Homeostatic control of neural activity: from phenomenology to molecular design. Annu Rev Neurosci 29:307–323

Davis GW, Goodman CS (1998a) Genetic analysis of synaptic development and plasticity: homeostatic regulation of synaptic efficacy. Curr Opin Neurobiol 8(1):149–156

Davis GW, Goodman CS (1998b) Synapse-specific control of synaptic efficacy at the terminals of a single neuron. Nature 392(6671):82–86

Desai NS, Rutherford LC, Turrigiano GG (1999) Plasticity in the intrinsic excitability of cortical pyramidal neurons. Nat Neurosci 2(6):515–520

DiAntonio A et al (1999) Glutamate receptor expression regulates quantal size and quantal content at the Drosophila neuromuscular junction. J Neurosci 19(8):3023–3032

Farrant M, Kaila K (2007) The cellular, molecular and ionic basis of GABA(A) receptor signalling. Prog Brain Res 160:59–87

Fedirchuk B et al (1999) Spontaneous network activity transiently depresses synaptic transmission in the embryonic chick spinal cord. J Neurosci 19(6):2102–2112

Feller MB (1999) Spontaneous correlated activity in developing neural circuits. Neuron 22(4):653–656

Fong MF et al (2015) Upward synaptic scaling is dependent on neurotransmission rather than spiking. Nat Commun 6:6339

Fortin G et al (1995) Rhythm generation in the segmented hindbrain of chick embryos. J Physiol Lond 486(Pt 3):735–744
Frank CA et al (2006) Mechanisms underlying the rapid induction and sustained expression of synaptic homeostasis. Neuron 52(4):663–677
Frank CA, Pielage J, Davis GW (2009) A presynaptic homeostatic signaling system composed of the Eph receptor, ephexin, Cdc42, and CaV2.1 calcium channels. Neuron 61(4):556–569
Frank CA, James TD, Muller M (2020) Homeostatic control of Drosophila neuromuscular junction function. Synapse 74(1):e22133
Garcia-Bereguiain MA et al (2013) In vivo synaptic scaling is mediated by GluA2-lacking AMPA receptors in the embryonic spinal cord. J Neurosci 33(16):6791–6799
Garcia-Bereguiain MA et al (2016) Spontaneous release regulates synaptic scaling in the embryonic spinal network in vivo. J Neurosci 36(27):7268–7282
Gavino MA et al (2015) Homeostatic synaptic depression is achieved through a regulated decrease in presynaptic calcium channel abundance. elife 4
Genc O, Davis GW (2019) Target-wide induction and synapse type-specific robustness of presynaptic homeostasis. Curr Biol 29(22):3863–3873 e2
Giachello CN, Baines RA (2015) Inappropriate neural activity during a sensitive period in embryogenesis results in persistent seizure-like behavior. Curr Biol 25(22):2964–2968
Giachello CN, Baines RA (2017) Regulation of motoneuron excitability and the setting of homeostatic limits. Curr Opin Neurobiol 43:1–6
Goel P, Dickman D (2021) Synaptic homeostats: latent plasticity revealed at the Drosophila neuromuscular junction. Cell Mol Life Sci
Goel P, Li X, Dickman D (2017) Disparate postsynaptic induction mechanisms ultimately converge to drive the retrograde enhancement of presynaptic efficacy. Cell Rep 21(9):2339–2347
Goel P et al (2019) Homeostatic scaling of active zone scaffolds maintains global synaptic strength. J Cell Biol 218(5):1706–1724
Gonzalez-Islas C, Wenner P (2006) Spontaneous network activity in the embryonic spinal cord regulates AMPAergic and GABAergic synaptic strength. Neuron 49(4):563–575
Gonzalez-Islas C, Chub N, Wenner P (2009) NKCC1 and AE3 appear to accumulate chloride in embryonic motoneurons. J Neurophysiol 101(2):507–518
Gonzalez-Islas C et al (2010) GABAergic synaptic scaling in embryonic motoneurons is mediated by a shift in the chloride reversal potential. J Neurosci 30(39):13016–13020
Gonzalez-Islas C et al (2016) Tonic nicotinic transmission enhances spinal GABAergic presynaptic release and the frequency of spontaneous network activity. Dev Neurobiol 76(3):298–312
Gonzalez-Islas C, Garcia-Bereguiain MA, Wenner P (2020) Homeostatic recovery of embryonic spinal activity initiated by compensatory changes in resting membrane potential. eNeuro 7(4)
Gratz SJ et al (2019) Endogenous tagging reveals differential regulation of Ca(2+) channels at single active zones during presynaptic homeostatic potentiation and depression. J Neurosci 39(13):2416–2429
Gummer AW, Mark RF (1994) Patterned neural activity in brain stem auditory areas of a prehearing mammal, the tammar wallaby (Macropus eugenii). Neuroreport 5(6):685–688
Haghighi AP et al (2003) Retrograde control of synaptic transmission by postsynaptic CaMKII at the Drosophila neuromuscular junction. Neuron 39(2):255–267
Hall BK, Herring SW (1990) Paralysis and growth of the musculoskeletal system in the embryonic chick. J Morphol 206(1):45–56
Hanson MG, Landmesser LT (2004) Normal patterns of spontaneous activity are required for correct motor axon guidance and the expression of specific guidance molecules. Neuron 43(5):687–701
Hauswirth AG et al (2018) A postsynaptic PI3K-cII dependent signaling controller for presynaptic homeostatic plasticity. elife 7
Ho SM, Waite PM (1999) Spontaneous activity in the perinatal trigeminal nucleus of the rat. Neuroreport 10(3):659–664

Hou Q et al (2008) Homeostatic regulation of AMPA receptor expression at single hippocampal synapses. Proc Natl Acad Sci U S A 105(2):775–780
Itaya SK, Fortin S, Molotchnikoff S (1995) Evolution of spontaneous activity in the developing rat superior colliculus. Can J Physiol Pharmacol 73(9):1372–1377
Kikuma K et al (2019) Cul3 and insomniac are required for rapid ubiquitination of postsynaptic targets and retrograde homeostatic signaling. Nat Commun 10(1):2998
Kilman V, van Rossum MC, Turrigiano GG (2002) Activity deprivation reduces miniature IPSC amplitude by decreasing the number of postsynaptic GABA(A) receptors clustered at neocortical synapses. J Neurosci 22(4):1328–1337
Kim EZ et al (2017) Nonreciprocal homeostatic compensation in Drosophila potassium channel mutants. J Neurophysiol 117(6):2125–2136
Kiragasi B et al (2017) A presynaptic glutamate receptor subunit confers robustness to neurotransmission and homeostatic potentiation. Cell Rep 19(13):2694–2706
Kiragasi B et al (2020) The auxiliary glutamate receptor subunit dSol-1 promotes presynaptic neurotransmitter release and homeostatic potentiation. Proc Natl Acad Sci U S A 117(41):25830–25839
Knogler LD, Liao M, Drapeau P (2010) Synaptic scaling and the development of a motor network. J Neurosci 30(26):8871–8881
Kuner T, Augustine GJ (2000) A genetically encoded ratiometric indicator for chloride: capturing chloride transients in cultured hippocampal neurons. Neuron 27(3):447–459
Lee KY et al (2015) N-methyl-D-aspartate receptors mediate activity-dependent down-regulation of potassium channel genes during the expression of homeostatic intrinsic plasticity. Mol Brain 8:4
Li X et al (2018a) Synapse-specific and compartmentalized expression of presynaptic homeostatic potentiation. elife 7
Li X et al (2018b) A glutamate homeostat controls the presynaptic inhibition of neurotransmitter release. Cell Rep 23(6):1716–1727
Lin WH et al (2012) Activity-dependent alternative splicing increases persistent sodium current and promotes seizure. J Neurosci 32(21):7267–7277
Lindsly C, Gonzalez-Islas C, Wenner P (2014) Activity blockade and GABAA receptor blockade produce synaptic scaling through chloride accumulation in embryonic spinal motoneurons and interneurons. PLoS One 9(4):e94559
Lindsly C, Gonzalez-Islas C, Wenner P (2017) Elevated intracellular Na(+) concentrations in developing spinal neurons. J Neurochem 140(5):755–765
Lippe WR (1995) Relationship between frequency of spontaneous bursting and tonotopic position in the developing avian auditory system. Brain Res 703(1-2):205–213
Lissin DV et al (1998) Activity differentially regulates the surface expression of synaptic AMPA and NMDA glutamate receptors. Proc Natl Acad Sci U S A 95(12):7097–7102
MacLean JN et al (2003) Activity-independent homeostasis in rhythmically active neurons. Neuron 37(1):109–120
Maghsoodi B et al (2008) Retinoic acid regulates RARalpha-mediated control of translation in dendritic RNA granules during homeostatic synaptic plasticity. Proc Natl Acad Sci U S A 105(41):16015–16020
Marley R, Baines RA (2011) Increased persistent Na+ current contributes to seizure in the slamdance bang-sensitive Drosophila mutant. J Neurophysiol 106(1):18–29
Mee CJ et al (2004) Regulation of neuronal excitability through pumilio-dependent control of a sodium channel gene. J Neurosci 24(40):8695–8703
Muller M, Davis GW (2012) Transsynaptic control of presynaptic Ca(2)(+) influx achieves homeostatic potentiation of neurotransmitter release. Curr Biol 22(12):1102–1108
Muller M et al (2012) RIM controls homeostatic plasticity through modulation of the readily-releasable vesicle pool. J Neurosci 32(47):16574–16585
Muraro NI et al (2008) Pumilio binds para mRNA and requires Nanos and Brat to regulate sodium current in Drosophila motoneurons. J Neurosci 28(9):2099–2109

Newman ZL et al (2017) Input-specific plasticity and homeostasis at the drosophila larval neuromuscular junction. Neuron 93(6):1388–1404 e10

O'Brien RJ et al (1998) Activity-dependent modulation of synaptic AMPA receptor accumulation. Neuron 21(5):1067–1078

O'Donovan MJ, Chub N, Wenner P (1998) Mechanisms of spontaneous activity in developing spinal networks. J Neurobiol 37(1):131–145

Paradis S, Sweeney ST, Davis GW (2001) Homeostatic control of presynaptic release is triggered by postsynaptic membrane depolarization. Neuron 30(3):737–749

Petersen SA et al (1997) Genetic analysis of glutamate receptors in Drosophila reveals a retrograde signal regulating presynaptic transmitter release. Neuron 19(6):1237–1248

Picton LD et al (2017) Sodium pumps mediate activity-dependent changes in mammalian motor networks. J Neurosci 37(4):906–921

Picton LD, Sillar KT, Zhang HY (2018) Control of xenopus tadpole locomotion via selective expression of IH in excitatory interneurons. Curr Biol 28(24):3911–3923 e2

Pilarski JQ et al (2011) Developmental nicotine exposure alters neurotransmission and excitability in hypoglossal motoneurons. J Neurophysiol 105(1):423–433

Plomp JJ (2017) Trans-synaptic homeostasis at the myasthenic neuromuscular junction. Front Biosci (Landmark Ed) 22:1033–1051

Plomp JJ, van Kempen GT, Molenaar PC (1992) Adaptation of quantal content to decreased postsynaptic sensitivity at single endplates in alpha-bungarotoxin-treated rats. J Physiol 458:487–499

Preyer W (1937) Embryonic motility and sensitivity. Monogr Soc Res Child Dev 2(6):1–115

Rich MM, Wenner P (2007) Sensing and expressing homeostatic synaptic plasticity. Trends Neurosci 30(3):119–125

Rivera C et al (1999) The K+/Cl- co-transporter KCC2 renders GABA hyperpolarizing during neuronal maturation. Nature 397(6716):251–255

Rosenberg SS, Spitzer NC (2011) Calcium signaling in neuronal development. Cold Spring Harb Perspect Biol 3(10):a004259

Ruano-Gil D, Nardi-Vilardaga J, Tejedo-Mateu A (1978) Influence of extrinsic factors on the development of the articular system. Acta Anat (Basel) 101(1):36–44

Sutton MA et al (2006) Miniature neurotransmission stabilizes synaptic function via tonic suppression of local dendritic protein synthesis. Cell 125(4):785–799

Swapna I, Borodinsky LN (2012) Interplay between electrical activity and bone morphogenetic protein signaling regulates spinal neuron differentiation. Proc Natl Acad Sci U S A 109(40):16336–16341

Thiagarajan TC, Lindskog M, Tsien RW (2005) Adaptation to synaptic inactivity in hippocampal neurons. Neuron 47(5):725–737

Toutant JP et al (1979) Enzymatic differentiation of muscle fibre types in embryonic latissimus dorsii of the chick: effects of spinal cord stimulation. Cell Diff 8(5):375–382

Turrigiano GG (2008) The self-tuning neuron: synaptic scaling of excitatory synapses. Cell 135(3):422–435

Turrigiano G (2012) Homeostatic synaptic plasticity: local and global mechanisms for stabilizing neuronal function. Cold Spring Harb Perspect Biol 4(1):a005736

Turrigiano G, Abbott LF, Marder E (1994) Activity-dependent changes in the intrinsic properties of cultured neurons. Science 264(5161):974–977

Turrigiano GG et al (1998) Activity-dependent scaling of quantal amplitude in neocortical neurons. Nature 391(6670):892–896

Wang X, Rich MM (2018) Homeostatic synaptic plasticity at the neuromuscular junction in myasthenia gravis. Ann N Y Acad Sci 1412(1):170–177

Wenner P, O'Donovan MJ (1999) Identification of an interneuronal population that mediates recurrent inhibition of motoneurons in the developing chick spinal cord. J Neurosci 19:7557–7567

Weyhersmuller A et al (2011) Rapid active zone remodeling during synaptic plasticity. J Neurosci 31(16):6041–6052

Wilhelm JC, Wenner P (2008) GABAA transmission is a critical step in the process of triggering homeostatic increases in quantal amplitude. Proc Natl Acad Sci U S A 105(32):11412–11417

Wilhelm JC, Rich MM, Wenner P (2009) Compensatory changes in cellular excitability, not synaptic scaling, contribute to homeostatic recovery of embryonic network activity. Proc Natl Acad Sci U S A 106(16):6760–6765

Wollman LB, Levine RB, Fregosi RF (2018a) Developmental nicotine exposure alters glycinergic neurotransmission to hypoglossal motoneurons in neonatal rats. J Neurophysiol 120(3):1135–1142

Wollman LB, Levine RB, Fregosi RF (2018b) Developmental plasticity of GABAergic neurotransmission to brainstem motoneurons. J Physiol 596(23):5993–6008

Wong RO (1999) Retinal waves and visual system development. Annu Rev Neurosci 22:29–47

Younger MA et al (2013) A presynaptic ENaC channel drives homeostatic plasticity. Neuron 79(6):1183–1196

Zhang HY, Sillar KT (2012) Short-term memory of motor network performance via activity-dependent potentiation of Na+/K+ pump function. Curr Biol 22(6):526–531

Part II
Motoneuron Connectivity and Function

Homeostatic Plasticity of the Mammalian Neuromuscular Junction

Kathrin L. Engisch, Xueyong Wang, and Mark M. Rich

Abstract The mammalian neuromuscular junction (NMJ) is an ideal preparation to study synaptic plasticity. Its simplicity- one input, one postsynaptic target- allows experimental manipulations and mechanistic analyses that are impossible at more complex synapses. Homeostatic synaptic plasticity attempts to maintain normal function in the face of perturbations in activity. At the NMJ, 3 aspects of activity are sensed to trigger 3 distinct mechanisms that contribute to homeostatic plasticity:

- Block of presynaptic action potentials triggers increased quantal size secondary to increased release of acetylcholine from vesicles.
- Simultaneous block of pre- and postsynaptic action potentials triggers an increase in the probability of vesicle release.
- Block of acetylcholine binding to acetylcholine receptors during spontaneous fusion of single vesicles triggers an increase in the number of releasable vesicles as well as increased motoneuron excitability.

Understanding how the NMJ responds to perturbations of synaptic activity informs our understanding of its response to diverse neuromuscular diseases.

Keywords Neuromuscular junction · Endplate · Acetylcholine receptor · Trophic · Motoneuron · Muscle · Plasticity · Homeostatic · Motoneuron

1 The NMJ Allows for Detailed Measurement of Parameters Governing Synaptic Function

The neuromuscular junction (NMJ) serves as the connection between lower motoneurons and skeletal muscle. It is made up of a presynaptic nerve terminal from a lower motoneuron, which is apposed by acetylcholine receptors that are expressed

K. L. Engisch · X. Wang · M. M. Rich (✉)
Department of Neuroscience, Cell Biology and Physiology, Wright State University, Dayton, OH, USA
e-mail: mark.rich@wright.edu

M. J. O'Donovan, M. Falgairolle (eds.), *Vertebrate Motoneurons*, Advances in Neurobiology 28, https://doi.org/10.1007/978-3-031-07167-6_5

by the postsynaptic muscle fiber (Fig. 1). Its role in force generation is to reliably translate presynaptic axonal action potentials into action potentials in the postsynaptic muscle fiber. Acetylcholine is released from the presynaptic terminal, diffuses across the synaptic cleft, and binds to acetylcholine receptors on the muscle. The opening of muscle acetylcholine receptors depolarizes the fiber and triggers firing of an action potential, completing the process of synaptic transmission. The NMJ is easily accessible and because there is only one axon entering the adult NMJ, changes in synaptic strength cannot be due to gain or loss of inputs. Details of technical considerations and various recording configurations relating to use of the NMJ as a model synapse are included at the end of this review.

Two types of synaptic events occur at the NMJ. The first occurs spontaneously and is termed either a spontaneous miniature endplate potential (MEPP) or a

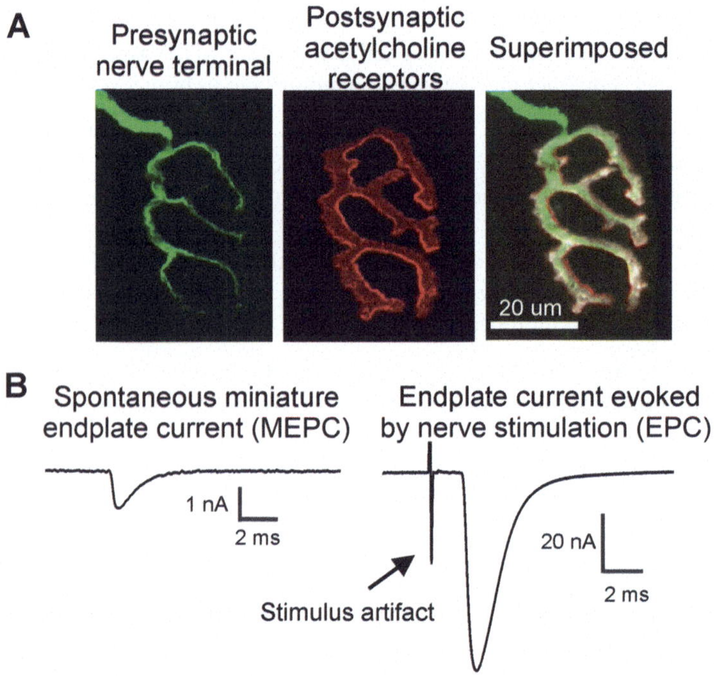

Fig. 1 The mammalian NMJ. (**a**) Shown is neurofilament staining of an axon entering the NMJ from the top left (in green) and SV2 staining of synaptic vesicles at the NMJ (also in green) as well as α-bungarotoxin (BTX) staining of acetylcholine receptors (AChRs, in red). When the green and red images are superimposed the alignment of presynaptic vesicles with AChRs is apparent. (**b**) Shown are voltage clamp recordings from a single NMJ of both a spontaneous miniature endplate current (MEPC) and the endplate current (EPC) evoked by nerve stimulation. The miniature endplate current is thought to result from release of acetylcholine from a single vesicle. The endplate current is triggered by a nerve action potential and is roughly 50 times larger than the spontaneously occurring miniature endplate current, suggesting release of acetylcholine from 50 vesicles. The stimulus artifact preceding the endplate current is caused by the current pulse used to trigger the nerve action potential

spontaneous miniature endplate current (MEPC) depending on whether the recording is done in the current clamp or voltage clamp configuration, respectively (Fig. 1). The second type of response occurs following nerve stimulation and is termed either an endplate potential (EPP) or an endplate current (EPC, Fig. 1). Study of these events and the mechanisms underlying them generated hypotheses that are the foundation of our understanding of synaptic function across the nervous system.

One of the central tenets of synaptic transmission, the "vesicle hypothesis," was generated at the NMJ. This hypothesis is that packets of neurotransmitter released via fusion of synaptic vesicles are the basic unit of synaptic transmission. One reason the vesicle hypothesis was generated at the NMJ rather than CNS synapses is that quantal amplitude (the postsynaptic response to release of a single vesicle) is much more variable at CNS synapses (Korn and Faber 1991). The variability in quantal amplitude at CNS synapses makes it difficult to determine quantal content (the number of vesicles released) following stimulation of the presynaptic input (evoked release). In contrast, at the NMJ, quantal amplitude has a relatively narrow distribution which is close to a normal distribution (Wang et al. 2011). This advantage of the NMJ is critical when studying synaptic function and plasticity of function.

The vesicle hypothesis was generated by recording at the NMJ in solutions with very low extracellular calcium concentrations (with accompanying raised magnesium to prevent hyperexcitability). When extracellular calcium is low, evoked release becomes so small that quantal content varies between 0, 1 or 2. Under these conditions (a low probability of vesicle release), the frequency of 0s, 1s and 2s can be predicted by the Poisson distribution, and the average quantal content, *m*, calculated from the Poisson formula (m = ln number of stimuli/number of failures), matches the quantal content determined by dividing EPP amplitude by MEPP amplitude (Del Castillo and Katz 1954; Boyd and Martin 1956). These experiments provided a key piece of evidence that evoked responses were due to the combined effect of packets of transmitter that were probabilistically released. They further demonstrated that the MEPP was the basic unit of synaptic transmission. MEPPs were subsequently proposed to correspond to the release of acetylcholine from one vesicle (Heuser et al. 1979; Heuser and Reese 1981).

Two difficulties with using the Poisson distribution are that it is generally only useful when probability of release is low (such that there are failures) and it only has one parameter: *m*, leaving two critical parameters, *p* and *n*, hidden. *p* is the probability of an event happening, such as flipping a coin and getting heads. *n* is the number of chances for the event to happen, for example how many times the coins is flipped. It is generally agreed that *p* represents the probability of release of the contents of one synaptic vesicle (Miyamoto 1975; Redman 1990; Korn and Faber 1991; Provan and Miyamoto 1993). Evidence from the frog NMJ indicates that *n* represents the number of active zones (Wernig 1975; Kelly and Robbins 1987). Alternatively, *n* may represent the number of releasable (docked) vesicles (Searl and Silinsky 2002; Stevens 2003). The use of binomial statistics to derive *p* and *n* at synapses (as well as the use of Poisson statistics) is well described in a review by Korn and Faber (Korn and Faber 1991).

At the NMJ p and n can be estimated by measuring the variance of quantal content when probability is varied by varying extracellular calcium concentration (Wang et al. 2010a, b). Quantal content is measured by repeatedly evoking EPCs at a slow rate of simulation (0.5 Hz) to avoid the depression or facilitation of EPC amplitude that occurs with high rates of stimulation. Quantal content of each stimulation is determined by dividing EPC amplitude by the mean MEPC amplitude. Mean quantal content and variance of quantal content (the square of standard deviation) can then be calculated. Plotting the variance of quantal content versus mean quantal content (varied by changing calcium concentration) generates a plot that allows for graphical estimation of p and n (Wang et al. 2010a, b) (Fig. 2). When p is either very high or very low, the variance in quantal content becomes small (each releasable vesicle is either always being released or never being released, respectively). When probability of release for each vesicle is 50%, variance of quantal content is maximal.

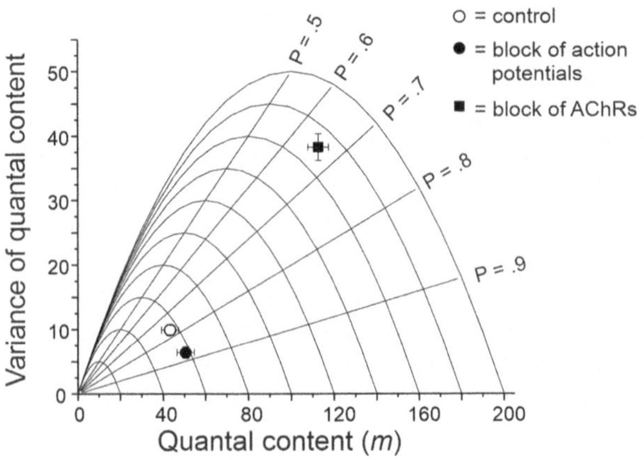

Fig. 2 Determination p and n using analysis of variance of quantal content at the NMJ. To determine p and n at the NMJ one can plot the variance (standard deviation squared) of quantal content versus quantal content. Each parabola represents the plot for a given n as p is increased from 0 to 1.0, assuming uniform p for all synaptic sites. Included in the plot shown are the parabolas for $n = 20$ up to $n = 200$. All the parabolas start with $\mathrm{Var}(m) = 0$ when $m = 0$ and increase to a maximum $\mathrm{Var}(m)$ when $p = 0.5$. When p reaches 1.0 for each n, $\mathrm{Var}(m)$ again equals 0 and $n = m$ as every vesicle that can be released is released. Intersecting the parabolas are straight lines representing the theoretical plot for a given p as n is increased. Each point on the plot has a unique p and n allowing for simultaneous measurement of both parameters. One limitation of this analysis is that when p is low, the parabolas for various values of n run together, making determination of p and n difficult. Superimposed on the theoretical lines for values of p and n are the mean value and standard error of $\mathrm{Var}(m)$ plotted versus m for control NMJs, NMJs in which pre- and postsynaptic action potentials were blocked by placement of a tetrodotoxin (TTX) containing cuff on the sciatic nerve for 1 week and NMJs in which AChRs were partially blocked with α-bungarotoxin (BTX) for 4–5 days. All recordings were performed in solution containing 1 mM Ca^{2+}. Following block of action potentials there was an increase in m that was accompanied by reduction in $\mathrm{Var}(m)$: indicating an increase in p while n remained constant. Following block of AChRs there was an increase in m that was accompanied by a dramatic increase in $\mathrm{Var}(m)$: indicating p may have decreased slightly while n increased dramatically. (Wang et al. 2010b)

2 Homeostatic Synaptic Plasticity at the Mouse NMJ

Homeostatic synaptic plasticity was initially demonstrated in vitro by adding drugs which either increased or decreased action potentials in the neuronal networks for several days. It was found that the networks responded in ways that opposed the manipulation of network activity (O'Brien et al. 1998; Turrigiano et al. 1998). Because the plasticity works towards restoring normal activity it was termed "homeostatic." A central question in studies of activity-dependent synaptic plasticity is how changes in synaptic activity are sensed (Rich and Wenner 2007). Potential ways that loss of synaptic activity might be sensed include loss of signals generated by presynaptic action potentials, such as loss of the rise in intracellular calcium or depolarization in the presynaptic terminal; loss of those same signals in the postsynaptic cell; or, lack of binding of acetylcholine to acetylcholine receptors (AChRs). After considering the ways in which alterations in synaptic activity are sensed to trigger homeostatic plasticity, we will discuss the alteration in synaptic function triggered by the various disruptions of activity.

Homeostatic plasticity at the mouse NMJ has been examined following two distinct ways of manipulating neuromuscular activity. In the first, action potentials of both nerve and muscle were blocked using placement of a cuff around the sciatic nerve in vivo to slowly release tetrodotoxin (TTX) and inhibit Na channels. This manipulation also blocks the release of acetylcholine triggered by presynaptic action potentials but does not affect acetylcholine release from spontaneous vesicle fusion. In the second, drugs which block muscle AChRs were injected in vivo or were acutely applied to muscle in vitro, which spares presynaptic action potentials and release of acetylcholine. The homeostatic responses triggered by these two manipulations of neuromuscular activity differ and will be considered separately.

3 Homeostatic Plasticity of Synaptic Function Triggered by Block of Action Potentials

Regulation of Quantal Amplitude

Action potentials at the NMJ can be blocked in vivo for days via placement of a cuff around the sciatic nerve to cause slow release of TTX to block Na channels. Following chronic (~1 week) loss of action potentials in both nerve and muscle, the EPC was increased. The cause of the increase in EPC amplitude was an increase in the MEPC amplitude with no increase in quantal content (Wang et al. 2005).

Block of nerve action potentials with a TTX cuff also blocks muscle action potentials. Therefore, it was not clear whether block of action potentials in the nerve or muscle was responsible for triggering the homeostatic increase in quantal amplitude. To address this question, a TTX cuff was placed on the sciatic nerve of mice with myotonia congenita. These mice have a loss of function mutation in the muscle chloride channel; the muscle is hyperexcitable and spontaneously fires action potentials, and this spontaneous action potential activity in the muscle continues even when nerve action potentials are blocked with a TTX cuff (Wang et al. 2005). MEPC amplitude in mice with myotonia congenita was larger following block of nerve activity despite the presence of continued muscle activity. These data suggest block of presynaptic action potentials triggers the increase in quantal amplitude (Wang et al. 2005).

Changes in quantal amplitude can be caused by a presynaptic change in the amount of acetylcholine released from a single vesicle, a change in the breakdown of acetylcholine by acetylcholinesterase, or a change in the current generated by postsynaptic AChRs (either number of AChRs or single channel conductance) (Wang et al. 2005). If a signal associated with presynaptic action potentials is the sensor for regulation of quantal amplitude, it seems more likely that a presynaptic change, such as increased acetylcholine release from vesicles, would be the underlying mechanism, rather than an increase in number of postsynaptic AChRs (although the latter could be a result of anterograde signaling across the synapse). Consistent with this idea, the increase in quantal amplitude was not associated with increased AChRs (Wang et al. 2005). The role of altered acetylcholinesterase function was also ruled out, because the increase in quantal amplitude persisted after complete block of acetylcholinesterase (Wang et al. 2005). This leaves either a presynaptic increase in the amount of acetylcholine release per quantum or an increase in single channel conductance of AChRs as the mechanism responsible.

Support for a presynaptic mechanism underlying the homeostatic increase in MEPC amplitude comes from absence of the increase in mice with a point mutation in the synaptic vesicle protein Rab3A (Wang et al. 2011). Disruption in Rab3A function has been shown to affect the time course of release from a single vesicle in both adrenal chromaffin cells, where it causes an increase in the duration of the fusion pore preceding full fusion (Wang et al. 2008), and at the neuromuscular junction, where it leads to an increase in abnormal, long duration MEPCs (Wang et al. 2008). These studies led to the following hypothesis: reduction in presynaptic action potentials is sensed by the presynaptic terminal, altering Rab3A function, which triggers an increase in the amount of acetylcholine released with fusion of a single vesicle. Generation of this hypothesis benefitted from the ability to selectively manipulate presynaptic and postsynaptic activity at the NMJ as well as the lack of variability of quantal amplitude.

Regulation of Probability of Release (p)

When extracellular Ca^{2+} was normal, the only contributor to increased EPC amplitude following block of action potentials was the increase in MEPC amplitude. However, when extracellular Ca^{2+} was lowered, there was a doubling in the number of vesicles released by a presynaptic action potential (quantal content) (Wang et al. 2004, 2010b). Estimation of p and n using the variance analysis described above revealed an increase in *p* with no change in *n* as the mechanism underlying the increase (Wang et al. 2010b) (Fig. 2). Because *p* is near maximal at the NMJ in normal physiologic solution, the increase had no effect on quantal content when extracellular Ca^{2+} was normal (Wang et al. 2010a, b). The only way to appreciate that synaptic plasticity had occurred was to lower extracellular Ca, reducing *p* below maximal.

Although as described above, block of action potentials at the NMJ in wild type mice triggered an increase in *p* (Wang et al. 2010b), selective block of presynaptic action potentials in mice with myotonia congenita did not increase *p* (Wang and Rich unpublished). Block of AChRs, which selectively blocks muscle action potentials, was also insufficient to increase *p* (Wang et al. 2010b). These data suggest selective block of presynaptic or postsynaptic action potentials alone is insufficient to trigger the increase. The increase is only triggered when both pre- and postsynaptic action potentials are blocked simultaneously. How the simultaneous block of pre- and postsynaptic action potentials is sensed is unknown, but it suggests there is communication between the presynaptic nerve and postsynaptic muscle.

The mechanism responsible for the increase in *p* is an increase in Ca^{2+} entry through P/Q type Ca^{2+} channels. When P/Q type Ca^{2+} channels were blocked with ω-agatoxin during the recording of EPCs, the increase in quantal content triggered by placement of a TTX cuff around the nerve was eliminated (Wang et al. 2004). A similar mechanism appears to contribute to upregulation of quantal content at the drosophila NMJ following partial block of postsynaptic glutamate receptors (Muller and Davis 2012). A point mutation in the presynaptic Ca^{2+} channel of drosophila, which prevents Ca^{2+} entry through the channel, prevented the upregulation in presynaptic Ca^{2+} entry responsible for the upregulation of quantal content after block of glutamate receptors (Muller and Davis 2012). Note that although a similar mechanism (increased Ca^{2+} entry) is involved in increasing quantal content in both systems, what triggers the increase appears to differ, being induced by only block of receptors in drosophila but requiring disruption of action potentials in both the pre- and postsynaptic elements in mouse. One possible explanation for the difference is that block of action potentials at the mammalian NMJ might block release of neurotransmitters other than acetylcholine. It has been shown that activation of NMDA receptors by glutamate at the NMJ affects developmental synaptic plasticity (Personius et al. 2016) and functional NMDA receptors appear to be present at the adult mammalian NMJ (Mays et al. 2009; Proskurina et al. 2018).

Summary of Changes in Synaptic Function Following Block of Action Potentials

The changes in NMJ function triggered by block of action potentials at the NMJ are summarized in Fig. 3. When both pre- and postsynaptic action potentials are blocked (as would normally occur in vivo) there is both an increase in quantal amplitude due to increased release of acetylcholine, and an increase in *p* which is likely due to

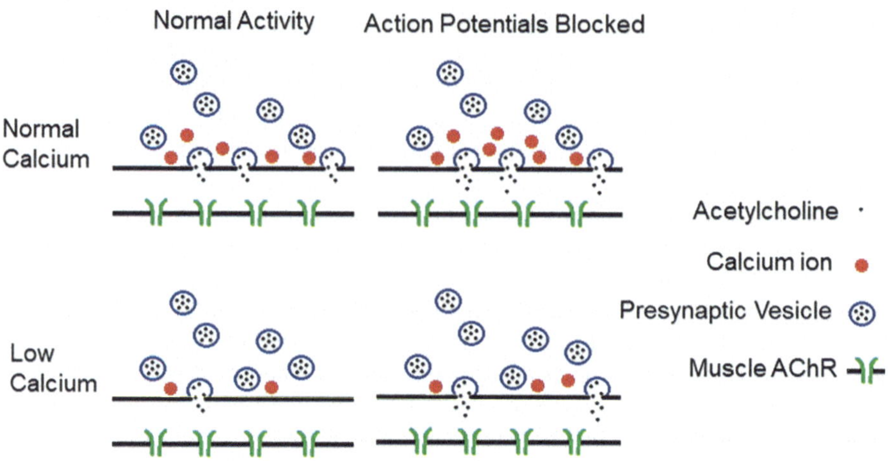

Fig. 3 Homeostatic synaptic plasticity at the mouse NMJ triggered by prolonged block of pre- and postsynaptic action potentials. In the top row, synaptic function is shown in solution containing normal external Ca^{2+}. Following block of action potentials there are two changes in synaptic function. The first is an increase in the release of acetylcholine during fusion of individual synaptic vesicles (illustrated as increased acetylcholine in the synaptic cleft). The second change is an increase in probability of release of vesicles that appears to be due to increased entry of Ca^{2+} into the presynaptic terminal during an action potential. When extracellular Ca^{2+} is normal, the increase in Ca^{2+} entry has no significant effect on the number of synaptic vesicles released (quantal content) as each releasable vesicle is already released with each action potential. In the bottom row is shown the situation when extracellular Ca^{2+} is lowered. In this case Ca^{2+} entry during the action potential limits quantal content. When Ca^{2+} entry is increased following block of action potentials, there is an increase in quantal content. In addition, there is increased acetylcholine release from each vesicle. This combines with the increase in quantal content to cause a dramatic increase in synaptic strength. This is illustrated as an increase from 2 to 8 acetylcholine molecules in the synaptic cleft. AChR = acetylcholine receptor. (This figure is reproduced from Wang and Rich 2018)

increased Ca^{2+} entry through P/Q channels. While the increase in p has little effect on the function of the NMJ in physiologic solution in vivo, it might be beneficial in diseases such Lambert Eaton Myasthenic Syndrome, where reduction of presynaptic Ca^{2+} entry causes pathologic reduction in probability of release (Vincent et al. 1989; Engisch et al. 1999).

4 Homeostatic Plasticity of Synaptic Function Triggered by Partial Block of AChRs

Regulation of the Number of Releasable Vesicles (n)

Many studies have found that when postsynaptic AChRs are partially blocked at the NMJ, there is an increase in quantal content (for review see (Plomp 2017; Wang and Rich 2018)). The first report of increased quantal content at the mammalian NMJ following reduced number of functional postsynaptic AChRs came from studies of patients with myasthenia gravis, an autoimmune disease directed against AChRs (Molenaar et al. 1979; Cull-Candy et al. 1980). A similar increase in quantal content occurs following block of AChRs in multiple species including rat (Plomp et al. 1992, 1994), snake (Harborne et al. 1988), cat (Blaber 1972), frog (Magleby et al. 1981) and mouse (Wang et al. 2016, 2018). Variance analysis of quantal content revealed that the mechanism underlying the increase in quantal content is an increase in *n* with no associated increase in *p* (Tian et al. 1994; Wang et al. 2010b) (Fig. 2). The conservation of this mechanism across multiple species suggests it serves an important function in vivo.

The increase in quantal content is present the first time that the nerve is stimulated after applying blockers of AChRs, ruling out loss of presynaptic or postsynaptic action potentials as the signal that is sensed. How can partial block of ACh receptors be sensed in the absence of action potential evoked release of acetylcholine? The answer appears to be that the muscle is sensing the reduced binding of acetylcholine to AChRs during MEPPs (Wang et al. 2010b, 2018). This finding complements studies of synaptic plasticity in the developing chick spinal cord, in which it was found that pharmacologically altering the frequency of spontaneous vesicle release was able to fully account for homeostatic changes in quantal amplitude whereas dramatic changes in spiking activity associated with spontaneous network activity had little effect on quantal amplitude (Garcia-Bereguiain et al. 2016; Gonzalez-Islas et al. 2018). Together, our work at the NMJ and the work in the chick spinal cord demonstrate an important role for spontaneous neurotransmitter release in triggering synaptic plasticity in vivo- something that has been debated since its discovery close to 60 years ago (for review see (Kavalali et al. 2011; Kavalali 2015)).

The increase in quantal content occurred even when muscle fibers were voltage clamped at a constant membrane potential, indicating that the signal could not be the loss of depolarization during MEPPs. Current flow during MEPCs was also found not to be involved as changing the muscle membrane potential (which changes MEPC amplitude) via voltage clamp had no effect on quantal content (Wang et al. 2018). The only change that correlated with the increase in quantal content was an approximately 15% shortening in the duration of MEPCs that occurred following partial block of AChRs with 5 different AChR blocking drugs (Wang et al. 2018). Surprisingly, treatment with acetylcholine and succinylcholine, which both reduced MEPC amplitude due to desensitization of AChRs, did not shorten MEPC duration and did not trigger the increase in quantal content. Our interpretation is that blockers of AChRs trigger a change in the state of AChRs such that their open time is decreased (Wang et al. 2018). However, such a change has not, to our knowledge, ever been demonstrated. The change in AChR state is somehow communicated to the presynaptic terminal to cause an increase in *n*. We propose that in addition to functioning in synaptic transmission, AChRs at the NMJ serve as signaling molecules that trigger synaptic plasticity.

The increase in quantal content triggered by partial block of AChRs occurs in physiologic solution, but not in a solution in which extracellular Ca^{2+} is lowered (Gallant 1982; Matzner et al. 1988; Molenaar et al. 1991; Plomp et al. 1992, 1994,

1995, Tian et al. 1994, 1997; Wilson et al. 1995; Wang et al. 2010b, 2016). The finding that the increase in quantal content was dependent on adequate extracellular Ca^{2+} led to the question of whether Ca^{2+} entry during action potentials is involved. This possibility was tested by application of 3,4 diaminopyridine when extracellular Ca^{2+} was reduced. 3,4 diaminopyridine blocks voltage gated K channels in the presynaptic terminal, widening the action potential and allowing greater Ca^{2+} entry (Thomsen and Wilson 1983). Under these conditions, the increase in quantal content following block of AChRs was present when extracellular Ca^{2+} was low (Wang et al. 2016). We conclude that a certain minimal amount of Ca^{2+} entry during action potentials is necessary for this synaptic plasticity to be observed.

The other feature of the increase in quantal content triggered by partial block of AChRs is that it is highly dependent on stimulation frequency. The increase in quantal content was only present at low frequency stimulation or during the first few pulses of a train of action potentials (Magleby et al. 1981; Wilson 1982; Hong and Chang 1991; Wilson and Thomsen 1991, Tian et al. 1994, 1997). The depression of EPC amplitude during trains of stimuli was shown not to be due to reduction of postsynaptic sensitivity to acetylcholine in two studies using different methods. In the first, iontophoretic application of acetylcholine was used and the response did not decrease during a train of stimuli (Gibb and Marshall 1984). In the second, MEPP amplitude was followed and did not decrease in parallel with EPP amplitude during repetitive stimulation (Hong and Chang 1991). Both studies suggest reduced action potential-evoked release of acetylcholine is the cause of the depression occurring with repetitive stimulation.

Our hypothesis is that homeostatic upregulation of quantal content is accomplished by mobilizing a unique pool of synaptic vesicles which are able to be replenished during low, but not high rates of stimulation (Wang et al. 2016). Furthermore, this pool of vesicles requires relatively high levels of Ca^{2+} entry during action potentials to be released; they do not participate in release when Ca^{2+} entry is reduced. Figure 4 provides a summary of the hypothesis. The hypothesis illustrated in Fig. 4 was supported by estimating the number of releasable vesicles before and after partial block of AChRs. Two methods were used to estimate the size of the pool of vesicles recruited to participate in release following partial block of AChRs. In the first method, rapid repetitive stimulation was applied and depression of the EPC measured. High frequency stimulation does not allow enough time for replenishment of the readily releasable vesicle pool (Schneggenburger et al. 1999; Ruiz et al. 2011). By measuring the difference in quantal content at the beginning and at the end of high frequency stimulation, one can estimate the number of vesicles in the pool that was depleted. Using high frequency stimulation we estimated there was an increase of 244 in the pool of readily releasable vesicles following partial block of AChRs (Wang et al. 2016). In the second method, refilling of synaptic vesicles was inhibited using the drug vesamicol. Any vesicles that were released could not be refilled after membrane recycling and so would no longer participate in synaptic transmission. When vesicle refilling was prevented, partial block of AChRs triggered an increase in quantal content that was transient. With repeated stimulation, the increase in quantal content disappeared, whereas when vesicle refilling was

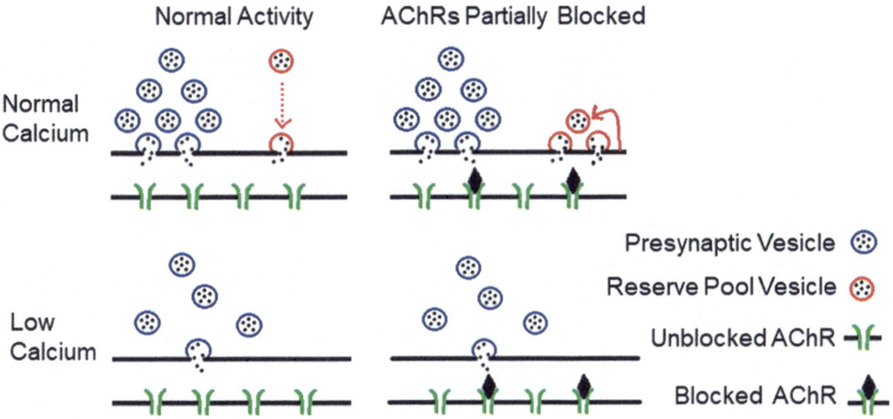

Fig. 4 The increase in *n* following block of AChRs. In the top row is shown synaptic function in solution containing normal external Ca^{2+}. Two types of synaptic vesicles are present. The blue vesicles represent normal vesicles that participate in synaptic transmission at baseline. The red vesicles represent a special pool of synaptic vesicles that normally do not play an important role in synaptic transmission. A few vesicles in this pool are released late in the course of the synaptic current (indicated by the long dotted red arrow). Following partial block of AChRs, the special pool of synaptic vesicles is released more rapidly (synchronous with the normal pool). In addition, the special pool of vesicles rapidly recycles (indicated by the curved red arrow) to increase quantal content. In solution containing low extracellular Ca^{2+}, Ca^{2+} entry during action potentials is insufficient for the special pool of vesicles to participate in release. (This figure is reproduced from Wang and Rich 2018)

allowed, the increase was maintained (Wang et al. 2016). Vesamicol causes a steady decline in quantal content as vesicles become depleted and no longer contribute. We were able to estimate the additional vesicles after receptor blockade by comparing the decline before and after receptor block. The "extra" vesicles released gave an estimate of 316 vesicles recruited to the releasable pool by partial block of AChRs. This estimate was similar to the estimate obtained by using high rates of repetitive stimulation (Wang et al. 2016).

The molecular signaling pathway involved in the upregulation of *n* is not well understood. Inhibition of Ca^{2+}/calmodulin-dependent protein kinase II (CaMKII) and Trk tyrosine kinases prevented the upregulation of *n* at the rat NMJ after partial block of AChRs with α-bungarotoxin (Plomp and Molenaar 1996). A number of proteins have been identified at the drosophila NMJ that participate in the upregulation of quantal content following block of postsynaptic glutamate receptors (Davis and Muller 2015). A member of the peptidoglycan pattern recognition receptor family as well as endostatin (an antiangiogenesis signaling factor) and semaphorin-plexin signaling have all been proposed to be involved in the retrograde signaling from muscle to nerve (Wang et al. 2014; Harris et al. 2015; Orr et al. 2017). Other molecules demonstrated to be required using genetic manipulations in drosophila include a presynaptic Na channel (Younger et al. 2013), dysbindin (a gene linked to schizophrenia) (Dickman and Davis 2009), Rab3A interacting molecule (RIM)

(Muller et al. 2012) as well as several innate immune signaling genes (Harris et al. 2018). Whether any of these proteins are involved in the increase in *n* or *p* during homeostatic plasticity at the mammalian NMJ is unknown.

Homeostatic Plasticity of Motoneuron Excitability Following Partial Block of AChRs

In addition to triggering the increase in quantal content, block of AChRs also triggers an increase in motoneuron excitability, which manifests as a reduction in the amount of current necessary to bring the cell to threshold (rheobase current) (Nakanishi et al. 2005). This suggests that activation of AChRs is involved in a retrograde trophic signaling from muscle that regulates excitability of the motoneuron.

Loss of this retrograde trophic signal may be the mechanism responsible for an increase in motoneuron excitability previously described following peripheral nerve injury (Foehring et al. 1986a, b; Pinter and Vanden Noven 1989). The increase in excitability following nerve injury could be a consequence of injury per se, or loss of a trophic signal from the NMJ, or a combination of the two. The post-injury changes in excitability persist if motoneurons remain disconnected from muscle and recover when muscle is reinnervated (Foehring et al. 1986a, b; Pinter and Vanden Noven 1989). These data argue against injury being the signal triggering the increase in excitability and instead favor loss of a trophic signal as the mechanism. Further evidence that the changes are not due to injury come from the finding that block of vesicle fusion by botulinum toxin can trigger the change (Pinter et al. 1991). The botulinum data and the finding that block of AChRs triggers increased motoneuron excitability suggest that either loss activation of AChRs or loss of a signal deriving during muscle action potentials is the trigger.

To determine whether muscle action potentials were necessary for the signal, motor axons were crushed and then allowed to regenerate to muscle through TTX containing cuffs that prevented nerve and muscle action potentials. Despite the block of evoked release of acetylcholine and muscle action potentials, motoneuron excitability returned to normal when nerve regenerated into muscle (Bichler et al. 2007). A signal that remains when action potentials are blocked is the spontaneous occurrence of MEPPs. Combining these findings with previous studies suggests the presence of a retrograde trophic signal from muscle mediated by activation of AChRs during MEPPs. Loss of this signal following nerve injury, botulinum toxin or block of AChRs triggers an increase in motoneuron excitability. One possibility is that ACh released during MEPPs could have its effect on both quantal content and motoneuron excitability by binding presynaptic α-7 subunit containing AChRs. However, specific block of α-1 subunit containing AChRs on muscle triggered the increase quantal content and the increase still occurred following block of AChRs in mice lacking α-7 subunit containing AChRs (Wang et al. 2018). While this does not rule out involvement of presynaptic AChRs in motoneuron plasticity, it does not favor the possibility.

Summary of Homeostatic Changes in Synaptic Function and Their Implications for Disease

Shown in Fig. 5 is a summary of homeostatic synaptic plasticity at the mammalian NMJ. Perturbation of distinct aspects of synaptic activity are sensed to trigger distinct changes is synaptic function and motoneuron excitability. Block of nerve action potentials triggers increased quantal size secondary to increased release of acetylcholine from vesicles and, combined with the loss of muscle action potentials

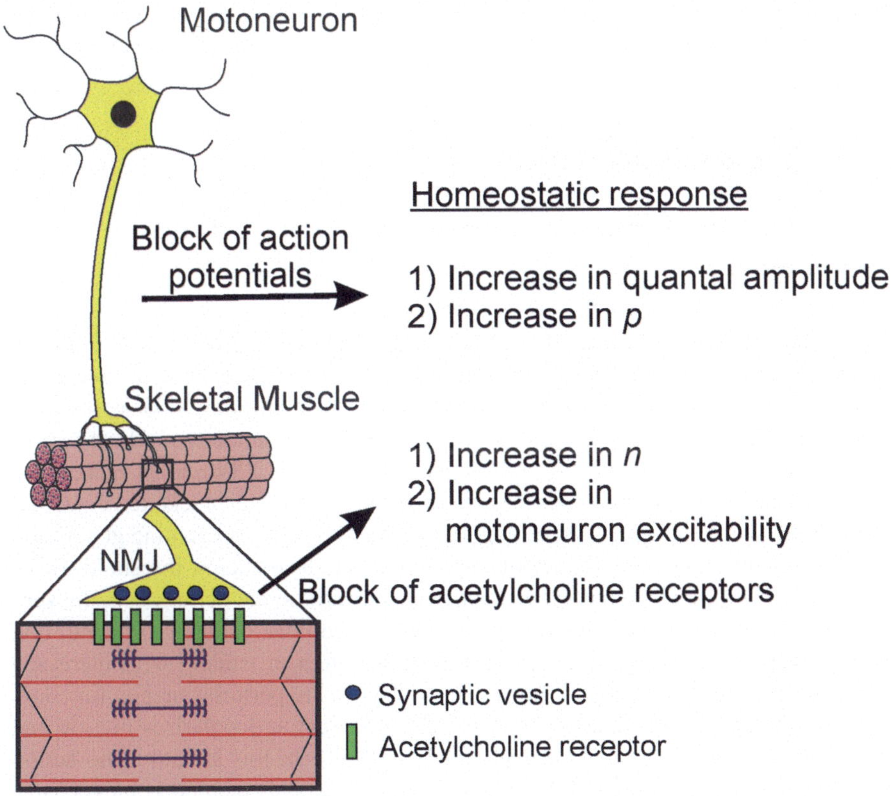

Fig. 5 Summary of homeostatic synaptic plasticity and the mammalian NMJ. Block of action potentials in the presynaptic axon, which also causes block of muscle action potentials, triggers to an increase in quantal amplitude as well as an increase in p. Block of postsynaptic AChRs on muscle causes a retrograde signal(s) that increases both n at the nerve terminal and an increase in motoneuron excitability that is present in the cell body in the spinal cord

caused by loss of nerve action potentials, also triggers an increase in the probability of vesicle release. Block of acetylcholine binding to AChRs during MEPPs triggers an increase in *n* as well as an increase in motoneuron excitability.

The understanding of homeostatic synaptic plasticity achieved by studies described here identifies mechanisms that could be involved in the response of the NMJ and motoneuron to various diseases. In myasthenia gravis, block and destruction of AChRs likely triggers homeostatic responses in the NMJ and motoneuron (Wang and Rich 2018). The increased release of acetylcholine may combat failure of neuromuscular transmission while the effects of increased motoneuron excitability are more difficult to predict. In the motoneuron disease spinal muscular atrophy, there is a reduction in synaptic inputs to motoneurons (Fletcher et al. 2017) that may lessen action potential firing and trigger an increase in quantal amplitude and probability of vesicle release. As we continue to increase our understanding of how the NMJ responds to perturbations in activity, we may one day be able to harness homeostatic responses in order to provide effective therapies for diseases affecting the motoneuron and NMJ. Furthermore, homeostatic mechanisms that turn out to be conserved at central synapses might be enlisted to treat neurological disorders.

5 Technical Considerations

Preventing Muscle Contraction and Activation of Muscle Na Channels

While the NMJ has the advantages of having a single input and less variability in quantal amplitude than CNS synapses (Korn and Faber 1991), there are technical challenges that must be overcome. One challenge of using the NMJ as a model synapse is that its normal function is to trigger muscle contraction, which causes loss of impalement of the postsynaptic muscle fiber (Rich 2006). Four approaches have been taken to eliminate muscle contraction. (1) Partial blockade of AChRs with a toxin such as curare, lessens depolarization of muscle, preventing generation of action potentials because the endplate potential is reduced below threshold (Boyd and Martin 1956). Unfortunately, this approach perturbs synaptic function, the very thing one wants to study. (2) Reduction of extracellular calcium concentration reduces release below action potential threshold, often in tandem with increased extracellular magnesium. Again, however, one has had to perturb synaptic function. (3) Crushing muscle fibers away from the endplate region depolarizes muscle and inactivates Na channels (Barstad and Lilleheil 1968). In this situation, no action potential can be generated and there is no contraction. This method is inexpensive and does not alter the number of vesicles released, but has the disadvantage that release of potassium from crushed muscle fibers can cause block of conduction of presynaptic action potentials. It takes close to an hour for the preparation to recover prior to recording, and then recording time is generally limited to less than 2 h because muscle fibers gradually repolarize and regain the ability to contract. (4) Nav1.4 type sodium channels expressed by skeletal muscle can be selectively blocked using μ-conotoxin GIIIB (Robitaille and Charlton 1992; Wood and Slater 1997). At the correct dose, μ-conotoxin spares presynaptic action potentials, while eliminating contraction by blocking postsynaptic action potentials. In this case one has not altered function and the preparation is stable over time. The disadvantage is that it is expensive.

Pros and Cons of Recording Configurations

In many studies of neuromuscular transmission, a single electrode is inserted into the muscle fiber and the membrane potential of the fiber measured as the attached motor nerve is stimulated. This type of recording produces an "endplate potential" (EPP), typically ~50 mV in physiologic solution (Khedraki et al. 2017). This method is easy to perform; it does not require visualization of the electrode, and only a single electrode has to be inserted into a muscle fiber. However, one issue is that changes in muscle fiber properties such as input resistance and capacitance can affect the amplitude of the postsynaptic depolarization caused by synaptic current, which can lead to incorrect conclusions about changes in synaptic function (Khedraki et al. 2017). An example of a disease in which there are profound changes in muscle properties in addition to changes in NMJ function is spinal muscular atrophy. In spinal muscular atrophy, muscle fibers are much smaller and there is a reduction in quantal content due to a reduction in probability of release (Kong et al. 2009; Ruiz et al. 2010; Martinez et al. 2012). The magnitude of the reduction in quantal content can be underestimated when measuring the EPP, as the EPP amplitude is increased by the high input resistance and low capacitance of the small fibers. A second disease in which muscle fiber properties and NMJ function are both altered is Huntington's disease in which there is increased input resistance and reduced capacitance of muscle fibers along with a reduction in quantal content (Waters et al. 2013; Khedraki et al. 2017; Miranda et al. 2017).

A second issue with using a single electrode to record endplate potentials is that when the endplate potential gets close to the reversal potential of current flow through acetylcholine receptors, each additional vesicle that releases acetylcholine triggers less depolarization. This is known as non-linear summation of the endplate potential and requires correction in order to more accurately estimate the number of synaptic vesicles released (McLachlan and Martin 1981).

The use of two electrode voltage clamp eliminates the confounds introduced by differences in muscle fiber resistance and capacitance, but is more technically challenging to perform. In this method, two electrodes are inserted into a single muscle fiber, one to sample the difference between the membrane potential and the desired potential; the other to inject current to clamp the membrane potential at the desired level. Because two electrodes must be inserted into the same fiber, this method requires visualization of the electrode tips and the NMJ. This issue can be addressed by using fluorescent imaging of electrodes loaded with sulforhodamine and NMJs labeled with the fluorescent vital dye 4-Di-2-ASP, or by expression of a fluorescent protein in the motor nerve terminals (Wang et al. 2004). Because voltage is constant during voltage clamp of muscle, factors altered by muscle fiber size such as input resistance and capacitance do not affect the amplitude of the postsynaptic response. Also, because voltage is constant there is no need to correct for non-linear summation of the EPP (Glavinovic 1979; Khedraki et al. 2017). Thus, if possible, it is preferable to use voltage clamp to record from the NMJ.

Acknowledgements This work was supported by NIH grant AR074985 (M.M.R.).

References

Barstad JA, Lilleheil G (1968) Transversaly cut diaphragm preparation from rat. An adjuvant tool in the study of the physiology and pbarmacology of the myoneural junction. Arch Int Pharmacodyn Ther 175(2):373–390

Bichler EK, Nakanishi ST, Wang QB, Pinter MJ, Rich MM, Cope TC (2007) Enhanced transmission at a spinal synapse triggered in vivo by an injury signal independent of altered synaptic activity. J Neurosci 27(47):12851–12859

Blaber LC (1972) The mechanism of the facilitatory action of edrophonium in cat skeletal muscle. Br J Pharmacol 46(3):498–507

Boyd IA, Martin AR (1956) The end-plate potential in mammalian muscle. J Physiol 132(1):74–91

Cull-Candy SG, Miledi R, Trautmann A, Uchitel OD (1980) On the release of transmitter at normal, myasthenia gravis and myasthenic syndrome affected human end-plates. J Physiol 299:621–638

Davis GW, Muller M (2015) Homeostatic control of presynaptic neurotransmitter release. Annu Rev Physiol 77:251–270

Del Castillo J, Katz B (1954) Quantal components of the end-plate potential. J Physiol 124(3):560–573

Dickman DK, Davis GW (2009) The schizophrenia susceptibility gene dysbindin controls synaptic homeostasis. Science 326(5956):1127–1130

Engisch KL, Rich MM, Cook N, Nowycky MC (1999) Lambert-Eaton antibodies inhibit Ca2+ currents but paradoxically increase exocytosis during stimulus trains in bovine adrenal chromaffin cells. J Neurosci 19(9):3384–3395

Fletcher EV, Simon CM, Pagiazitis JG, Chalif JI, Vukojicic A, Drobac E, Wang X, Mentis GZ (2017) Reduced sensory synaptic excitation impairs motor neuron function via Kv2.1 in spinal muscular atrophy. Nat Neurosci 20(7):905–916

Foehring RC, Sypert GW, Munson JB (1986a) Properties of self-reinnervated motor units of medial gastrocnemius of cat. I. Long-term reinnervation. J Neurophysiol 55(5):931–946

Foehring RC, Sypert GW, Munson JB (1986b) Properties of self-reinnervated motor units of medial gastrocnemius of cat. II. Axotomized motoneurons and time course of recovery. J Neurophysiol 55(5):947–965

Gallant PE (1982) The relationship between anti-acetylcholine receptor antibody levels and neuromuscular function in chronically myasthenic rats. J Neurol Sci 54(1):129–141

Garcia-Bereguiain MA, Gonzalez-Islas C, Lindsly C, Wenner P (2016) Spontaneous release regulates synaptic scaling in the embryonic spinal network in vivo. J Neurosci 36(27):7268–7282

Gibb AJ, Marshall IG (1984) Pre-and post-junctional effects of tubocurarine and other nicotinic antagonists during repetitive stimulation in the rat. J Physiol 351:275–297

Glavinovic MI (1979) Voltage clamping of unparalysed cut rat diaphragm for study of transmitter release. J Physiol 290(2):467–480

Gonzalez-Islas C, Bulow P, Wenner P (2018) Regulation of synaptic scaling by action potential-independent miniature neurotransmission. J Neurosci Res 96(3):348–353

Harborne AJ, Bowman WC, Marshall IG (1988) Effects of tubocurarine on end-plate current run-down and quantal content during rapid nerve stimulation in the snake. Clin Exp Pharmacol Physiol 15(6):479–490

Harris N, Braiser DJ, Dickman DK, Fetter RD, Tong A, Davis GW (2015) The innate immune receptor PGRP-LC controls presynaptic homeostatic plasticity. Neuron 88(6):1157–1164

Harris N, Fetter RD, Brasier DJ, Tong A, Davis GW (2018) Molecular interface of neuronal innate immunity, synaptic vesicle stabilization, and presynaptic homeostatic plasticity. Neuron 100(5):1163–1179 e1164

Heuser JE, Reese TS (1981) Structural changes after transmitter release at the frog neuromuscular junction. J Cell Biol 88(3):564–580

Heuser JE, Reese TS, Dennis MJ, Jan Y, Jan L, Evans L (1979) Synaptic vesicle exocytosis captured by quick freezing and correlated with quantal transmitter release. J Cell Biol 81(2):275–300

Hong SJ, Chang CC (1991) Run-down of neuromuscular transmission during repetitive nerve activity by nicotinic antagonists is not due to desensitization of the postsynaptic receptor. Br J Pharmacol 102(4):817–822

Kavalali ET (2015) The mechanisms and functions of spontaneous neurotransmitter release. Nat Rev Neurosci 16(1):5–16

Kavalali ET, Chung C, Khvotchev M, Leitz J, Nosyreva E, Raingo J, Ramirez DM (2011) Spontaneous neurotransmission: an independent pathway for neuronal signaling? Physiology (Bethesda) 26(1):45–53

Kelly SS, Robbins N (1987) Statistics of neuromuscular transmitter release in young and old mouse muscle. J Physiol 385:507–516

Khedraki A, Reed EJ, Romer SH, Wang Q, Romine W, Rich MM, Talmadge RJ, Voss AA (2017) Depressed synaptic transmission and reduced vesicle release sites in Huntington's disease neuromuscular junctions. J Neurosci 37(34):8077–8091

Kong L, Wang X, Choe DW, Polley M, Burnett BG, Bosch-Marce M, Griffin JW, Rich MM, Sumner CJ (2009) Impaired synaptic vesicle release and immaturity of neuromuscular junctions in spinal muscular atrophy mice. J Neurosci 29(3):842–851

Korn H, Faber DS (1991) Quantal analysis and synaptic efficacy in the CNS. Trends Neurosci 14(10):439–445

Magleby KL, Pallotta BS, Terrar DA (1981) The effect of (+)-tubocurarine on neuromuscular transmission during repetitive stimulation in the rat, mouse, and frog. J Physiol 312:97–113

Martinez TL, Kong L, Wang X, Osborne MA, Crowder ME, Van Meerbeke JP, Xu X, Davis C, Wooley J, Goldhamer DJ, Lutz CM, Rich MM, Sumner CJ (2012) Survival motor neuron protein in motor neurons determines synaptic integrity in spinal muscular atrophy. J Neurosci 32(25):8703–8715

Matzner H, Parnas H, Parnas I (1988) Presynaptic effects of d-tubocurarine on neurotransmitter release at the neuromuscular junction of the frog. J Physiol 398:109–121

Mays TA, Sanford JL, Hanada T, Chishti AH, Rafael-Fortney JA (2009) Glutamate receptors localize postsynaptically at neuromuscular junctions in mice. Muscle Nerve 39(3):343–349

McLachlan EM, Martin AR (1981) Non-linear summation of end-plate potentials in the frog and mouse. J Physiol 311:307–324

Miranda DR, Wong M, Romer SH, McKee C, Garza-Vasquez G, Medina AC, Bahn V, Steele AD, Talmadge RJ, Voss AA (2017) Progressive Cl- channel defects reveal disrupted skeletal muscle maturation in R6/2 Huntington's mice. J Gen Physiol 149(1):55–74

Miyamoto MD (1975) Binomial analysis of quantal transmitter release at glycerol treated frog neuromuscular junctions. J Physiol 250(1):121–142

Molenaar PC, Polak RL, Miledi R, Alema S, Vincent A, Newsom-Davis J (1979) Acetylcholine in intercostal muscle from myasthenia gravis patients and in rat diaphragm after blockade of acetylcholine receptors. Prog Brain Res 49:449–458

Molenaar PC, Oen BS, Plomp JJ, Van Kempen GT, Jennekens FG, Hesselmans LF (1991) A non-immunogenic myasthenia gravis model and its application in a study of transsynaptic regulation at the neuromuscular junction. Eur J Pharmacol 196(1):93–101

Muller M, Davis GW (2012) Transsynaptic control of presynaptic Ca(2)(+) influx achieves homeostatic potentiation of neurotransmitter release. Curr Biol 22(12):1102–1108

Muller M, Liu KS, Sigrist SJ, Davis GW (2012) RIM controls homeostatic plasticity through modulation of the readily-releasable vesicle pool. J Neurosci 32(47):16574–16585

Nakanishi ST, Cope TC, Rich MM, Carrasco DI, Pinter MJ (2005) Regulation of motoneuron excitability via motor endplate acetylcholine receptor activation. J Neurosci 25(9):2226–2232

O'Brien RJ, Kamboj S, Ehlers MD, Rosen KR, Fischbach GD, Huganir RL (1998) Activity-dependent modulation of synaptic AMPA receptor accumulation. Neuron 21(5):1067–1078

Orr BO, Fetter RD, Davis GW (2017) Retrograde semaphorin-plexin signalling drives homeostatic synaptic plasticity. Nature 550(7674):109–113

Personius KE, Slusher BS, Udin SB (2016) Neuromuscular NMDA receptors modulate developmental synapse elimination. J Neurosci 36(34):8783–8789

Pinter MJ, Vanden Noven S (1989) Effects of preventing reinnervation on axotomized spinal motoneurons in the cat. I. Motoneuron electrical properties. J Neurophysiol 62(2):311–324

Pinter MJ, Vanden Noven S, Muccio D, Wallace N (1991) Axotomy-like changes in cat motoneuron electrical properties elicited by botulinum toxin depend on the complete elimination of neuromuscular transmission. J Neurosci 11(3):657–666

Plomp JJ (2017) Trans-synaptic homeostasis at the myasthenic neuromuscular junction. Front Biosci (Landmark Ed) 22:1033–1051

Plomp JJ, Molenaar PC (1996) Involvement of protein kinases in the upregulation of acetylcholine release at endplates of alpha-bungarotoxin-treated rats. J Physiol 493(Pt 1):175–186

Plomp JJ, van Kempen GT, Molenaar PC (1992) Adaptation of quantal content to decreased postsynaptic sensitivity at single endplates in alpha-bungarotoxin-treated rats. J Physiol 458:487–499

Plomp JJ, van Kempen GT, Molenaar PC (1994) The upregulation of acetylcholine release at endplates of alpha-bungarotoxin-treated rats: its dependency on calcium. J Physiol 478(Pt 1):125–136

Plomp JJ, Van Kempen GT, De Baets MB, Graus YM, Kuks JB, Molenaar PC (1995) Acetylcholine release in myasthenia gravis: regulation at single end-plate level. Ann Neurol 37(5):627–636

Proskurina SE, Petrov KA, Nikolsky EE (2018) Influence of the activation of NMDA receptors on the resting membrane potential of the Postsynaptic cell at the neuromuscular junction. Acta Nat 10(3):100–102

Provan SD, Miyamoto MD (1993) Unbiased estimates of quantal release parameters and spatial variation in the probability of neurosecretion. Am J Phys 264(4 Pt 1):C1051–C1060

Redman S (1990) Quantal analysis of synaptic potentials in neurons of the central nervous system. Physiol Rev 70(1):165–198

Rich MM (2006) The control of neuromuscular transmission in health and disease. Neuroscientist 12(2):134–142

Rich MM, Wenner P (2007) Sensing and expressing homeostatic synaptic plasticity. Trends Neurosci 30(3):119–125

Robitaille R, Charlton MP (1992) Presynaptic calcium signals and transmitter release are modulated by calcium-activated potassium channels. J Neurosci 12(1):297–305

Ruiz R, Casanas JJ, Torres-Benito L, Cano R, Tabares L (2010) Altered intracellular Ca2+ homeostasis in nerve terminals of severe spinal muscular atrophy mice. J Neurosci 30(3):849–857

Ruiz R, Cano R, Casanas JJ, Gaffield MA, Betz WJ, Tabares L (2011) Active zones and the readily releasable pool of synaptic vesicles at the neuromuscular junction of the mouse. J Neurosci 31(6):2000–2008

Schneggenburger R, Meyer AC, Neher E (1999) Released fraction and total size of a pool of immediately available transmitter quanta at a calyx synapse. Neuron 23(2):399–409

Searl TJ, Silinsky EM (2002) Evidence for two distinct processes in the final stages of neurotransmitter release as detected by binomial analysis in calcium and strontium solutions. J Physiol 539(Pt 3):693–705

Stevens CF (2003) Neurotransmitter release at central synapses. Neuron 40(2):381–388

Thomsen RH, Wilson DF (1983) Effects of 4-aminopyridine and 3,4-diaminopyridine on transmitter release at the neuromuscular junction. J Pharmacol Exp Ther 227(1):260–265

Tian L, Prior C, Dempster J, Marshall IG (1994) Nicotinic antagonist-produced frequency-dependent changes in acetylcholine release from rat motor nerve terminals. J Physiol 476(3):517–529

Tian L, Prior C, Dempster J, Marshall IG (1997) Hexamethonium- and methyllycaconitine-induced changes in acetylcholine release from rat motor nerve terminals. Br J Pharmacol 122(6):1025–1034

Turrigiano GG, Leslie KR, Desai NS, Rutherford LC, Nelson SB (1998) Activity-dependent scaling of quantal amplitude in neocortical neurons. Nature 391(6670):892–896

Vincent A, Lang B, Newsom-Davis J (1989) Autoimmunity to the voltage-gated calcium channel underlies the Lambert-Eaton myasthenic syndrome, a paraneoplastic disorder. Trends Neurosci 12(12):496–502

Wang X, Rich MM (2018) Homeostatic synaptic plasticity at the neuromuscular junction in myasthenia gravis. Ann N Y Acad Sci 1412(1):170–177

Wang X, Engisch KL, Li Y, Pinter MJ, Cope TC, Rich MM (2004) Decreased synaptic activity shifts the calcium dependence of release at the mammalian neuromuscular junction in vivo. J Neurosci 24(47):10687–10692

Wang X, Li Y, Engisch KL, Nakanishi ST, Dodson SE, Miller GW, Cope TC, Pinter MJ, Rich MM (2005) Activity-dependent presynaptic regulation of quantal size at the mammalian neuromuscular junction in vivo. J Neurosci 25(2):343–351

Wang X, Thiagarajan R, Wang Q, Tewolde T, Rich MM, Engisch KL (2008) Regulation of quantal shape by Rab3A: evidence for a fusion pore-dependent mechanism. J Physiol 586(16):3949–3962

Wang X, Pinter MJ, Rich MM (2010a) Ca2+ dependence of the binomial parameters p and n at the mouse neuromuscular junction. J Neurophysiol 103(2):659–666

Wang X, Wang Q, Engisch KL, Rich MM (2010b) Activity-dependent regulation of the binomial parameters p and n at the mouse neuromuscular junction in vivo. J Neurophysiol 104(5):2352–2358

Wang X, Wang Q, Yang S, Bucan M, Rich MM, Engisch KL (2011) Impaired activity-dependent plasticity of quantal amplitude at the neuromuscular junction of Rab3A deletion and Rab3A earlybird mutant mice. J Neurosci 31(10):3580–3588

Wang T, Hauswirth AG, Tong A, Dickman DK, Davis GW (2014) Endostatin is a trans-synaptic signal for homeostatic synaptic plasticity. Neuron 83(3):616–629

Wang X, Pinter MJ, Rich MM (2016) Reversible recruitment of a homeostatic reserve pool of synaptic vesicles underlies rapid homeostatic plasticity of quantal content. J Neurosci 36(3):828–836

Wang X, McIntosh JM, Rich MM (2018) Muscle Nicotinic Acetylcholine receptors may mediate trans-synaptic signaling at the mouse neuromuscular junction. J Neurosci 38(7):1725–1736

Waters CW, Varuzhanyan G, Talmadge RJ, Voss AA (2013) Huntington disease skeletal muscle is hyperexcitable owing to chloride and potassium channel dysfunction. Proc Natl Acad Sci U S A 110(22):9160–9165

Wernig A (1975) Estimates of statistical release parameters from crayfish and frog neuromuscular junctions. J Physiol 244(1):207–221

Wilson DF (1982) Influence of presynaptic receptors on neuromuscular transmission in rat. Am J Physiol 242(5):C366–C372

Wilson DF, Thomsen RH (1991) Nicotinic receptors on the rat phrenic nerve: evidence for negative feedback. Neurosci Lett 132(2):163–166
Wilson DF, West AE, Lin Y (1995) Inhibitory action of nicotinic antagonists on transmitter release at the neuromuscular junction of the rat. Neurosci Lett 186(1):29–32
Wood SJ, Slater CR (1997) The contribution of postsynaptic folds to the safety factor for neuromuscular transmission in rat fast- and slow-twitch muscles. J Physiol 500(Pt 1):165–176
Younger MA, Muller M, Tong A, Pym EC, Davis GW (2013) A presynaptic ENaC channel drives homeostatic plasticity. Neuron 79(6):1183–1196

Diversity of Mammalian Motoneurons and Motor Units

Marcin Bączyk, Marin Manuel, Francesco Roselli, and Daniel Zytnicki

Abstract Although they share the common function of controlling muscle fiber contraction, spinal motoneurons display a remarkable diversity. Alpha-motoneurons are the "final common pathway", which relay all the information from spinal and supraspinal centers and allow the organism to interact with the outside world by controlling the contraction of muscle fibers in the muscles. On the other hand, gamma-motoneurons are specialized motoneurons that do not generate force and instead specifically innervate muscle fibers inside muscle spindles, which are proprioceptive organs embedded in the muscles. Beta-motoneurons are hybrid motoneurons that innervate both extrafusal and intrafusal muscle fibers. Even among alpha-motoneurons, there exists an exquisite diversity in terms of motoneuron electrical and molecular properties, physiological and structural properties of their neuromuscular junctions, and molecular and contractile properties of the innervated muscle fibers. This diversity, across species, across muscles, and across muscle fibers in a given muscle, underlie the vast repertoire of movements that one individual can perform.

Authors Marcin Bączyk, Marin Manuel, Francesco Roselli, Daniel Zytnicki have equally contributed to this chapter.

M. Bączyk
Department of Neurobiology, Poznań University of Physical Education, Poznań, Poland

M. Manuel (✉) · D. Zytnicki
SPPIN – Saints-Pères Paris Institute for the Neurosciences, CNRS, Université de Paris, Paris, France
e-mail: marin.manuel@neurobio.org

F. Roselli
Department of Neurology, Ulm University, Ulm, Germany

Institute of Anatomy and Cell Biology, Ulm University, Ulm, Germany

German Center for Neurodegenerative Diseases (DZNE)-Ulm, Ulm, Germany

Neurozentrum Ulm, Ulm, Germany

M. J. O'Donovan, M. Falgairolle (eds.), *Vertebrate Motoneurons*, Advances in Neurobiology 28, https://doi.org/10.1007/978-3-031-07167-6_6

Keywords Motoneuron · Motor unit · Physiological type · Electrophysiology · Neuromuscular junctions · Muscle fibers · Contractile properties

1 Introduction

Mammals display an extraordinary diversity of muscle fibers and motoneurons. Indeed, the majority of motoneurons (alpha-motoneurons) innervate ordinary (extrafusal) muscle fibers which produce the force necessary for all movements, whereas other motoneurons (gamma-motoneurons) innervate a special type of muscle fibers (intrafusal fibers) located inside muscle spindles, which are proprioceptive sensory organs. A third motoneuron type (beta-motoneurons) innervates both intra- and extrafusal muscle fibres. All motoneurons and muscle fibers types can be further divided in several subtypes with different anatomical and functional properties. In this chapter we will characterize the alpha-, beta- and gamma-motoneurons and then we will focus on the alpha-motoneurons to show how their electrical properties fit with the anatomical and contractile properties of the innervated muscle fibers to form well-adapted functional motor units, which are the elementary constituent of all motor acts.

2 Alpha, Beta and Gamma Motoneurons

All muscles acting on the skeleton (so-called "skeletal muscles") are composed of a heterogeneous population of muscle fibers (Fig. 1). Most of them are "ordinary" fibers, whose contraction produces the forces exerted by muscles on the skeleton. Ordinary fibers are thus responsible for postural activities and movements. Other specialized muscle fibers play a very different role: they are located inside sensory organs called "muscle spindles" which play a crucial role in proprioception. These so-called "intrafusal" fibers do not contribute to muscle force. Instead, their contraction modulates the transduction properties of muscle spindles (i.e., how spindles translate a mechanical input—muscle stretch—to firing of their sensory afferents) (Matthews 1972; Hunt 1990). In adults, intrafusal fibers are shorter and thinner than ordinary fibers (Gregory and Proske 1991). Like "ordinary" extrafusal fibers, all intrafusal fibers originate from the fusion of multiple embryonic myoblasts and they therefore contain multiple nuclei. In intrafusal fibers, the nuclei are gathered together at the middle of the fiber ("equatorial" region) whereas the contractile material (myosin and actin) lies in the two polar regions on each side of the equator. In some intrafusal fibers, the nuclei are lined up like a chain ("chain" fibers) whereas in other intrafusal fibers they are packed like in a bag ("bag" fiber) (Milburn 1984). Furthemore, the bag fibers subdivide into bag1 fiber (there is only one bag1 fiber per

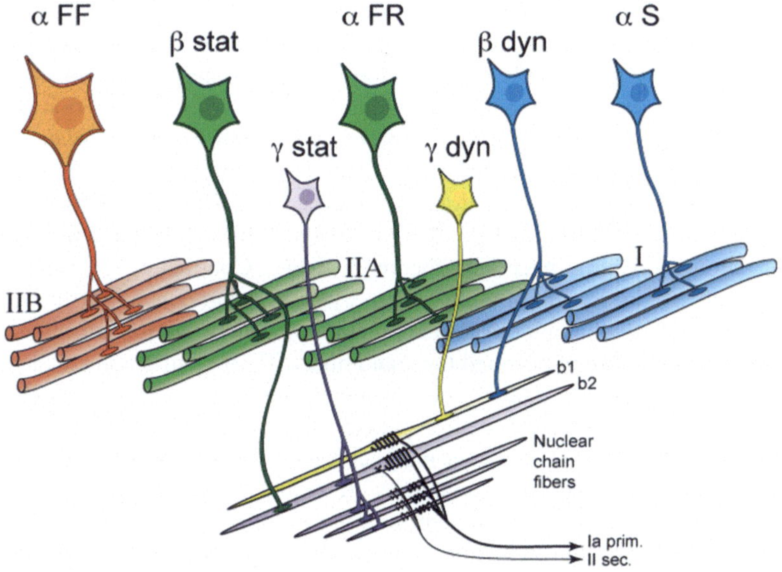

Fig. 1 Schema illustrating the different types of motoneurons and the muscle fibers they innervate. Gamma-motoneurons innervate the intrafusal motoneurons of the muscle spindles. Their action (dynamic vs. static) depends on which intrafusal fiber they innervate. Among alpha-motoneurons, S motor units have a small motoneuron soma, a small diameter axon and innervate a small number of type I fibers. FR motor units tend to have larger motoneuron cell bodies, larger diameter axons, and innervate a larger number of type IIA or type IIX muscle fibers. FF motor units have the largest motoneurons, the largest diameter axons, and innervate a very large number of type IIB muscle fibers. Beta-motoneurons are hybrids that innervate both extra- and intrafusal muscle fibers. Schema not to scale. The full length of the extrafusal muscle fibers is truncated for clarity. (From Manuel and Zytnicki (2011))

spindle) and bag2 fibers (Thornell et al. 2015). The anatomical organization of "ordinary" fibers is very different. The contractile material and nuclei are regularly spaced along the fiber and are located at the periphery between the sarcolemma and the contractile material (Roman and Gomes 2018).

The motoneuron population reflects the diversity of muscle fibers they innervate (Manuel and Zytnicki 2011). The largest fraction of motoneurons, so-called alpha-motoneurons, innervate the ordinary muscle fibers. Each motoneuron innervates a separate set of muscle fibers because, in adults, each muscle fiber is innervated by only one motoneuron. All the muscle fibers innervated by a motoneuron contract concurrently, hence the name "motor unit", which encompass the motor neuron (whose cell body is located in the spinal cord or the brainstem), its axon, and the muscle fibers that it innervates. As shown in the next section (classification of muscle fibers and motor units), alpha-motoneurons are subdivided in three main types depending on the functional properties of their respective muscle fibers (Burke et al. 1971). A characteristic feature of mammals is that intrafusal fibers are innervated by a particular class of motoneurons, so-called gamma motoneurons, whose only function is to control the activity of intrafusal fibers (Kuffler et al. 1951). Contrary to the ordinary fibers, one intrafusal fiber is innervated by several gamma-motoneurons (poly-innervated) even in adult animals (Banks 1981). In intrafusal fibers, each neuromuscular junction exerts a focal action (the contraction is focal and does not spread along the fiber). Gamma-motoneurons may innervate several fibers in the same spindle or in different spindles from the same muscle. However the innervation pattern is not random (Matthews 1972; Brown and Butler 1973; Bessou and Pagés 1975; Boyd et al. 1977). Actually, a subclass of gamma motoneurons (gamma dynamic) innervates the bag1 fibers. Contraction of bag1 fibers increases the sensitivity of spindles to the "dynamic" components of the muscle stretch (rapid length changes). Another subclass of gamma-motoneurons (gamma static) innervate the bag2 fibers and the chain fibers. Their contraction enhances the spindle's sensitivity to the "static" components of the stretch. Finally, a fraction of motoneurons have a double innervation pattern: they form neuromuscular junctions both on ordinary fibers (like alpha-motoneurons) and on intrafusal fibers (like gamma-motoneurons). These motoneurons are called "beta-motoneurons". It was long believed that these beta-motoneurons were present only in lower vertebrates (amphibians, reptiles, birds), while in mammals, spindles were controlled by gamma-motoneurons only, which allows spindles to be activated independently of the activity of alpha-motoneurons. The discovery of beta-motoneurons in mammals required challenging electrophysiological investigations (Bessou et al. 1963a). Their presence is not a vestigial remnant of evolution since the innervation pattern of beta-motoneurons is very specific. Like gamma-motoneurons, we distinguish "dynamic" and "static" beta-motoneurons that innervate the bag1 or the bag2/chain fibers in the spindles, respectively (Barker et al. 1977). But the "extrafusal" innervation (ordinary muscle fibers) is also very specific since the "dynamic" beta motoneurons specifically innervate the slow-contracting ordinary fibers whereas the "static" beta-motoneurons specifically innervate the fast-contracting (fatigable or fatigue resistant) fibers (Jami et al. 1982a; Emonet-Dénand et al. 1992). To this day, the specific function of beta-motoneurons is still largely unknown.

3 Classification of Muscles Fibers and Motor Units

The ordinary muscle fibers do not constitute a homogeneous population. They display a variety of biochemical and contractile properties. This was apparent very early on when it was observed that some muscles appear redder than others, and this difference of color correlated with contraction speed (Ranvier 1873; Needham 1926; Denny-Brown 1929). The red muscles are more vascularized, contain a larger quantity of myoglobin but contract more slowly than white muscles. Later, analyses based on myosin ATPase activity, as well as glycolytic and oxidative enzymes revealed that muscle fibers could be classified in 3 major groups (with different nomenclatures according to authors, (Brooke and Kaiser 1970; Ariano et al. 1973)): (i) slow-twitch oxidative (SO), also called type I fibers. These fibers contract slowly and are rich in oxidative enzymes, allowing them to sustain long-duration energy demands. Consequently, these fibers are highly resistant to fatigue. (ii) Fast-twitch oxidative glycolytic (FOG), also called type IIA fibers, that are resistant to fatigue, albeit to a lesser degree than type I fibers. (iii) Fast-twitch glycolytic (FG), also called type IIB fibers, which are poor in oxidative enzymes and thereby fatigable (Burke 1981; Kernell 2006; Schiaffino and Reggiani 2011). In the 1980s–1990s, there was a shift to using monoclonal antibodies directed towards several isoform of the myosin heavy-chain (MyHC) protein to type the muscle fibers. Although type I, IIA and IIB fibers were found to express different MyHC isoforms (MyHC-slow, MyHC-2A and MyHC-2B, respectively), a fourth myosin isoform, MyHC-2X, was identified (Schiaffino et al. 1989), as well as some isoforms that are expressed selectively in specific muscles (extra-ocular and jaw muscles in particular; see Schiaffino and Reggiani 2011 for a review).

The diversity of muscle fibers is a property that is highly conserved across vertebrates, from zebrafish, mice, rats, cats, to humans (although it should be noted that humans do not express the MyHC-2B isoform, but only MyHC-2X (Smerdu et al. 1994)). This suggests that this diversity is essential to generate a vast repertoire of movements and behaviors with a finite number of muscles. Most interestingly, all muscles throughout the body contain a mosaic of muscle fiber types (Ariano et al. 1973), however, the proportion of each type varies across muscles depending on the muscle function (Fig. 2). Muscles involved in postural activity tend to contain high concentration of Slow (Type I) muscle fibers, while flexor muscles tend to contain more Type II fibers. However, there is considerable inter-species difference, owing to the difference in size, lifestyle, locomotion type, etc.... Overall, muscles from smaller animals contain a larger proportion of IIX and IIB fibers, while large mammals, such as humans, possess more type I and IIA fibers (Schiaffino and Reggiani 2011). For instance, the ankle extensor muscle soleus consists of a mix of type I and type IIA/IIX fibers in mice and rats (albeit with slightly different proportions (Armstrong and Phelps 1984; Augusto et al. 2004; Sawano et al. 2016)). In the cat, however, this muscle is exclusively constituted of Type I muscle fibers (Ariano et al. 1973; Burke et al. 1974).

Elegant in vivo experiments allowed correlating the physiological properties of motor units with the histochemical profile of their muscle fibers. The contraction of a single motor unit can be studied by stimulating either the cell body of a motoneuron (via an intracellular electrode) or a filament of a ventral root (split until the filament contains only one axon innervating a specific muscle attached to a strain gauge). After long-duration repetitive stimulation, muscle fibers constituting the stimulated motor unit can be depleted of their reserve in glycogen, and these fibers can be identified on serial histological slices of the muscle by their absence of coloration to the periodic acid-Schiff staining. At the same time, the histochemical properties of the same muscle fibers can be studied on adjacent serial slices. Pioneering studies using this technique in rats (Edström and Kugelberg 1968) and cats (Burke et al. 1971) have demonstrated that motor units are homogenous in terms of the muscle fiber types they contain (at least as far as their oxidative and ATPase activity is concerned), and that their contractile properties match the

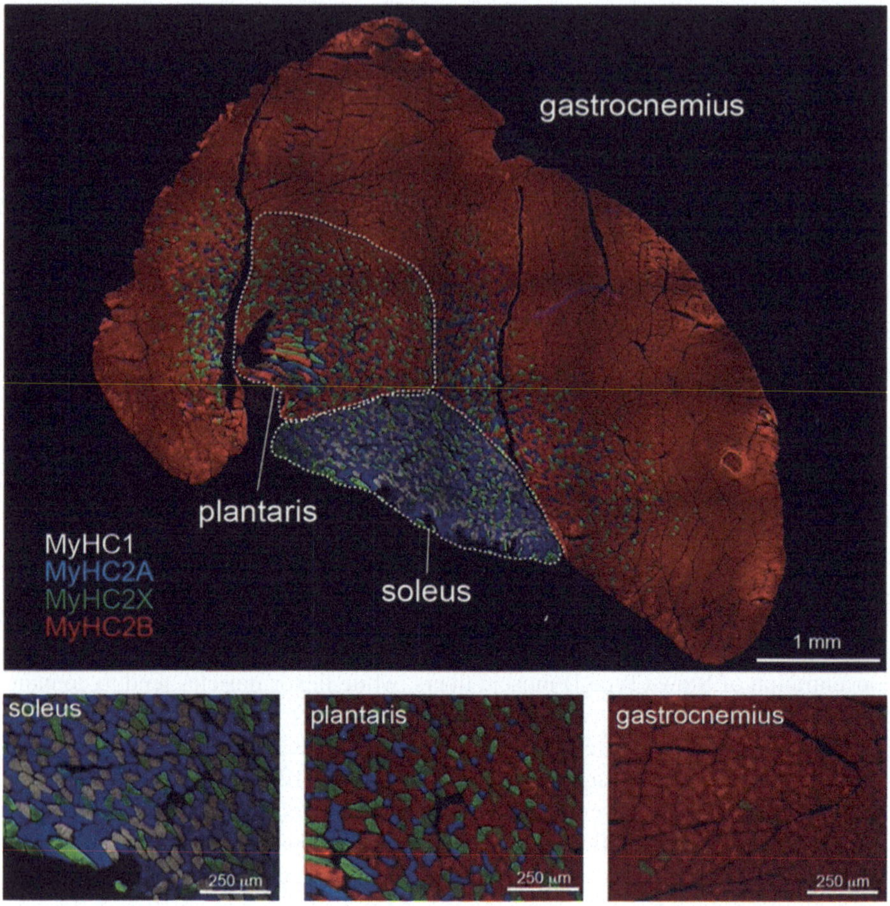

Fig. 2 Muscles are composed of a mosaic of muscle fiber types. Histological image of the cross-section of a mouse calf muscle (including the gastrocnemius medialis and lateralis, the soleus and the plantaris muscles), labeled using antibodies against the MyHC isoforms MyHC1 (grey), MyHC2A (blue), MyHC2X (green), and MyHC2B (red). Most muscles contain muscle fibers expressing either of these isoforms, but the relative abundance of each muscle fiber type varies across muscles. (Image adapted from Sawano et al. (2016) licensed under CC-BY)

contractile properties of the fibers, leading to the identification of three types of motor units: slow-contracting (S-type) motor units, which are comprised of type I fibers, contract slowly, develop a small amount of force, but are highly resistant to fatigue; fast-contracting, fatigue resistant (FR) motor units contain type 2A fibers, develop more force than S motor units and are also resistant to fatigue, albeit to a lesser degree than type S; and finally, fast-contracting, fatigable (FF) motor units contain type 2B fibers, develop the largest amount of force, but are highly fatigable. These results were further confirmed in humans, using controlled intramuscular microstimulation, glycogen depletion and muscle biopsy (Garnett et al. 1979).

It should be noted, however, that further studies have shown that muscle fibers could express several MyHC isoforms (Caiozzo et al. 2003), and that there is, in fact, a continuous spectrum of fiber types. Similar conclusions can be reached when looking at the contractile properties of single motor units, which tend to exhibit continuous distributions of properties (Burke 1981; Macefield et al. 1996; McNulty et al. 2000; Schiaffino and Reggiani 2011; Heckman and Enoka 2012). One possible exception is the property of the unfused tetanus of some motor units to drop slightly after reaching a peak of force ("sag property"; Burke et al. 1971; Celichowski et al. 1999; Drzymała-Celichowska et al. 2016), while other units do not exhibit such sag in their force. Units that exhibit sag are fast-contracting, while units that do not are slow-contracting (Burke et al. 1971; Celichowski et al. 1999; Drzymała-Celichowska et al. 2016). Even in this case, there is still a significant variability among motor units in the shape of the unfused tetanus and the range of frequencies at which this sag is visible.

When discussing motor units, their properties, and their susceptibility to disease, it nonetheless remains more practical to group them under the generic terms S, FR and FF (Fig. 1), as we will do in the rest of this chapter, while keeping in mind that there are no strict boundaries between these groups and that intermediate motor unit exist.

4 Contractile Properties of Motor Unit Types

The different muscle fibers composition of fast and slow motor units is reflected in their contractile properties. In general, the fast fatigable (FF) motor units develop the highest contraction force; have the fastest contraction and relaxation times, while in parallel being highly susceptible to fatigue. On the other end of the spectrum, the slow motor units (S) display high resistance to fatigue, slow contraction and relaxation times, and small capability for developing high contractile force. Between the FF and S, the intermediate region of contractile properties is occupied by the fast fatigable (FR) motor units with short contraction times and moderate force production capabilities. Throughout the years, many attempts were made to determine cut-off values of selected contractile parameters which would enable the precise classification of motor unit type without the necessity to perform muscle fiber histochemical analysis (Burke et al. 1974; Jami et al. 1982b; Kernell et al. 1983; Bawa et al. 1984; Zengel et al. 1985; Gardiner 1993). Already very early on, researchers noticed that, during unfused tetanus, some motor units display an early peak and subsequent decline in tension toward a lower plateau level (Burke et al. 1973). This phenomenon, coined "sag" property is highly correlated with motor unit half-relaxation time and could be used to distinguish between fast and slow motor unit type in medial gastrocnemius muscle of cats (Burke et al. 1973) and rats (Grottel and Celichowski 1990). However it does not allow the identification of FF and FR motor units. These units can be segregated based on their ability to withstand a fatiguing stimulation protocol and measuring a fatigue index (FI), describing the ratio of motor unit force between the last and the first tetanic contraction of the stimulation protocol (lasting generally 2 min). In cats, motor units with FI < 0.25 were considered FF while these with FI > 0.75 were considered FR (Burke et al. 1973). In rats and mice, a threshold value of 0.5 was chosen instead (Grottel and Celichowski 1990; Martínez-Silva et al. 2018). Unfortunately the aforementioned criteria are not infallible and Burke et al. (1973) acknowledged that 2.5% of cat MG motor units were considered "intermediate" between FF and FR. Still, as mentioned before, the motor unit classification as FF, FR and S based on those parameters remains relevant when discussing the motor units susceptibility to disease.

The major determinant of motor unit force is its size, i.e. the number of muscle fibers it contains. In the cat MG, FF motor units typically contain 400–700 muscle fibers, while S motor units are composed of ~200 muscle fibers (Rafuse and Gordon 1996; Rafuse et al. 1997). Differences in muscle fiber cross-sectional areas also contribute to the force difference (Burke 1981; Kernell 2006). Nevertheless, it is important to keep in mind that force transmission in situ is a highly non-linear process, as most of the force transmission happens through the extra-cellular matrix, connective tissue and myofascia (Maas 2019; Ward et al. 2020). Of course the developed force and contraction times of FF, FR and S motor unit types vary significantly between different species. For example the twitch force of an average medial gastrocnemius FF motor unit is about 230 mN in cats, 30 mN in rats and 15 mN in mice (Krutki et al. 2006; Martínez-Silva et al. 2018). Motor unit contraction time depends essentially on the type of Myosin isoform (which determines the cross-bridges detachment rate through the rate of ADP release, Schiaffino and Reggiani 2011) as well as the calcium dynamics inside the fibers (Schiaffino and Reggiani 2011). On average, there is a 2–3-fold difference in contraction time between S and F motor units, regardless of species (Burke et al. 1973; Krutki et al. 2006; Łochyński et al. 2016; Martínez-Silva et al. 2018).

5 Motoneuron Electrophysiological Properties Depend on Their Motor Unit Type

Since motor units are homogeneous in their muscle fiber composition, it is essential that the firing of the motoneuron be precisely matched to the contractile properties of its muscle fibers in order to take full advantage of their contraction capabilities.

First, motor units need to be activated at the right moment. S-type motor units can sustain very long duration contractions thanks to their high resistance to fatigue, but cannot generate large amounts of forces. On the other hand, although FF motor units can generate a lot of force, they should not be active for long periods, as their force output will collapse very quickly. Experiments in animal models have shown that the motor units tend to be recruited in a specific order according to their size, hence the name "size principle" (Henneman et al. 1965). During postural activities or movements, S motor units are activated first. If the motor act requires more force than can be generated by the S motor units alone, then FR motor units become recruited as well. It is only in the cases where very large amounts of forces are required that the FF motor units will be recruited. Hennig and Lømo (1985) have followed the activity of several motor units across extended periods of the daily activity of laboratory rats and have indeed shown that motor units presumed to be S-type fire continuously for many hours during the day, while presumed FF motor units only fire bursts of action potentials for only 0.5–3 min per 24 h (Hennig and Lømo 1985).

Many properties of the motoneurons contribute to this orderly recruitment, which have been extensively reviewed before (Kernell 2006; Kanning et al. 2009; Manuel and Zytnicki 2011; Heckman and Enoka 2012; Manuel et al. 2019). The recent development of motor unit recordings in mice have shown that, for the most part, the same differences in the properties of the motor unit types are conserved between mice, rats and cats (Martínez-Silva et al. 2018; Manuel et al. 2019).

One of the major determinants of recruitment is the input conductance of the motoneuron, which conditions the amount of current required to reach firing threshold (rheobase). In motoneurons, the input conductance varies according to the total surface area of the cell body (Burke 1981). S motoneurons are small, and have fewer dendritic branches, resulting in a small input conductance, a low rheobase, and are therefore the easiest to recruit. On the other end of the spectrum, FF motoneurons have the largest cell bodies, have a highly developed dendritic arborization, and the largest input conductance, which makes them the least excitable (Burke 1981; Kernell 2006). The reasons for this difference in cell body sizes is not entirely clear, but it might be directly related to the size of their axons. FF motoneurons have the largest diameter axons due to the fact that they innervate the highest number of muscle fibers, requiring that their axon bifurcate extensively. On the other hand, S motoneurons innervate a smaller number of muscle fibers, and have a thinner axon (Cullheim and Ulfhake 1979; Burke 1982).

Recruitment is also dependent on the expression of voltage-dependent channels that are active below the firing threshold. For instance, in 6–10 day old mice, about one third of lumbar motoneurons start to discharge immediately during a long current pulse at rheobase intensity (immediate firing pattern) whereas the remaining motoneurons display a slow depolarizing ramp and start to discharge a few seconds only after the pulse onset (delayed firing pattern) (Leroy et al. 2014) (see also Russier et al. 2003; Pambo-Pambo et al. 2009). The delay before firing onset is caused by the interaction of two potassium currents acting at two different time scales. First, at the onset of a current pulse, the fast potassium current I_A prevents the motoneuron from firing (Leroy et al. 2015). Then, a slow depolarization is apparent, which slowly brings the membrane potential to the firing threshold. This slow depolarization is caused by a slowly-inactivating potassium current (Leroy et al. 2015) that was later identified as being mediated by Kv1.2 channels (Bos et al. 2018). Although these experiments were conducted in vitro and in young animals, making the identification of the motor unit type based on its contractile properties impossible, several lines of evidence suggest that the immediate firing motoneurons innervate S-type motor units whereas the delayed firing ones innervate F-type motor units (Leroy et al. 2014). Indeed, the delayed firing motoneurons displayed a lower input resistance, a higher voltage for spiking, a higher rheobase, a shorter spike and a shorter AHP than the immediate firing motoneurons. In addition, the delayed firing motoneurons are larger (larger soma, longer dendrites) than the immediate firing ones. Furthermore, the immediate firing motoneurons express estrogen-related receptor beta (ERRβ), a candidate marker of slow motoneurons (Enjin et al. 2010), but not matrix metalloproteinase-9 (MMP9), proposed to be a marker of large F-type motoneurons (Kaplan et al. 2014). In sharp contrast, the delayed firing motoneurons are ERRβ negative and the largest ones express MMP9 (Leroy et al. 2014) Although Kv1.2 expression persists in adult mice (Bos et al. 2018) whether its expression remains different between S and F motoneurons is still unknown.

Not only the different types of motor units need to be activated at the opportune time, the firing rate of the motoneuron needs to be adapted to the speed of contraction of the muscle fibers. Since type II muscle fibers have a shorter contraction time and shorter relaxation time than type I muscle fibers (Burke 1981) F-type motoneurons need to fire at higher rates to allow individual muscle twitches to fuse and thereby increase the force developed by the motor unit. The firing rate of a motoneuron is controlled in large part by the after-hyperpolarization, a period of hyperpolarization following each action potential caused by the activation of calcium-dependent potassium channels (SK channels, (Sah and McLachlan 1992)). S motoneurons tend to have a bigger and longer AHP than F motoneurons (Burke 1981; Zengel et al. 1985). Consequently, several labs use the half-decay time of AHP to distinguish between F-fast and S-slow rat motoneurons (F > 20 ms, S < 20 ms, (Gardiner 1993)) but the difference between the FR and FF motoneurons is not well defined.

In cat motor units, the duration of the AHP matches the duration of the twitch ("speed matching", Kernell 2006) and determines the minimal repetitive-firing frequency of the motoneuron (Kernell 2006). This adaptation seemingly ensures that force output increases smoothly with the increase in firing frequency (rate modulation). As reviewed by Meehan et al. (2010) both the AHP duration and the twitch duration decrease in parallel with the size of the animal, indicating that this adaptation is conserved across species and must serve an important role in force gradation (Meehan et al. 2010). However, recordings experiments in mice (Manuel and Heckman 2011) and in rats (Turkin et al. 2010) show that motoneurons fire preferentially at rates at which the force output is saturated, suggesting that these species rely more heavily on recruitment of new motor units rather than rate modulation for force gradation (Manuel et al. 2019).

The firing rate is not the only parameter that must be matched to the properties of the motor unit. An FF-type motoneuron must drive its muscle fibers at high frequency to generate force, but it can only do so for a short period of time before the muscle fibers fatigue and the force collapses. That is likely why FF motoneurons have a tendency to fire in bursts of action potentials, while S motoneurons can discharge tonically for long periods of time (Granit et al. 1956; Hennig and Lømo 1985). This difference in firing characteristics is due to multiple currents. First, motoneurons exhibit a property called *spike-frequency adaptation*. In response to a long supra-threshold pulse of current, motoneuron firing starts at high frequency and then decreases progressively (Kernell 1965; Sawczuk et al. 1997; Miles et al. 2005; Brownstone 2006). High rheobase, high input conductance (likely FF motoneurons) tend to have a stronger spike frequency adaptation than smaller motoneurons (Zengel et al. 1985; Button et al. 2007), and are therefore likely to stop firing earlier than S-type motoneurons. Motoneurons also possess a type of slowly-inactivating currents called *persistent inward currents* (PICs) that have been extensively studied (reviewed in Heckman et al. 2005; ElBasiouny et al. 2010; Heckman and Enoka 2012). One of the functions of these currents would be to generate sustained tonic firing ("bistability") in motoneurons, which could contribute to muscle tone in postural muscles (Heckman and Enoka 2012; Lømo et al. 2020). However, PICs in large motoneurons inactivate faster than in smaller motoneurons. Consequently, large motoneurons tend to be only partially-bistable, and cannot sustain a tonic firing for long periods of time (Lee and Heckman 1998a, b). PICs are also important for the amplification of synaptic inputs (Bennett et al. 1998; Heckman et al. 2005). Because PICs can interact with the *h*-current, which is different between S and FF motoneurons (Gustafsson and Pinter 1984; Zengel et al. 1985), PICs tend to amplify tonic input in small motoneurons, while the amplification is more efficient for phasic inputs in larger motoneurons (Manuel et al. 2007).

6 Anatomical and Synaptic Properties of the Neuromuscular Junctions Depend on Their Motor Unit Type

In order to evoke muscle contraction, action potentials generated by motoneurons have to travel orthodromically along the axon, and reach the innervated muscle and activate the actin-myosin complex. This is achieved through the neuromuscular junction (NMJ), a fundamental meeting place of muscle and nervous system. Motoneuron axons reaching the muscle bifurcate to innervate a large number of muscle fibers, and terminate at each sarcolemma endplate, juxtaposed to postsynaptic junction folds highly enriched in nicotinic ACh receptors and voltage-dependent channels. An action potential reaching the boutons opens voltage gated calcium channels enabling Ca^{2+} influx and docking of the acetylcholine containing synaptic vesicles to the boutons' active zones and ACh exocytosis into the synaptic cleft. Following activation of nicotinic ACh channel by the secreted neurotransmitter, a rapid sodium influx evokes an endplate action potential which then propagates along the muscle fiber (Schiaffino and Reggiani 2011).

Although this basic description is shared among all motor units, some type-specific differences in NMJ structure and functioning are evident. Similarly to the motor units properties highlighted above, NMJ properties tend to form a continuum from fast to slow motor unit type. The axons terminals innervating type I muscle fibers are round, the juxtaposed sarcolemma is highly enriched in mitochondria and the overall terminal size is smaller than those innervating IIB or IIA fibers (Fig. 3)

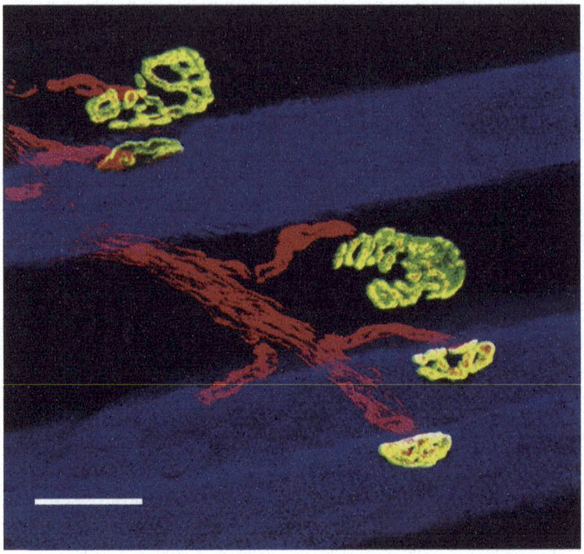

Fig. 3 Different morphologies of neuromuscular junctions on different muscle fiber types. Image from a rat diaphragm showing motor axons (labeled by incubation with FM4-64, red), motor end-plates (labeled with Alexa Fluor 488-conjugated α-bungarotoxin, green) and muscle fibers (stained with specific antibodies to MHC isoforms Slow and 2A, blue). Notice the difference in neuromuscular junction morphology between NMJs on type I/IIA fibers (blue) and IIX/IIB fibers (black). Scale bar = 20 μm. (Image reproduced from Mantilla et al. (2007) with permission from Elsevier)

(Mantilla et al. 2007; Tremblay et al. 2017). The terminals of IIA and IIB innervating fibers are oval, cover larger areas and their postsynaptic sarcolemma contains fewer mitochondria. In addition the postsynaptic folds of type IIB innervating terminals are larger than those of type IIA (Lømo and Wærhaug 1985; Ogata 1988; Wærhaug and Lømo 1994; Seene et al. 2017). Importantly, type I terminals contain fewer synaptic vesicles than type II terminals, and this is reflected in their smaller quantal content (number of vesicles released) following single pulse stimulation (Reid et al. 1999; Tremblay et al. 2017).

However, the smaller quantal content is not reflected in miniature end plate potentials, which tend to be the largest in NMJs of S motor units, and then progressively decrease from FR to FF motor units (Tremblay et al. 2017). Actually, the electrophysiological properties of NMJ types are closely related to the physiological demand imposed on them by muscle activity patterns. As slow motor units are active in the majority of long and short lasting motor tasks (Hennig and Lømo 1985) their respective NMJs have to withstand continuous activation in order to provide the necessary muscle contraction. To meet these requirements, in comparison to NMJs of FF or FR motor units the NMJ of S type motor units show stronger paired-pulse facilitation (Tremblay et al. 2017) and a slower decrease in quantal content upon repetitive stimulation (Reid et al. 1999), effectively increasing their ability to evoke repeated sarcolemma depolarization over a long time period. This is a very important mechanism, as the synaptic vesicle trafficking dynamics seem to be the same across the entire motor unit pool (Reid et al. 1999). On the other hand, the NMJs of rarely activated FF motor units release massive amounts of ACh in a short time, enabling high frequency muscle activation, which, however, will be quite limited in duration.

Importantly, the three subtypes of NMJs innervating type I, IIA and IIB muscle fibers not only possess different morphological and electrophysiological properties, but also exhibit different capacities to compensate for motoneuron loss in aging, disease, and injury, a process called "sprouting", where a new nerve terminal will form from an intact axon in order to re-innervate its own, or adjacent NMJs left vacant (Tam and Gordon 2009). Experiments conducted in cats following peripheral nerve injury demonstrated that axonal sprouting can compensate for the loss of up to 85% of motor units (Rafuse and Gordon 1996). When less than 20% of functional motor units remain, the maximal capacity of axonal sprouting is exceeded, reinnervation of all denervated muscle fibers fails, and muscle weakness becomes evident (Tam and Gordon 2009). As efficient as this process can be, it strongly depends on the type of the motor unit. In particular, the efficacy of sprouting is tightly controlled by electrical activity (which differs between motor units, see above). Curiously, after partial denervation, axonal sprouting is strongly inhibited by both high neuromuscular activity (elicited through chronic electrical stimulation or forced exercise) (Tam and Gordon 2003) and blockade of neuromuscular activity (using TTX or α-bungarotoxin) (Connold and Vrbová 1991). Sprouting can also be triggered in the absence of injury with chronic blockade of neurotransmission by Botulinum toxin (a toxin that blocks the release of ACh from the nerve terminals, (Pamphlett 1989)). In all these cases, axons from "slow" muscles exhibit more/faster sprouting than axons from "fast" muscles (Duchen and Tonge 1973; Brown et al. 1980).

To summarize, it is clear that the neuromuscular system is constituted of multiple types of motor units, each exhibiting their own particular properties (contractile properties at the level of the muscle fibers, synaptic properties at the level of the NMJs, electrical properties at the level of the motoneurons; Fig. 4). This diversity is conserved across species, including humans, and must play a fundamental role in our ability to generate highly complex and adaptable behaviors.

7 Conclusion

Motoneurons display a wide diversity in terms of types of muscle fibers they are innervating. Importantly, the innervation patterns are not random but are very specific, providing different functions to the different motoneuron types. Furthermore, intrinsic electrical and molecular properties of motoneurons, synaptic properties of their neuromuscular junctions, and contractile properties of the innervated muscle fibers are all precisely matched to form motor units with highly tuned functional properties. The diversity of the resulting motor units allows the motor system to execute a large repertoire of motor tasks.

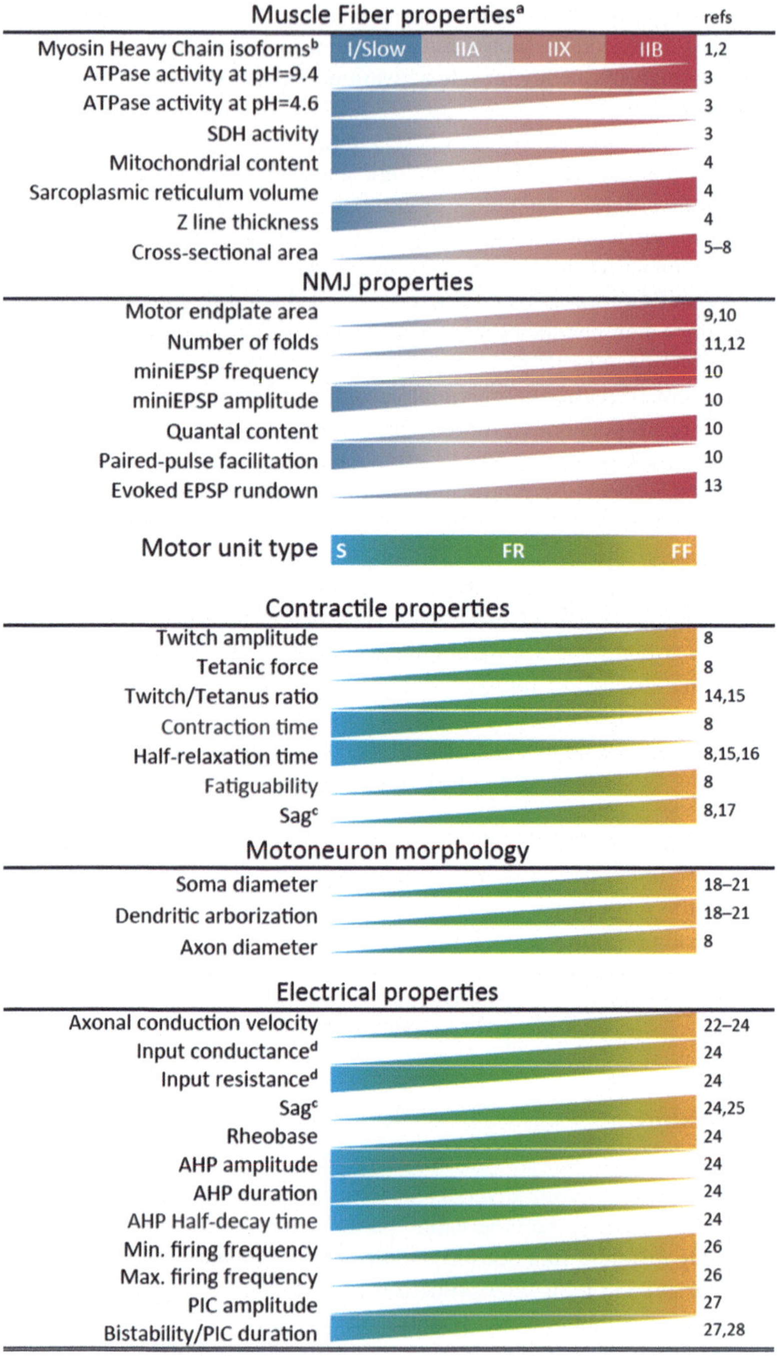

Fig. 4 Summary of the main differences in properties across motor unit types. This table shows the overall relationship between various properties and the type of muscle fibers and motor

Acknowledgments MB is supported by National Science Centre 2017/26/D/NZ7/00728. MM and DZ are supported by NIH-NINDS R01NS110953, the Thierry Latran Foundation project "TRiALS", Association pour la Recherche sur la SLA et autres maladies du motoneurone (ARSLA), the Association Française contre les Myopathies (AFM) project "HYPERTOXIC", Radala Foundation for ALS Research, and Programme Hubert Curien "Polonium" for scientific exchanges. MM would like to thank Alexandra Elbakyan for her help with bibliography. FR is supported by the Thierry Latran Foundation (projects "Trials" and "Hypothals"), by the Radala Foundation, by the Deutsche Forschungsgemeinschaft (DFG) as part of the SFB1149 and with the individual grant no. 431995586 (RO-5004/8-1) and no. 443642953 (RO5004/9-1), by the Cellular and Molecular Mechanisms in Aging (CEMMA) Research Training Group and by BMBF (FKZ 01EW1705A, as member of the ERANET-NEURON consortium "MICRONET").

Conflict of Interest The authors declare no conflict of interest.

References

Ariano MA, Edgerton VR, Amstrong RB (1973) Hindlimb muscle fiber populations of five mammals. J Histochem Cytochem 21:51–55. https://doi.org/10.1177/21.1.51

Armstrong RB, Phelps RO (1984) Muscle fiber type composition of the rat hindlimb. Am J Anat 171:259–272. https://doi.org/10.1002/aja.1001710303

Augusto V, Padovani C, Campos G (2004) Skeletal muscle fiber types in C57BL6J mice. Braz J Morphol Sci 21:89–94

Banks RW (1981) A histological study of the motor innervation of the cat's muscle spindle. J Anat 133:571–591

Barker D, Emonet-Dénand F, Harker DW et al (1977) Types of intra- and extrafusal muscle fibre innervated by dynamic skeleto-fusimotor axons in cat peroneus brevis and tenuissimus muscles, as determined by the glycogen-depletion method. J Physiol 266:713–726. https://doi.org/10.1113/jphysiol.1977.sp011789

Barnard RJ, Edgerton VR, Furukawa T, Peter JB (1971) Histochemical, biochemical, and contractile properties of red, white, and intermediate fibers. Am J Phys 220:410–414. https://doi.org/10.1152/ajplegacy.1971.220.2.410

Fig. 4 (continued) units. Note that these relationships are not necessarily linear, and the graphical representation used here is merely a cartoon showing the overall direction of the trend. Notes: (a) only a few differences are presented here. See (Schiaffino and Reggiani 2011) for a detailed review. (b) There exists intermediate types expressing multiple MyHC ex: I/IIA, IIA/IIX, IIX/IIB (Caiozzo et al. 2003; Schiaffino and Reggiani 2011). (c) the "sag" on the unfused tetanus of a motor unit corresponds to decline in the force toward a lower plateau level after an initial peak (Burke et al. 1973). When speaking about electrical properties of a motoneuron, the "sag" refers to a decline in the membrane potential after an initial peak upon injection of a square pulse of current, due to the activation of the Ih current (Gustafsson and Pinter 1984; Zengel et al. 1985). (d) Input resistance is the inverse of input conductance. References: 1. Schiaffino and Reggiani (2011) 2. Caiozzo et al. (2003) 3. Barnard et al. (1971) 4. Schiaffino et al. (1970) 5. Rafuse and Gordon (1996) 6. Staron et al. (1999) 7. Augusto et al. (2004) 8. Burke (1981) 9. Mantilla et al. (2007) 10. Tremblay et al. (2017) 11. Ogata (1988) 12. Padykula and Gauthier (1970) 13. Reid et al. (1999) 14. Stephens and Stuart (1975) 15. Łochyński et al. (2016) 16. Hammarberg and Kellerth (1975) 17. Grottel and Celichowski (1990) 18. Conradi et al. (1979) 19. Kellerth et al. (1979) 20. Kellerth et al. (1983) 21. Cullheim et al. (1987) 22. Burke et al. (1973) 23. Bessou et al. (1963b) 24. Zengel et al. (1985) 25. Gustafsson and Pinter (1984) 26. Kernell (2006) 27. Lee and Heckman (1998a) 28. Lee and Heckman (1998b)

Bawa P, Binder MD, Ruenzel P, Henneman E (1984) Recruitment order of motoneurons in stretch reflexes is highly correlated with their axonal conduction velocity. J Neurophysiol 52:410–420. https://doi.org/10.1152/jn.1984.52.3.410

Bennett DJ, Hultborn H, Fedirchuk B, Gorassini MA (1998) Short-term plasticity in hindlimb motoneurons of decerebrate cats. J Neurophysiol 80:2038–2045

Bessou P, Pagés B (1975) Cinematographic analysis of contractile events produced in intrafusal muscle fibres by stimulation of static and dynamic fusimotor axons. J Physiol 252:397–427

Bessou P, Emonet-Denand F, Laporte Y (1963a) Occurrence of intrafusal muscle fibres innervation by branches of slow α motor fibres in the cat. Nature 198:594–595. https://doi.org/10.1038/198594a0

Bessou P, Emonet-Dénand F, Laporte Y (1963b) Relation entre la vitesse de conduction des fibres nerveuses motrices et le tempe de contraction de leurs unites motrices. CR Acad Sci Paris 256:5625–5627

Bos R, Harris-Warrick RM, Brocard C et al (2018) Kv1.2 channels promote nonlinear spiking motoneurons for powering up locomotion. Cell Rep 22:3315–3327. https://doi.org/10.1016/J.CELREP.2018.02.093

Boyd IA, Gladden MH, McWilliam PN, Ward J (1977) Control of dynamic and static nuclear bag fibres and nuclear chain fibres by gamma and beta axons in isolated cat muscle spindels. J Physiol 265:133–162

Brooke MH, Kaiser KK (1970) Muscle fiber types: how many and what kind? Arch Neurol 23:369–379. https://doi.org/10.1001/archneur.1970.00480280083010

Brown MC, Butler RG (1973) Studies on the site of termination of static and dynamic fusimotor fibres within muscle spindles of the tenuissimus muscle of the cat. J Physiol 233:553–573

Brown MC, Holland RL, Ironton R (1980) Nodal and terminal sprouting from motor nerves in fast and slow muscles of the mouse. J Physiol 306:493–510. https://doi.org/10.1113/jphysiol.1980.sp013410

Brownstone RM (2006) Beginning at the end: repetitive firing properties in the final common pathway. Prog Neurobiol 78:156–172. https://doi.org/10.1016/J.PNEUROBIO.2006.04.002

Burke RE (1981) Motor units: anatomy, physiology, and functional organization. In: Handbook of physiology. The nervous system. Motor control. American Physiological Society, Bethesda, pp 345–422

Burke RE (1982) Motor units in cat muscles: anatomical considerations in relation to motor unit types. Adv Neurol 36:31–45

Burke RE, Levine DN, Zajac FE (1971) Mammalian motor units: physiological-histochemical correlation in three types in cat gastrocnemius. Science 174:709–712

Burke RE, Levine DN, Tsairis P, Zajac FE (1973) Physiological types and histochemical profiles in motor units of the cat gastrocnemius. J Physiol 234:723–748

Burke RE, Levine DN, Salcman M, Tsairis P (1974) Motor units in cat soleus muscle: physiological, histochemical and morphological characteristics. J Physiol 238:503–514. https://doi.org/10.1113/jphysiol.1974.sp010540

Button DC, Kalmar JM, Gardiner K et al (2007) Spike frequency adaptation of rat hindlimb motoneurons. J Appl Physiol 102:1041–1050. https://doi.org/10.1152/japplphysiol.01148.2006

Caiozzo VJ, Baker MJ, Huang K et al (2003) Single-fiber myosin heavy chain polymorphism: how many patterns and what proportions? Am J Physiol Regul Integr Comp Physiol 285:R570–R580. https://doi.org/10.1152/ajpregu.00646.2002

Celichowski J, Grottel K, Bichler E (1999) Differences in the profile of unfused tetani of fast motor units with respect to their resistance to fatigue in the rat medial gastrocnemius muscle. J Muscle Res Cell Motil 20:681–685. https://doi.org/10.1023/A:1005541013209

Connold AL, Vrbová G (1991) Temporary loss of activity prevents the increase of motor unit size in partially denervated rat soleus muscles. J Physiol 434:107–119. https://doi.org/10.1113/jphysiol.1991.sp018461

Conradi S, Kellerth JO, Berthold CH, Hammarberg C (1979) Electron microscopic studies of serially sectioned cat spinal alpha-motoneurons. IV. Motoneurons innervating slow-twitch (type S) units of the soleus muscle. J Comp Neurol 184:769–782. https://doi.org/10.1002/cne.901840409

Cullheim S, Ulfhake B (1979) Relations between cell body size, axon diameter and axon conduction velocity of triceps surae alpha montoneurons during the postnatal development in the cat. J Comp Neurol 188:679–686. https://doi.org/10.1002/cne.901880410

Cullheim S, Fleshman JW, Glenn LL, Burke RE (1987) Membrane area and dendritic structure in type-identified triceps surae alpha motoneurons. J Comp Neurol 255:68–81. https://doi.org/10.1002/cne.902550106

Denny-Brown DE (1929) The histological features of striped muscle in relation to its functional activity. Proc R Soc Lond Ser B Contain Pap Biol Character 104:371–411. https://doi.org/10.1098/rspb.1929.0014

Drzymała-Celichowska H, Raikova R, Krutki P (2016) Prolonged activity evokes potentiation and the "sag" phenomenon in slow motor units of rat soleus. Acta Neurobiol Exp (Warsz) 76:152–157. https://doi.org/10.21307/ane-2017-014

Duchen LW, Tonge DA (1973) The effects of tetanus toxin on neuromuscular transmission and on the morphology of motor end-plates in slow and fast skeletal muscle of the mouse. J Physiol 228:157–172. https://doi.org/10.1113/jphysiol.1973.sp010078

Edström L, Kugelberg E (1968) Properties of motor units in the rat anterior tibial muscle. Acta Physiol Scand 73:543–544. https://doi.org/10.1111/j.1365-201X.1968.tb10894.x

ElBasiouny SM, Schuster JE, Heckman CJ (2010) Persistent inward currents in spinal motoneurons: important for normal function but potentially harmful after spinal cord injury and in amyotrophic lateral sclerosis. Clin Neurophysiol Off J Int Fed Clin Neurophysiol 121:1669–1679. https://doi.org/10.1016/j.clinph.2009.12.041

Emonet-Dénand F, Petit J, Laporte Y (1992) Comparison of skeleto-fusimotor innervation in cat peroneus brevis and peroneus tertius muscles. J Physiol 458:519–525

Enjin A, Rabe N, Nakanishi ST et al (2010) Identification of novel spinal cholinergic genetic subtypes disclose Chodl and Pitx2 as markers for fast motor neurons and partition cells. J Comp Neurol 518:2284–2304. https://doi.org/10.1002/cne.22332

Gardiner PF (1993) Physiological properties of motoneurons innervating different muscle unit types in rat gastrocnemius. J Neurophysiol 69:1160–1170. https://doi.org/10.1152/jn.1993.69.4.1160

Garnett RA, O'Donovan MJ, Stephens JA, Taylor A (1979) Motor unit organization of human medial gastrocnemius. J Physiol 287:33–43. https://doi.org/10.1113/jphysiol.1979.sp012643

Granit R, Henatsch H-D, Steg G (1956) Tonic and phasic ventral horn cells differentiated by post-tetanic potentiation in cat extensors. Acta Physiol Scand 37:114–126. https://doi.org/10.1111/j.1748-1716.1956.tb01347.x

Gregory JE, Proske U (1991) Extrafusal and intrafusal motor units in the kitten. Int J Dev Neurosci Off J Int Soc Dev Neurosci 9:101–109. https://doi.org/10.1016/0736-5748(91)90077-y

Grottel K, Celichowski J (1990) Division of motor units in medial gastrocnemius muscle of the rat in the light of variability of their principal properties. Acta Neurobiol Exp (Warsz) 50:571–587

Gustafsson B, Pinter MJ (1984) Relations among passive electrical properties of lumbar alpha-motoneurones of the cat. J Physiol 356:401–431

Hammarberg C, Kellerth JO (1975) Studies of some twitch and fatigue properties of different motor unit types in the ankle muscles of the adult cat. Acta Physiol Scand 95:231–242

Heckman CJ, Enoka RM (2012) Motor Unit Compr Physiol 2:2629–2682. https://doi.org/10.1002/cphy.c100087

Heckman CJ, Gorassini MA, Bennett DJ (2005) Persistent inward currents in motoneuron dendrites: implications for motor output. Muscle Nerve 31:135–156. https://doi.org/10.1002/MUS.20261

Henneman E, Somjen G, Carpenter DO (1965) Functional significance of cell size in spinal motoneurons. J Neurophysiol 28:560–580. https://doi.org/10.1152/JN.1965.28.3.560

Hennig R, Lømo T (1985) Firing patterns of motor units in normal rats. Nature 314:164–166. https://doi.org/10.1038/314164A0

Hunt CC (1990) Mammalian muscle spindle: peripheral mechanisms. Physiol Rev 70:643–663. https://doi.org/10.1152/physrev.1990.70.3.643

Jami L, Murthy KS, Petit J (1982a) A quantitative study of skeletofusimotor innervation in the cat peroneus tertius muscle. J Physiol 325:125–144

Jami L, Murthy KS, Petit J, Zytnicki D (1982b) Distribution of physiological types of motor units in the cat peroneus tertius muscle. Exp Brain Res Exp Hirnforsch Exp Cerebrale 48:177–184. https://doi.org/10.1007/BF00237213

Kanning KC, Kaplan A, Henderson CE (2009) Motor neuron diversity in development and disease. Annu Rev Neurosci 33:409–440

Kaplan A, Spiller KJ, Towne C et al (2014) Neuronal matrix Metalloproteinase-9 is a determinant of selective neurodegeneration. Neuron 81:333–348. https://doi.org/10.1016/J.NEURON.2013.12.009

Kellerth JO, Berthold CH, Conradi S (1979) Electron microscopic studies of serially sectioned cat spinal alpha-motoneurons. III. Motoneurons innervating fast-twitch (type FR) units of the gastrocnemius muscle. J Comp Neurol 184:755–767

Kellerth JO, Conradi S, Berthold CH (1983) Electron microscopic studies of serially sectioned cat spinal-motoneurons: V. motoneurons innervating fast-twitch (type FF) units of the gastrocnemius muscle. J Comp Neurol 214:451–458

Kernell D (1965) The adaptation and the relation between discharge frequency and current strength of cat lumbosacral motoneurones stimulated by long-lasting injected currents. Acta Physiol Scand 65:65–73

Kernell D (2006) The motoneurone and its muscle fibres. Oxford University Press, Oxford

Kernell D, Eerbeek O, Verhey BA (1983) Motor unit categorization on basis of contractile properties: an experimental analysis of the composition of the cat's m. peroneus longus. Exp Brain Res Exp Hirnforsch Exp Cerebrale 50:211–219. https://doi.org/10.1007/BF00239185

Krutki P, Celichowski J, Lochyński D et al (2006) Interspecies differences of motor units properties in the medial gastrocnemius muscle of cat and rat. Arch Ital Biol 144:11–23

Kuffler SW, Hunt CC, Quilliam JP (1951) Function of medullated small-nerve fibers in mammalian ventral roots; efferent muscle spindle innervation. J Neurophysiol 14:29–54. https://doi.org/10.1152/jn.1951.14.1.29

Lee RH, Heckman CJ (1998a) Bistability in spinal motoneurons in vivo: systematic variations in persistent inward currents. J Neurophysiol 80:583–593

Lee RH, Heckman CJ (1998b) Bistability in spinal motoneurons in vivo: systematic variations in rhythmic firing patterns. J Neurophysiol 80:572–582

Leroy F, Lamotte d'Incamps B, Imhoff-Manuel RD, Zytnicki D (2014) Early intrinsic hyperexcitability does not contribute to motoneuron degeneration in amyotrophic lateral sclerosis. elife 3. https://doi.org/10.7554/eLife.04046

Leroy F, Lamotte d'Incamps B, Zytnicki D (2015) Potassium currents dynamically set the recruitment and firing properties of F-type motoneurons in neonatal mice. J Neurophysiol 114:1963–1973. https://doi.org/10.1152/JN.00193.2015

Łochyński D, Kaczmarek D, Mrówczyński W et al (2016) Contractile properties of motor units and expression of myosin heavy chain isoforms in rat fast-type muscle after volitional weight-lifting training. J Appl Physiol 121:858–869. https://doi.org/10.1152/japplphysiol.00330.2016

Lømo T, Wærhaug O (1985) Motor endplates in fast and slow muscles of the rat: what determines their difference? J Physiol Paris 80:290–297

Lømo T, Eken T, Bekkestad Rein E, Njå A (2020) Body temperature control in rats by muscle tone during rest or sleep. Acta Physiol 228:e13348. https://doi.org/10.1111/apha.13348

Maas H (2019) Significance of epimuscular myofascial force transmission under passive muscle conditions. J Appl Physiol Bethesda Md 1985. https://pubmed.ncbi.nlm.nih.gov/30605398/?from_term=Maas%2C+Huub&from_sort=date & from_pos=8. Accessed 16 Apr 2020

Macefield VG, Fuglevand AJ, Bigland-Ritchie B (1996) Contractile properties of single motor units in human toe extensors assessed by intraneural motor axon stimulation. J Neurophysiol 75:2509–2519. https://doi.org/10.1152/jn.1996.75.6.2509

Mantilla CB, Rowley KL, Zhan W-Z et al (2007) Synaptic vesicle pools at diaphragm neuromuscular junctions vary with motoneuron soma, not axon terminal, inactivity. Neuroscience 146:178–189. https://doi.org/10.1016/j.neuroscience.2007.01.048

Manuel M, Heckman CJ (2011) Adult mouse motor units develop almost all of their force in the subprimary range: a new all-or-none strategy for force recruitment? J Neurosci 31:15188–15194. https://doi.org/10.1523/JNEUROSCI.2893-11.2011

Manuel M, Zytnicki D (2011) Alpha, beta and gamma motoneurons: functional diversity in the motor system's final pathway. J Integr Neurosci 10:243–276. https://doi.org/10.1142/S0219635211002786

Manuel M, Meunier C, Donnet M, Zytnicki D (2007) Resonant or not, two amplification modes of proprioceptive inputs by persistent inward currents in spinal motoneurons. J Neurosci 27:12977–12988. https://doi.org/10.1523/JNEUROSCI.3299-07.2007

Manuel M, Chardon M, Tysseling V, Heckman CJ (2019) Scaling of motor output, from mouse to humans. Physiology 34:5–13. https://doi.org/10.1152/PHYSIOL.00021.2018

Martínez-Silva M d L, Imhoff-Manuel RD, Sharma A et al (2018) Hypoexcitability precedes denervation in the large fast-contracting motor units in two unrelated mouse models of ALS. elife 7:e30955. https://doi.org/10.7554/ELIFE.30955

Matthews PBC (1972) Mammalian muscle receptors and their central actions. Edward Arnold, London

McNulty PA, Falland KJ, Macefield VG (2000) Comparison of contractile properties of single motor units in human intrinsic and extrinsic finger muscles. J Physiol 526(Pt 2):445–456. https://doi.org/10.1111/j.1469-7793.2000.t01-2-00445.x

Meehan CF, Sukiasyan N, Zhang M et al (2010) Intrinsic properties of mouse lumbar motoneurons revealed by intracellular recording in vivo. J Neurophysiol 103:2599–2610. https://doi.org/10.1152/jn.00668.2009

Milburn A (1984) Stages in the development of cat muscle spindles. J Embryol Exp Morphol 82:177–216

Miles GB, Dai Y, Brownstone RM (2005) Mechanisms underlying the early phase of spike frequency adaptation in mouse spinal motoneurones. J Physiol 566:519–532. https://doi.org/10.1113/JPHYSIOL.2005.086033

Needham DM (1926) Red and white muscle. Physiol Rev 6:1–27. https://doi.org/10.1152/physrev.1926.6.1.1

Ogata T (1988) Structure of motor endplates in the different fiber types of vertebrate skeletal muscles. Arch Histol Cytol 51:385–424. https://doi.org/10.1679/aohc.51.385

Padykula HA, Gauthier GF (1970) The ultrastructure of the neuromuscular junctions of mammalian red, white, and intermediate skeletal muscle fibers. J Cell Biol 46:27–41. https://doi.org/10.1083/jcb.46.1.27

Pambo-Pambo A, Durand J, Gueritaud J-P (2009) Early excitability changes in lumbar motoneurons of transgenic SOD1G85R and SOD1G(93A-Low) mice. J Neurophysiol 102:3627–3642. https://doi.org/10.1152/jn.00482.2009

Pamphlett R (1989) Early terminal and nodal sprouting of motor axons after botulinum toxin. J Neurol Sci 92:181–192. https://doi.org/10.1016/0022-510x(89)90135-4

Rafuse VF, Gordon T (1996) Self-reinnervated cat medial gastrocnemius muscles. I. Comparisons of the capacity for regenerating nerves to form enlarged motor units after extensive peripheral nerve injuries. J Neurophysiol 75:268–281. https://doi.org/10.1152/jn.1996.75.1.268

Rafuse VF, Pattullo MC, Gordon T (1997) Innervation ratio and motor unit force in large muscles: a study of chronically stimulated cat medial gastrocnemius. J Physiol 499:809–823. https://doi.org/10.1113/jphysiol.1997.sp021970

Ranvier LA (1873) Propriétés et structures différentes des muscles rouges et des muscles blancs-chez les lapins et chez les raies. CR Acad Sci Paris 77:1030–1034

Reid B, Slater CR, Bewick GS (1999) Synaptic vesicle dynamics in rat fast and slow motor nerve terminals. J Neurosci 19:2511–2521. https://doi.org/10.1523/JNEUROSCI.19-07-02511.1999

Roman W, Gomes ER (2018) Nuclear positioning in skeletal muscle. Semin Cell Dev Biol 82:51–56. https://doi.org/10.1016/j.semcdb.2017.11.005
Russier M, Carlier E, Ankri N et al (2003) A-, T-, and H-type currents shape intrinsic firing of developing rat abducens motoneurons. J Physiol 549:21–36
Sah P, McLachlan EM (1992) Potassium currents contributing to action potential repolarization and the afterhyperpolarization in rat vagal motoneurons. J Neurophysiol 68:1834–1841. https://doi.org/10.1152/JN.1992.68.5.1834
Sawano S, Komiya Y, Ichitsubo R et al (2016) A one-step immunostaining method to visualize rodent muscle fiber type within a single specimen. PLoS One 11:e0166080. https://doi.org/10.1371/journal.pone.0166080
Sawczuk A, Powers RK, Binder MD (1997) Contribution of outward currents to spike-frequency adaptation in hypoglossal motoneurons of the rat. J Neurophysiol 78:2246–2253. https://doi.org/10.1152/JN.00831.2004
Schiaffino S, Reggiani C (2011) Fiber types in mammalian skeletal muscles. Physiol Rev 91:1447–1531. https://doi.org/10.1152/physrev.00031.2010
Schiaffino S, Hanzlíková V, Pierobon S (1970) Relations between structure and function in rat skeletal muscle fibers. J Cell Biol 47:107–119. https://doi.org/10.1083/jcb.47.1.107
Schiaffino S, Gorza L, Sartore S et al (1989) Three myosin heavy chain isoforms in type 2 skeletal muscle fibres. J Muscle Res Cell Motil 10:197–205. https://doi.org/10.1007/bf01739810
Seene T, Umnova M, Kaasik P (2017) Morphological peculiarities of neuromuscular junctions among different fiber types: effect of exercise. Eur J Transl Myol 27. https://doi.org/10.4081/ejtm.2017.6708
Smerdu V, Karsch-Mizrachi I, Campione M et al (1994) Type IIx myosin heavy chain transcripts are expressed in type IIb fibers of human skeletal muscle. Am J Phys 267:C1723–C1728. https://doi.org/10.1152/ajpcell.1994.267.6.C1723
Staron RS, Kraemer WJ, Hikida RS et al (1999) Fiber type composition of four hindlimb muscles of adult Fisher 344 rats. Histochem Cell Biol 111:117–123. https://doi.org/10.1007/s004180050341
Stephens JA, Stuart DG (1975) The motor units of cat medial gastrocnemius. Twitch potentiation and twitch-tetanus ratio. Pflugers Arch 356:359–372. https://doi.org/10.1007/BF00580008
Tam SL, Gordon T (2003) Neuromuscular activity impairs axonal sprouting in partially denervated muscles by inhibiting bridge formation of perisynaptic Schwann cells. J Neurobiol 57:221–234. https://doi.org/10.1002/neu.10276
Tam SL, Gordon T (2009) Axonal sprouting in health and disease. In: Binder MD, Hirokawa N, Windhorst U (eds) Encyclopedia of neuroscience. Springer, Berlin/Heidelberg, pp 322–328
Thornell L-E, Carlsson L, Eriksson P-O et al (2015) Fibre typing of intrafusal fibres. J Anat 227:136–156. https://doi.org/10.1111/joa.12338
Tremblay E, Martineau É, Robitaille R (2017) Opposite synaptic alterations at the neuromuscular junction in an ALS mouse model: when motor units matter. J Neurosci 37:8901–8918. https://doi.org/10.1523/JNEUROSCI.3090-16.2017
Turkin VV, O'Neill D, Jung R et al (2010) Characteristics and organization of discharge properties in rat hindlimb motoneurons. J Neurophysiol 104:1549–1565. https://doi.org/10.1152/JN.00379.2010
Wærhaug O, Lømo T (1994) Factors causing different properties at neuromuscular junctions in fast and slow rat skeletal muscles. Anat Embryol (Berl) 190:113–125. https://doi.org/10.1007/BF00193409
Ward SR, Winters TM, O'Connor SM, Lieber RL (2020) Non-linear scaling of passive mechanical properties in fibers, bundles, fascicles and whole rabbit muscles. Front Physiol 11. https://doi.org/10.3389/fphys.2020.00211
Zengel JE, Reid SA, Sypert GW, Munson JB (1985) Membrane electrical properties and prediction of motor-unit type of medial gastrocnemius motoneurons in the cat. J Neurophysiol 53:1323–1344. https://doi.org/10.1152/jn.1985.53.5.1323

Synaptic Projections of Motoneurons Within the Spinal Cord

Marco Beato and Gary Bhumbra

Abstract Motoneurons have long been considered as the final common pathway of the nervous system, transmitting the neural impulses that are transduced into action.

While many studies have focussed on the inputs that motoneurons receive from local circuits within the spinal cord and from other parts of the CNS, relatively few have investigated the targets of local axonal projections from motoneurons themselves, with the notable exception of those contacting Renshaw cells or other motoneurons.

Recent research has not only characterised the detailed features of the excitatory connections between motoneurons and Renshaw cells but has also established that Renshaw cells are not the only target of motoneurons axons within the spinal cord. Motoneurons also form synaptic contacts with other motoneurons as well as with a subset of ventrally located V3 interneurons. These findings indicate that motoneurons cannot be simply viewed as the last relay station delivering the command drive to muscles, but perform an active role in the generation and modulation of motor patterns.

Keywords Synaptic transmission · Co-release · Glutamate · Acetylcholine · Synaptic connectivity

There are notable exceptions to Sherrington's statement of motoneurons as the final common pathway (Sherrington 1906). For instance, the autonomic, enteric, and neuroendocrine systems each constitute components of the nervous system whose output pathways evade somatic motoneurons. The magnocellular neurosecretory cells of the ventral hypothalamus for example project infundibular axons to the posterior pituitary from where they secrete peptide hormones directly into the bloodstream. While the associated neuronal circuitry is not as well characterised as that of the somatic motor system where the first recurrent circuit was discovered (Renshaw 1946), it has long been known that magnocellular neurosecretory cells

M. Beato (✉) · G. Bhumbra
Department of Neuroscience, Physiology and Pharmacology, University College London, London, UK
e-mail: m.beato@ucl.ac.uk

M. J. O'Donovan, M. Falgairolle (eds.), *Vertebrate Motoneurons*, Advances in Neurobiology 28, https://doi.org/10.1007/978-3-031-07167-6_7

also send recurrent fibres, which project back to the ventral hypothalamus (Dreifuss and Kelly 1972).

Motifs of recurrent connectivity may thus be a feature common to the different efferent components of the nervous system. It has been proposed that the intercommunication between magnocellular neurosecretory cells performs an important role in modulating hypothalamic outputs for physiological neurohormone release (Leng and Dyball 1983). By analogy, it is highly likely that modulation of motoneuron activity by recurrent circuitry is essential for shaping their outputs for normal motor function (Hultborn et al. 1979).

1 Projections of Motoneurons to Renshaw Cells

Renshaw cells were first identified in the cat and distinctively respond with prolonged bursts of spikes in response to antidromic ventral root or nerve stimulation (Eccles et al. 1954). While there have been numerous experimental recordings from motoneurons, intracellular access to the Renshaw cells have proved more difficult mostly because of their smaller size. However, field recordings have confirmed the presence of prolonged bursts of activation following motoneuron firing (Ryall and Piercey 1971). Extracellular recordings in cat preparations have provided evidence of strong excitatory motoneuron synapses onto Renshaw cells. Continuous firing in a motoneuron can increase spiking in Renshaw cells (Ross et al. 1975) and drive discharge rates up to ≥40–60 Hz (Ross et al. 1976). Ventral root stimulation in decerebrated cats and rabbit preparations (Renshaw 1946) can elicit bursts of discharges from putative Renshaw cells and stimulation of a single motoneuron alone can be sufficient to evoke multiple spikes (Van Keulen 1981). On the basis of these observations, the motoneuron to Renshaw cell synapse has always been postulated to be a strong synapse with Renshaw cells reliably propagating the firing in motoneurons. The motoneuron to Renshaw cell synapse might thus be considered as a 'relay synapse', giving a faithful image of motoneuron firing.

The strength of this synapse has been confirmed using paired recordings (Moore et al. 2015 Fig. 1a) performed in neonatal (P8–14) mice spinal cord slices. Quantal analysis of these recordings revealed that each motoneuron forms 5–10 contacts with its post-synaptic Renshaw cell, with high vesicular release probabilities (> ~ 0.5) following a single nerve impulse. In many of such recordings, firing in a single motoneuron could evoke a spike in the Renshaw cell.

In addition to the unitary conductance, the degree of convergence from within a motoneuron pool onto a single Renshaw cell will determine the size of excitation.

Immunohistochemical studies have have showed that in rats each Renshaw cell receives 20–140 synaptic contacts from VAChT immunoreactive terminals originating from motoneurons (Alvarez et al. 1999). This would correspond to tens of motoneurons converging onto a single Renshaw cell. A functional gauge of convergence was estimated using electrophysiological recordings and quantal analysis of ventral root stimulation induced responses in Renshaw cells (Moore et al. 2015), that

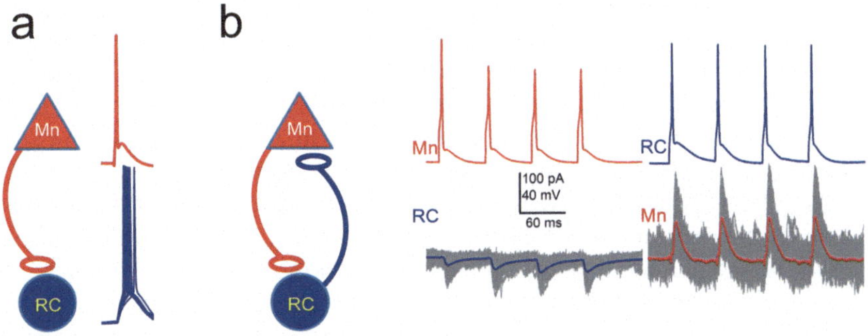

Fig. 1 Synapses between motoneurons and Renshaw cells: In a paired Motoneuron-Renshaw cell recording, a spike evoked in the motoneuron always induces firing in the connected Renshaw cell (panel **a**). Reciprocal connections between individual motoneurons and Renshaw cells (panel **b**) occurred in ~30% of paired recordings (adapted from Moore et al. 2015). Excitatory and inhibitory currents are consistently evoked by a train of spikes induced in the motoneuron or in the Renshaw cell

indicated that 5–10 motoneurons contact a single Renshaw cell. While the results may appear somewhat discrepant, the electrophysiological studies were performed in a different species (mouse) and, more importantly, in a reduced 'oblique slice' spinal cord preparation with the ventral root attached, but with a thickness less than that of a single segment. Since motoneurons axon collaterals can extend for more than one segment, contacting more distant Renshaw cells, estimates obtained from electrophysiological recordings from slices should be considered as a lower bound for the degree of convergence. A similar caveat is applicable to the estimate of the number of reciprocal connections between single motoneurons and Renshaw cells (Fig. 1b): Moore et al. (2015) found that within the spinal slice, ~1/3 of motoneuron to Renshaw cell synapses were reciprocal. Given the inevitable severance of connections in the slice preparation as highlighted above, this proportion is likely to be much larger in an intact system.

In addition to the synaptic strength of individual connections between motoneuron and Renshaw cells, the degree of convergence and divergence between each cell type are a key determinants of their function. While we have estimates of the convergence, with respect to the number of Renshaw cells contacting each motoneuron and the number of motoneurons contacting each Renshaw cell (Moore et al. 2015), we do not know the level of divergence. Given the strong effect of a single Renshaw cell input on a motoneuron (Bhumbra et al. 2014), if the same Renshaw cell were to contact several motoneurons, it is likely that this configuration would increase synchronicity of motoneuron firing.

A consequence of the combination of a considerable degree of convergence and high reliability of the motoneuron to Renshaw cell synapse is that the Renshaw cell can follow the firing of motoneurons with high fidelity. Indeed, even with ventral root stimulation at 100 Hz, near the upper range of physiological firing frequencies in motoneurons, the Renshaw cells show just ~50% of synaptic depression and can follow reliably the firing in the motoneuron pool (Moore et al. 2015).

The precise role of Renshaw cells in the control of motor outputs is unknown. While they could just prevent excessive motoneuron firing, several other hypotheses were raised about their function (Windhorst 1996), including a role as a variable gain regulator (Hultborn and Pierrot-Deseilligny 1979) or in correlating or decorrelating (Maltenfort et al. 1998) motoneuron firing. In a recent report (Enjin et al. 2017), selective ablation of Renshaw cells input to motoneurons was achieved using a knockout strategy, taking advantage of the selective expression of the neuronal nicotinic receptor subunit α2 in Renshaw cells (Perry et al. 2015), Constitutive ablation of the vesicular inhibitory amino acid transporter (VIAAT) from Renshaw cells terminals however did not result in any apparent motor phenotype, but these results are potentially confounded by compensatory changes occurring within the motor circuits.

A more recent model (Brownstone et al. 2015) has suggested that the combination of Renshaw cell connectivity and components of spinal circuitry projecting to motoneurons themselves may contribute to sensorimotor learning. It is proposed that an 'internal model' of muscle action can learn as a result of differences between feed-forward predictive (recurrent) information and feedback reactive (proprioceptive) information. The advantage of tuning the internal model to feedforward information is that by nature it is more immediately available compared to the feedback information that inherently can only be transmitted later in time (Brownstone et al. 2015).

Co-transmission to Renshaw Cells

When Dale's principle was initially proposed, it stated only that the chemical function of a neuron was the same at all its terminals. It is important to note that only two putative transmitters were known at the time: acetylcholine and noradrenaline (at the time thought to be adrenaline, Dale 1935). Consequently, Dale could not have envisaged co-release of both transmitters. This statement was subsequently re-appropriated and misinterpreted in a form that states that all synaptic terminals of the same neuron release the same transmitter onto all its different targets, perhaps in part due to Eccles' early formulation (Eccles et al. 1954) that used the word 'transmitter' in the singular. The large number of cases in which this statement is clearly inaccurate led many to describe the discovery of co-release (of different inhibitory amino-acids, or neurotransmitter and peptides or gases) as a violation of Dale's principle. However, we owe it to Eccles to clarify his more faithful interpretation of Dale's principle in which he reformulated the originally loose statement by Dale in more precise terms: 'Dale's Principle be defined as stating that at all the axonal branches of a neuron, there was liberation of the same transmitter substance or substances' (Eccles et al. 1976). With this definition, it is clear that the many instances in which co-release has been demonstrated do not constitute any violation of Dale's principle.

It is however important to emphasise that where there is co-release (meaning the release of vesicles containing two or more transmitters), it does not necessarily follow that transmission at that synapse is mediated by both transmitters. While both transmitters could be released, their corresponding detections depend on the repertoire of post-synaptic receptors (Chery and De Koninck 1999; Lu et al. 2008). Eccles' observation that ACh blockers do not completely abolish the Renshaw cell discharge elicited by antidromic nerve stimulation is consistent with a possible second transmitter released from motoneurons. At the neuromuscular junction, some studies reported the presence of glutamate *(W*ærhaug and Ottersen 1993; Meister et al. 1993) while others failed to detect it (Herzog et al. 2004; Kraus et al. 2004). Similarly conflicting results were reported from studies of immunoreactivity for vesicular glutamate transporters in the motoneurons terminals opposed to Renshaw cells, with some authors detecting the expression of Vglut2 (Oliveira et al. 2003; Herzog et al. 2004; Nishimaru et al. 2005) or Vglut RNA (Schäfer et al. 2002) and others failing to detect its presence (Mentis et al. 2005).

The issue of a second transmitter was finally resolved using direct electrophysiological recordings of the excitatory response evoked by ventral root stimulations in the neonatal spinal cord. Two groups independently observed a non cholinergic component in the ventral root evoked EPSC recorded in Renshaw cells (Mentis et al. 2005; Nishimaru et al. 2005) that could be blocked by glutamatergic antagonists. These recordings confirmed that motoneurons could release an excitatory amino acid from their central synapses. Without unequivocal identification of mechanisms for excitatory amino acid loading at the motoneuron terminals, glutamate remains only a putative transmitter at this synapse. Since motoneuron terminals opposed to Renshaw cells are enriched with aspartate (Richards et al. 2014), glutamate might not be the only feasible candidate.

The situation is potentially more complex on the post-synaptic side, where at least 4 different subtypes of receptors contribute to the post-synaptic response (Lamotte d'Incamps and Ascher 2008). A fast α7 mediated response is followed by a slower nicotinic response mediated by heteromeric receptors of unknown subunit composition, although it has been suggested that two classes of receptors with different αβ stoichiometry might be involved (Lamotte d'Incamps and Ascher 2014). Furthermore, both AMPA and NMDA receptors are activated post-synaptically. The combined cholinergic and glutamatergic contributions to the connections to Renshaw cells results in a synapse with two transmitters and at least four different types of post-synaptic receptors. Bursts of spikes observed in response to ventral root stimulation may result from a 'priming' depolarisation, mediated by AMPA and nicotinic receptors, relieving magnesium blockade and prolonging activation of NMDA receptors (Lamotte d'Incamps and Ascher 2008).

Definitive confirmation of the synaptic activation of AMPA receptors is somewhat inconsistent with aspartate as the second transmitter, since aspartate, while being an agonist at NMDA receptors, does not activate AMPA receptors (Patneau and Mayer 1990), even though some reports suggest that AMPA receptors can be weekly activated by direct application of aspartate, at least in a subset of dopaminergic neurons (Krashia et al. 2016). Motoneurons thus appear to communicate using only ACh at the neuromuscular junction, and using both ACh and glutamate (or similar excitatory amino-acid) at the motoneuron to Renshaw cell synapse.

Contrary to GABA and glycine co-transmission in the spinal cord (Jonas et al. 1998; Singer et al. 1998), that is largely confined to the early developmental stage (Bhumbra et al. 2012; Jiang and Alstermark 2015) co-transmission of ACh and glutamate is preserved at a mature age (Lamotte d'Incamps et al. 2017). Most of the electrophysiological evidence for co-transmission was obtained from recordings of Renshaw cell responses to ventral root stimulation, that activates a large number of motoneurons. Compound responses in single Renshaw cells could thus result from mixed transmission whereby some motoneurons released ACh only and others released a glutamate like substance. Paired recordings from connected motoneurons and Renshaw cells (Lamotte d'Incamps et al. 2017) have excluded this possibility by unmasking of a non-nicotinic component following blockade of ACh receptors. Since responses to single motoneurons were recorded, the residual non-cholinergic component confirms the occurrence of both modes of transmission at the level of single motoneurons.

Kinetic analysis of spontaneously occurring miniature synaptic currents and of asynchronous mEPSC originating from the motoneurons revealed a considerable degree of segregation between the two transmitters systems: namely, miniature events were either mediated by the two nicotinic receptors or by the two glutamate receptors, but no mixed cholinergic and glutamatergic mEPSC were observed (Lamotte d'Incamps et al. 2017), suggesting that transmission occurs with only one transmitter at each individual terminal.

It is possible that all four receptors are present at the post-synaptic Renshaw cell membrane, but each terminal of any given motoneuron can release vesicles containing either ACh or glutamate, but not both (Fig. 2c). Alternatively, single synaptic vesicles can contain both transmitters, but the post-synaptic receptors expressed Renshaw cells are either GluRs or AChRs, thus precluding mixed transmission at individual contacts (Fig. 2d). The function of any such arrangements is unclear, since at the level of individual synapses between motoneurons and Renshaw cells, that are made up of several functional contacts (Moore et al. 2015), transmission is always mixed. In order to distinguish between the pre-synaptic and post-synaptic configurations for segregation, it will be necessary to perform experiments aimed at evoking release from single motoneuron terminals onto Renshaw cells. Such experiments might employ spatially restricted light activation of excitatory opsins or of caged calcium compounds loaded into the pre-synaptic cell.

2 Recurrent Excitation Between Motoneurons

Inter-communication between motoneurons in vertebrates has often been attributed to gap-junctions (Fulton et al. 1980). While gap junctions are expressed during early mammalian development (Fulton et al. 1980; Hinckley and Ziskind-Conhaim 2006), their density tend to decrease in more mature animals (Walton and Navarrete

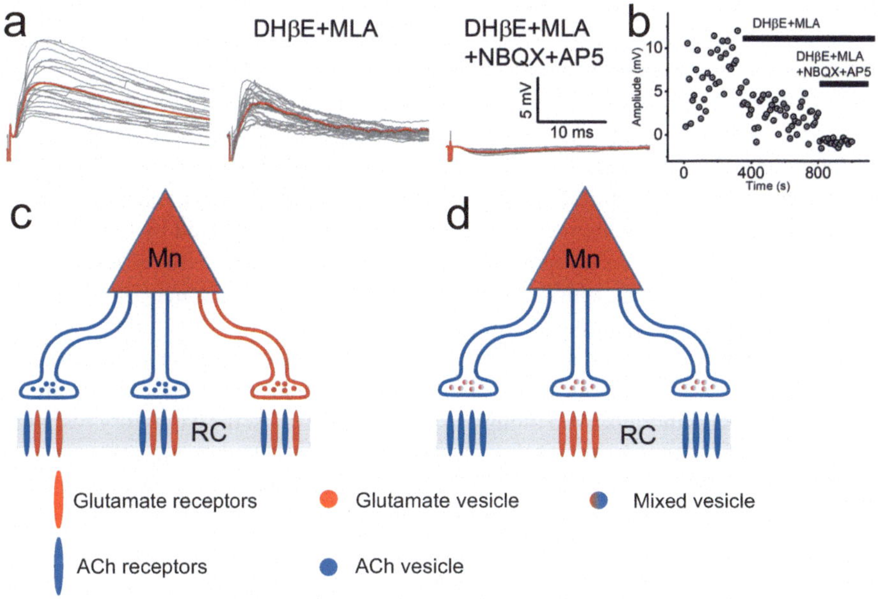

Fig. 2 Modes of transmissions between motoneurons and Renshaw cells. An example of a ventral root evoked response in a Renshaw cell from a mature (P18) spinal cord slice, showing the progressive decrease in the evoked potential following application of cholinergic (middle panel) and glutamatergic (right panel) antagonists. The full time course of the experiment is shown in panel **b**. Segregation of glutamate and acetylcholine neurotransmission at the motoneuron to Renshaw cell synapse can occur either at the pre-synaptic level (panel **c**), with individual terminals from a single motoneuron containing only one of the two transmitters, or at the post-synaptic level (panel **d**), if all pre-synaptic terminals contain mixed content vesicles, but the post-synaptic membranes opposed to each terminal contain either glutamate or nicotinic receptors

1991; Chang et al. 1999; Personius et al. 2007), even though the presence of mixed electrical and chemical synapses in the spinal cord has been detected also in adult rats motoneurons (Rash et al. 1996).

Most motor pools, with the exception of those innervating the more distal paw muscles (Cullheim and Kellerth 1978; McCurdy and Hamm 1992) show extensive axonal branching. Anatomical studies in the cat lumbar spinal cord have revealed that motoneurons axon collaterals also invade the motor nuclei, potentially forming synapses onto other motoneurons (Cullheim et al. 1977; Cullheim et al. 1987). The first functional synapses between motoneurons were observed in frog embryos (Perrins and Roberts 1995). Such synapses are often mixed exhibiting electrical and chemical components, with the latter mediated by ACh and transmitted across spinal segments. However, there are no reports of similar connectivity in adult frogs. In mammals, reports of synaptic connectivity between motoneurons were scarce and somewhat contradictory. In neonatal rats, ventral root simulation not only evoked the expected di-synaptic inhibition of motoneurons mediated via Renshaw cells, but also a smaller response, sensitive to glutamate receptors antagonists (Schneider and

Fyffe 1992) and whose reversal potential was compatible with a mixed cation mediated current (Jiang et al. 1991). The presence of this excitatory input was attributed to the invasion of primary afferent fibers through the ventral roots. This view was supported by some anatomical evidence (Coggeshall 1980) and by the capacity of glutamate receptors antagonists to block the excitation (Jiang et al. 1991). More recent studies (Mentis et al. 2005) however have excluded the presence of primary afferent in ventral roots, indicating that the source of glutamatergic excitation could be from the motoneurons themselves. This suggestion is supported by the observed close juxtaposition between motoneurons terminals loaded with different dyes through adjacent ventral roots (Mentis et al. 2005) and evidence of glutamate enrichment in a number of these terminals.

Ventral root evoked excitatory responses in motoneurons were also described in neonatal mice, P0–4 (Nishimaru et al. 2005). Such responses were mixed glutamatergic-cholinergic, even though in 7/9 recorded cells, the excitatory response was dominated (>80%) by the glutamatergic component. On the contrary, recurrent excitation measured in neonatal rats (Ichinose and Miyata 1998) was abolished in 3/5 cases by cholinergic antagonist, but glutamate antagonists were not tested in the remaining cases.

A more recent systematic study of recurrent excitation between motoneurons (Bhumbra and Beato 2018) confirmed the presence of recurrent excitatory connectivity between motoneurons. This was shown not only by ventral root stimulation, but also from paired recording between synaptically coupled motoneurons. Both paired recordings and ventral root stimulation established that excitation between motoneurons is entirely mediated by glutamate and that excitatory inputs from motoneurons can propagate across neighbouring segments. Furthermore, while previous studies were limited to neonatal mouse preparations (up to 10 days old), (Bhumbra and Beato 2018) showed that recurrent excitation is not a transient phenomenon in early development, but it is present also in older animals up to P20, when motor systems are considered mature and the animal is capable of executing most motor tasks normally. In summary, the presence of recurrent excitation between motoneurons has been consistently observed in different species and ages, but while one recent study suggested that transmission is purely glutamatergic (Bhumbra and Beato 2018), previous works (Ichinose and Miyata 1998; Nishimaru et al. 2005) point at mixed glutamatergic-cholinergic trnasmission, a view that is also supported by the presence of cholinergic boutons originating from motoneurons onto other motoneurons (Mentis et al. 2005).

One remarkable feature of recurrent excitation is that putative fast motoneurons, identified by their firing properties (Leroy et al. 2014), receive almost 10 times more excitation than putative slow motoneurons, characterized by their early firing behaviour (Fig. 3a–d). However, it is not known whether the pre-synaptic origins of excitation of fast motoneurons are predominantly slow or fast units (Fig. 3e, f). Hennemann's size principle (Henneman 1957) states that motor units are recruited in a specific order, with slow units being recruited first, and progressively more involvement of fast units as the required force increases. If the preferential pattern of connectivity favoured connections between slow to fast units (Fig. 3e), recurrent excitation could be a further mechanism for the implementation of the size principle and could mediate the progressive recruitment of faster fibers in synergy with progressively increasing input strength received by different motor units. Preferential connectivity from pre-synaptic slow motoneurons to post-synaptic fast motoneurons would favour the recruitment of fast units in a graded way, facilitating a coordinated increase in the strength of muscle contractions.

If on the other hand, the preferential pattern of connectivity only involves fast units (Fig. 3f), recurrent excitation would lead to a closed loop amplification that could increase firing rates, and potentially coherence between fast units after a sufficient proportion are activated. In order to distinguish between these two possibilities, a complete electrophysiological characterization of pre and post-synaptic partners is necessary, and the interpretation of results should be supported by detection of genetic (Muller et al. 2014) or immunohistochemical (Enjin et al. 2010; Leroy et al. 2014) markers of fast and slow motoneurons.

While synaptic connectivity between individual motoneurons has been demonstrated through paired recordings between motoneurons belonging to the same motor nucleus (lateral gastrocnemius), it remains to be ascertained whether this connectivity extends also across different nuclei. Experiments using Ca^{2+} imaging

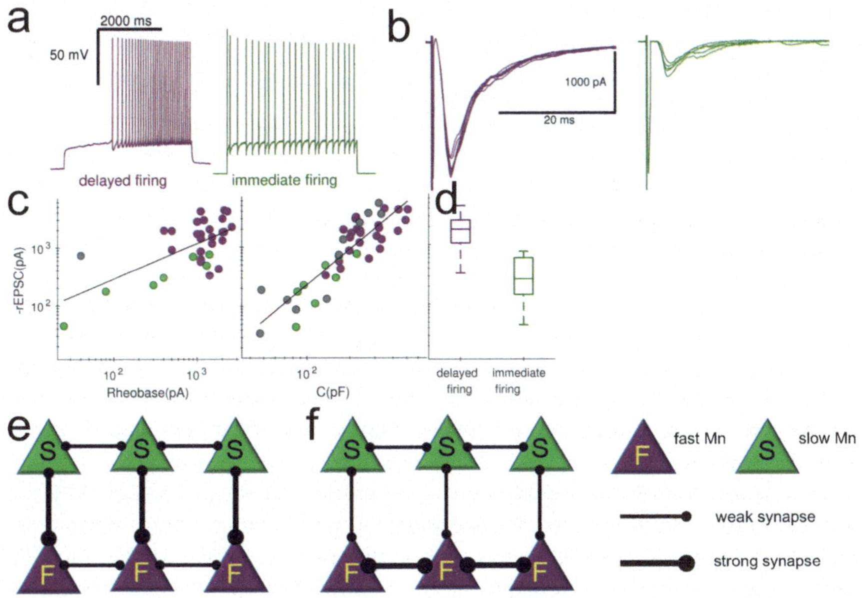

Fig. 3 Motoneurons can be distinguished based on their firing properties, with delayed firing and immediate firing motoneurons associated with fast and slow units respectively (panel **a**). Recurrent excitation is larger in delayed firing motoneurons (panel **b** and **d**) and its size positively correlates with rheobase and cell capacitance (panel **c**, adapted from (Bhumbra and Beato 2018). Larger recurrent excitation in fast units could be due to stronger connectivity of slow to fast motoneurons (panel **e**) or to stronger connectivity between fast motoneurons themselves (panel **f**)

across two segments showed that the wave of excitation evoked by ventral root stimulation could be observed throughout the scanned segments in virtually all motoneurons, that presumably belonged to different motor nuclei. However, motoneurons also make synaptic contact with V3 interneurons, and the poor time resolution of Ca^{2+} imaging does not allow to distinguish between direct motoneuron to motoneuron excitation and excitation mediated by a potential excitatory disynaptic loop through V3 interneurons (Chopek et al. 2018). Therefore, to date, there is no conclusive evidence that motoneuron connectivity extends across different nuclei.

While the presence of connectivity among synergist muscles would again be consistent with a role of recurrent excitation in amplifying the muscle contractions by providing positive interactions between muscles that are normally co-activated during movements, the presence of connectivity extending across antagonist units is more difficult to interpret. Since recurrent excitation might be effective only when a critical number of motor units are activated, it is possible that excitation between antagonist motor units might play a role in the preparation and execution of ballistic movements, such as throwing, jumping, or lifting heavy weights, wherein an explosive movement is characteristically preceded by isometric co-contraction of antagonist muscles.

Investigation of these possibilities will require exploration in the future, possibly using electrophysiological recordings from labelled motor units from different muscles or using trans-synaptic tracing methods (Stepien et al. 2010; Tripodi et al. 2011) or selective expression of excitatory opsins in different motor nuclei.

Even once the connectivity pattern is unravelled, establishing the exact role of recurrent excitation will not be straightforward.

3 Other Synaptic Targets of Motoneurons

It is generally accepted that in many invertebrate species motoneurons actively contribute to the generation and execution of motor patterns through either gap junctions or chemical synaptic connections, in vertebrate species the only identified post-synaptic targets were Renshaw cells and other motoneurons. The recurrent inhibitory and excitatory loops, could down- or up-scale firing in motoneurons, thus altering the strength of muscle contraction. A seminal paper however showed that antidromic activation of motoneurons through ventral root stimulation could elicit episodes of fictive locomotion in an intact neonatal spinal cord in vitro (Mentis et al. 2005). The consequence of this observation is that motoneurons terminal must be capable of activating some elements of the central pattern generator, resulting in locomotor like pattern throughout the spinal cord at least at a developmental stage. Further confirmation came from studies in neonatal rats (Machacek and Hochman 2006, Fig. 4b), showing that ventral root stimulation can entrain rhythmic bursting activity following blockade of synaptic inhibition. These experiments also show that noradrenaline unmasks a connectivity pattern between motoneurons and an unidentified class of interneurons. Similarly, during drug-activated fictive locomotion, light induced activation of motoneurons, following selective expression of the excitatory opsin channelrhodopsin, increases the bursting frequency of the locomotor pattern (Falgairolle et al. 2017).

The initiation of locomotion, entrainment of disinhibited bursting activity, and up-regulation of the frequency of the locomotor patterns are incompatible with an effect mediated by recurrent excitation alone or by activation of Renshaw cells, due to their inhibitory nature, even accounting for projections from Renshaw cells to Ia inhibitory interneurons and to ventral spinocerebellar neurons (Jankowska and Hammar 2013). Furthermore, even if Renshaw cells had synaptic projections to some elements of the CPG, their effect would most likely be inhibitory, especially since, already at the neonatal stage, the chloride reversal potential is more negative than the membrane potential (Delpy et al. 2008).

Evidence of motoneuron-induced effects on the CPG suggests that motoneuron axon collaterals directly contact an unspecified group of interneurons associated with the CPG. It is possible that modulation of motoneuron activity on fictive locomotion result from excitation of large motor pools, through antidromic or light stimulation, activating the CPG not through direct synaptic contacts. Instead, the modulatory effects may be mediated through either ephaptic transmission (Jefferys 1995), or through local increases in extracellular potassium, that is known to contribute to the initiation of locomotor bursts following dorsal root stimulation (Marchetti et al. 2001a, b). However, this explanation is inconsistent with results from optogenetic experiments (Falgairolle et al. 2017) that show that during drug-activated fictive locomotion, light-induced inhibition of motoneurons, selectively expressing the inhibitory opsin halorhodopsin, results in a remarkable slowing down of the rhythmic pattern. Furthermore, this effect is strongly attenuated following blockade of AMPA receptors, but unchanged by partial block of gap junctions. This elegant set of experiments indicates not only that motoneurons communicate with interneurons other than Renshaw cells and that such interneurons are associated with the CPG, but also that such communication is not mediated through gap junctions but through an excitatory amino acid that activates AMPA receptors, such as glutamate.

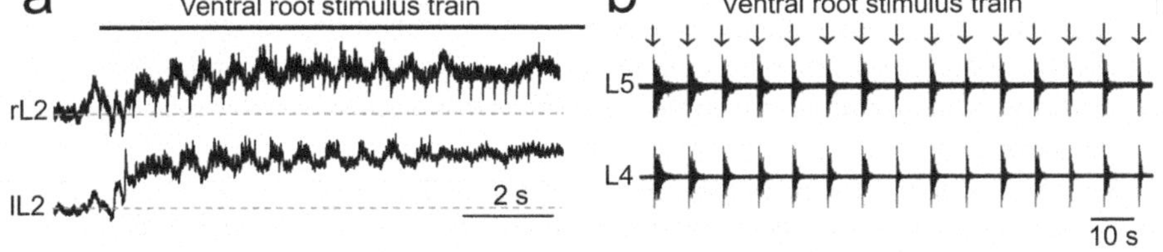

Fig. 4 Motoneurons influence the activity of central pattern generators: a train of ventral root stimulations can induce a locomotor like pattern recorded in the lumbar ventral roots (panel **a**, adapted from Bonnot et al. 2009). Similarly, single ventral root stimulations can entrain the spontaneous bursting pattern induced by block of inhibition (panel **b**, adapted from Machacek and Hochman 2006. Copyright 2006, Society for Neuroscience)

Direct evidence of communication between motoneurons and other CPG related interneurons came initially from zebrafish preparations (Song et al. 2016), where it was shown that activation of motoneurons perturbs the frequency of the swimming pattern. The study identified a novel motoneuron synaptic target in the population of V2a interneurons, with connections that are either electrical, chemical, or a combination.

Mammalian V2a interneurons however do not receive any input from motoneurons, whether chemical or electrical (Bhumbra and Beato 2018), raising the possibility that other classes of interneuron might be responsible for the motoneuron induced modulation of locomotor activity. Potential alternative candidates have been identified among V3 interneurons, that are known to send commissural axons and contact contralateral motoneurons (Zhang et al. 2008). It has been recently shown that within a spinal cord slice, V3 interneurons also send direct ipsilateral projection to motor pools (Chopek et al. 2018). Remarkably in the subset of the most ventrally located V3 interneurons, this pattern of connectivity appears to be reciprocal, with at least a proportion of interneurons receiving excitatory glutamatergic inputs from motoneurons (Chopek et al. 2018). This was the first direct evidence of connectivity between motoneurons and an identified class of excitatory interneurons within the mammalian spinal cord.

While identification of these excitatory projections does not fully explain the modulatory effects of motoneuron activity on the frequency of the locomotor rhythm, it at least introduces a candidate class of interneurons that could feasibly mediate this effect.

V3 interneurons contribute to the CPG for locomotion through contralateral connections (Danner et al. 2017) and their acute or chronic ablation (Zhang et al. 2008) causes changes the regularity, though do not abolish, the motor pattern.

New methods establishing the connectivity patterns of cells in the CNS may reveal further classes of neurons post-synaptic to motoneurons that could modulate the central pattern generators.

4 Conclusions

It seems that the era of perceiving motoneurons confined to a purely passive role integrating inputs from the brain and local spinal cord circuits is coming to an end as evidence emerges of motoneurons as active protagonists in the generation of motor patterns.

Recent evidence concerning the nature of the different post-synaptic targets of motoneurons have unveiled a peculiar diversity of neurotransmitter phenotypes. Activation at the neuromuscular junction is purely mediated by ACh, motoneurons excite Renshaw cells via a combination of ACh and glutamate, while transmission from motoneurons to themselves and to V3 interneurons is purely glutamatergic motoneurons thus communicate in different languages according to their specific post-synaptic target (Fig. 5). This arrangement is not unique in the central nervous

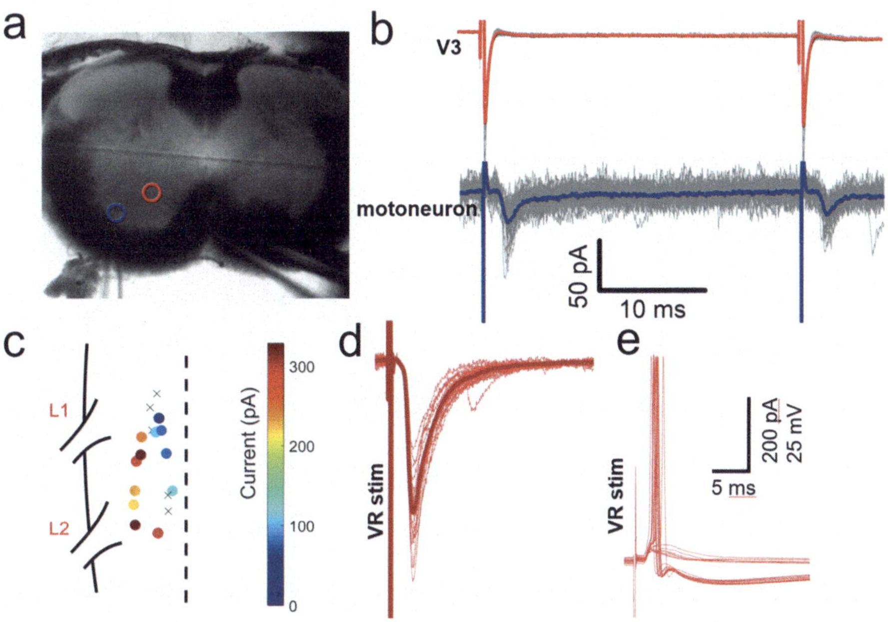

Fig. 5 Reciprocal connectivity between motoneurons and V3 interneurons. V3 interneurons are monosynaptically connected to ipsilateral motoneurons. An example of a paired recording is shown in panels **a** and **b**, with the location of the recorded cells indicated by an open circle. A spike evoked in the V3 interneuron in the loose cell-attached configuration (panel **b**, upper trace) evokes a post-synaptic response in the recorded motoneuron (panel **b**, lower trace). In a longitudinal spinal cord preparation with dorsal horn ablated (panel **c**), some V3 interneurons respond to ventral root stimulation (size of the postsynaptic response is colour coded in panel **c**, crosses correspond to not responding cells) with a large postsynaptic current (panel **d**) that can bring the V3 interneurons to threshold for the generation of an action potential (panel **e**). (Adapted from Chopek et al. 2018)

system, but as far as we know it is certainly rare. To the best of our knowledge, the only other case of such neurotransmitter dissociation is at the synapse between Golgi cells and unipolar brush cells and granule cells in the cerebellum, where the first transmitting with glycine and the latter with GABA (Dugue et al. 2005). In both this case and for motoneurons, it is difficult to envisage a rationale or a function for such differentiation in the transmitter used, especially since there seems to be no relation between the size of the post-synaptic target and the kinetics of the post-synaptic receptors involved. It is yet to be determined whether segregation of the different neurotransmitter systems occurs at the pre or post-synaptic sites. Attempts of answering this question are challenged by the disparate immunohistochemistry evidence at motoneurons terminals and by our ignorance of the exact mechanism of glutamate loading in the vesicles.

It is humbling to admit that we do not know the exact role of the recurrent inhibitory loop in the control and execution of motor tasks, despite the fact that it is one of the first characterized closed loop circuits in the CNS and 70 years of subsequent research. The recent discovery of two further recurrent loops, both excitatory, one

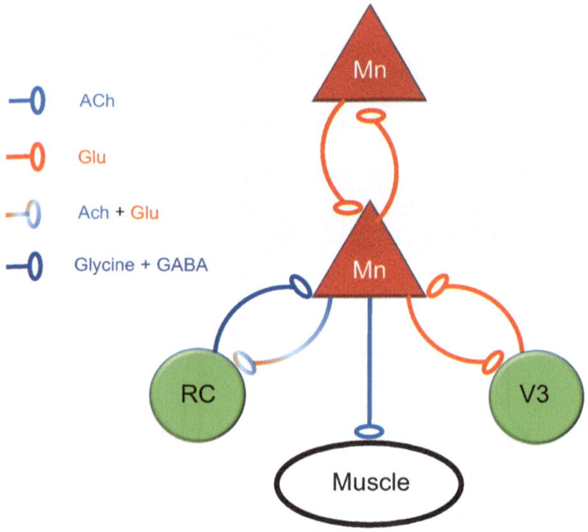

Fig. 6 Modes of synaptic communication from motoneurons: synaptic transmission is entirely cholinergic at the neuromuscular junction, glutamatergic at the motoneuron to motoneuron and motoneuron to V3 interneurons synapses, but mixed at the motoneuron to Renshaw cell synapse

between motoneurons themselves and one between motoneurons and V3 interneurons, raises the question of which one of these feedback loops is dominant during the execution of motor tasks. Addressing this question requires knowledge of the specific connectivity patterns between each element of the feedback loops (Fig. 6).

While it is accepted that recurrent inhibition is largely confined to homonymous or synergist motor nuclei (McCurdy and Hamm 1994), it is known that there is differential degree of convergence between fast and slow motoneurons: slow motoneurons receive more inhibition than fast ones (Hultborn et al. 1988a), and contact fewer Renshaw cells (Hultborn et al. 1988b). Similarly, within the recurrent excitatory loop, fast motoneurons receive 10 times greater recurrent excitation than slow ones, while the relative strengths of synaptic connections in the loop between motoneurons and V3 interneurons is unknown. It is tempting to speculate that the differential pattern of connectivity within each loop might relate to the differential order of recruitment of slow and fast motor units, conferring dominance to one circuit or the other depending on the task being performed.

Whatever the relationship between the recurrent loops, it is clear that motoneurons must assume a central active role in the control of motor tasks, well beyond their postulated role within the final common pathway as simple passive units relaying information from the CNS to the muscles.

References

Alvarez FJ, Dewey DE, McMillin P, Fyffe RE (1999) Distribution of cholinergic contacts on Renshaw cells in the rat spinal cord: a light microscopic study. J Physiol 515(Pt 3):787–797

Bhumbra GS, Beato M (2018) Recurrent excitation between motoneurones propagates across segments and is purely glutamatergic. PLoS Biol 16. https://doi.org/10.1371/journal.pbio.2003586

Bhumbra GS, Moore NJ, Moroni M, Beato M (2012) Co-release of GABA does not occur at glycinergic synapses onto lumbar motoneurons in juvenile mice. https://www.ncbi.nlm.nih.gov/pmc/articles/PMC3309924/. Accessed 25 Oct 2019

Bhumbra GS, Bannatyne BA, Watanabe M, Todd AJ, Maxwell DJ, Beato M (2014) The recurrent case for the Renshaw cell. J Neurosci 34:12919–12932

Bonnot A, Chub N, Pujala A, O'Donovan MJ (2009) Excitatory actions of ventral root stimulation during network activity generated by the disinhibited neonatal mouse spinal cord. J Neurophysiol 101:2995–3011. https://doi.org/10.1152/jn.90740.2008

Brownstone RM, Bui TV, Stifani N (2015) Spinal circuits for motor learning. Curr Opin Neurobiol 33:166–173. https://doi.org/10.1016/j.conb.2015.04.007

Chang Q, Gonzalez M, Pinter MJ, Balice-Gordon RJ (1999) Gap junctional coupling and patterns of connexin expression among neonatal rat lumbar spinal motor neurons. J Neurosci 19:10813–10828

Chery N, De Koninck Y (1999) Junctional versus extrajunctional glycine and GABA(A) receptor-mediated IPSCs in identified lamina I neurons of the adult rat spinal cord. J Neurosci 19:7342–7355

Chopek JW, Nascimento F, Beato M, Brownstone RM, Zhang Y (2018) Sub-populations of spinal V3 interneurons form focal modules of layered pre-motor microcircuits. Cell Rep 25. https://doi.org/10.1016/j.celrep.2018.08.095

Coggeshall RE (1980) Law of separation of function of the spinal roots. Physiol Rev 60:716–755

Cullheim S, Kellerth JO (1978) A morphological study of the axons and recurrent axon collaterals of cat alpha-motoneurones supplying different functional types of muscle unit. J Physiol 281:301–313. https://doi.org/10.1113/jphysiol.1978.sp012423

Cullheim S, Kellerth JO, Conradi S (1977) Evidence for direct synaptic interconnections between cat spinal alpha-motoneurons via the recurrent axon collaterals: a morphological study using intracellular injection of horseradish peroxidase. Brain Res 132:1–10

Cullheim S, Fleshman JW, Glenn LL, Burke RE (1987) Three-dimensional architecture of dendritic trees in type-identified alpha-motoneurons. J Comp Neurol 255:82–96. https://doi.org/10.1002/cne.902550107

Dale H (1935) Pharmacology and nerve-endings (Walter Ernest Dixon Memorial Lecture). Proc R Soc Med 28:319–332

Danner SM, Shevtsova NA, Alain F, Rybak IA (2017) Computational modeling of spinal circuits controlling limb coordination and gaits in quadrupeds. eLife 6. Cambridge. https://doi.org/10.7554/eLife.31050

Delpy A, Allain A-E, Meyrand P, Branchereau P (2008) NKCC1 cotransporter inactivation underlies embryonic development of chloride-mediated inhibition in mouse spinal motoneuron. J Physiol 586:1059–1075. https://doi.org/10.1113/jphysiol.2007.146993

Dreifuss JJ, Kelly JS (1972) Recurrent inhibition of antidromically identified rat supraoptic neurones. J Physiol 220:87–103. https://doi.org/10.1113/jphysiol.1972.sp009696

Dugue GP, Dumoulin A, Triller A, Dieudonne S (2005) Target-dependent use of co-released inhibitory transmitters at central synapses. J Neurosci 25:6490–6498

Eccles JC, Fatt P, Koketsu K (1954) Cholinergic and inhibitory synapses in a pathway from motor-axon collaterals to motoneurones. J Physiol Lond 126:524–562

Eccles JC, Jones RV, Paton WDM (1976) From electrical to chemical transmission in the central nervous system: the closing address of the Sir Henry Dale Centennial Symposium Cambridge, 19 September 1975. Notes Rec R Soc Lond 30:219–230. https://doi.org/10.1098/rsnr.1976.0015

Enjin A, Rabe N, Nakanishi ST, Vallstedt A, Gezelius H, Memic F, Lind M, Hjalt T, Tourtellotte WG, Bruder C, Eichele G, Whelan PJ, Kullander K (2010) Identification of novel spinal cholinergic genetic subtypes disclose Chodl and Pitx2 as markers for fast motor neurons and partition cells. J Comp Neurol 518:2284–2304. https://doi.org/10.1002/cne.22332

Enjin A, Perry S, Hilscher MM, Nagaraja C, Larhammar M, Gezelius H, Eriksson A, Leão KE, Kullander K (2017) Developmental disruption of recurrent inhibitory feedback results in compensatory adaptation in the Renshaw cell–motor neuron circuit. J Neurosci 37:5634–5647. https://doi.org/10.1523/JNEUROSCI.0949-16.2017

Falgairolle M, Puhl JG, Pujala A, Liu W, O'Donovan MJ (2017) Motoneurons regulate the central pattern generator during drug-induced locomotor-like activity in the neonatal mouse. Elife 6:e26622–e26622. https://doi.org/10.7554/eLife.26622
Fulton BP, Miledi R, Takahashi T (1980) Electrical synapses between motoneurons in the spinal cord of the newborn rat. Proc R Soc Lond B Biol Sci 208:115–120
Henneman E (1957) Relation between size of neurons and their susceptibility to discharge. Science 126:1345–1347. https://doi.org/10.1126/science.126.3287.1345
Herzog E, Landry M, Buhler E, Bouali-Benazzouz R, Legay C, Henderson CE, Nagy F, Dreyfus P, Giros B, El MS (2004) Expression of vesicular glutamate transporters, VGLUT1 and VGLUT2, in cholinergic spinal motoneurons. Eur J Neurosci 20:1752–1760. https://doi.org/10.1111/j.1460-9568.2004.03628.x
Hinckley CA, Ziskind-Conhaim L (2006) Electrical coupling between locomotor-related excitatory interneurons in the mammalian spinal cord. J Neurosci 26:8477–8483. https://doi.org/10.1523/JNEUROSCI.0395-06.2006
Hultborn H, Pierrot-Deseilligny E (1979) Input-output relations in the pathway of recurrent inhibition to motoneurones in the cat. J Physiol 297:267–287
Hultborn H, Lindstrom S, Wigstrom H (1979) On the function of recurrent inhibition in the spinal cord. Exp Brain Res 37:399–403
Hultborn H, Katz R, Mackel R (1988a) Distribution of recurrent inhibition within a motor nucleus. II. Amount of recurrent inhibition in motoneurones to fast and slow units. Acta Physiol Scand 134:363–374. https://doi.org/10.1111/j.1748-1716.1988.tb08502.x
Hultborn H, Lipski J, Mackel R, Wigstrom H (1988b) Distribution of recurrent inhibition within a motor nucleus. I. Contribution from slow and fast motor units to the excitation of Renshaw cells. Acta Physiol Scand 134:347–361. https://doi.org/10.1111/j.1748-1716.1988.tb08503.x
Ichinose T, Miyata Y (1998) Recurrent excitation of motoneurons in the isolated spinal cord of newborn rats detected by whole-cell recording. Neurosci Res 31:179–187
Jankowska E, Hammar I (2013) Interactions between spinal interneurons and ventral spinocerebellar tract neurons. J Physiol 591:5445–5451. https://doi.org/10.1113/jphysiol.2012.248740
Jefferys JG (1995) Nonsynaptic modulation of neuronal activity in the brain: electric currents and extracellular ions. Physiol Rev 75:689–723. https://doi.org/10.1152/physrev.1995.75.4.689
Jiang J, Alstermark B (2015) Not GABA but glycine mediates segmental, propriospinal, and bulbospinal postsynaptic inhibition in adult mouse spinal forelimb motor neurons. J Neurosci 35:1991–1998. https://doi.org/10.1523/JNEUROSCI.1627-14.2015
Jiang ZG, Shen E, Wang MY, Dun NJ (1991) Excitatory postsynaptic potentials evoked by ventral root stimulation in neonate rat motoneurons in vitro. J Neurophysiol 65:57–66
Jonas P, Bischofberger J, Sandkuhler J (1998) Corelease of two fast neurotransmitters at a central synapse. Science 281:419–424
Krashia P, Ledonne A, Nobili A, Cordella A, Errico F, Usiello A, D'Amelio M, Mercuri NB, Guatteo E, Carunchio I (2016) Persistent elevation of D-aspartate enhances NMDA receptor-mediated responses in mouse substantia nigra pars compacta dopamine neurons. Neuropharmacology 103:69–78. https://doi.org/10.1016/j.neuropharm.2015.12.013
Kraus T, Neuhuber WL, Raab M (2004) Vesicular glutamate transporter 1 immunoreactivity in motor endplates of striated esophageal but not skeletal muscles in the mouse. Neurosci Lett 360:53–56. https://doi.org/10.1016/j.neulet.2004.02.039
Lamotte d'Incamps B, Ascher P (2008) Four excitatory postsynaptic ionotropic receptors coactivated at the motoneuron-Renshaw cell synapse. J Neurosci 28:14121–14131. https://doi.org/10.1523/JNEUROSCI.3311-08.2008
Lamotte d'Incamps B, Ascher P (2014) High affinity and low affinity heteromeric nicotinic acetylcholine receptors at central synapses. J Physiol Lond 592:4131–4136. https://doi.org/10.1113/jphysiol.2014.273128
Lamotte d'Incamps B, Bhumbra GSS, Foster JDD, Beato M, Ascher P, Lamotte d'Incamps B, Bhumbra GSS, Foster JDD, Beato M, Ascher P (2017) Segregation of glutamatergic and

cholinergic transmission at the mixed motoneuron Renshaw cell synapse. Sci Rep 7:4037–4037. https://doi.org/10.1038/s41598-017-04266-8
Leng G, Dyball REJ (1983) Intercommunication in the rat supraoptic nucleus. Q J Exp Physiol 68:493–504. https://doi.org/10.1113/expphysiol.1983.sp002742
Leroy F, d'Incamps BL, Imhoff-Manuel RD, Zytnicki D (2014) Early intrinsic hyperexcitability does not contribute to motoneuron degeneration in amyotrophic lateral sclerosis. elife 3:e04046–e04046. https://doi.org/10.7554/eLife.04046
Lu T, Rubio ME, Trussell LO (2008) Glycinergic transmission shaped by the corelease of GABA in a mammalian auditory synapse. Neuron 57:524–535
Machacek DW, Hochman S (2006) Noradrenaline unmasks novel self-reinforcing motor circuits within the mammalian spinal cord. J Neurosci 26:5920–5928. https://doi.org/10.1523/JNEUROSCI.4623-05.2006
Maltenfort MG, Heckman CJ, Rymer WZ (1998) Decorrelating actions of Renshaw interneurons on the firing of spinal motoneurons within a motor nucleus: a simulation study. J Neurophysiol 80:309–323
Marchetti C, Beato M, Nistri A (2001a) Alternating rhythmic activity induced by dorsal root stimulation in the neonatal rat spinal cord in vitro. J Physiol 530:105–112. https://doi.org/10.1111/j.1469-7793.2001.0105m.x
Marchetti C, Beato M, Nistri A (2001b) Evidence for increased extracellular K+ as an important mechanism for dorsal root induced alternating rhythmic activity in the neonatal rat spinal cord in vitro. Neurosci Lett 304:77–80. https://doi.org/10.1016/S0304-3940(01)01777-3
McCurdy ML, Hamm TM (1992) Recurrent collaterals of motoneurons projecting to distal muscles in the cat hindlimb. J Neurophysiol 67:1359–1366. https://doi.org/10.1152/jn.1992.67.5.1359
McCurdy ML, Hamm TM (1994) Topography of recurrent inhibitory postsynaptic potentials between individual motoneurons in the cat. J Neurophysiol 72:214–226. https://doi.org/10.1152/jn.1994.72.1.214
Meister B, Arvidsson U, Zhang X, Jacobsson G, Villar MJ, Hökfelt T (1993) Glutamate transporter mRNA and glutamate-like immunoreactivity in spinal motoneurones. Neuroreport 5:337–340. https://doi.org/10.1097/00001756-199312000-00040
Mentis GZ, Alvarez FJ, Bonnot A, Richards DS, Gonzalez-Forero D, Zerda R, O'Donovan MJ (2005) Noncholinergic excitatory actions of motoneurons in the neonatal mammalian spinal cord. Proc Natl Acad Sci 102:7344–7349
Moore NJ, Bhumbra GS, Foster JD, Beato M (2015) Synaptic connectivity between Renshaw cells and motoneurons in the recurrent inhibitory circuit of the spinal cord. J Neurosci. https://doi.org/10.1523/jneurosci.2541-15.2015
Muller D, Cherukuri P, Henningfeld K, Poh CH, Wittler L, Grote P, Schluter O, Schmidt J, Laborda J, Bauer SR, Brownstone RM, Marquardt T (2014) Dlk1 promotes a fast motor neuron biophysical signature required for peak force execution. Science 343:1264–1266. https://doi.org/10.1126/science.1246448
Nishimaru H, Restrepo CE, Ryge J, Yanagawa Y, Kiehn O (2005) Mammalian motor neurons corelease glutamate and acetylcholine at central synapses. Proc Natl Acad Sci 102:5245–5249
Oliveira ALR, Hydling F, Olsson E, Shi T, Edwards RH, Fujiyama F, Kaneko T, Hökfelt T, Cullheim S, Meister B (2003) Cellular localization of three vesicular glutamate transporter mRNAs and proteins in rat spinal cord and dorsal root ganglia. Synapse 50:117–129. https://doi.org/10.1002/syn.10249
Patneau DK, Mayer ML (1990) Structure-activity relationships for amino acid transmitter candidates acting at N-methyl-D-aspartate and quisqualate receptors. J Neurosci 10:2385–2399. https://doi.org/10.1523/JNEUROSCI.10-07-02385.1990
Perrins R, Roberts A (1995) Cholinergic and electrical synapses between synergistic spinal motoneurones in the Xenopus laevis embryo. J Physiol 485(Pt 1):135–144
Perry S, Gezelius H, Larhammar M, Hilscher MM, Lamotte d'Incamps B, Leao KE, Kullander K (2015) Firing properties of Renshaw cells defined by Chrna2 are modulated by hyperpolarizing and small conductance ion currents Ih and ISK. Eur J Neurosci 41:889–900. https://doi.org/10.1111/ejn.12852

Personius KE, Chang Q, Mentis GZ, O'Donovan MJ, Balice-Gordon RJ (2007) Reduced gap junctional coupling leads to uncorrelated motor neuron firing and precocious neuromuscular synapse elimination. Proc Natl Acad Sci U S A 104:11808–11813. https://doi.org/10.1073/pnas.0703357104
Rash JE, Dillman RK, Bilhartz BL, Duffy HS, Whalen LR, Yasumura T (1996) Mixed synapses discovered and mapped throughout mammalian spinal cord. Proc Natl Acad Sci U S A 93:4235–4239
Renshaw B (1946) Central effects of centripetal impulses in axons of spinal ventral roots. J Neurophysiol 9:191–204
Richards DS, Griffith RW, Romer SH, Alvarez FJ (2014) Motor axon synapses on renshaw cells contain higher levels of aspartate than glutamate. PLoS One 9:e97240–e97240. https://doi.org/10.1371/journal.pone.0097240
Ross HG, Cleveland S, Haase J (1975) Contribution of single motoneurons to renshaw cell activity. Neurosci Lett 1:105–108
Ross HG, Cleveland S, Haase J (1976) Quantitative relation between discharge frequencies of a Renshaw cell and an intracellularly depolarized motoneuron. Neurosci Lett 3:129–132
Ryall RW, Piercey MF (1971) Excitation and inhibition of Renshaw cells by impulses in peripheral afferent nerve fibers. J Neurophysiol 34:242–251. https://doi.org/10.1152/jn.1971.34.2.242
Schäfer MK-H, Varoqui H, Defamie N, Weihe E, Erickson JD (2002) Molecular cloning and functional identification of mouse vesicular glutamate transporter 3 and its expression in subsets of novel excitatory neurons. J Biol Chem 277:50734–50748. https://doi.org/10.1074/jbc.M206738200
Schneider SP, Fyffe RE (1992) Involvement of GABA and glycine in recurrent inhibition of spinal motoneurons. J Neurophysiol 68:397–406
Sherrington C (1906) The integrative action of the nervous system. Yale University Press, New Haven
Singer JH, Talley EM, Bayliss DA, Berger AJ (1998) Development of glycinergic synaptic transmission to rat brain stem motoneurons. J Neurophysiol 80:2608–2620
Song J, Ampatzis K, Björnfors ER, El Manira A (2016) Motor neurons control locomotor circuit function retrogradely via gap junctions. Nature. https://doi.org/10.1038/nature16497
Stepien AE, Tripodi M, Arber S (2010) Monosynaptic rabies virus reveals premotor network organization and synaptic specificity of cholinergic partition cells. Neuron 68:456–472. https://doi.org/10.1016/j.neuron.2010.10.019
Tripodi M, Stepien AE, Arber S (2011) Motor antagonism exposed by spatial segregation and timing of neurogenesis. Nature 479:61–66. https://doi.org/10.1038/nature10538
Van Keulen L (1981) Autogenetic recurrent inhibition of individual spinal motoneurones of the cat. Neurosci Lett 21:297–300
Wærhaug O, Ottersen OP (1993) Demonstration of glutamate-like immunoreactivity at rat neuromuscular junctions by quantitative electron microscopic immunocytochemistry. Anat Embryol 188:501–513. https://doi.org/10.1007/BF00190144
Walton KD, Navarrete R (1991) Postnatal changes in motoneurone electrotonic coupling studied in the in vitro rat lumbar spinal cord. J Physiol 433:283–305
Windhorst U (1996) On the role of recurrent inhibitory feedback in motor control. Prog Neurobiol 49:517–587
Zhang Y, Narayan S, Geiman E, Lanuza GM, Velasquez T, Shanks B, Akay T, Dyck J, Pearson K, Gosgnach S, Fan CM, Goulding M (2008) V3 spinal neurons establish a robust and balanced locomotor rhythm during walking. Neuron 60:84–96

Recruitment of Motoneurons

Vatsala Thirumalai and Urvashi Jha

Abstract Beginning about half a century ago, the rules that determine how motor units are recruited during movement have been deduced. These classical experiments led to the formulation of the 'size principle'. It is now clear that motoneuronal size is not the only indicator of recruitment order. In fact, motoneuronal passive, active and synaptic conductances are carefully tuned to achieve sequential recruitment. More recent studies, over the last decade or so, show that the premotor circuitry is also functionally specialized and differentially recruited. Modular sub networks of interneurons and their post-synaptic motoneurons have been shown to drive movements with varying intensities. In addition, these modular networks are under the influence of neuromodulators, which are capable of acting upon multiple motor and premotor targets, thereby altering behavioral outcomes. We discuss the recruitment patterns of motoneurons in light of these new and exciting studies.

Keywords Size principle · Intrinsic properties · Synaptic properties · Neuromodulation · Interneurons · Force · Circuits · Dopamine · Serotonin · Bistability · Excitability

1 Introduction

A motor unit consists of a motoneuron and all the muscle fibers it innervates. Within every motor unit, force generated by muscle fiber contraction is graded by two factors: the firing rate of the motoneuron and the transform function that determines how electrical excitation is transduced to mechanical contractility (Kernell 2003). When a motoneuron fires an action potential, the electrical excitation is transduced to mechanical contractility, i.e. a twitch, in the muscle fiber. As motoneuronal firing rate increases, individual twitches begin to summate and the peak contractile force

V. Thirumalai (✉) · U. Jha
National Centre for Biological Sciences, Bangalore, India
e-mail: vatsala@ncbs.res.in

M. J. O'Donovan, M. Falgairolle (eds.), *Vertebrate Motoneurons*, Advances in Neurobiology 28, https://doi.org/10.1007/978-3-031-07167-6_8

generated by the muscle fiber increases. Beyond a threshold firing rate, the individual twitches fuse, the fiber is said to be in tetanus and no further increase in force is possible. Thus, as motoneuronal firing rate increases, the force generated by the fiber increases linearly over a range and then saturates. Rate modulation of twitch force in single motor units is a useful way of regulating forces generated by the whole muscle. However, relying on only rate modulation will only lead to a linear increase in force and a smaller range over which force can be modulated.

Muscles comprise several motor units and therefore every muscle receives input from several motoneurons. Therefore muscle force can also be varied by changing the numbers of motoneurons/units activated. In addition, motor units are not all the same. Motor units can be classified based on the time to achieve peak force, the magnitude of force generated and their fatigability, properties that are determined by the biochemical composition of the muscle fibers in that unit. Fast fatigue (FF) motor units generate large forces but fatigue quickly, slow (S) units generate small forces but are fatigue-resistant and fast, fatigue resistant (FR) units generate intermediate forces. Every muscle is composed of a mixture of these three types of motor units. Muscle force is a direct function of the number and type of motor units activated and the intensity with which each of those motor units are activated. A non-linear gradation of force generated is achieved by selecting which motoneurons in a pool will fire action potentials and at what rate, referred to as motoneuronal recruitment. Below, we examine the rules that determine motoneuronal recruitment order and the mechanisms by which these rules operate.

2 Size Principle

The earliest demonstration that motor units are differentially recruited to generate incremental force came from experiments performed nearly a century ago (Adrian and Bronk 1929; Denny-Brown and Pennybacker 1938). While measuring motor unit activity from the ventral root upon stimulation of sciatic nerve in cats, Elwood Henneman (Henneman 1957) observed that motoneurons continued to fire a few spikes after the initial reflex response. He showed that at low stimulation intensity, small amplitude spikes were seen and that with increased stimulation intensity, while the small amplitude spikes persisted, increasingly larger amplitude spikes were seen (Fig. 1a). Since the amplitude of the extracellularly recorded spike varies directly as the axonal diameter, it followed that small spikes were generated by thin axons and large spikes by thicker axons. Further, since soma size and axon diameter are likely correlated, thin axons are likely to belong to small somata and thick axons to large somata (Henneman et al. 1965a). These experiments suggested that as stimulation intensity is gradually increased, small motoneurons are first recruited to fire followed by increasingly larger motoneurons.

Henneman and colleagues surmised that if size were the primary determinant of excitability then the order of recruitment would remain the same regardless of the source of excitation. They tested this idea using various reflexes such as

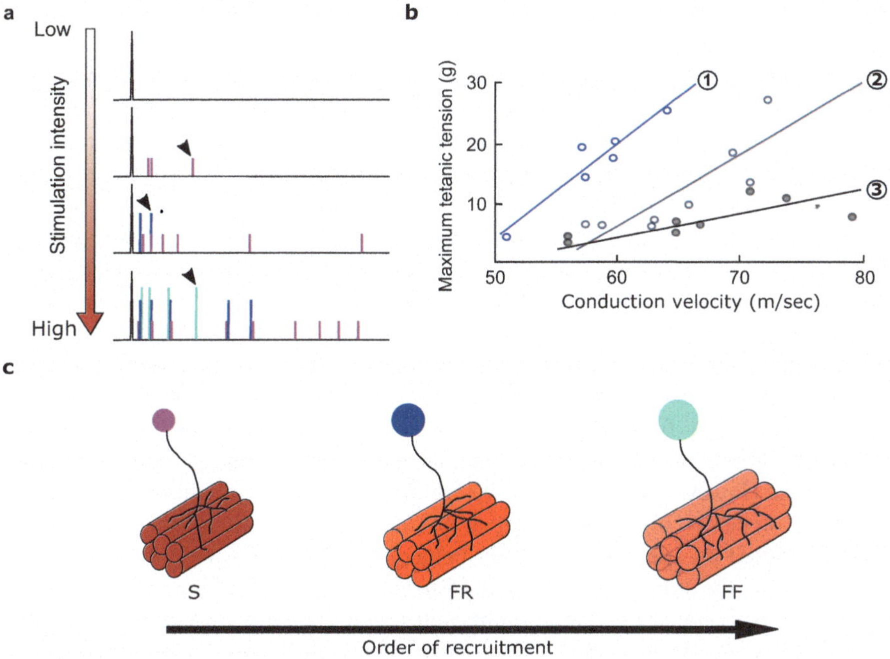

Fig. 1 Motor units are recruited in a precise order from smallest to largest. (**a**) Diagrammatic representation of ventral root recordings with increasing stimulation of sciatic nerve. Responses in magenta, blue and cyan represent activity of three different alpha motoneurons recruited at progressively higher stimulation intensity respectively. Note the difference in amplitudes of these responses (marked by arrowhead). (Adapted with permission from Henneman 1957). (**b**) Relationship between maximum tension developed by individual motor units and conduction velocity of their motoneuronal axons in three different soleus muscles numbered 1–3. In all cases, maximum tension developed increases linearly with axonal conduction velocity. (Figure modified with permission from Mcphedran et al. 1965). (**c**) Schematic diagram showing three different motor units and their recruitment order. Left: Smaller-sized, slow conducting motoneuron innervating weaker, fatigue-resistant muscles. Middle: Intermediate-sized motoneuron innervating muscles with intermediate contractile properties. Right: Larger-sized, fast conducting motoneuron innervating stronger and fatigable muscles

crossed-extension reflex, flexor reflex, and monosynaptic stretch reflex and also upon electrical stimulation of the muscle nerve. Irrespective of the source of excitation, the order of recruitment remained the same (Henneman et al. 1965a, b; Bawa et al. 1984).

Further experiments established the relationships between motoneuron size and the muscle fibers they innervate (Henneman and Olson 1965; Mcphedran et al. 1965; Wuerker et al. 1965). Slowly conducting axons tended to innervate the weaker fatigue resistant Type I muscle fibers. In contrast, axons with faster conduction velocity were shown to innervate the stronger, fatigable Type IIB muscle fibers (Fig. 1b). Since conduction velocity varies inversely with axon diameter, it follows that small motoneurons with thin, slowly conducting axons likely innervate weak fatigue resistant fibers. The converse is true for large motoneurons and they innervate the fast fatigable fibers.

The above results form the basis of the 'size principle' which states that gradation of force is achieved by first recruiting the slow motor units via the activation of small motoneurons and then by sequentially activating the fast fatigue resistant and finally the fast fatigable motor units via the recruitment of increasingly larger motoneurons (Mendell 2005) (Fig. 1c).

Size-ordered, sequential recruitment has been observed to be true across systems (Davis 1971; Milner-Brown et al. 1973; McLean et al. 2007; Hill and Cattaert 2008; Ampatzis et al. 2013; Azevedo et al. 2019) and across motor behaviors (Seven et al. 2014). Since many of these studies were done on paralyzed or isolated spinal cord preparations, the frequency of fictive locomotor rhythms was used to classify neurons as being recruited during slow or fast movements. Thus slow motor units were shown to drive low frequency movements and continued to be active during fast movements while fast motor units became recruited only during high frequency movements. However, a few notable exceptions, of selective recruitment also exist. Zebrafish, on perceiving a threatening or noxious stimulus, initiate a ballistic motor response called escape response. This behavior involves a powerful tail bend to quickly propel the animal away from danger. Such high-performance motor behavior involves activation of fast motor units and suppression of slow motor units in the adult zebrafish (Song et al. 2015). Even in larval zebrafish, some slow motoneurons are selectively silenced during very fast swims (Menelaou and McLean 2012). Selective activation of high-threshold motor units with simultaneous inhibition of low-threshold motor units has also been observed in cats (Kanda et al. 1977) and humans (Stephens et al. 1978) upon stimulation of cutaneous afferents.

Irrespective of sequential or selective recruitment of motoneurons, the underlying mechanisms responsible for orderly recruitment are interesting to explore and have been investigated systematically. We describe these findings below broadly under intrinsic excitability, synaptic connectivity and neuromodulatory mechanisms driving motoneuronal recruitment.

3 Intrinsic Properties

Spinal motoneuronal intrinsic properties have been studied in several vertebrate organisms in slice, *en bloc* or *in vivo* preparations. From these studies, we now have a good understanding of the conductances present on the motoneuronal membrane (McLarnon 1995; Heckman et al. 2005; Binder et al. 2019). Here we will focus on what is known about how motoneuronal intrinsic properties regulate their recruitment order. Henneman proposed that the size-ordered recruitment of motoneurons could simply be a reflection of their excitability, in turn determined by the input resistance, where smaller S motoneurons have higher input resistance in comparison to larger F units (Henneman et al. 1965a) (Fig. 2a-1). Subsequently, such gradations in input resistance were shown to be due to differences in specific membrane

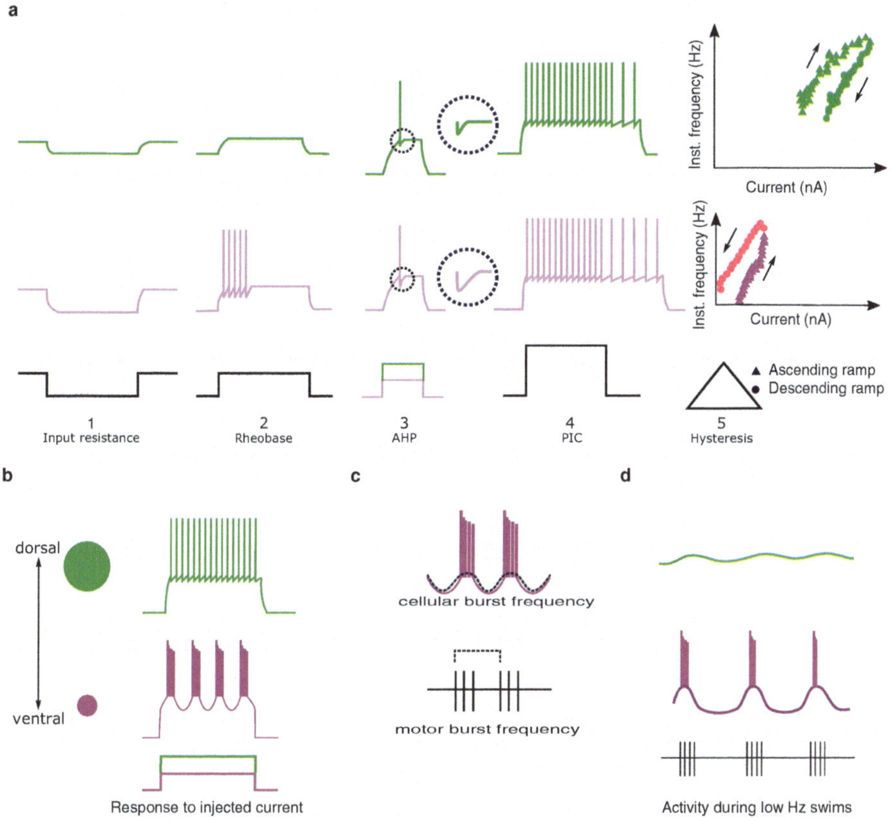

Fig. 2 Systematic variations in intrinsic properties of fast and slow motoneurons. (**a**) Schematic representation of responses of fast (green) and slow (magenta) motoneurons to five different current injection protocols showing their differences in (1) Input resistance (2) Rheobase (3) After-hyperpolarization (AHP). Magnified view of AHP is shown on the right inside the dashed circle. (4) PIC mediated sustained firing (5) Hysteresis. (Adapted with permission from Turkin et al. 2010). (**b**) Left: Schematic representation of topographical recruitment pattern of spinal motoneurons in larval zebrafish. Dorsally located, low input resistance motoneuron is labelled in green while ventrally located, high-input resistance motoneuron is labelled in magenta. Right: Tonic and bursting responses of fast and slow motoneurons respectively in response to injected current. (**c**) Top: Schematic representation of rhythmic bursts in slow motoneurons in response to injected current. Rhythmic membrane oscillations underlying action potentials are highlighted with dashed black line. Bottom: Schematic representation of ventral root recordings. Start of each motor burst is highlighted with a dashed line on top. (**d**) Schematic representation of membrane potential changes in fast and slow motoneurons during fictive swims. Depicted at bottom is schematic of ventral root recordings. (Figure 2b–d adapted from Menelaou and McLean 2012)

resistivity (Gustafsson and Pinter 1984a), and not arising solely out of changes in size.

It is now becoming clear that both passive and active properties of motoneurons systematically vary within a motoneuronal pool to aid in the orderly recruitment.

For example, motoneurons in the S-motor units have considerably smaller dendritic elaborations and therefore smaller membrane capacitance compared to F-motoneurons (Cullheim et al. 1987). Moreover, rheobase (the current required to bring a neuron to threshold), increases in the order S < FR < FF (Fig. 2a-2). However, this could not be explained based on input resistance alone, as units that had similar resistances showed very different rheobases (Fleshman et al. 1981; Gustafsson and Pinter 1984b). These early findings in cat are supported by studies in other vertebrates (Bakels and Kernell 1993; Leroy et al. 2014, 2015) and invertebrates (Davis 1971; Hill and Cattaert 2008; Sasaki and Burrows 1998; Azevedo et al. 2019; Gabriel et al. 2003) showing a gradation of motoneuronal size, active and passive properties to generate slow/weak, and fast/strong muscle action.

Resting membrane potential is also an important regulator of recruitment order. In the brainstem respiratory network of neonatal rats, phrenic motoneurons that had a higher resting membrane potential and input resistance fired during inspiratory cycles. In contrast, motoneurons having a lower resting membrane potential and input resistance were silent. Though the synaptic current during inspiration was the same, synaptic drive potentials were smaller in the silent neurons due to the larger leak conductance. Further, the hyperpolarized resting membrane potential meant that the synaptic potentials were not large enough to reach threshold (Su et al. 1997). In stick insect also, differences in resting membrane potential between slow and fast flexor motoneurons lead to their differential recruitment during walking (Gabriel et al. 2003).

The spike after-hyperpolarization (AHP) is another property that varies systematically among motoneurons to set their recruitment order (Fig. 2a-3). S-type motoneurons have longer and larger AHP's compared to F-type ones (Eccles et al. 1957, 1958; Gustafsson and Pinter 1984a, b; Zengel et al. 1985; Bakels and Kernell 1993). As AHP's are often generated by calcium-dependent potassium currents, these results point to differences in calcium-dependent potassium channel densities and kinetics in S-type and F-type motoneurons.

Besides these, motoneurons express a range of ionic conductances, which differ between motoneuron types. For example, motoneurons exhibit persistent inward currents (PICs) mediated by voltage-dependent L-type calcium channels, calcium-activated non-selective cation currents and/or voltage-gated sodium channels that inactivate slowly and are located predominantly on dendrites (Schwindt and Crill 1977; Zhang et al. 1995; Lee and Heckman 1998a, b; Hamm et al. 2010). In decerebrate cats, low-threshold motoneurons belonging to S-type motor units show long-lasting depolarization in response to brief synaptic inputs due to the activation of PICs. In contrast, high-threshold motoneurons likely belonging to F-type motor units, though they have strong PICs, generate shorter spiking responses, probably due to differences in the voltage-dependence of PICs (Lee and Heckman 1998a, b) (Fig. 2a-4). These differences result in altered responses to current injections or synaptic input such that the low-threshold motoneurons show immediate acceleration of firing rate that is sustained even after the input has ended, while high-threshold cells show delayed acceleration of firing and do not sustain firing rates over longer periods of time (Heckman et al. 2005; Bhumbra and Beato 2018;

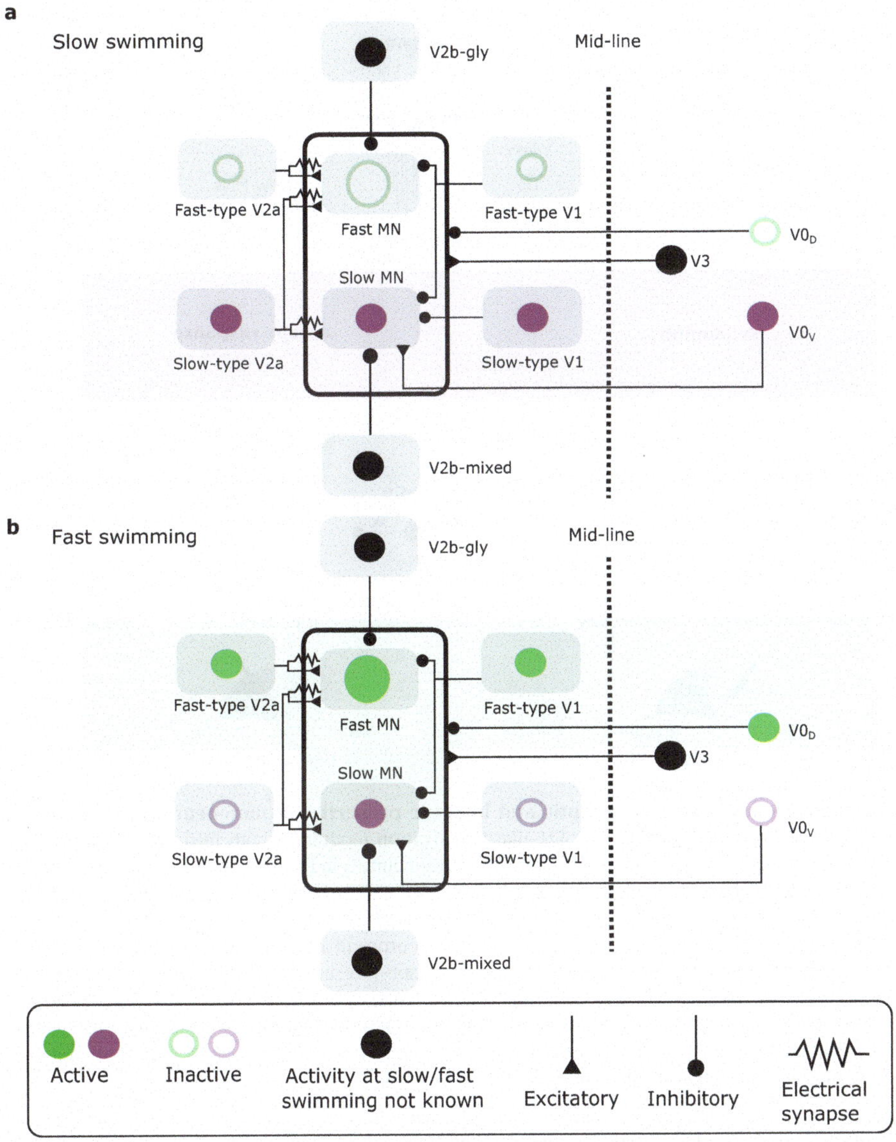

Fig. 3 Fast and slow motoneurons receive differential inputs from spinal interneurons that show speed dependent recruitment patterns. (**a**) At slow speeds slow motoneurons (magenta) are active (filled circle) while fast motoneurons (green) are not recruited. Slow motoneurons receive inputs from spinal interneurons that also show speed-dependent recruitment pattern (**b**) At fast swim speeds most slow motoneurons continue to fire and fast motoneurons are further recruited. Spinal interneurons that were active at slow speed are inhibited and a different set of spinal interneurons become active at fast speeds in larval zebrafish. All cells active at slow swimming speed are highlighted in magenta bar and those active at fast speeds are highlighted in green bar

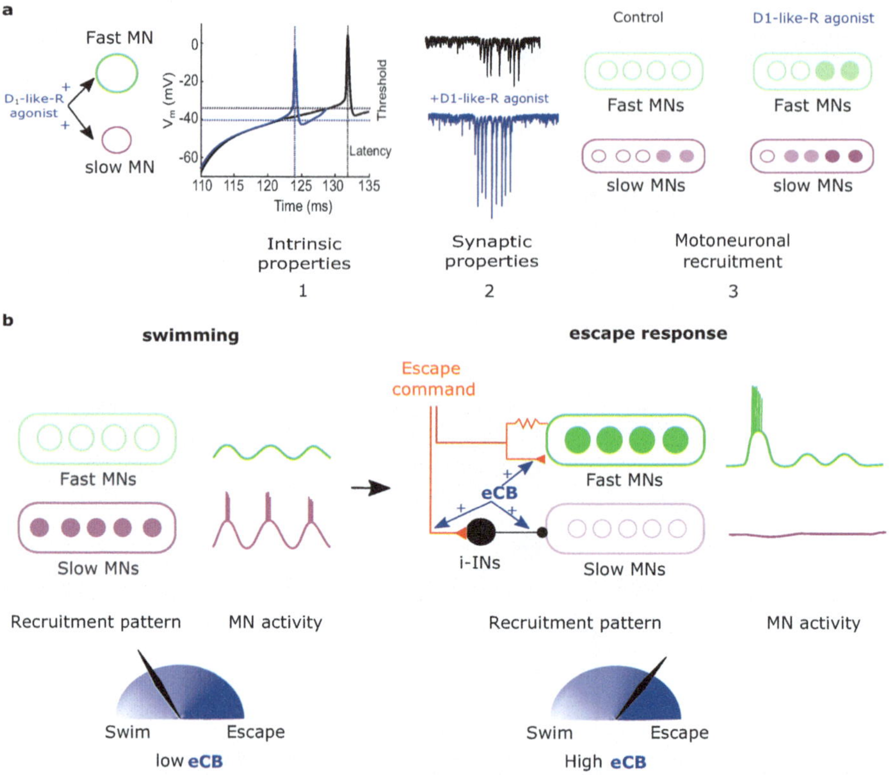

Fig. 4 Neuromodulation of synaptic and intrinsic properties of motoneurons alters motoneuronal recruitment pattern. (**a**) D1-like-R activation modulates both intrinsic properties (1) and excitatory synaptic drive (2) in fast and slow motoneurons in larval zebrafish during optomotor response. These changes result in enhanced activation of slow motoneurons as well as additional recruitment of fast motorneurons that are typically silent during this behavior. (Figure adapted from Jha and Thirumalai 2020). (**b**) Left: Slow motoneurons in adult zebrafish are recruited during slow swimming while fast motoneurons only receive sub-threshold synaptic inputs. Right: During escape response, the Mauthner system indirectly inhibits slow motoneurons via inhibitory interneurons (i-INs) while fast motoneurons are activated via mixed synapses. This selection of fast motoneurons over slow motoneurons during escape is positively modulated by endocannabinoids (eCB, blue arrows). (Based on Song et al. 2015)

Manuel and Zytnicki 2019). Presence of PICs leads to a negative slope region in the current-voltage relationship of neurons and induces several non-linear characteristics including membrane potential bistability in the form of plateau potentials (Binder et al. 2019). PICs are major targets of neuromodulation and serve to amplify dendritic synaptic currents (Heckman et al. 2009). The neuromodulation of PICs by monoamines and the induction of bistability are discussed in greater detail in a later section.

The characteristics of the AHP, PICs and spike frequency adaptation determine the neuronal gain, i.e. the slope of the relationship between current injected versus firing rate of a neuron (f-I curve) (Powers and Binder 2001). The f-I curves of mammalian motoneurons show a linear relationship over a range of injected current values, their 'primary' range. Beyond this, a 'secondary' and sometimes a 'tertiary' region with distinct slopes are present (Schwindt 1973). In rat and mouse motoneurons, a 'sub-primary' region, where spiking is irregular and the f-I curve is non-linear, could be discerned (Manuel et al. 2009; Turkin et al. 2010). In addition, presence of PICs and spike frequency adaptation leads to hysteresis in f-I relationships, i.e., ascending and descending ramp currents generate different responses. Neurons with spike frequency adaptation fire less during descending ramps than during ascending ramps and therefore have 'clockwise' hysteresis (e.g., Fig. 2a-5, top panel). Neurons with strong PICs fire more during descending ramps due to the activation of these inward currents leading to 'anti-clockwise' hysteresis (Fig. 2a-5, bottom panel). Interestingly, the nature of hysteresis depended on input resistance, such that high resistance neurons showed anti-clockwise hysteresis while low resistance neurons showed clockwise hysteresis (Manuel et al. 2009; Hamm et al. 2010; Turkin et al. 2010) (Fig. 2a-5). The net result of these differences in hysteresis is that the threshold current for recruitment in slow motoneurons is higher than that during de-recruitment. The opposite is true in fast motoneurons: the threshold current for recruitment is lower than that during de-recruitment. These properties imply that as locomotor synaptic drive decreases at the end of a motor episode, slow motoneurons continue to fire while fast motoneurons switch off quickly.

Further, motoneuronal classes show distinct frequency-dependent responses making them amplify inputs at certain frequencies better. The presence of the hyperpolarization activated 'sag' current (I_h) induces resonance in some motoneurons with a peak at around 10 Hz on average. The amplitude of I_h is different in slow and fast motoneurons such that slow motoneurons likely have little to no I_h while fast motoneurons show a strong I_h component (Gustafsson and Pinter 1984a; Manuel et al. 2007; Turkin et al. 2010). Monosynaptic excitatory post synaptic potentials (EPSPs) from Ia afferents that signal muscle stretch consist of static and dynamic components. In a modeling study, it was shown that the interaction between PICs and I_h-induced resonance in fast motoneurons could lead to greater amplification of fast, dynamic EPSPs (Manuel et al. 2007). In turn, this means that during strong, fast muscle stretches such as those occurring during fast movements, the dynamic components of Ia monosynaptic input generate much larger EPSPs in fast motoneurons increasing their probability of spiking. In non-resonant slow motoneurons, fast PICs led to amplification of the static component of Ia EPSPs leading to plateau potentials and sustained firing responses (Manuel et al. 2007; Turkin et al. 2010; Binder et al. 2019).

Larval zebrafish afford a convenient model system in which mechanisms of motoneuronal recruitment patterns can be studied in a whole organismal context. This has led to investigations of motoneuronal recruitment with respect to birth order of motoneurons and their passive, active and synaptic properties. Primary motoneurons are specified soon after gastrulation ends, occupy a relatively dorsal location in the spinal cord and are also larger in size (Eisen 1991; Berg et al. 2018). Primary motoneurons innervate fast muscle and are recruited during escapes, struggles and fast swims (McLean et al. 2007, 2008; Gabriel et al. 2011; Ampatzis et al. 2013; Song et al. 2015; Berg et al. 2018). In contrast, secondary motoneurons are born later, occupy relatively ventral locations in the spinal cord and are smaller in size (Myers 1985). Secondary motoneurons are recruited at lower swim speeds compared to primary motoneurons. In general, recruitment order proceeds from ventral to dorsal soma position such that the most ventrally located secondary motoneurons are the first ones to be recruited at very low swim speeds, followed by dorsal secondary motoneurons and finally the primary motoneurons at the fastest frequencies (McLean et al. 2007, 2008; Gabriel et al. 2011; Ampatzis et al. 2013). These topographic gradients in motoneuronal recruitment pattern match the differences in intrinsic properties. In larval zebrafish, primary motoneurons were the largest and had the lowest input resistance and the highest rheobase. The most ventrally located secondary motoneurons had the lowest rheobases while an intermediate population showed properties in between the two groups (McLean et al. 2007; Menelaou and McLean 2012). These classes of motoneurons also responded differently to injected depolarizing current steps where smaller, ventral motoneurons were shown to be intrinsically rhythmic in both the larva and the adult (Fig. 2b) (Gabriel et al. 2011; Menelaou and McLean 2012). Current-evoked bursting frequencies were similar to burst frequencies observed during fictive slow swims (Fig. 2c). This suggests that these ventral secondary motoneurons are tuned to be endogenously rhythmic near natural swim frequencies and could be easily recruited during slow swims (Gabriel et al. 2011; Menelaou and McLean 2012) (Fig. 2d) or that these motoneurons participate in generating the locomotor rhythm at these frequencies (Falgairolle and O'Donovan 2019b).

Taken together, all of the above studies suggest that recruitment order is set by the careful tuning of passive and active properties and are not only dependent on size. The net effect of such tuning is to give rise to smaller, more excitable motoneurons that generate weak movements and larger, less excitable motoneurons recruited for generating strong forces.

4 Synaptic Properties

Motoneurons receive thousands of excitatory and inhibitory synaptic inputs all along their extensive dendritic arbors. Synaptic inputs to motoneurons arise from diverse sources such as group Ia and group II sensory afferents, excitatory and inhibitory spinal interneurons and descending projections. In general, three classes of synaptic inputs have been identified: (1) Post synaptic potentials (PSPs) that scale with input resistance; (2) PSPs that are independent of input resistance; and (3) PSPs that are predominantly inhibitory to low threshold motoneurons but predominantly excitatory to high threshold motoneurons (Powers and Binder 2001).

Class 1: The synapses made by group Ia afferents onto motoneurons are probably the best studied amongst all synaptic inputs to motoneurons. The monosynaptic

EPSP from Ia afferents to homonymous motoneurons scales in amplitude with motoneuronal input resistance. In addition, voltage clamp studies of motoneurons in adult cat show that the somatically recorded synaptic current amplitude scales directly with input resistance. Thus, in S-type motoneurons, high input resistance and a large synaptic current combine to generate an even larger EPSP while in F-type motoneurons, a much smaller EPSP is seen (Heckman and Binder 1988).

Class 2: EPSPs generated by group II afferents did not vary in amplitude across the different types of motor units (Munson et al. 1982). It then follows that to compensate for the low resistance of the F-type motoneurons, group II afferent synaptic currents will have to scale inversely in order to generate similar sized EPSPs. These findings also imply that while sensory input from Ia afferents might be particularly important for sustained movements of small forces, group II afferent input drives all motoneurons equally.

Class 3: Stimulation of the rubrospinal tract from the red nucleus, the pyramidal tract or group Ib afferents, predominantly excites the F-type and inhibits the S-type motoneurons. These actions are thought to derive from their connections to last order spinal excitatory and inhibitory interneurons (Powers and Binder 2001).

These early studies suggest that spinal pre-motor networks could be segregated based on the motoneuronal types that they target, an idea that has recently been investigated in zebrafish and mice (Goulding 2009; Arber 2012; McLean and Dougherty 2015; Gosgnach et al. 2017; Berg et al. 2018). Indeed, in zebrafish, fast motoneurons have been shown to receive strong phasic excitatory synaptic input preferentially during fast swims (Kishore et al. 2014). The ability to label distinct spinal interneuronal types in zebrafish and mice has enabled investigations into motoneuron type-specific connectivity within the spinal cord. Such studies have asked whether specific groups of interneurons are recruited during slow or fast movements and if they are preferentially connected to slow and fast motoneurons respectively. The view that has emerged from these studies is that in contrast to motoneurons, interneurons are selectively activated or silenced at different locomotor speeds such that distinct populations of interneurons drive motoneuronal activation at these speeds (McLean et al. 2007, 2008; Kimura and Higashijima 2019; Satou et al. 2020) (Fig. 3). However, incremental recruitment of interneurons as speed increases, similar to the recruitment of motoneurons has also been reported (Ausborn et al. 2012; Ampatzis et al. 2014; Björnfors and El Manira 2016).

Molecular analyses of embryonic mammalian spinal cords have revealed domains along the dorso-ventral axis, where distinct transcription factors are expressed. Neuronal progenitors within these transcription factor domains give rise to neurons that are broadly classified into different classes of dorsal and ventral interneuronal, and motoneuronal populations. The ventral interneurons, members of spinal CPG networks, are in turn sub-divided into V0, V1, V2, and V3 populations, with V0 being the dorsal-most and V3 being the ventral-most (Jessell 2000; Alaynick et al. 2011; Catela et al. 2015). However, these groups are collections of diverse neuronal types in terms of neurotransmitter phenotype, projection patterns, molecular composition and function during locomotion (see below). All four groups of interneurons are active during locomotor episodes and serve distinct functions to generate the rhythm (Gosgnach et al. 2017). For some populations, preferential recruitment during slow or fast locomotion and preferential connectivity to slow and fast motoneurons have been demonstrated. These studies are briefly discussed below.

V0 interneurons can be identified and genetically labeled based on their expression of the transcription factor Dbx1 (Pierani et al. 2001). V0 neurons are commissural, and are divided into two types: excitatory and ventrally located $V0_V$ and inhibitory and more dorsally located $V0_D$ (Lanuza et al. 2004; Satou et al. 2012). A small population of cholinergic $V0_C$ neurons is also present, which projects to ipsilateral motoneurons and may modulate their firing rates (Zagoraiou et al. 2009). In general, V0 neurons regulate left-right alternation but $V0_V$ regulate left-right alternation during high frequency locomotor rhythms while $V0_D$ are required for left-right alternation at low frequencies (Talpalar et al. 2013; Bellardita and Kiehn 2015). Recent work in adult zebrafish shows that even within the $V0_V$ class, there are two types: those that are rhythmic during swim bouts and those that are not. The rhythmic $V0_V$ neurons show preferential recruitment at slow, intermediate or fast swim speeds where the recruitment order of $V0_V$ interneurons is determined by a combination of intrinsic and synaptic properties (Björnfors and El Manira 2016). Study of V0 interneurons in larval zebrafish also shows them to be important for left-right alternation, while differing with adult zebrafish and mice in the frequencies at which they are recruited (McLean et al. 2008; Satou et al. 2020). Monosynaptic connections between V0 interneurons and motoneurons have been demonstrated (McLean et al. 2008; Satou et al. 2020), however specific connectivity to different motoneuronal classes has not yet been comprehensively determined.

V1 interneurons are ipsilaterally projecting inhibitory interneurons marked by the expression of the transcription factor Engrailed1 (Higashijima et al. 2004; Li et al. 2004; Gosgnach et al. 2006). These neurons give rise to several distinct classes of ipsilateral inhibitory interneurons (Alvarez et al. 2005; Bikoff et al. 2016). In larval zebrafish and tadpoles, V1 interneurons provide in-phase inhibition to ipsilateral premotor CPG neurons and motoneurons (Higashijima et al. 2004; Li et al. 2004; Kimura and Higashijima 2019). Studies in zebrafish show that V1 neurons can also be sub-divided based on the frequencies at which they are active. Fast-type V1 interneurons fire during fast swims and provide strong in-phase inhibition to both slow and fast type motoneurons. In-phase inhibition provided to fast motoneurons is critical for terminating the burst and therefore increases the frequency of bursts in fast motoneurons. Slow-type V1s are silent during fast swims and fire during slow swims, terminating the burst in slow motoneurons (Kimura and Higashijima 2019). In juvenile mice also, V1 interneurons restrict the burst duration of flexor motoneurons and promote the transition from swing to stance (Zhang et al. 2014; Britz et al. 2015). Consistent with these results, removing V1 input leads to slowing of the motor rhythm in mammals and fish (Gosgnach et al. 2006; Falgairolle and O'Donovan 2019a; Kimura and Higashijima 2019). Further investigations into patterns of connectivity between V1 neurons and distinct motoneuronal classes in mammals are necessary to delineate how they aid in selective recruitment of motoneurons.

V2 interneurons are ipsilaterally projecting mixed type neurons, expressing the transcription factor Lhx3 and further subdivided into excitatory V2a and inhibitory V2b interneurons (Karunaratne et al. 2002). The role of the V2a class in motoneuronal recruitment is better understood compared to other ventrally located interneurons (Dougherty and Kiehn 2010a; Berg et al. 2018). V2a interneurons are the primary source of excitation to motoneurons and are genetically labelled by the transcription factor Chx10 (Kimura et al. 2006; Al-Mosawie et al. 2007; Lundfald et al. 2007). They contribute to important functions in the spinal cord like maintaining left-right coordination (Crone et al. 2008, 2009; Dougherty and Kiehn 2010b; Zhong et al. 2010) and providing excitatory locomotor drive (Eklöf-Ljunggren et al. 2012, 2014; Sternberg et al. 2016).

The heterogeneity of V2a interneurons, in terms of their molecular composition, morphology, intrinsic electrical properties and connectivity (Ausborn et al. 2012; Kimura et al. 2006; Dougherty and Kiehn 2010b; Zhong et al. 2010; Menelaou et al. 2014; Hayashi et al. 2018; Menelaou and McLean 2019), results in different populations of V2a being recruited at different locomotor frequencies (McLean et al. 2007, 2008; Crone et al. 2009; Zhong et al. 2010; Ausborn et al. 2012). V2a interneurons in mice and zebrafish were shown to belong to type I or type II with type I neurons having descending axons while type II neurons have a bifurcating axon that projects in the ascending and descending directions (Dougherty and Kiehn 2010b; Menelaou et al. 2014; Hayashi et al. 2018; Menelaou and McLean 2019). In addition, the two types differ in their expression levels of Chx10, passive and active membrane properties, and in connectivity to other V2a neurons, V0 neurons and motoneurons. Type I neurons seem to provide stronger inputs to V0 interneurons, while type II provide strong excitation to motoneurons (Hayashi et al. 2018; Song et al. 2018; Menelaou and McLean 2019).

In zebrafish, V2a neurons were also shown to fall into 'slow', 'intermediate' or 'fast' classes depending on their preferential firing during slow, intermediate and fast rhythms respectively. Within each of these pools, type I and type II V2a neurons were present at varying proportions. The slow module consisted predominantly of type I neurons while the fast module consisted predominantly of type II neurons (Song et al. 2020). Slow V2a neurons are preferentially connected to slow motoneurons, intermediate V2as to intermediate motoneurons and fast V2as to fast motoneurons (Ampatzis et al. 2014). Recurrent excitation via electrical, mixed and/or glutamatergic synapses between V2a neurons is also pool specific with preferential connectivity within but not across pools (Menelaou and McLean 2019; Song et al. 2020). These results suggest that the V2a and motoneuronal speed modules comprise distinct microcircuits that are engaged to generate different speeds of movement. In addition, several recent studies show that motoneurons are able to send feedback to the CPG and regulate the rhythm (Song et al. 2016; Falgairolle et al. 2017; Falgairolle and O'Donovan 2019b). In zebrafish, this is likely mediated via bidirectional gap junctions between motoneurons and their respective V2a pools

(Kimura et al. 2006; Song et al. 2016). In contrast, studies in neonatal mice have shown that motoneurons and V2as are not electrically coupled (Bhumbra and Beato 2018) and other mechanisms are probably at play for motoneuronal modulation of locomotor frequency (Falgairolle et al. 2017; Falgairolle and O'Donovan 2019b).

V2b neurons express GATA2/3 and are GABA/glycinergic inhibitory interneurons (Karunaratne et al. 2002). Together with V1 interneurons, V2b ensure alternation between ipsilateral flexors and extensors in the neonatal mouse spinal cord (Zhang et al. 2014; Britz et al. 2015). In larval zebrafish, two different populations of V2b were identified: ones that were purely glycinergic and those that were GABA and glycinergic. V2bs used both GABA and glycine to inhibit slow motoneurons while the V2b input to fast motoneurons was pure glycinergic. V2b input was shown to regulate swim speed by reducing tail beat frequency (Callahan et al. 2019).

The V3 class of interneurons are an interesting group of commissural neurons that express the transcription factor Sim1 and make monosynaptic excitatory connections with contralateral motoneurons. They have been shown to be important for maintaining a stable locomotor pattern (Zhang et al. 2008) and in controlling left-right asymmetry of motion such as those required to make turns (Danner et al. 2019). However their recruitment patterns or differential innervation of motoneuronal types have not yet been investigated.

In sum, it is clear that synaptic drive to motoneurons varies across classes and is critical for determining their recruitment order. The above studies clearly show that premotor excitatory and inhibitory drive is not the same for all motoneurons and in fact motoneurons and their premotor synaptic cohort could be organized into distinct modules. These modules are differentially recruited depending on the speed of locomotion (Fig. 3).

Output properties are also distinct in slow and fast type motoneurons. This was demonstrated in zebrafish, where it was observed that connections made by fast and slow type motoneurons with their target muscles differed in strength and activity-dependent plasticity. Fast-type motoneurons formed stronger and more reliable synapses with muscle fibers and these synapses underwent depression during repetitive high frequency motoneuronal spiking. In contrast, slow-type motoneurons formed weaker synapses with highly variable output. These synapses underwent potentiation making the response of slow type motoneurons consistent after repetitive firing as would occur during continuous swimming (Wang and Brehm 2017). Thus, differences at the level of both input and output of motoneuronal classes match motoneuronal recruitment patterns.

5 Neuromodulation

From the above, it can be seen that the recruitment order of motoneurons is a function of their passive and active properties as well as the excitatory and inhibitory synaptic drive they receive. These properties are under the constant influence of neuromodulators such as monoamines and peptides (Marder and Thirumalai 2002;

Marder 2012). In particular, the monoamines serotonin, norepinephrine and dopamine have received much attention with regards to their effects on locomotor networks and the cellular mechanisms involved (Miles and Sillar 2011). Serotonergic and dopaminergic neuromodulatory nuclei with known projections to the spinal cord show activity that is correlated with movement (Veasey et al. 1995; Jay et al. 2015). Both serotonin and norepinephrine lower the action potential threshold in spinal motoneurons (Fedirchuk and Dai 2004). Serotonin also modulates several ion channel and synaptic receptor classes in spinal motoneurons including leak channels, inward rectifying potassium channels, the 'sag' current I_h and NMDAR currents (Takahashi and Berger 1990; Maclean et al. 1998; Kjaerulff and Kiehn 2001; MacLean and Schmidt 2001). In addition to these, PICs, described in detail in Sect. 3, are also an important target of neuromodulators like serotonin and norepinephrine (Heckman et al. 2009). Serotonin induces plateau potentials in motoneurons via the activation of PICs (Hounsgaard et al. 1988; Hounsgaard and Kiehn 1989; Perrier and Hounsgaard 2003) leading to bistability of resting membrane potential. As seen earlier, motoneurons of different classes exhibit bistability to varying degrees (Lee and Heckman 1998b). Motoneuronal dendrites are richly innervated by serotonergic fibers (Alvarez et al. 1998) and are also where PICs are located (Heckman et al. 2003). This enables serotonergic input to amplify synaptic input in the dendrites via the activation of PICs, in turn implying that neuromodulators like serotonin can amplify synaptic input in S-type motoneurons to a greater extent than they do in F-type motoneurons. This would be important for motoneuronal recruitment as this would allow fatigue resistant S type units to generate long duration self-sustained firing.

Serotonin can also affect motoneuronal firing indirectly by modulating the premotor network. Serotonin depolarizes V2a neurons, increases their input resistance and induces membrane bistability in them (Zhong et al. 2010; Husch et al. 2015). Serotonin also increases the excitability of ascending commissural interneurons, some of whom are members of the V0 pool, by depolarizing these neurons and by altering the shape of the action potential and the spike after-hyperpolarization (Zhong et al. 2006).

Dopamine increases the excitability of motoneurons (Han et al. 2007; Jha and Thirumalai 2020) and reduces the first spike latency by acting on multiple classes of potassium channels (Han et al. 2007). Further, dopamine strengthens glutamatergic synaptic input arriving in motoneurons (Han and Whelan 2009). Thus, at least for monoamines, the evidence that neuromodulation alters motoneuronal firing properties is strong.

Though these studies describe the effects of neuromodulators at a cellular or network level, how neuromodulators like dopamine act on motoneuronal firing properties in a behavioral context was not clearly understood. Recently, dopamine was shown to alter motoneuronal firing rates and recruitment patterns during the optomotor response in larval zebrafish. During this behavior, larvae generate slow swims by engaging many slow motoneurons while some slow motoneurons and all fast motoneurons remain silent. Activation of D1-like dopamine receptors (D1-like-R) decreased spike latency and action potential threshold in both classes of motoneuron. Further, D1-like-R activation also increased the excitatory drive to motoneurons (Fig. 4a). All of these changes at the cellular level enhanced action potential firing in slow motoneurons. Additionally, the silent slow motoneurons and fast motoneurons were also recruited to fire during the optomotor response, ultimately causing the fish to make larger tail bends and therefore, swim faster (Jha and Thirumalai 2020) (Fig. 4a). It follows from this study that the recruitment of motoneurons during a given task is dependent on the neuromodulatory context. This study also provides evidence that neuromodulators like dopamine can affect the force generated by involving both rate modulation and recruitment patterns of motoneurons.

Finally, neuromodulators can switch behavioral outputs by biasing motoneurons towards one set of inputs. During escape response in adult zebrafish, fast motoneurons receive strong excitatory inputs while slow motoneurons are inhibited from spiking as stated above (Ampatzis et al. 2013). This hard-wired circuit is acted upon by the endocannabinoid 2-arachidonoyl glycerol (2-AG) that differentially modulates both slow and fast units. 2-AG promotes potentiation of both excitatory input to fast motoneurons and inhibitory inputs to slow motoneurons (Fig. 4b). Endocannabinoid modulation of the escape and swim modules promotes preferential selection of less excitable fast units and suppression of slow units. Thus under conditions of high endocannabinoid modulation, the fast motor units are easily activated and fish are biased to execute escapes even at low stimulus intensities (Song et al. 2015) (Fig. 4b). In sum, though recruitment pattern is dictated by the carefully tuned intrinsic and synaptic properties of motoneurons, neuromodulators can alter these to dramatically alter motoneuronal recruitment patterns during behavior.

6 Conclusion

Investigations over the last 100 years have now led to a deeper understanding of the rules that govern motoneuron excitability and recruitment patterns. We now appreciate that pre-motor circuits are heterogenous and likely deliver precise patterns of excitation and inhibition to motoneurons depending on the motor command. However, there are still many open questions such as how the recruitment patterns are set-up during development or how descending motor command is translated into distinct recruitment patterns. The availability of modern tools to label neuronal types and image and manipulate their activity during locomotor behavior will allow us to answer many of these open questions in the future.

Acknowledgements The authors would like to thank the following sources of funding support: Wellcome Trust-DBT India Alliance Intermediate (VT; 500040/Z/09/Z) and Senior fellowships (VT; IA/S/17/2/503297), Department of Biotechnology (VT; BT/PR4983/MED/30/790/2012), Science and Engineering Research Board, Department of Science and Technology (VT; EMR/2015/000595), and CSIR-UGC fellowship (UJ). We acknowledge support of the Department of Atomic Energy, Government of India, under project no. 12-R & D-TFR-5.04-0800. We thank Mr. Aalok Varma for comments on this manuscript.

References

Adrian ED, Bronk DW (1929) The discharge of impulses in motor nerve fibres. J Physiol 67:i3–i151
Alaynick WA, Jessell TM, Pfaff SL (2011) SnapShot: spinal cord development. Cell 146: 178–178.e1
Al-Mosawie A, Wilson JM, Brownstone RM (2007) Heterogeneity of V2-derived interneurons in the adult mouse spinal cord. Eur J Neurosci 26:3003–3015
Alvarez FJ, Pearson JC, Harrington D, Dewey D, Torbeck L, Fyffe RE (1998) Distribution of 5-hydroxytryptamine-immunoreactive boutons on alpha-motoneurons in the lumbar spinal cord of adult cats. J Comp Neurol 393:69–83
Alvarez FJ, Jonas PC, Sapir T, Hartley R, Berrocal MC, Geiman EJ, Todd AJ, Goulding M (2005) Postnatal phenotype and localization of spinal cord V1 derived interneurons. J Comp Neurol 493:177–192
Ampatzis K, Song J, Ausborn J, El Manira A (2013) Pattern of innervation and recruitment of different classes of motoneurons in adult zebrafish. J Neurosci 33:10875–10886
Ampatzis K, Song J, Ausborn J, El Manira A (2014) Separate microcircuit modules of distinct v2a interneurons and motoneurons control the speed of locomotion. Neuron 83:934–943
Arber S (2012) Motor circuits in action: specification, connectivity, and function. Neuron 74:975–989
Ausborn J, Mahmood R, El Manira A (2012) Decoding the rules of recruitment of excitatory interneurons in the adult zebrafish locomotor network. Proc Natl Acad Sci U S A 109:E3631–E3639
Azevedo AW, Dickinson ES, Gurung P, Venkatasubramanian L, Mann R, Tuthill JC (2019) A size principle for leg motor control in Drosophila. bioRxiv:730218
Bakels R, Kernell D (1993) Matching between motoneurone and muscle unit properties in rat medial gastrocnemius. J Physiol 463:307–324
Bawa P, Binder MD, Ruenzel P, Henneman E (1984) Recruitment order of motoneurons in stretch reflexes is highly correlated with their axonal conduction velocity. J Neurophysiol 52:410–420
Bellardita C, Kiehn O (2015) Phenotypic characterization of speed-associated gait changes in mice reveals modular organization of locomotor networks. Curr Biol 25:1426–1436
Berg EM, Björnfors ER, Pallucchi I, Picton LD, El Manira A (2018) Principles governing locomotion in vertebrates: lessons from zebrafish. Front Neural Circuits 12. Available at https://www.ncbi.nlm.nih.gov/pmc/articles/PMC6146226/. Accessed 22 Apr 2020
Bhumbra GS, Beato M (2018) Recurrent excitation between motoneurones propagates across segments and is purely glutamatergic. PLoS Biol 16:e2003586
Bikoff JB, Gabitto MI, Rivard AF, Drobac E, Machado TA, Miri A, Brenner-Morton S, Famojure E, Diaz C, Alvarez FJ, Mentis GZ, Jessell TM (2016) Spinal inhibitory interneuron diversity delineates variant motor microcircuits. Cell 165:207–219
Binder MD, Powers RK, Heckman CJ (2019) Nonlinear input-output functions of motoneurons. Physiology 35:31–39
Björnfors ER, El Manira A (2016) Functional diversity of excitatory commissural interneurons in adult zebrafish. elife 5:e18579
Britz O, Zhang J, Grossmann KS, Dyck J, Kim JC, Dymecki S, Gosgnach S, Goulding M (2015) A genetically defined asymmetry underlies the inhibitory control of flexor-extensor locomotor movements. elife 4:e13038
Callahan RA, Roberts R, Sengupta M, Kimura Y, Higashijima S-I, Bagnall MW (2019) Spinal V2b neurons reveal a role for ipsilateral inhibition in speed control. elife 8:e47837
Catela C, Shin MM, Dasen JS (2015) Assembly and function of spinal circuits for motor control. Annu Rev Cell Dev Biol 31:669–698
Crone SA, Quinlan KA, Zagoraiou L, Droho S, Restrepo CE, Lundfald L, Endo T, Setlak J, Jessell TM, Kiehn O, Sharma K (2008) Genetic ablation of V2a ipsilateral interneurons disrupts left-right locomotor coordination in mammalian spinal cord. Neuron 60:70–83
Crone SA, Zhong G, Harris-Warrick R, Sharma K (2009) In mice lacking V2a interneurons, gait depends on speed of locomotion. J Neurosci 29:7098–7109

Cullheim S, Fleshman JW, Glenn LL, Burke RE (1987) Membrane area and dendritic structure in type-identified triceps surae alpha motoneurons. J Comp Neurol 255:68–81
Danner SM, Zhang H, Shevtsova NA, Borowska-Fielding J, Deska-Gauthier D, Rybak IA, Zhang Y (2019) Spinal V3 interneurons and left-right coordination in mammalian locomotion. Front Cell Neurosci 13:516
Davis WJ (1971) Functional significance of motorneuron size and soma position in swimmeret system of the lobster. J Neurophysiol 34:274–288
Denny-Brown D, Pennybacker JB (1938) Fibrillation and fasciculation in voluntary muscle. Brain 61:311–312
Dougherty KJ, Kiehn O (2010a) Functional organization of V2a-related locomotor circuits in the rodent spinal cord. Ann N Y Acad Sci 1198:85–93
Dougherty KJ, Kiehn O (2010b) Firing and cellular properties of V2a interneurons in the rodent spinal cord. J Neurosci 30:24–37
Eccles JC, Eccles RM, Lundberg A (1957) Durations of after-hyperpolarization of motoneurones supplying fast and slow muscles. Nature 179:866–868
Eccles JC, Eccles RM, Lundberg A (1958) The action potentials of the alpha motoneurones supplying fast and slow muscles. J Physiol 142:275–291
Eisen JS (1991) Motoneuronal development in the embryonic zebrafish. Dev Suppl Suppl 2:141–147
Eklöf-Ljunggren E, Haupt S, Ausborn J, Dehnisch I, Uhlén P, Higashijima S, El Manira A (2012) Origin of excitation underlying locomotion in the spinal circuit of zebrafish. Proc Natl Acad Sci U S A 109:5511–5516
Eklöf-Ljunggren E, Haupt S, Ausborn J, Ampatzis K, El Manira A (2014) Optogenetic activation of excitatory premotor interneurons is sufficient to generate coordinated locomotor activity in larval zebrafish. J Neurosci 34:134–139
Falgairolle M, O'Donovan MJ (2019a) V1 interneurons regulate the pattern and frequency of locomotor-like activity in the neonatal mouse spinal cord. PLoS Biol 17:e3000447
Falgairolle M, O'Donovan MJ (2019b) Feedback regulation of locomotion by motoneurons in the vertebrate spinal cord. Curr Opin Physiol 8:50–55
Falgairolle M, Puhl JG, Pujala A, Liu W, O'Donovan MJ (2017) Motoneurons regulate the central pattern generator during drug-induced locomotor-like activity in the neonatal mouse. elife 6:e26622
Fedirchuk B, Dai Y (2004) Monoamines increase the excitability of spinal neurones in the neonatal rat by hyperpolarizing the threshold for action potential production. J Physiol Lond 557:355–361
Fleshman JW, Munson JB, Sypert GW, Friedman WA (1981) Rheobase, input resistance, and motor-unit type in medial gastrocnemius motoneurons in the cat. J Neurophysiol 46:1326–1338
Gabriel JP, Scharstein H, Schmidt J, Büschges A (2003) Control of flexor motoneuron activity during single leg walking of the stick insect on an electronically controlled treadwheel. J Neurobiol 56:237–251
Gabriel JP, Ausborn J, Ampatzis K, Mahmood R, Eklöf-Ljunggren E, El Manira A (2011) Principles governing recruitment of motoneurons during swimming in zebrafish. Nat Neurosci 14:93–99
Gosgnach S, Lanuza GM, Butt SJB, Saueressig H, Zhang Y, Velasquez T, Riethmacher D, Callaway EM, Kiehn O, Goulding M (2006) V1 spinal neurons regulate the speed of vertebrate locomotor outputs. Nature 440:215–219
Gosgnach S, Bikoff JB, Dougherty KJ, El Manira A, Lanuza GM, Zhang Y (2017) Delineating the diversity of spinal interneurons in locomotor circuits. J Neurosci 37:10835–10841
Goulding M (2009) Circuits controlling vertebrate locomotion: moving in a new direction. Nat Rev Neurosci 10:507–518
Gustafsson B, Pinter MJ (1984a) Relations among passive electrical properties of lumbar alpha-motoneurones of the cat. J Physiol Lond 356:401–431
Gustafsson B, Pinter MJ (1984b) An investigation of threshold properties among cat spinal alpha-motoneurones. J Physiol Lond 357:453–483

Hamm TM, Turkin VV, Bandekar NK, O'Neill D, Jung R (2010) Persistent currents and discharge patterns in rat hindlimb motoneurons. J Neurophysiol 104:1566–1577
Han P, Whelan PJ (2009) Modulation of AMPA currents by D(1)-like but not D(2)-like receptors in spinal motoneurons. Neuroscience 158:1699–1707
Han P, Nakanishi ST, Tran MA, Whelan PJ (2007) Dopaminergic modulation of spinal neuronal excitability. J Neurosci 27:13192–13204
Hayashi M, Hinckley CA, Driscoll SP, Moore NJ, Levine AJ, Hilde KL, Sharma K, Pfaff SL (2018) Graded arrays of spinal and supraspinal V2a interneuron subtypes underlie forelimb and hindlimb motor control. Neuron 97:869–884.e5
Heckman CJ, Binder MD (1988) Analysis of effective synaptic currents generated by homonymous Ia afferent fibers in motoneurons of the cat. J Neurophysiol 60:1946–1966
Heckman CJ, Lee RH, Brownstone RM (2003) Hyperexcitable dendrites in motoneurons and their neuromodulatory control during motor behavior. Trends Neurosci 26:688–695
Heckman CJ, Gorassini MA, Bennett DJ (2005) Persistent inward currents in motoneuron dendrites: implications for motor output. Muscle Nerve 31:135–156
Heckman CJ, Mottram C, Quinlan K, Theiss R, Schuster J (2009) Motoneuron excitability: the importance of neuromodulatory inputs. Clin Neurophysiol 120:2040–2054
Henneman E (1957) Relation between size of neurons and their susceptibility to discharge. Science 126:1345–1347
Henneman E, Olson CB (1965) Relations between structure and function in the design of skeletal muscles. J Neurophysiol 28:581–598
Henneman E, Somjen G, Carpenter DO (1965a) Functional significance of cell size in spinal motoneurons. J Neurophysiol 28:560–580
Henneman E, Somjen G, Carpenter DO (1965b) Excitability and inhibitibility of motoneurons of different sizes. J Neurophysiol 28:599–620
Higashijima S, Masino MA, Mandel G, Fetcho JR (2004) Engrailed-1 expression marks a primitive class of inhibitory spinal interneuron. J Neurosci 24:5827–5839
Hill AAV, Cattaert D (2008) Recruitment in a heterogeneous population of motor neurons that innervates the depressor muscle of the crayfish walking leg muscle. J Exp Biol 211:613–629
Hounsgaard J, Kiehn O (1989) Serotonin-induced bistability of turtle motoneurones caused by a nifedipine-sensitive calcium plateau potential. J Physiol Lond 414:265–282
Hounsgaard J, Hultborn H, Jespersen B, Kiehn O (1988) Bistability of alpha-motoneurones in the decerebrate cat and in the acute spinal cat after intravenous 5-hydroxytryptophan. J Physiol Lond 405:345–367
Husch A, Dietz SB, Hong DN, Harris-Warrick RM (2015) Adult spinal V2a interneurons show increased excitability and serotonin-dependent bistability. J Neurophysiol 113:1124–1134
Jay M, De Faveri F, McDearmid JR (2015) Firing dynamics and modulatory actions of supraspinal dopaminergic neurons during zebrafish locomotor behavior. Curr Biol 25:435–444
Jessell TM (2000) Neuronal specification in the spinal cord: inductive signals and transcriptional codes. Nat Rev Genet 1:20–29
Jha U, Thirumalai V (2020) Neuromodulatory selection of motor neuron recruitment patterns in a visuomotor behavior increases speed. Curr Biol 30:788–801.e3
Kanda K, Burke RE, Walmsley B (1977) Differential control of fast and slow twitch motor units in the decerebrate cat. Exp Brain Res 29:57–74
Karunaratne A, Hargrave M, Poh A, Yamada T (2002) GATA proteins identify a novel ventral interneuron subclass in the developing chick spinal cord. Dev Biol 249:30–43
Kernell D (2003) Principles of force gradation in skeletal muscles. Neural Plast 10:69–76
Kimura Y, Higashijima S (2019) Regulation of locomotor speed and selection of active sets of neurons by V1 neurons. Nat Commun 10:1–12
Kimura Y, Okamura Y, Higashijima S (2006) alx, a zebrafish homolog of Chx10, marks ipsilateral descending excitatory interneurons that participate in the regulation of spinal locomotor circuits. J Neurosci 26:5684–5697

Kishore S, Bagnall MW, McLean DL (2014) Systematic shifts in the balance of excitation and inhibition coordinate the activity of axial motor pools at different speeds of locomotion. J Neurosci 34:14046–14054
Kjaerulff O, Kiehn O (2001) 5-HT modulation of multiple inward rectifiers in motoneurons in intact preparations of the neonatal rat spinal cord. J Neurophysiol 85:580–593
Lanuza GM, Gosgnach S, Pierani A, Jessell TM, Goulding M (2004) Genetic identification of spinal interneurons that coordinate left-right locomotor activity necessary for walking movements. Neuron 42:375–386
Lee RH, Heckman CJ (1998a) Bistability in spinal motoneurons in vivo: systematic variations in rhythmic firing patterns. J Neurophysiol 80:572–582
Lee RH, Heckman CJ (1998b) Bistability in spinal motoneurons in vivo: systematic variations in persistent inward currents. J Neurophysiol 80:583–593
Leroy F, Lamotte d'Incamps B, Imhoff-Manuel RD, Zytnicki D (2014) Early intrinsic hyperexcitability does not contribute to motoneuron degeneration in amyotrophic lateral sclerosis Kiehn O, ed. elife 3:e04046
Leroy F, Lamotte d'Incamps B, Zytnicki D (2015) Potassium currents dynamically set the recruitment and firing properties of F-type motoneurons in neonatal mice. J Neurophysiol 114:1963–1973
Li W-C, Higashijima S, Parry DM, Roberts A, Soffe SR (2004) Primitive roles for inhibitory interneurons in developing frog spinal cord. J Neurosci 24:5840–5848
Lundfald L, Restrepo CE, Butt SJB, Peng C-Y, Droho S, Endo T, Zeilhofer HU, Sharma K, Kiehn O (2007) Phenotype of V2-derived interneurons and their relationship to the axon guidance molecule EphA4 in the developing mouse spinal cord. Eur J Neurosci 26:2989–3002
MacLean JN, Schmidt BJ (2001) Voltage-sensitivity of motoneuron NMDA receptor channels is modulated by serotonin in the neonatal rat spinal cord. J Neurophysiol 86:1131–1138
Maclean JN, Cowley KC, Schmidt BJ (1998) NMDA receptor-mediated oscillatory activity in the neonatal rat spinal cord is serotonin dependent. J Neurophysiol 79:2804–2808
Manuel M, Zytnicki D (2019) Molecular and electrophysiological properties of mouse motoneuron and motor unit subtypes. Curr Opin Physiol 8:23–29
Manuel M, Meunier C, Donnet M, Zytnicki D (2007) Resonant or not, two amplification modes of proprioceptive inputs by persistent inward currents in spinal motoneurons. J Neurosci 27:12977–12988
Manuel M, Iglesias C, Donnet M, Leroy F, Heckman CJ, Zytnicki D (2009) Fast kinetics, high-frequency oscillations, and subprimary firing range in adult mouse spinal motoneurons. J Neurosci 29:11246–11256
Marder E (2012) Neuromodulation of neuronal circuits: back to the future. Neuron 76:1–11
Marder E, Thirumalai V (2002) Cellular, synaptic and network effects of neuromodulation. Neural Netw 15:479–493
McLarnon JG (1995) Potassium currents in motoneurones. Prog Neurobiol 47:513–531
McLean DL, Dougherty KJ (2015) Peeling back the layers of locomotor control in the spinal cord. Curr Opin Neurobiol 33:63–70
McLean DL, Fan J, Higashijima S, Hale ME, Fetcho JR (2007) A topographic map of recruitment in spinal cord. Nature 446:71–75
McLean DL, Masino MA, Koh IYY, Lindquist WB, Fetcho JR (2008) Continuous shifts in the active set of spinal interneurons during changes in locomotor speed. Nat Neurosci 11:1419–1429
Mcphedran AM, Wuerker RB, Henneman E (1965) Properties of motor units in a homogeneous red muscle (soleus) of the cat. J Neurophysiol 28:71–84
Mendell LM (2005) The size principle: a rule describing the recruitment of motoneurons. J Neurophysiol 93:3024–3026
Menelaou E, McLean DL (2012) A gradient in endogenous rhythmicity and oscillatory drive matches recruitment order in an axial motor pool. J Neurosci 32:10925–10939
Menelaou E, McLean DL (2019) Hierarchical control of locomotion by distinct types of spinal V2a interneurons in zebrafish. Nat Commun 10:1–12

Menelaou E, VanDunk C, McLean DL (2014) Differences in the morphology of spinal V2a neurons reflect their recruitment order during swimming in larval zebrafish. J Comp Neurol 522:1232–1248
Miles GB, Sillar KT (2011) Neuromodulation of vertebrate locomotor control networks. Physiology (Bethesda) 26:393–411
Milner-Brown HS, Stein RB, Yemm R (1973) The orderly recruitment of human motor units during voluntary isometric contractions. J Physiol Lond 230:359–370
Munson JB, Sypert GW, Zengel JE, Lofton SA, Fleshman JW (1982) Monosynaptic projections of individual spindle group II afferents to type-identified medial gastrocnemius motoneurons in the cat. J Neurophysiol 48:1164–1174
Myers PZ (1985) Spinal motoneurons of the larval zebrafish. J Comp Neurol 236:555–561
Perrier J-F, Hounsgaard J (2003) 5-HT2 receptors promote plateau potentials in turtle spinal motoneurons by facilitating an L-type calcium current. J Neurophysiol 89:954–959
Pierani A, Moran-Rivard L, Sunshine MJ, Littman DR, Goulding M, Jessell TM (2001) Control of interneuron fate in the developing spinal cord by the progenitor homeodomain protein Dbx1. Neuron 29:367–384
Powers RK, Binder MD (2001) Input-output functions of mammalian motoneurons. Rev Physiol Biochem Pharmacol 143:137–263
Sasaki null, Burrows null (1998) Innervation pattern of a pool of nine excitatory motor neurons in the flexor tibiae muscle of a locust hind leg. J Exp Biol 201(Pt 12):1885–1893
Satou C, Kimura Y, Higashijima S (2012) Generation of multiple classes of V0 neurons in zebrafish spinal cord: progenitor heterogeneity and temporal control of neuronal diversity. J Neurosci 32:1771–1783
Satou C, Sugioka T, Uemura Y, Shimazaki T, Zmarz P, Kimura Y, Higashijima S (2020) Functional diversity of glycinergic commissural inhibitory neurons in larval zebrafish. Cell Rep 30:3036–3050.e4
Schwindt PC (1973) Membrane-potential trajectories underlying motoneuron rhythmic firing at high rates. J Neurophysiol 36:434–439
Schwindt P, Crill WE (1977) A persistent negative resistance in cat lumbar motoneurons. Brain Res 120:173–178
Seven YB, Mantilla CB, Sieck GC (2014) Recruitment of rat diaphragm motor units across motor behaviors with different levels of diaphragm activation. J Appl Physiol 117:1308–1316
Song J, Ampatzis K, Ausborn J, El Manira A (2015) A hardwired circuit supplemented with endocannabinoids encodes behavioral choice in zebrafish. Curr Biol 25:2610–2620
Song J, Ampatzis K, Björnfors ER, El Manira A (2016) Motor neurons control locomotor circuit function retrogradely via gap junctions. Nature 529:399–402
Song J, Dahlberg E, El Manira A (2018) V2a interneuron diversity tailors spinal circuit organization to control the vigor of locomotor movements. Nat Commun 9:3370
Song J, Pallucchi I, Ausborn J, Ampatzis K, Bertuzzi M, Fontanel P, Picton LD, El Manira A (2020) Multiple rhythm-generating circuits act in tandem with pacemaker properties to control the start and speed of locomotion. Neuron 105:1048–1061.e4
Stephens JA, Garnett R, Buller NP (1978) Reversal of recruitment order of single motor units produced by cutaneous stimulation during voluntary muscle contraction in man. Nature 272:362–364
Sternberg JR, Severi KE, Fidelin K, Gomez J, Ihara H, Alcheikh Y, Hubbard JM, Kawakami K, Suster M, Wyart C (2016) Optimization of a neurotoxin to investigate the contribution of excitatory interneurons to speed modulation in vivo. Curr Biol 26:2319–2328
Su C-K, Mellen NM, Feldman JL (1997) Intrinsic and extrinsic factors affecting phrenic motoneuronal excitability in neonatal rats. Brain Res 774:62–68
Takahashi T, Berger AJ (1990) Direct excitation of rat spinal motoneurones by serotonin. J Physiol Lond 423:63–76
Talpalar AE, Bouvier J, Borgius L, Fortin G, Pierani A, Kiehn O (2013) Dual-mode operation of neuronal networks involved in left-right alternation. Nature 500:85–88

Turkin VV, O'Neill D, Jung R, Iarkov A, Hamm TM (2010) Characteristics and organization of discharge properties in rat hindlimb motoneurons. J Neurophysiol 104:1549–1565

Veasey SC, Fornal CA, Metzler CW, Jacobs BL (1995) Response of serotonergic caudal raphe neurons in relation to specific motor activities in freely moving cats. J Neurosci 15:5346–5359

Wang W-C, Brehm P (2017) A gradient in synaptic strength and plasticity among motoneurons provides a peripheral mechanism for locomotor control. Curr Biol 27:415–422

Wuerker RB, McPhedran AM, Henneman E (1965) Properties of motor units in a heterogeneous pale muscle (m. gastrocnemius) of the cat. J Neurophysiol 28:85–99

Zagoraiou L, Akay T, Martin JF, Brownstone RM, Jessell TM, Miles GB (2009) A cluster of cholinergic premotor interneurons modulates mouse locomotor activity. Neuron 64:645–662

Zengel JE, Reid SA, Sypert GW, Munson JB (1985) Membrane electrical properties and prediction of motor-unit type of medial gastrocnemius motoneurons in the cat. J Neurophysiol 53:1323–1344

Zhang B, Wootton JF, Harris-Warrick RM (1995) Calcium-dependent plateau potentials in a crab stomatogastric ganglion motor neuron. II. Calcium-activated slow inward current. J Neurophysiol 74:1938–1946

Zhang Y, Narayan S, Geiman E, Lanuza GM, Velasquez T, Shanks B, Akay T, Dyck J, Pearson K, Gosgnach S, Fan C-M, Goulding M (2008) V3 spinal neurons establish a robust and balanced locomotor rhythm during walking. Neuron 60:84–96

Zhang J, Lanuza GM, Britz O, Wang Z, Siembab VC, Zhang Y, Velasquez T, Alvarez FJ, Frank E, Goulding M (2014) V1 and v2b interneurons secure the alternating flexor-extensor motor activity mice require for limbed locomotion. Neuron 82:138–150

Zhong G, Díaz-Ríos M, Harris-Warrick RM (2006) Serotonin modulates the properties of ascending commissural interneurons in the neonatal mouse spinal cord. J Neurophysiol 95:1545–1555

Zhong G, Droho S, Crone SA, Dietz S, Kwan AC, Webb WW, Sharma K, Harris-Warrick RM (2010) Electrophysiological characterization of V2a interneurons and their locomotor-related activity in the neonatal mouse spinal cord. J Neurosci 30:170–182

Electrical Properties of Adult Mammalian Motoneurons

Calvin C. Smith and Robert M. Brownstone

Abstract Motoneurons are the 'final common path' between the central nervous system (that intends, selects, commands, and organises movement) and muscles (that produce the behaviour). Motoneurons are not passive relays, but rather integrate synaptic activity to appropriately tune output (spike trains) and therefore the production of muscle force. In this chapter, we focus on studies of mammalian motoneurons, describing their heterogeneity whilst providing a brief historical account of motoneuron recording techniques. Next, we describe adult motoneurons in terms of their passive, transition, and active (repetitive firing) properties. We then discuss modulation of these properties by somatic (C-boutons) and dendritic (persistent inward currents) mechanisms. Finally, we briefly describe select studies of human motor unit physiology and relate them to findings from animal preparations discussed earlier in the chapter. This interphyletic approach to the study of motoneuron physiology is crucial to progress understanding of how these diverse neurons translate intention into behaviour.

Keywords Repetitive firing · Spike frequency adaptation · C-boutons · Persistent inward currents · Modulation

1 Introduction

The primary role of a pool of motoneurons is to work in concert to effect muscle contraction. To do so, each motoneuron of the pool involved in the behaviour must fire trains of action potentials at frequencies that cause its innervated muscle fibres

---

This chapter is dedicated to two friends who both lost their lives in April, 2019, Doug Stuart and Tom Jessell. Our knowledge of motoneurons would not be where it is today without their relentless quests to understand these fascinating neurons.

C. C. Smith · R. M. Brownstone (✉)
Department of Neuromuscular Diseases, University College London, London, UK
e-mail: calvin.smith@ucl.ac.uk; r.brownstone@ucl.ac.uk

M. J. O'Donovan, M. Falgairolle (eds.), *Vertebrate Motoneurons*, Advances in Neurobiology 28, https://doi.org/10.1007/978-3-031-07167-6_9

to contract with the required force. To do this, a motoneuron integrates inputs from spinal premotor neurons (in some cases including other motoneurons), brain stem, and forebrain, as well as proprioceptive afferents. These inputs inform the motoneuron's output (spike train), and are comprised of excitatory, inhibitory, and modulatory types. Thus to understand how motor commands are transduced into behaviour, it is necessary to understand how motoneurons integrate these various inputs.

Motoneuron integrative properties have primarily been studied using electrophysiological techniques in which glass micropipettes are used to obtain access to the interior of the cell soma by either impalement (sharp microelectrodes) or negative pressure following membrane seal formation (whole cell patch). Electrical responses can be classified into passive (e.g. whole cell capacitance, input resistance), transition (e.g. rheobase, single action potential characteristics), and repetitive firing properties, and are influenced by many factors. One important factor in integration is geometry. Although variable between cells, motoneurons can be large, with inputs arriving at dendrites that extend over millimetres in larger mammals (Fig. 1). The location of these inputs relative to the integrative site at the axon initial segment influences their "value" in the production of spike trains.

A motoneuron is defined by its output: these are neurons that innervate muscle fibres. (While there are indeed other post-synaptic targets, a key role is to cause muscle contraction; other targets are considered in other chapters.) Somatic motoneurons, to be considered here, innervate skeletal muscle fibres. While this may seem to be a sufficiently restricted definition, there is great diversity within motoneurons within any given animal. That is, there are many subgroups of motoneurons

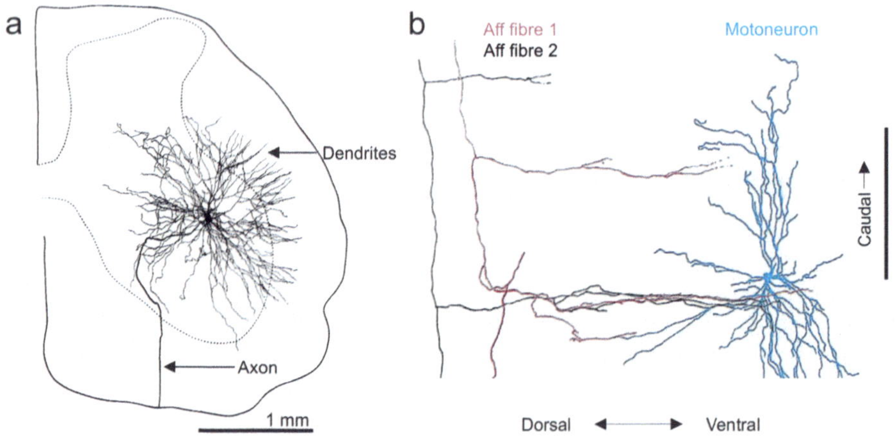

Fig. 1 Anatomy of a motoneuron. (**a**) Intracellularly filled, 3D reconstructed, adult cat triceps surae motoneuron shown in a transverse section of the lumbar spinal cord. *Arrows* show axon (no tapering) and dendritic field, which can extend more that 1 mm from the soma. (**b**) Intracellularly filled, 3D reconstructed, medial gastrocnemius motoneurons (blue) and invading proprioceptive afferent fibres (*Aff,* pink and black). Longitudinal/oblique spinal cord slice. Both scale bars are 1 mm. (**a**) from Ulfhake et al. (1988), Fig. 5d. (**b**) from Burke and Glenn (1996), Fig. 1

within both the brain stem and spinal cord, and these subgroups have different integrative properties.

Somatic motoneurons in the brain stem include those that innervate the extraocular muscles (Evinger 1988), the muscles of mastication (Yamada et al. 2005), of the face (including whisking motoneurons in some species; Kleinfeld et al. 2014), of the oropharynx (Gestreau et al. 2005), of the larynx, and the tongue. Some of these populations have been studied more thoroughly than others, but it is clear that their properties differ from each other. Oculomotor neurons, for example, can fire at rates of up to 400 Hz (Tsuzuki et al. 1995), far faster than trigeminal motoneurons (Chandler et al. 1994), for example.

Motoneurons are located throughout the spinal cord, where they organise into columns during development. The phrenic motor column is in cervical C3–5 segments and innervates the diaphragm. In the cervical and lumbar spinal cord, medial motor column (MMC) motoneurons innervate epaxial (including, for example, erector spinae, multifidus, semispinalis, and splenius) muscle fibres (Tsuchida et al. 1994), and the lateral motor column (LMC) motoneurons are limb-innervating. Motoneurons innervating external urethral and anal sphincter muscles, and those innervating tail muscles are in the sacral spinal cord. LMC motoneurons have been most extensively studied and thus form the basis of much of our knowledge of motoneuron properties.

The LMC itself can be divided into columns (medial (LMC_m) and lateral (LMC_l)), which in turn can be subdivided into columels that contain motor pools (for review see, Stifani 2014). A motor pool is a population of motoneurons that innervates a single muscle (Kanning et al. 2010). There is further granularity within motor pools themselves, as most are comprised of motoneurons with different functions. γ-motoneurons, which are about 30% of many pools, innervate the contractile components of the sensory organs of muscles, muscle spindles (Granit 1975), whereas α-motoneurons innervate extrafusal fibres responsible for force production. In addition, β-motoneurons innervate both spindles and extrafusal fibres. α-motoneurons within a pool can be further sub-divided based on the muscle fibre type they innervate (type I, or slow twitch, type IIa or fast twitch fatigue-resistant, type IIb and IIx, fast twitch fatigable). An α-motoneuron and the muscle fibres it innervates are together called a motor unit (Liddell and Sherrington 1924). These motoneurons have electrophysiological properties that correspond to the fibre types they innervate, and are termed S, FR, and FF (slow, fast fatigue resistant, and fast fatigable) motoneurons (Burke et al. 1973).

Given their diversity, it is no wonder that motoneurons have diverse electrophysiological properties and it is thus impossible to define these properties for an "idealised" motoneuron. In this chapter, we will present a brief history of motoneuron recordings from Lord Adrian to the present day, and then discuss passive, transition, and repetitive firing properties, focusing on the properties and subtypes of spinal LMC α-motoneurons. We also discuss motoneuron modulation by somatic (C-boutons) and dendritic (persistent inward currents) pathways, and finally, human studies of motor unit physiology.

Foundational Studies of Motoneuron Properties

While Sherrington studied motor output from the behavioural viewpoint, it was Lord Adrian who first recorded mammalian motor axons (Adrian and Bronk 1929). These experiments revealed repetitive spike trains from phrenic motor nerves at frequencies which seemingly covaried with the contraction force. Eccles and Hoff (1932) followed this up with a study in which they modified these spike trains using electrical stimulation, leading them to propose that the neurons of origin had a "central excitatory state" that was depressed after each spike. They further concluded that these depressions could summate – this turned out to be prescient of the phenomenon of the summation of post-spike afterhyperpolarisations described by Ito and Oshima (1962).

With the advent of the paraphernalia required for intracellular recording that followed on studies of invertebrate neuronal excitability (Hodgkin and Huxley 1939), two groups began to record cat motoneurons in anaesthetised preparations (Brock et al. 1951; Brownstone 2006 review for a full account). These seminal studies led to an understanding of the basic electrical properties of motoneurons, which were then formalised by Rall and co-workers (Rall 1960; Eccles 1961). Meanwhile, Kernell and colleagues outlined repetitive firing properties, defining the relationships between motoneuron input and output (Granit et al. 1966a, b). Thus, by the end of the 1960s, we had a good understanding of the fundamental passive, transition, and repetitive firing properties of motoneurons.

All of the above experiments were done in vivo, mostly in anaesthetised preparations although some in decerebrate preparations. It became evident that motoneuron properties are "state-dependent" – i.e. different if recorded under anaesthesia vs in the decerebrate preparation, and different during induced locomotor activity than at rest (Brownstone et al. 1992). Most of these early experiments were in cat (with a few in rats), and all were done with sharp electrodes, "blindly" seeking ventral horn neurons that could be identified by their antidromic responses to stimulation of muscle nerves. All were in motoneurons innervating the hindlimb (usually left), and most innervated extensor muscles such as gastrocnemius. And because they were done with sharp electrodes, the probability was in favour of recording the largest motoneurons (type F), although some studies intentionally sampled a wide range (Zengel et al. 1985). In other words, much of our early knowledge of motoneuron electrophysiological properties has been derived from a rather limited selection of these diverse neurons.

The advent of slice preparations and patch clamping led to the study of membrane channels responsible for neuronal properties (Edwards et al. 1989; Dodt and Zieglgänsberger 1990; Sakmann and Neher 1984), which in turn led to the use of spinal cord slice preparations to study motoneuron properties (Takahashi 1978; Konishi and Otsuka 1974). Unfortunately, these studies have largely been confined to neonatal MNs, neurons that have properties that are still under development during their critical period.

2 Passive Properties of Motoneurons and Elementary Cable Theory

As the "final common path" that functions to cause effective muscle contraction, the ultimate goal of a motoneuron is to fire trains of action potentials in response to synaptic inputs to its soma and extensive dendritic tree. These inputs must be integrated to provide sustained depolarisation that is sufficient to initiate repetitive firing at a particular rate. In this section, we will look at the basic factors that govern this integration, which are the passive electrical properties of motoneurons: input resistance, whole cell capacitance, membrane time constant, and length constant. For a more thorough discussion of these properties and their contribution to integration, the reader is directed to Rall (2011).

The input resistance (R_N) is usually measured by injecting small amounts of current to cause voltage deflections that do not activate voltage-sensitive channels (Fig. 2a), and is calculated using Ohm's law:

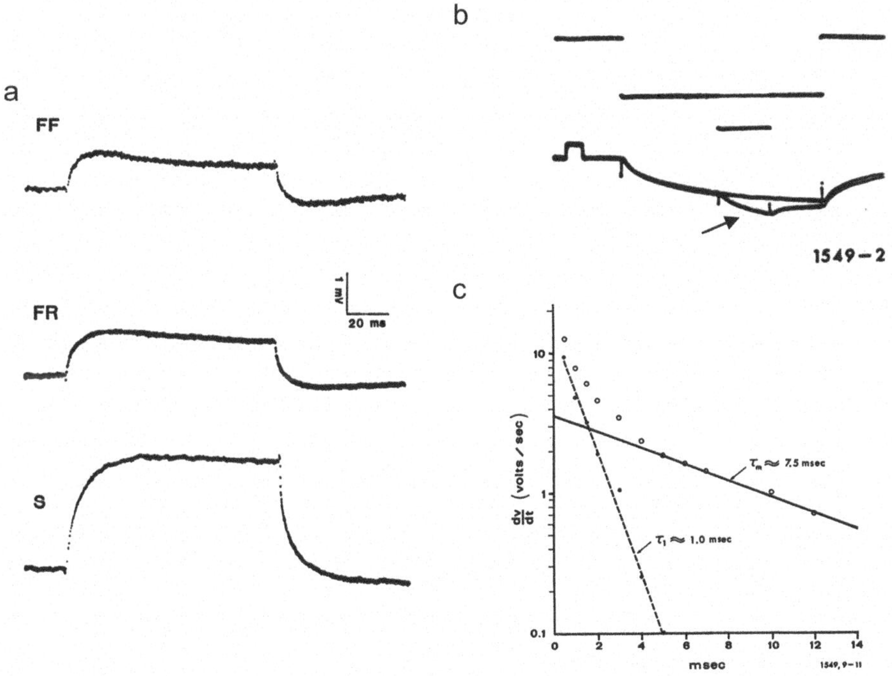

Fig. 2 Passive properties of an adult cat motoneuron. (**a**) Input resistance: Voltage response to current injection in slow (*S*), fast fatigue resistant (*FR*) and fast fatigable (*FF*) medial gastrocnemius motoneurons. (**b**) Time constant: Response recorded intracellularly to hyperpolarising current injection, current in *upper trace*, voltage in *lower trace*. Voltage calibration *pulse* is 10 mV, 1 ms. *Arrow* indicates transient analysed in c. (**c**) Slope vs time of voltage transient in b reveals more than one exponential. By subtracting the long exponential, a second linear component is revealed, showing the data can be approximated by the sum of two exponentials. The ratio of the two time constants is related to the electrotonic length of the dendritic tree. **a** is from Zengel et al. (1985), Fig. 3. (**b** and **c**) from Nelson and Lux (1970), Fig. 5d, 6

$$R_N = V / I \tag{1}$$

where V is the voltage deflection produced by the current (I) injected.

There are several important factors that will impact these measurements:

(a) Leak around the electrode, which is greater for sharp than patch electrodes – this can lead to an order of magnitude difference in measurement (Li et al. 2004a);
(b) "state" during which the measurements are taken, e.g. in vitro vs in vivo, the type of anaesthesia, decerebration (see, for example, Kernell 1999); and
(c) Intactness of the dendritic tree, which will be affected in slice preparations (typically sampled at 300–350 μm thick; Smith and Brownstone 2020), disproportionately affecting larger (e.g. more mature and/or faster) motoneurons. Juvenile mouse motoneuron dendrites can extend at least 0.5–1 mm from the soma (Leroy et al. 2014; Fukuda et al. 2020).

Furthermore, there is a high degree of biological variance between motoneurons, as small (type S) motoneurons have much higher input resistances than large (FF) motoneurons. These differences will also lead to systematic differences in measurements between studies: for example, in vivo studies that "blindly" target motoneurons will tend to bias towards the largest, lowest resistance motoneurons, and slice experiments generally tend to favour the smaller (S) motoneurons as they more readily survive in vitro (Mitra and Brownstone (2012) reported a mean R_N of 123 MΩ for lumbar motoneurons in postnatal day (P)43+ slice preparations, whereas Smith and Brownstone (2020) reported a mean R_N of 31 MΩ at P14–21). Nevertheless, we can make some general statements about input resistance: cat lumbar motoneurons (with sharp electrodes) range from 0.1 to 4 MΩ (Zengel et al. 1985), mouse motoneurons in vivo range from 1 to 12 MΩ (Meehan et al. 2010), and motoneurons from the mature (>P15) mouse lumbar spinal cord in slices (patch recordings) range from 2 to 200 MΩ (Smith and Brownstone 2020; Bhumbra and Beato 2018; Mitra and Brownstone 2012). These differences arise from the factors listed above.

Whole cell capacitance reflects the charge that the neuronal membrane can store, which depends on specific membrane capacity and membrane surface area. Since the former, a property of the lipid bilayer, is the same across all neurons, capacitance measurements reflect the size of the neuron, or the amount of membrane area affected by the current pulse.

Whole cell capacitance (C) is calculated from measurements of membrane time constant, τ and input resistance, R_N:

$$\tau = R_N * C \tag{2}$$

The time constant can be measured by fitting an exponential to the decay of membrane potential following a subthreshold pulse (Fig. 2b, c). The time course of a voltage change depends on τ such that at time t,

$$V = V_0 * e^{-t/\tau} \tag{3}$$

where V_0 is the initial voltage, and assuming that there is perfect space clamp (that is, that voltage is uniform throughout the cell). It can be seen that the time constant represents the time at which the voltage decays by 1/e, or to about 37% of the initial voltage. Typical measurements for capacitance and time constant are, respectively: 1.5 to 15 nF and 3 to 15 ms for cat motoneurons, and 0.2 to 1 nF and 2 to 20 ms for mature mouse motoneurons.

It is important to consider the membrane time constant when examining integration of synaptic inputs: the shorter the time constant, the faster a depolarising potential will rise and decay, thus affecting temporal summation. Temporal summation, or the summation of EPSPs close to each other in time, depends largely on the decay time of the EPSPs as rise times are very fast (see Iansek and Redman 1973a). This decay is more dependent on the time constant of the synaptic conductance and/or the membrane time constant, whichever is greater. Efficient temporal summation requires a frequency of excitatory inputs that at least matches the time constant of EPSP decay. In FF MNs, which have membrane and AMPA receptor time constants in the range of 1–2 ms, proximal unitary EPSPs correspondingly decay rapidly (Burke 1967). This fast τ would thus minimise temporal summation in these neurons, suggesting that close to concurrent excitatory inputs or activation of longer time constant synaptic conductances (such as those mediated by NMDA receptors) are needed for effective summation. But note also that temporal summation at a single location would at best be sublinear: the driving force for the second EPSP would be reduced, and the local conductance would be increased, both effects contributing to sublinear summation.

The basic properties above are, in the first instance, particularly relevant for spherical neurons. But motoneurons have extensive dendritic trees, and thus other factors come into play. Wilfred Rall [1922–2018] applied cable theory to motoneurons in order to put these factors together, so as to understand passive motoneuron properties and how they affect motoneuron integration. The strength of Rall's work on cable theory was its close relationship with experimental neuroscience, each informing the other.

One example of the interaction of theory and experiment arose from the analysis of τ. When measuring τ as above, it can be seen – particularly when very brief current pulses are delivered – that more than one exponential can be fit to the voltage decay (Fig. 2c). This multi-exponential decay results from a lack of space clamp: neurons are not spheres, and the current injected does not lead to a membrane that is isopotential throughout the dendritic tree. That is, there is voltage decay due to current flow both across the membrane and through the cell, along dendritic branches. This latter current flow is dependent on axial resistance, R_i, which in turn depends on cross-sectional area of the process in which the current flows.

The length constant, λ, is an indication of the decay of voltage along the length of dendrite, such that

$$V = V_0 * e^{-x/\lambda} \tag{4}$$

where x is the distance along the dendrite from the voltage change V_0. Thus, a voltage decays to 1/e (~37%) of its value at distance λ. In other words, the larger the length constant, the less decay of the voltage for any given length, and thus the more "effective" that voltage would be at a distance.

The length constant is dependent on the axial resistance and the membrane resistance such that λ can be approximated by:

$$\lambda = \left(r_m / r_i\right)^{1/2} \tag{5}$$

or

$$\lambda = \left[\left(R_m / R_i\right)\left(d / 4\right)\right]^{1/2} \tag{6}$$

where r_m is the resistance across a unit length of membrane (Ω-cm), r_i is the axial resistance per unit length (Ω/cm), R_m is the membrane resistivity (resistance across a unit area of membrane, Ω-cm^2), R_i is the intracellular resistivity (Ω/cm), and d is the diameter of the dendrite (cm).

While it is thus difficult to measure λ, it is possible to measure its consequences. That is, the dendritic tree can be thought of in terms of its electrotonic length, L, or the length of a dendrite (ℓ) in terms of the number of length constants:

$$L = \ell / \lambda \tag{7}$$

L can be estimated by measuring the time constant of current spread within the neuron. By injecting a brief current pulse, two time constants can be peeled from the voltage decay (Nelson and Lux 1970; Rall 1967). The slowest τ (τ_0 or τ_m) would reflect the membrane time constant, and the next slowest, τ_1, and others (τ_2 and theoretically beyond) the time constant of the cable(s), reflecting the equalisation of voltage through the neuron. The ratio of τ_0/τ_1 reflects L. In cat motoneurons, L = 1.0–2.1 (mean 1.5; Iansek and Redman 1973a), indicating that the depolarisation in the soma would be about 22% ($1/e^{1.5}$) of the voltage produced by an excitatory post-synaptic potential (EPSP) at the dendritic tip.

Understanding these principles has proven to be very useful in understanding motoneuron integration. For example, using cable theory and meticulous measurements of cable properties in cat motoneurons, the shapes of EPSPs (rise times and half-widths) can be studied to reveal their location on the dendritic trees and the synaptic currents they produce there (Iansek and Redman 1973b). Doing so revealed that the quantal content of synapses on dendrites are far greater than those on somata.

Thus, in the soma, changes in voltage produced by synaptic inputs throughout the motoneuron are dependent on space and time, and can be expressed by this partial differential equation of cable theory:

$$\lambda^2 \partial^2 V / \partial x^2 - V - \tau \partial V / \partial t = 0 \tag{8}$$

The time constant τ influences temporal summation, and the length constant λ influences spatial summation.

It may seem that cable theory is limited by the underlying assumption that dendritic membranes are passive. However, even so, cable theory is foundational for understanding how active dendrites contribute to motoneuron integration (see below).

3 Motoneuron Transition Properties

Transition properties can be defined to include subthreshold voltage-gated conductances, as well as properties that influence single action potentials. These are the attributes that, while influenced by passive properties, are foundational for the ultimate goal of motoneurons: producing repetitive spike trains.

Sag and Post-inhibitory Rebound

A subthreshold phenomenon that can contribute to motoneuron recruitment is the sag potential. Sag potentials are seen when a rectangular hyperpolarising current is injected to produce a voltage response that after a delay (perhaps ~100 ms; Bayliss et al. 1994) starts to return (rectify) towards resting potential. Upon cessation of the hyperpolarisation, this slow inward current that has been activated causes the membrane voltage to rebound beyond rest, producing post-inhibitory rebound (PIR). The PIR is characterised by 'over-rectification' following inhibitory input, producing a depolarising 'hump'. Voltage clamp studies have shown that conductances mediating sag potentials (I_h) are non-specific cation conductances carried by a family of hyperpolarisation-activated cyclic nucleotide-gated channels (HCN; Zemankovics et al. 2010).

There is evidence that I_h can be modulated by monoaminergic inputs, and thus can contribute to motoneuron activity in a state-dependent manner (Takahashi and Berger 1990; Larkman and Kelly 1992). Activation of I_h during the hyperpolarised phase of the locomotor cycle may also contribute to PIR-induced firing to assist in initiating the active phase of the step cycle (Bertrand and Cazalets 1998; Hochman et al. 1994). Of note, as well as HCN-mediated I_h, a calcium current carried by low voltage-activated Ca^{2+} (Ca_V3) channels can contribute to PIR (Canto-Bustos et al. 2014).

Rheobase

When depolarisation of the axon initial segment is sufficient such that it reaches threshold, an action potential is initiated. The minimum current needed for this to occur, rheobase, depends on the passive properties of the motoneuron (see above), in particular R_N. Thus, smaller S-type motoneurons with higher R_N have lower rheobase currents than the larger, lower R_N FR and FF motoneurons.

In 1957, Henneman described the "size principle," in which motoneurons are recruited by synaptic inputs in order from the smallest to the largest size (Henneman 1957). This "orderly recruitment" of motoneurons in response to stimulation of spindle primary afferents fits with the distribution of rheobase currents across different motoneuron types. That is, given that each primary afferent projects to every motoneuron in its pool (at least in the adult cat; Baldissera et al. 1981), then if synaptic transmission at the fibre terminals produces approximately an equivalent synaptic current, higher currents (produced by increasing stimulation) would be necessary to recruit the higher rheobase, larger motoneurons. These neurons would thereby only be recruited in response to afferent activity greater than would activate the lower rheobase, smaller motoneurons.

Action Potential Characteristics

Changes to spike frequency and spike timing are influenced by changes in motoneuron action potential morphology (Fig. 3a). As all-or-none phenomena, action potentials, regardless of neuron type, have the same phases: threshold, depolarisation (rising), repolarisation (falling), fast after hyperpolarisation (fAHP), after depolarisation (ADP), and medium after hyperpolarisation (mAHP, Fig. 3a). However, due to the relative expression of constitutive ion channels, the duration and/or amplitude of these phases varies between neurons.

Threshold and Rising Phase

The voltage threshold is the voltage at which an action potential is generated; this voltage can change over time and with behaviour. When the membrane potential reaches threshold, voltage-gated sodium channels are activated, initiating a rapid depolarisation of the membrane potential constituting the rising phase of the spike, which usually results in an overshoot greater than 0 mV (Barrett and Crill 1980). Given the reliance of the spike on sodium channel availability, an increase in the proportion of inactivated channels will depolarise the voltage threshold. Therefore, the voltage threshold in response to slow depolarisations is less negative than following rapid depolarisations.

Initial assessments of motoneuron action potentials were done in cats and described 2 distinct 'spike potentials: an 'A' and a 'B' spike (Brock et al. 1951; Fuortes et al. 1957). The small spike is an initial depolarisation to an inflection point followed by a rapid depolarisation to peak. An elegant set of experiments by Coombs et al. (1957) showed that the size of the A and B spikes change relative to the position of the microelectrode between the soma-dendritic (SD) compartment and the axon initial segment (IS). When the electrode penetrates close to the initial axon

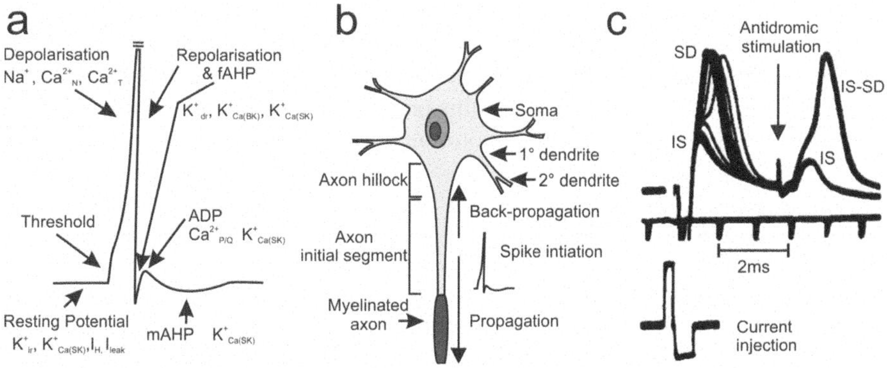

Fig. 3 Motoneuron action potential characteristics and generation. (**a**) Each phase of the action potential identified by *arrows* with a sampling of associated conductances indicated. (**b**) Schematic illustrating motoneuron proximal morphology and highlighting the axon initial segment, where spikes are generated. (**c**) Traces showing initial segment (IS) and somato-dendritic (SD) spikes. The top trace is the membrane potential and shows several large and small spikes evoked by either intracellular current injection (bottom trace, after 1 ms) or antidromic activation via peripheral nerve stimulation (artefact on top trace, after ~3.7 ms). Depolarising current pulses evoked high amplitude spikes, which represent *SD* spikes. When depolarising pulses are followed by a brief hyperpolarising pulse, the *SD* spike is inhibited and reveals the *IS* spike. When "intermediate sized" hyperpolarising pulses at threshold for suppression preceded antidromic stimulation, either *IS* or *IS-SD* spikes were evoked. (**a**) was adapted from Nordstrom et al. (2007), Fig. 1. (**b**) is from Li et al. (2004b), Fig. 2b, and (**c**) is from Coombs et al. (1957), Fig. 11f

segment (Fig. 3b), the A spike is larger than the B spike, whereas if well into the soma, the B spike is larger. Ultimately, this led to the adoption of the more specific nomenclature, with the early phase called the IS spike and the later the SD spike (Fig. 3c). Because the IS spike has a lower threshold and precedes the SD, it was concluded that action potentials are initiated in the axon IS.

The high excitability of the initial segment is due to the differential expression of ion channels, acquired during early embryogenesis (Le Bras et al. 2014). The boundaries of the IS are clearly defined by expression of the scaffolding protein ankyrin G (AnkG), which is crucial for assembly of the IS. The motoneuron IS expresses 2 main sodium channels, $Na_V1.6$ which is expressed throughout the IS, and Na_V 1.1 expressed at the axon hillock (Duflocq et al. 2008, 2011). A variety of potassium channels ($K_V7.2$, $K_V1.1$, $K_V1.2$ and $K_V\beta2$) are also expressed in the IS, ensuring a fast repolarisation of the membrane potential at this site (Garrido et al. 2003; Pan et al. 2006; Lorincz and Nusser 2008).

While many factors influence threshold, motoneuron type does not. Although threshold varies between motoneurons, there is no systematic difference in voltage threshold between fast and slow motoneurons, or between those from cervical and lumbar cord (Pinter et al. 1983; Smith and Brownstone 2020). Thus it is unlikely that threshold contributes to orderly recruitment of the size principle (Henneman 1957).

There is evidence that the voltage threshold of motoneurons is state-dependent, and thus likely subject to neuromodulatory control. During fictive locomotion in the cat, voltage threshold is hyperpolarised in relation to the control threshold (Krawitz et al. 2001). Therefore, modulation of this property is likely another tool available to the nervous system for regulating motor output.

Repolarisation Phase and fAHP

There is a greater diversity of ion channels contributing to the falling compared to the rising phase of the action potential. These include voltage activated delayed rectifier potassium channels, rapidly inactivating potassium channels, and small conductance calcium-dependent potassium (SK) channels (Sah and McLachlan 1992; Schwindt and Crill 1981; Barrett and Barret 1976). Action potentials tend to be wider in slow motoneurons compared to fast (Krutki et al. 2017), suggesting differential expression of the channels involved in the falling phase.

The fast after hyperpolarisation (*f*AHP) marks the end of the action potential and is largely determined by the same currents responsible for the repolarisation phase of the AP. The amplitude and duration of the *f*AHP varies with motoneuron type and species as it is quite prominent in smaller animals such as frogs (Barrett and Barret 1976), toads (Araki and Otani 1955), turtles (Hounsgaard et al. 1988b), and rodents, but less so in the cat. The trough voltage of the fAHP was found to be very constant regardless of how the spike was initiated (Kolmodin and Skoglund 1958), and is thought to represent a high conductance state due to the delayed rectifier potassium conductances (Krnjević et al. 1978; Nelson and Burke 1967).

Afterdepolarisation

Motoneuron action potentials may have a short depolarisation phase immediately following the *f*AHP, called the afterdepolarisation (ADP; Granit et al. 1963b; Kernell 1964). The ADP is thought to be dependent upon voltage-gated Ca^{2+} channels as it can be blocked by cadmium and enhanced by increasing extracellular Ca^{2+} concentrations (Kobayashi et al. 1997).

A prominent ADP can reach threshold and promote high frequency spike doublets or triplets at the start of an action potential train (Spielmann et al. 1993). These "additional" spikes that ride upon the ADP of the initial action potential were described as defining 'the catch property' of motoneurons by Burke et al. (1970). The group showed that while stimulating a motoneuron to produce a spike train, a single extra stimulus with an interval of <10 ms added to the start of the train increased the rate of force generation and induced long lasting enhancement of tension in the innervated muscle. Stein and Parmiggiani (1979) used extracellular stimulation to confirm that high initial frequencies followed by slower spiking is the most efficient for muscle force generation (Garland and Griffin 1999; Parmiggiani and Stein 1981). Thus, the ADP promotes high initial firing frequencies in motoneuron spike trains, thereby enhancing rate and magnitude of force generation. Interestingly, slow motoneurons tend to express greater amplitude ADPs and are therefore more likely to produce doublets than fast motoneurons (Spielmann et al. 1993).

Medium Afterhyperpolarisation

From the very first intracellular recordings from motoneurons it was noticed that the AHP was responsible for the refractory period and therefore regulated firing frequency (Brock et al. 1951). It is now well established that the medium afterhyperpolarisation (mAHP) amplitude and duration correlate inversely with motoneuron size (Eccles et al. 1958). Slower motoneurons have a larger amplitude, longer duration mAHP compared to fast motoneurons and fire at lower frequencies. The majority of the mAHP current is carried by SK channels, which can be selectively blocked by apamin (Kobayashi et al. 1997; Zhang and Krnjević 1987). As might be expected by the different characteristics of the mAHP in slow vs fast motoneurons, SK channel expression also differs between the 2 types (Deardorff et al. 2013). There are several different SK isoforms, but only SK2 and 3 have been identified in motoneurons to date. SK2 is expressed by all motoneurons, but the expression of SK3 channels is lower or absent in fast motoneurons, and dominates in slow motoneurons (Deardorff et al. 2013). It is thus likely the SK3 channels that are responsible for the larger mAHP.

The source of calcium that leads to SK channel activation is unknown, but several studies have shown that Ca_V2 (N and P/Q type Ca^{2+}) channels are necessary for generation of the mAHP (Li and Bennett 2007; Bayliss et al. 1995; Umemiya and Berger 1994). Given the clustering of SK channels at C-bouton synapses and their proximity to Ca^{2+} stores in the subsurface cisternae (see below), it is possible that local intracellular Ca^{2+} release also contributes to SK channel activation and the mAHP. The mAHP is the target of modulatory inputs to motoneurons, with reductions in the mAHP amplitude/duration leading to significant increases in excitability as determined by the f-I relationship. During locomotor activity, the mAHP is reduced and high frequency firing ensues (Brownstone et al. 1992).

4 Repetitive Firing Properties of Motoneurons

Ultimately, it is the role of a motoneuron to translate synaptic inputs into repetitive spike trains at a frequency appropriate to produce the contraction of its innervated muscle fibres required for a given task. Motoneurons fire action potentials repetitively in response to repetitive (Eccles and Hoff 1932) or sustained input (Barron and Matthews 1938), with a positive relationship between the frequency of output and the magnitude of current input. This frequency-current (f-I) relationship (Fig. 4a) has been extensively studied to quantify motoneuron excitability using both sustained rectangular current pulses and slow triangular ramps (Fig. 4b).

Between 2 and 4 different 'ranges' of firing have been demarcated, largely in studies of cat motoneurons, by changes in the slope of the f-I curve (Fig. 4a, b). The two main ranges have been designated as the primary and secondary ranges (Granit et al. 1966a, b), with a tertiary range (Schwindt 1973) sometimes seen. In addition, a sub-primary range (Manuel et al. 2009; Jensen et al. 2018), thought to be related to mixed mode membrane oscillations (high frequency, sub-threshold membrane potential oscillations) has also been identified in some conditions in mouse motoneurons (Iglesias et al. 2011; Fig. 4c, d).

Each of the ranges is more-or-less linear. In cat motoneurons, the primary range is characterised by a relatively low f-I slope, with transitions to the secondary range seen by a sharp increase in slope (Granit et al. 1966a, b). A third segment, the tertiary range, is usually demarcated by a sudden reduction in f-I slope (to below that of the primary range), although in some motoneurons it has the steepest slope (Schwindt 1973). In mouse motoneurons, the secondary range, if present, typically has a lower slope than the primary range. Possible mechanisms underlying this transition are discussed below (*Persistent inward currents*).

To understand how different ranges of the f-I relationship relate to force generation in the muscle, it is important to consider the contractile properties of the muscle fibres innervated. Motoneuron firing frequency is tuned to the force output of the muscle fibre innervated, with increases in frequency resulting in repetitive twitch contractions that summate. As firing frequency increases, the twitch contractions fuse until maximum force is generated, at which point further increases in spike frequency are superfluous. This relationship can be seen by the sigmoid shape of the relationship between tension and stimulation frequency (Kernell 1983; Cooper and Eccles 1930).

Contraction properties differ in different muscle fibre types; motoneuron properties are tuned to the muscle fibre types they innervate, and thus their firing frequencies differ. For example, the firing frequency at which maximum tetanic force is produced will differ depending not only on the motor unit assessed, but also the species (smaller animals have faster equivalent motor units). Our understanding of the relationship between motoneuron firing frequency and muscle force production comes mainly from studies in cats (Kernell 2006). However, contemporary studies of motor unit physiology use rodent models, particularly the mouse. The maximum muscle force output in rats, for example, is generally achieved at the start of the primary or early secondary range of motoneuron firing (Turkin et al. 2010), and in anaesthetised mice, there is evidence to suggest that near maximal force can be generated by sub-primary range firing (Manuel and Heckman 2011). It is unclear why, at least in smaller mammals, motoneuron firing ranges exceed the contractile force capabilities of the muscles they innervate.

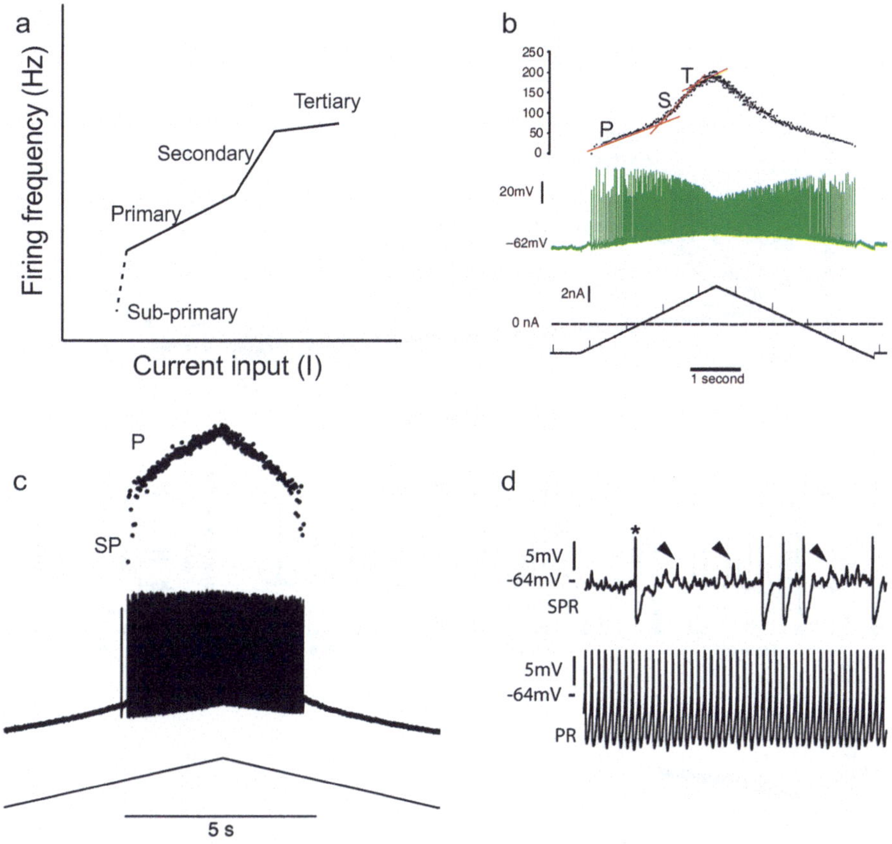

Fig. 4 Repetitive firing patterns in motoneurons. (**a**) Frequency-current plot schematic. The primary, secondary and tertiary ranges (*solid lines*) are labelled and the sub-primary range is the *dashed line*. (**b**) Voltage trace (*middle*) during current clamp experiment in an adult rat motoneuron shows firing in response to a triangular current ramp (*lower trace*). The *top trace* is a plot showing the instantaneous firing frequencies during the primary (P), secondary (S) and tertiary (T) ranges. (**c**) Similar layout as b showing the sub-primary (SP) and primary (P) ranges in an adult mouse motoneuron. (**d**) Irregular adult mouse motoneuron firing produced in sub-primary range due to mixed mode oscillations (*upper trace*). Firing in primary range is regular (*lower trace*). Spikes truncated for illustrative purposes. (**a**) adapted from Heckman et al. (2005), Fig. 1b. (**b**) adapted from Jensen et al. (2020), Fig. 6A. (**c**) from Manuel et al. (2009), Fig. 8b1. (**d**) from Iglesias et al. (2011), Fig. 1b1

Spike Frequency Adaptation

When long supra-threshold rectangular currents are injected into a motoneuron, inter-spike-intervals (ISI) become longer over the duration of the pulse (Fig. 5a). This slowing of instantaneous spike frequency is termed spike frequency adaptation (SFA). There are 2 main phases of spike frequency adaptation. Early SFA represents the slowing of spike frequency over the first hundreds of milliseconds (Granit et al.

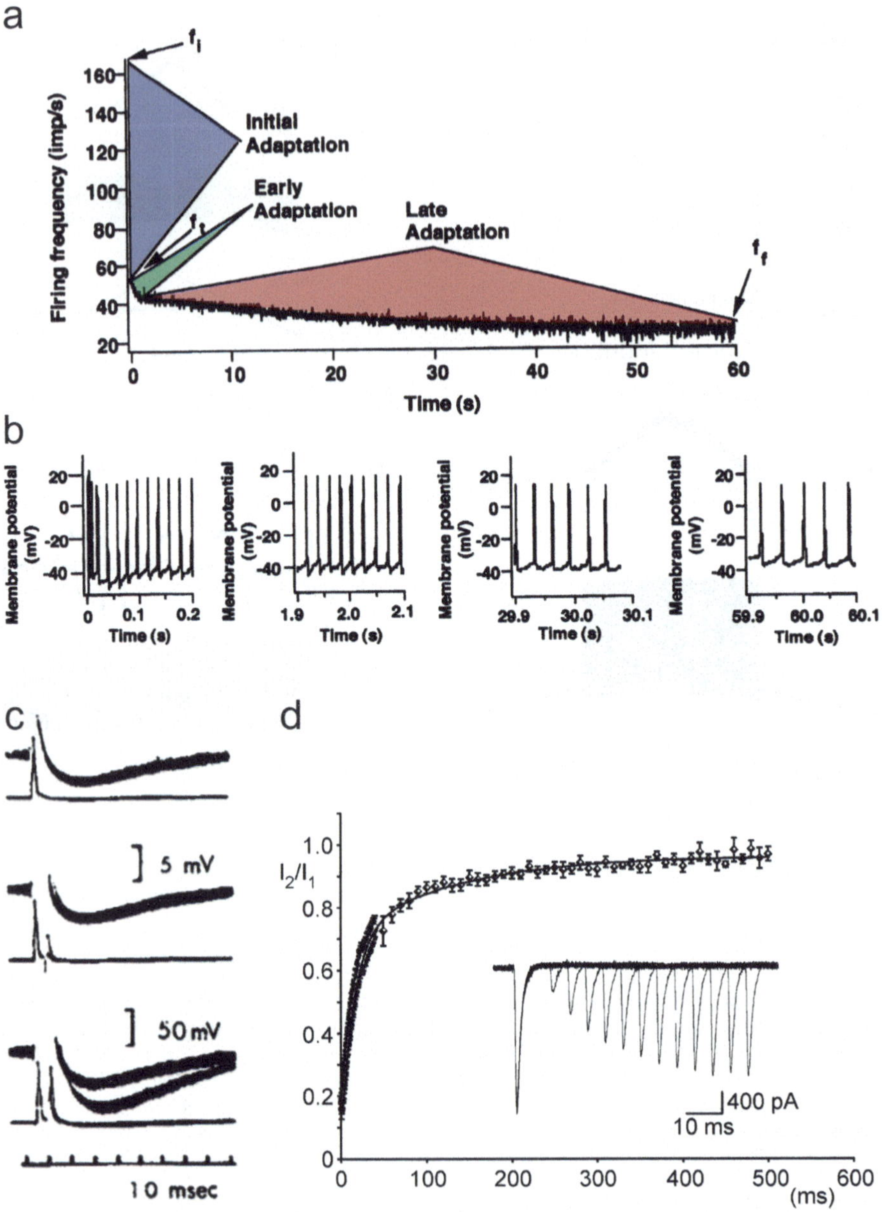

Fig. 5 Spike frequency adaptation and mAHP summation. (**a**) Firing rate in an adult rat hypoglossal motoneuron over the course of a 60s injected current pulse showing the initial frequency (f_I, *blue*), subsequent exponential decline (f_t, *green*) and slow decline (f_f, red) to final frequency, representing immediate (or initial), early, and late SFA. (**b**) 200 ms expanded samples of spiking show initial (first trace), early (second trace) and late (third-fourth traces) SFA. (**c**) AHP summation in an adult cat motoneuron. *Top panel*: AHP after single antidromic spike. *Middle panel*: no change

1963a; Kernell 1965), whereas late SFA occurs over many seconds to minutes (Kernell and Monster 1982; Fig. 5a, b). In addition, a third, earlier phase over a few spikes has been referred to as "immediate" SFA (Brownstone 2006). This phase results from the initial high instantaneous frequencies of initial doublets (see ADP section above), with subsequent intervals governed by the AHP (Sawczuk et al. 1995). SFA is more prominent in type F than type S motoneurons (Kernell 1972; Spielmann et al. 1993; Button et al. 2006).

Initially, early SFA was thought to arise from mAHP summation: the mAHP conductance is longer-lasting than the interspike interval, and would thus summate with that of the mAHP of the subsequent spike, leading to greater afterhyperpolarisation and a longer interspike interval (Fig. 5c; Ito and Oshima 1962; Baldissera and Gustafsson, 1971, 1974b). This concept was supported by computational studies (Baldissera and Gustafsson 1971, 1974a, b; Baldissera et al. 1978; Kernell and Sjöholm 1973; Kernell 1968, 1972).

On the other hand, it became clear that the mAHP was not completely responsible for SFA. Blocking calcium influx did not completely abolish early SFA (infact, it enhanced late SFA; Powers et al. 1999). These studies led to the suggestion that there are a number of potential mechanisms underlying SFA, including slow inactivation of sodium conductances and recruitment of other outward currents, such as an M-current (Powers et al. 1999). Further support for other mechanisms was found when blocking the AHP (SK) conductance in embryonic stem cell derived motoneurons did not alter early SFA (Miles et al. 2004). On noting that action potential amplitude progressively decreased, and duration increased throughout the spike train – both effects consistent with sodium channel inactivation – the time course of inactivation of Na^+ channels was studied and a slow component found, correlating with SFA (Fig. 5d; Miles et al. 2005). Furthermore, when incorporated into a computational model, the experimentally-determined slow inactivation parameters were sufficient to produce SFA (Miles et al. 2005). Thus, mechanisms underlying early SFA are likely multifactorial, with a slow time constant of sodium channel inactivation playing a significant role.

Late SFA, spanning seconds to minutes, is seen whether motoneurons are stimulated by continuous or repetitive stimulation (Stein and Parmiggiani 1979), and whether they are stimulated intracellularly (Kernell and Monster 1982) or extracellularly (Spielmann et al. 1993), suggesting that it is not an artefact produced by penetrating the membrane. The mechanisms contributing to late SFA are not clear, however slow inactivation of sodium channels is thought to contribute here as well (Brownstone 2006; Chen et al. 2006).

Fig 5 (continued) when a second spike fails to invade the soma. *Lower panel*: larger (summated) AHP when SD spike is elicited. In each panel, *upper trace* is amplification of *lower trace*. (**d**) Time course of recovery from sodium channel inactivation shown in a juvenile (P8–14) mouse motoneuron using a two pulse protocol. Note the (at least) double exponential of the recovery that parallels the phases of SFA. (**a** and **b**) Adapted from Sawczuk et al. (1995), Fig. 1. (**c**) from Ito and Oshima (1962), Fig. 1a–c. (**d**) from Miles et al. (2005), Fig. 5b

What are the functional consequences of SFA? On one hand, sustained high frequency firing is not necessary for optimal muscle contraction (Bigland-Ritchie et al. 1983a, b), and may even be detrimental to both the motoneuron and muscle fibre health (Vrbova 1983). Therefore, SFA could serve to protect motor units from excessive activation and subsequent damage. On the other hand, theories of motor unit rate coding suggest that SFA is a central mechanism contributing to motor fatigue (Kernell and Monster 1982; Gandevia 2001; Nordstrom et al. 2007). In this light, it is interesting that late SFA is reversed during fictive locomotion in cats and the reversal lasts even after the locomotor bout has finished, suggesting that it may be reversed by state dependent neuromodulatory inputs (Brownstone et al. 2011). If SFA does contribute to motor fatigue, a neuromodulatory system capable of supressing SFA would certainly be useful.

In summary, both early and late SFA represent fundamental characteristics of motoneuron repetitive firing and force control but their precise role in motor control remains unknown.

5 Modulation of Motoneuron Properties

Over the past several decades, it has become increasingly clear that motoneuron properties are not static, and that they can be modulated in a task-specific manner. While it is useful to study motoneuron properties in controlled, "quiescent," conditions to understand their basic properties and fundamental firing characteristics, these conditions do not reflect the state of motoneurons during behaviour when they are receiving both ionotropic and metabotropic (neuromodulatory) inputs.

We consider two main categories of modulation of motoneuron output: (a) output modulation, in which frequency of firing of the motoneuron in response to any given input is increased (i.e. agnostic to the specifics of the inputs), and (b) input modulation, which serves to alter the magnitude or duration of synaptic inputs received by the motoneuron. Thus, in broad terms, output modulation occurs at the soma, and in the case of motoneurons can be mediated by C-bouton synapses, and input modulation occurs at the dendrites and is often observed as persistent inward currents (PICs). Below we describe two best studied systems that contribute to output and input modulation of motoneuron activity.

Somatic Output Amplification: C-Bouton Synapses

C-boutons were first characterised anatomically as large synapses in apposition to α-motoneuron somata. They were named C-type boutons, or C-boutons, for the specialised endoplasmic reticulum, called sub surface cisternae (SSC), seen in proximity to the postsynaptic plasma membrane (Conradi 1969). C-boutons were later found to contain acetylcholine (Nagy et al. 1993). C-boutons were found to arise from medial partition neurons, just lateral to the central canal (Miles et al. 2007), and identified as $V0_C$ interneurons that express the transcription factor Paired-like homeodomain 2 (Pitx2; Zagoraiou et al. 2009). Here, we consider C-bouton circuits to be output modulators as they alter the biophysical properties of motoneurons to increase excitability and thus frequency of firing.

$V0_C$ interneurons are active during fictive locomotion in neonatal spinal cord preparations (Nascimento et al. 2020; Zagoraiou et al. 2009) and appear to modulate fictive locomotor motoneuron bursting. However, in adult behaving mice, knock down of C-bouton transmission has no observed effect on over-ground locomotion (Zagoraiou et al. 2009). To date, the only behavioural deficit observed following C-bouton knock out is reduced EMG amplification between walking and swimming (Landoni et al. 2019; Zagoraiou et al. 2009), suggesting that C-boutons amplify motor output in a task-dependent manner. In other words, it is likely that C-boutons are important for high force output tasks, of which swimming is one example.

Acetylcholine released by $V0_C$ interneurons at C-bouton terminals binds to type 2 muscarinic acetylcholine receptors (M2Rs), which form dense aggregations spanning the apposing postsynaptic membrane (Hellström et al. 2003; Deardorff et al. 2013, 2014). Activation of M2Rs leads to an increase in motoneuron excitability (increased slope of f-I relationship, Miles et al. 2007), thought to be mediated via a number of factors at the complex postsynaptic domain. Studies from many labs have contributed to our understanding of the diversity of proteins associated with the post-synaptic site of C-bouton synapses (Fig. 6). These include: M2Rs (Hellström et al. 2003), SK2, SK3 (Deardorff et al. 2013) and $K_V2.1$ potassium channels (Muennich and Fyffe 2004), Sigma1 receptors (Mavlyutov et al. 2010), neuregulin-1 (Gallart-Palau et al. 2014), and TMEM16F (Soulard et al. 2020). It has been challenging to determine the role of each of these proteins in C-bouton physiology.

C-boutons are found on all limb innervating α- but not γ-motoneurons (Lagerbäck et al. 1986). A series of papers using electron microscopy to study the synaptology of fast and slow motoneurons (identified by intracellular recording of properties) in cats showed a relatively lower density of C-boutons on slow compared to fast motoneurons (Kellerth et al. 1979, 1983; Conradi et al. 1979). Furthermore, on fast motoneurons C-boutons organise into clusters around the dendritic roots and can be found on the axon hillock, whereas on slow motoneurons they tend to be restricted to the soma. The functional relevance of this organisational discrepancy is not clear.

SK Channels

Motoneurons express at least 2 types of SK channels that are responsible for the mAHP (see above). SK2 and SK3 channels are clustered post-synaptically to C-boutons but in different proportions in fast and slow motoneurons: all motoneurons express SK2 but smaller, slow type motoneurons express a higher level of SK3 (Deardorff et al. 2013). Based on the correlation between AHP amplitude and SK3 expression, this finding may explain the larger amplitude, longer duration mAHP in slow motoneurons (Zhang and Krnjević 1987; Hounsgaard and Mintz 1988).

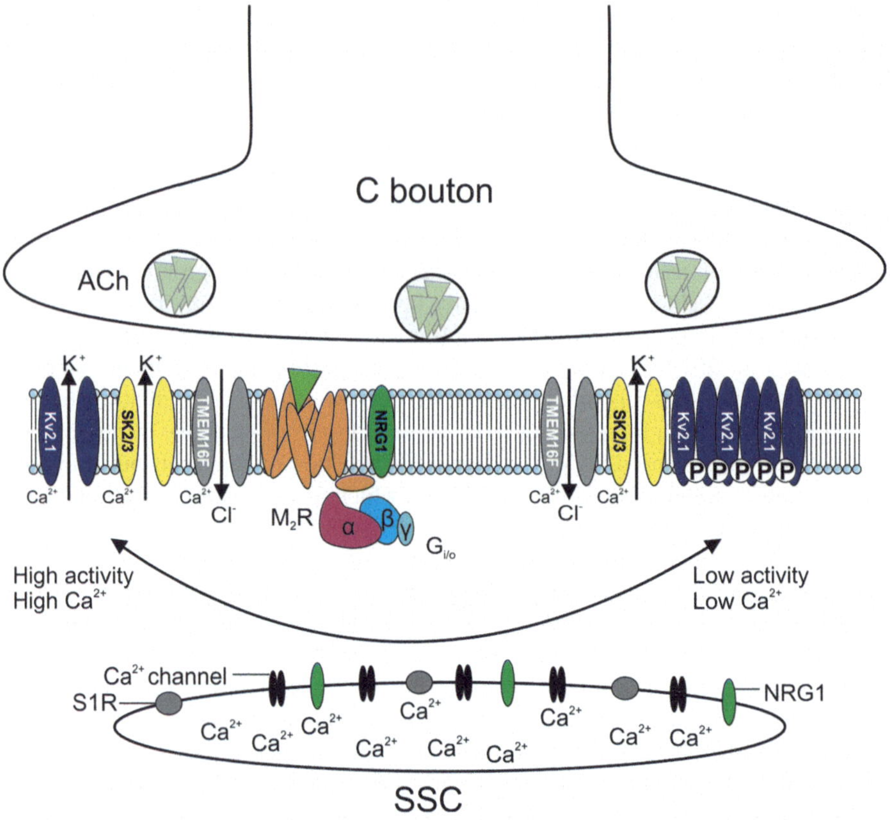

Fig. 6 C bouton organisation and hypothesised function. Presynaptic C-boutons release ACh which binds to M2 G-protein coupled receptors located in the motoneuron plasma membrane. Clustered together with M2 receptors are NRG1 proteins, SK2/3 and $K_V2.1$ potassium channels, and the chloride channel TMEM16F. Note that all channels post-synaptic to C boutons are calcium-dependent. $K_V2.1$ cycles between 2 states. During low activity, where intracellular Ca^{2+} is low, $K_V2.1$ channels are phosphorylated *(P)* and coalesce into large clusters, which are in a low or non-conducting state. With increasing activity leading to increases in intracellular Ca^{2+}, $K_V2.1$ is dephosphorylated and declustered, lowering its activation threshold and increasing K^+ conductance. SK2/3 and TMEM16F are activated by Ca^{2+} during high activity states. Just below the cell surface are the sub-surface cisterns (SSC), which are Ca^{2+} stores. Sigma-1 receptors and NRG1 proteins are primarily in the SSC membrane. M2 G-protein coupled receptor configurations is adapted Santiago and Abrol (2019), Fig. 1

Activation of M2Rs using pharmacological agonists in neonatal mouse spinal cord preparations reduces the mAHP, suggesting that the increase in *f*-I slope results from C-bouton-mediated reduction of SK conductance (Miles et al. 2007), either directly or indirectly (for example by reducing local calcium availability). It is not yet clear whether the source of local calcium is transmembrane, for example via Ca_V2-type Ca^{2+} currents (Li and Bennett 2007; Viana et al. 1993), or from an intracellular store like the adjacent SSC. In other neurons, ryanodine receptors have been found on SSC membranes and contribute to calcium induced calcium release

(Mandikian et al. 2014; Berridge 1998), but there is no evidence for their existence in motoneuronal SSC.

K_V2.1 Channels

K_V2 channels are delayed rectifier potassium channels that are widely expressed throughout the nervous system, contributing to the regulation of neuronal excitability (Misonou et al. 2005b; Du et al. 2000; Kihira et al. 2010). This family of voltage-gated potassium channels has 2 main isoforms, K_V2.1 and K_V2.2, that share many electrical properties. Likely due to a cloning artefact (see Kihira et al. 2010), K_V2.2 was initially thought to be restricted to distal dendritic compartments (Hwang et al. 1993), leaving K_V2.1 as the more studied isoform (Johnson et al. 2019).

In motoneurons, large clusters of K_V2.1 channels were discovered years ago (Muennich and Fyffe 2004), and are considered to be the main Kv2 isoform in these neurons (Deardorff et al. 2021; Romer et al. 2019). However, there is evidence that Kv2.2 is also expressed (Fletcher et al. 2017; Burger and Ribera 1996), although the exact membrane localisation and function has not been studied. Therefore, for the purposes of this chapter, we will focus on the Kv2.1 isoform.

Motoneuron K_V2.1 channels form large clusters at C-boutons, but their role in C-bouton physiology remains unclear (Muennich and Fyffe 2004). In the brain, due to the conductance's relatively slow activation/inactivation kinetics, K_V2.1 contributes significantly to the repolarisation of slow spiking neurons with broad action potentials (Liu and Bean 2014), but only minimally in faster spiking neurons (Guan et al. 2013; Du et al. 2000). On the other hand, K_V2.1 contributes to repetitive firing in all neuronal types studied, including motoneurons. All studies assessing K_V2.1 function in motoneurons so far use juvenile or neonatal in vitro preparations, before channels are fully clustered (Wilson et al. 2004), and none distinguishes between motoneuron type (e.g. fast vs. slow). A common finding is that inhibiting K_V2.1 channels with either Guangxitoxin 1-E (Fletcher et al. 2017; Nascimento et al. 2020) or stromatoxin (Romer et al. 2019) has no effect on the passive membrane properties of motoneurons. C-bouton-mediated increases in excitability are thought to be dependent on K_V2.1 channels (Nascimento et al. 2020), which may help to prevent gradual depolarisation of the interspike membrane voltage with repetitive firing, thereby reducing the depolarising block of sodium channels (Romer et al. 2019).

An interesting feature of K_V2.1 channels is that their state of clustering is dependent upon phosphorylation, which is regulated by Ca^{2+}/calcineurin dependent signalling mechanisms (Misonou et al. 2005a). In their clustered state, K_V2.1 channels either have higher activation thresholds (Murakoshi et al. 1997) and/or are non-conducting (Fox et al. 2013) – both effects decreasing potassium flux. An increase in neuronal activity and therefore calcium results in dephosphorylation of K_V2.1 channels, which increases their conductance by reducing activation threshold. One theory for K_V2.1 function in motoneurons is that C-boutons may act to inhibit local calcium activity, thus maintaining K_V2.1 clustering and maintaining basal rates of potassium conductance (Romer et al. 2019). This would be sufficient to support high frequency firing by preventing depolarising block (Liu and Bean 2014; Nascimento et al. 2020; Romer et al. 2019). But in states of high frequency firing, increased intracellular calcium would de-cluster $K_V2.1$ channels, thereby reducing their activation threshold and thus homeostatically reduce firing frequency (Romer et al. 2019). However, we do not yet fully appreciate how $K_V2.1$ clustering at C-bouton synapses contributes to the modulatory role of these synapses.

Other Proteins Located at C-Bouton Synapses

There is even less understanding of the function of other proteins at this site. Transmembrane protein 16F (TMEM16F) was found to be responsible for a Ca^{2+} activated Cl^- current in cultured motoneurons, contributing to lowering recruitment thresholds of fast motoneurons in neonatal slice preparations. TMEM16F knockout mice had reduced maximum speed and endurance capacity during a treadmill running task, suggesting that they may be involved in C-bouton amplification of motor output.

Neuregulin 1 (NRG1) is also located post-synaptically at the SSC (Casanovas et al. 2017; Gallart-Palau et al. 2014; Issa et al. 2010). The function of NRG1 is poorly understood, but it may play a role in synaptic retrograde communication (Mòdol-Caballero et al. 2018) or in maintaining protein clustering at the synapse. Also located at the SSC are sigma-1 receptors (Mavlyutov et al. 2010, 2012) and their ligand Indole-*N*-methyl transferase (INMT), but there is little knowledge on the roles these proteins may play in C-bouton function.

In contrast to the lower density of SK3 on fast motoneurons (Deardorff et al. 2013), these neurons express the $K_V2.1$ β-subunit KCNG4 whereas slow motoneurons do not (Müller et al. 2014). The density of NRG1 clusters is higher in tibialis anterior motoneurons (fast) compared to soleus (slow), but the cluster size was larger in soleus (Casanovas et al. 2017). While the implications of these differences are not known, they emphasise the importance of studying known motoneuron types in order to provide further understanding of C-bouton function.

C-Boutons and Intracellular Calcium Signalling

C-boutons form complex synapses, with more localised proteins being discovered before the function of those previously known are understood. However, it is clear that Ca^{2+} is a key player in C-bouton function. To date, four Ca^{2+}-dependent ion channels (SK2, SK3, $K_V2.1$, and TMEM16F) have been identified, and SSC Ca^{2+} stores are located in close proximity. Future work aimed at characterising local Ca^{2+} fluxes during M2 receptor activation will undoubtedly provide insight into the intracellular signalling cascades underlying C-bouton mediated amplification of motor output.

Dendritic Input Amplification: Persistent Inward Currents

In 1975, Hultborn and colleagues demonstrated that, in response to brief trains of low threshold sensory afferent stimulation, cat motoneurons can produce "long latency prolonged activity", as shown in EMG and tension recordings (Hultborn et al. 1975). They proposed that this persistent activity could be due to a spinal interneuron circuit creating a reverberating excitatory loop. Schwindt and Crill (1977), using voltage clamp recordings in cat motoneurons, showed that these neurons had non-inactivating "N-shaped" current-voltage relationships, and that this property reflects a persistent net inward current (PIC, Fig. 7a–c). The group later suggested that PICs were generated "on or near the soma", and were mediated by calcium conductances (Schwindt and Crill 1980b, c, 1981). Following the Copenhagen group's demonstration of motoneuron plateau potentials and bistability produced by PICs (see section "Dendritic input amplification: persistent inward currents"; Hounsgaard et al. 1988a; Crone et al. 1988), and the implications of this activity for motor control, work spread to many labs and PICs have since been well-studied.

Persistent inward currents were initially thought to be produced solely by dihydropyridine-sensitive, non-inactivating calcium currents mediated by Ca_V1 (likely $Ca_V1.3$) channels (Perrier and Hounsgaard 2003; Hounsgaard and Kiehn 1985, 1993; Carlin et al. 2000). It was later demonstrated that persistent sodium currents (INaP; likely Nav1.1 and 1.6; Schwindt and Crill 1980a; Li and Bennett 2003) contribute to repetitive firing as well (Miles et al. 2005). That is, PICs are mediated by a combination of persistent sodium and calcium currents, and are modulated by metabotropic inputs (Li and Bennett 2003; Lee and Heckman 1999; Hounsgaard and Kiehn 1985; See Binder et al. 2020 for review).

There is now little doubt that PICs are generated primarily in motoneuron dendrites (Fig. 7c; Carlin et al. 2000, although it is possible that cooperative clustering of CaV1 channels on the soma contributes, Moreno et al. 2016). Having active conductances in the dendrites near the sites of synaptic inputs to motoneurons means that dendrites are not passive cables and, furthermore, provides a widespread substrate such that the responses to synaptic inputs can be modulated.

Modulation of PICs: Neurotransmitter Systems

It was initially curious that PICs or their effects had not been noted in the first decades of investigations of motoneuron properties. The likely reason for this "blind spot" was that motoneurons were usually investigated in anaesthetised preparations, and PICs are state-dependent, relying on modulatory inputs that are depressed by anaesthesia (Hounsgaard et al. 1986; Button et al. 2006). PICs are activated by monoaminergic, metabotropic neuromodulators, with the level of activation setting the degree of modulatory drive. The main source of these modulatory inputs are brain stem nuclei: raphe nuclei for serotonin (5HT; Conway et al. 1988; Hounsgaard

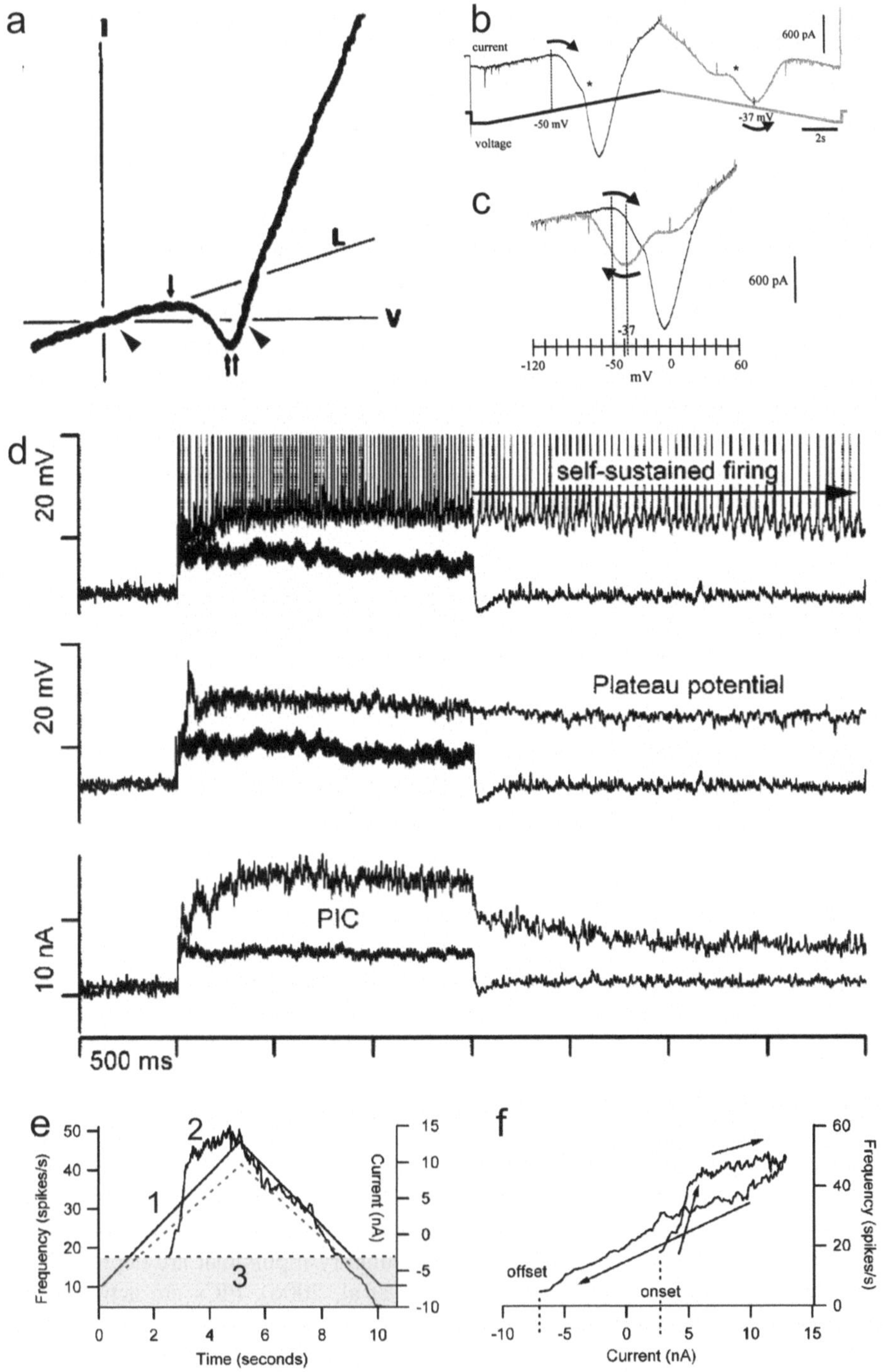
a
I
L
V
b
current
600 pA
-50 mV
-37 mV
voltage
2s
c
600 pA
-37
-120
-50
0
60
mV
d
20 mV
self-sustained firing
20 mV
Plateau potential
10 nA
PIC
500 ms
e
Frequency (spikes/s)
Current (nA)
Time (seconds)
1
2
3
f
Frequency (spikes/s)
Current (nA)
offset
onset

et al. 1988a; Alvarez et al. 1998), and the locus coeruleus for noradrenalin (NA; Björklund and Skagerberg 1982; Giroux et al. 1999; Conway et al. 1988). This is in contrast to output modulation by C-boutons, which is regulated by the evolutionarily older spinal cord. Not only are these systems depressed during anaesthesia, but they are modulated during changes in organismal state, increasing during wake as compared to sleep (Aston-Jones et al. 2000; Jacobs et al. 2002), and behavioural state, increasing during locomotion (Veasey et al. 1995). That is, these brain stem nuclei regulate PICs and thus dendritic integration in a state-dependent manner.

PICs, Repetitive Firing, and Synaptic Amplification

A key finding that led to our understanding of PICs arose from voltage clamp experiments in cats, in which N-shaped current-voltage (I–V) relations were found. The region of negative slope conductance in an N-shaped I–V relation is by definition unstable. But where the "N" crosses the abscissa with a positive slope (which can be at two different voltages depending on its position), the membrane potential will be stable – any small deviations will be counteracted by currents that bring the voltage back to this level (Schwindt and Crill 1980c). Furthermore, depolarisations from the more hyperpolarised "resting" membrane potential need only reach the "hump" of the N, or the unstable region, for the voltage to jump to the more depolarised stable

Fig. 7 (continued) N-shaped I–V relationship, PICs, and repetitive firing in motoneurons. (**a**) N-shaped I–V curve in an adult cat motoneuron revealed during slow depolarising ramp in voltage clamp. Slope of *line L* indicates passive leak conductance. *Single arrow* is first point of 0 slope conductance, and *double arrow* indicates maximum peak inward current. *Arrowheads* indicate abscissa crossings with positive slopes, which are stable voltages. (**b**) Current response to triangular voltage ramp in a juvenile (P8–15) mouse motoneuron in slice similarly reveals differences in 'N-shape' during ascending and descending ramps due to channel kinetics. * show inflection points, likely resulting from dendritic location of Ca_V1 channels. (**c**) These differences can be quantitatively appreciated when reflecting the response to the downward ramp (grey) on the upward ramp (black) to plot the I–V curve. The *dashed lines* indicate the onset of the region of negative slope conductance on the up-ramp, and the peak inward current on the down-ramp. (**d**) Voltage-dependent persistent inward currents generate plateau potentials and sustained firing in an adult cat motoneuron, with synaptic input activated by tendon vibration. The *top two panels* show voltage traces and the *lower panel* current traces. In each panel, the *lower trace* is when the cell held at hyperpolarised levels at which PICs are not activated, whereas the *top traces* are depolarised to voltages where PICs can be generated. The *top panel, top trace* illustrates self-sustained firing of a motoneuron activated by synaptic input. The *middle panel* shows the underlying plateau potential (*top trace*) in a motoneuron injected with QX-314 to block action potentials. In the *lower panel*, a PIC is initiated when the voltage is clamped at more depolarised levels. Note that the traces have been shifted along the y-axes, aligning the pre-pulse current/voltages to facilitate comparison. (**e**) Motoneuron model showing instantaneous firing frequency in response to a triangular current ramp in the absence (*red dashed line*) and presence (*solid, non-linear line*) of neuromodulation. *1–3* indicate major PIC phases. (**f**) *f*-I plot showing firing frequency hysteresis induced by PICs in an adult cat motoneuron. *Arrows* indicate direction of the change in current injection. (**a**) modified from Schwindt and Crill (1977), Fig. 1c. (**b** and **c**) from Carlin et al. (2000), Fig. 1b and c. (**d**) from Heckman et al. (2005), Fig. 2. (**e** and **f**) from Heckman et al. (2005), Fig. 4c–d crossing. Thus, bistability is produced. But bistability is a special case – the inward and outward currents must be such that the N crosses the abscissa with a positive slope twice. In many motoneurons, this is unlikely to be the case and bistability is not seen (Heckman et al. 2003). Nonetheless, the voltage-gated channels responsible for PICs will amplify EPSPs, and, given that they are either slowly- or non-inactivating, the EPSPs will be prolonged and can impact temporal summation (Fig. 7d); In other words, in modulating these channels, brain stem monoaminergic systems can significantly change the integrative properties of motoneuron dendrites.

Of note, the recruitment of PICs parallels the transition from primary to secondary range firing (Schwindt and Crill 1982). In cats, the f-I slope of secondary range firing is higher than primary range, but some motoneurons do not have a secondary range. It is thought that counteracting factors such as increases in both spike threshold and outward potassium currents reduce the effects of the inward currents in those motoneurons (Schwindt and Crill 1982). Perhaps these forces are greater in mouse motoneurons, eliminating any PIC-mediated increase in f-I slope for secondary range firing.

These conductances also mean that less synaptic drive – in both amplitude and time – would be sufficient to ensure repetitive firing of motoneurons and hence muscle contraction. PICs can be strong enough to maintain firing in the absence of synaptic drive, until sufficient inhibition is received (Hounsgaard et al. 1988a; Lee and Heckman 2000; Hultborn et al. 2003).

The effects of PICs on motoneuron firing are clearly demonstrated during triangular current ramps (Fig. 7e, f). In the absence of PICs, motoneuron firing increases and decreases linearly with current input (Fig. 7e, red dashed line). However, when PICs are activated, firing accelerates rapidly, producing a steep *f*-I curve (*1.* Fig. 7e, non-linear line), followed by a reduction in the slope as maximum firing is approached (*2.* Fig. 7e), and finally a linear reduction in firing. These firing rate changes can be seen as an onset-offset hysteresis (*3.* Fig. 7e, f) of the *f*-I relationship (Hounsgaard et al. 1988a). Furthermore, the de-recruitment current threshold for firing is often lower than the recruitment threshold (*3.* Fig. 7e, shaded area shows difference), presumably due to slow/non inactivating PICs (Note that AHP prolongation may contribute to this effect as well; Wienecke et al. 2009). These voltage-dependent effects can also be seen during ramp current injections during locomotor activity, indicating that PICs or other voltage-dependent conductances are activated during synaptic excitation as well (Brownstone et al. 1994).

Termination of PICs

While persistent sodium currents are subject to slow inactivation (Lee and Heckman 1999), persistent calcium currents require strong inhibitory input or hyperpolarising/outward current to terminate them – to move to the left beyond the unstable region of negative slope conductance (Moritz et al. 2007; Perrier and Hounsgaard 2003; Hultborn et al. 1975). Furthermore, PICs are dependent on the balance between inward and outward K^+ currents (Schwindt and Crill 1980a, c, 1981), so any increase in K^+ current, such as a slowly activating K^+ current, can lead to their termination.

This can be mediated by SK channels through AHP summation (Li and Bennett 2007), or other potassium currents that are slowly-activating and non-inactivating, such as those mediated by KCNQ channels (M-currents; Alaburda et al. 2002). Renshaw cell inhibition is also well placed to terminate PICs (Bui et al. 2008).

PICs and Motor Pools

PIC channels are not homogeneously expressed across a motor pool and therefore contribute differently to sustained firing in different motoneuron types (Grande et al. 2007). Bistable behaviour (self-sustained firing, Fig. 7d) is more prevalent in low threshold slow motoneurons, in keeping with their fatigue resistance and involvement in sustained contractions such as those required for postural control (Eken and Kiehn 1989). Self-sustained firing is seldom seen in the largest, fast motoneurons, but this does not mean that they do not express PICs. In fact, PICs increase the gain of the f-I slope and reduce the decruitment threshold in these neurons as well. Hence, in fast, high threshold motoneurons, PICs primarily serve to amplify synaptic inputs, whereas in slow motoneurons they can also maintain repetitive firing.

6 Other Voltage- and Time- Varying Currents

Other non-linear properties of motoneurons can contribute to motoneuron repetitive firing. For example, during repeated phasic inputs, similar to those that occur during locomotion, a cumulative effect can lead to increasing firing frequency in response to the same input. This history-dependent increase in frequency is known as 'windup,' and although initially thought to be related to L-type Ca^{2+} channel activity, is seen even in neonatal motoneurons in which L-type Ca^{2+} channel expression is low, suggesting that other channels are involved (Pambo-Pambo et al. 2009). Indeed, motoneuron windup was found to be dependent on the nifedipine-sensitive K^+ channel, $K_V1.2$ (Bos et al. 2018). $K_V1.2$ is a slowly inactivating channel concentrated in the axon initial segment. Although present in most motoneurons, its slow ramping membrane depolarisation in response to rectangular current injection is preferentially seen in fast, high threshold motoneurons (Leroy et al. 2014). Thus, in response to long depolarising, near threshold, current injection (lasting seconds), fast motoneurons exhibit delayed firing (Fig. 8a) compared to slow motoneurons, which fire immediately (Fig. 8a1). In considering synaptic integration, fast motoneurons would thus be oblivious to low amplitude, near threshold inputs unless they were prolonged. $K_V1.2$ channels could therefore contribute to the observed size principle of recruitment (see Sect. 2). Furthermore, this finding has provided a surrogate marker to identify motoneuron type: long, near threshold current injections can distinguish fast from slow motoneurons by their delay to repetitive firing (Leroy et al. 2014; Fig. 8a–c1).

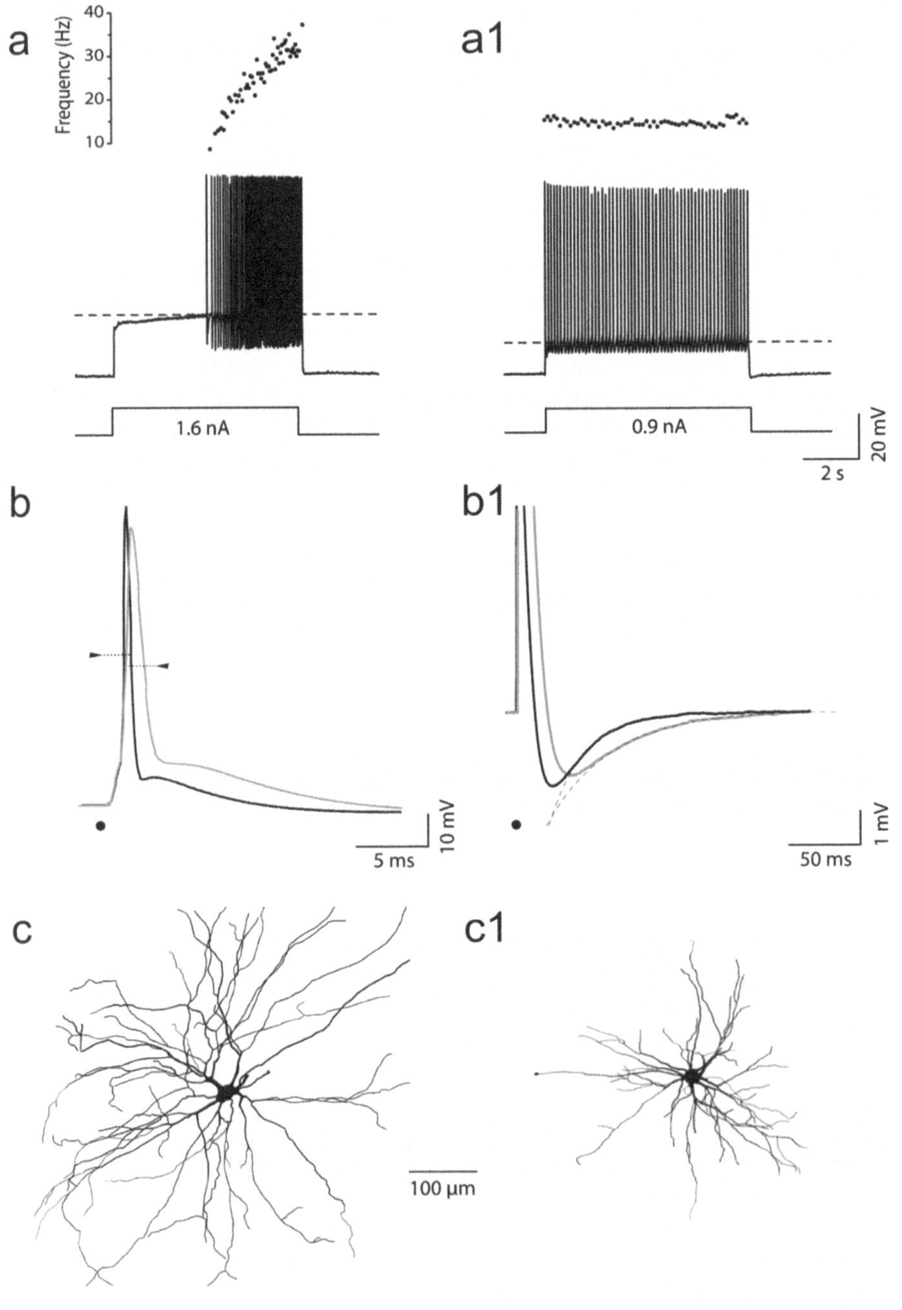

Fig. 8 Delayed and immediate firing motoneurons. Motoneurons may be categorized as fast or slow based on their threshold firing patterns during long duration rectangular current injections. (**a**) An example of a fast (6–10 day old) mouse motoneuron shows a ramping depolarisation in response to a stable near threshold current injection, resulting in delayed firing that increases in

The interaction of these slowly inactivating outward currents and SFA (see above) is interesting to consider, especially as both are more prominent in fast motoneurons (Spielmann et al. 1993; Button et al. 2006; Kernell 1972). Particularly in these motoneurons, it is therefore possible that for current injections in low suprathreshold ranges, the increases in firing rate resulting from inactivation of $K_V1.2$ could counterbalance decreases from SFA mechanisms (Brownstone et al. 2011). That is, at relatively low levels of current injections, SFA may not be seen (Fig. 8a; Leroy et al. 2014), whereas larger depolarisations would lead to clearer SFA (Sawczuk et al. 1995; Button et al. 2006).

7 Human Motor Unit Recordings

Our understanding of motoneuron physiology has been strengthened by the capacity to study motoneurons both in animals and in humans. Combining knowledge gained from experiments recording voluntary motor unit activity in humans with that gleaned from mechanistic studies in animal experiments has broadened our understanding of motoneuron physiology.

The first recordings of motor unit action potentials (MUAPs) from human muscle during voluntary contractions took place in the early 1900s (Wachholder 1928; see Duchateau & Enoka 2011 for comprehensive historical review). Adrian and Bronk (1929) used concentric needle electrodes to record single motor units, and provided an early description of rate coding, which describes the relationship between unit recruitment, spike frequency, and force output (Bigland and Lippold 1954). Development of fine wire electrodes for improving the signal to noise ratio of recordings (Stecko 1962) allowed principles of motoneuron physiology discovered in cat preparations, such as the size principle, to be studied during human voluntary contraction (Milner-Brown et al. 1973; Stein et al. 1972).

As humans can voluntarily produce different types of contractions with different magnitudes and rates of force generation, details about motor unit recruitment and rate coding could be readily studied (Büdingen and Freund 1976; Desmedt and Godaux 1977; Thomas et al. 1987). Subsequent investigations of motor unit firing patterns during different tasks enhanced our understanding of state-dependent motoneuron physiology (Collins et al. 2001). Thus, the fundamental knowledge from animal experiments could be interrogated in human motor units.

Fig 8 (continued) frequency over time. (**a1**) In slow motoneurons, action potential trains initiate at the onset of the input and spike frequencies do not accelerate. (**b**) Single action potentials in fast (*black*) motoneurons are shorter in duration (**b**) and have shorter mAHPs (**b1**) than slow (*grey*) motoneurons. These properties correspond to whether the motoneurons have delayed firing or not. (**c**) Analysis of motoneurons filled during recording demonstrates a larger dendritic tree in motoneurons with delayed (**c**) than those with immediate firing (**c1**), corresponding to the known characteristic that fast motoneurons are larger than slow motoneurons. (From Leroy et al. 2014)

Recently, much work has been directed toward recording multiple individual human motor units using intramuscular (Farina et al. 2008) and surface EMG electrode arrays (Holobar et al. 2009). These flexible, multi-channel EMG electrodes with tens of electrode contacts, combined with advances in signal processing algorithms led to the ability to record, decompose, and analyse most units in certain human muscles during isometric contractions (Farina et al. 2008; Muceli et al. 2018; Florestal et al. 2009; Marateb et al. 2011; Negro et al. 2016). This technique allowed for the comparison of activity in different motor units – for example early recruited (presumably S-type) vs late recruited (presumably F-type) units – which led to improved understanding of a motor pool. These studies revealed, for example, that while synaptic integration is non-linear in individual motoneurons, the population behaves linearly during these tasks (Farina and Negro 2015).

Development of these arrays and signal processing capacity to study human motor units has not only shed light on motor pool function in humans, but has allowed for the application of this technology in animals such as cat (Thompson et al. 2018) and song bird (Zia et al. 2018). This complementarity of human and animal studies is exemplary in demonstrating how the knowledge gained through such a combination is greater than the sum of each type of study independently.

Fatigue in Human Motoneurons

Muscle fatigue can be defined as the state occurring over time in which muscles have reduced contraction in response to a constant command, or in which an increase in command is needed to maintain a constant contraction. If that command is considered to originate at a pre-motoneuronal level, then a reduction in motoneuron responsiveness to that command could be considered a component of fatigue. Human studies of muscle fatigue have revealed that, during sustained voluntary contraction, there are two parallel components of fatigue: motor unit firing frequency decreases, and there is a concomitant decrease in the contractile properties of the innervated fibres. That is, the innervating motoneuron firing frequency reduces such that it does not fire at frequencies higher than that required for a fused tetanic contraction of its muscle fibre. Nonetheless, increasing the input to the motor pool by volitional control can increase motoneuron firing frequency during this fatigued state. The increase in command can recruit additional motor units in order to maintain the contraction, and likewise act on already recruited motoneurons to counterbalance reductions in firing frequency (Johnson et al. 2004). The degree to which descending modulatory activity changes and the roles that afferent input may play to enable motoneurons to increase their firing rate in these conditions are not known (Taylor et al. 2016).

But what are the cellular mechanisms behind this fatigue (Taylor et al. 2016; Enoka 2019)? Further studies on humans pointed to a central mechanism of fatigue, quite likely late spike frequency adaptation as seen in animal studies (Piotrkiewicz and Wilanowski 2012). As noted above, this late SFA can be reduced depending on behavioural state, such as when the animal is walking (Brownstone et al. 2011). As late SFA can be modulated, it may be that a strategy to "overcome" fatigue in the short term would be to alter neuromodulatory state (consider a fight or flight reaction). Furthermore, this concept could lead to the suggestion that long term strategies to reduce fatiguing of muscles, such as exercise, may act through metamodulation, or changing the baseline modulatory state of motoneurons.

PICs in Human Motoneurons

One example where knowledge gleaned from animal studies has directly led to corresponding investigations of human motor unit physiology is the phenomenon of PICs (see section "Dendritic input amplification: persistent inward currents"). As with animal motoneurons recordings, in which PICs can be studied by ramp increases and decreases in injected current, human PICs can be approximated using a surrogate of injected current: ramp increases (and decreases) in voluntary force production. This paradigm was first used to study human motor units to demonstrate orderly recruitment, and high threshold units were found to have nonlinear firing (likely secondary range; De Luca et al. 1982). Interestingly, although long before PICs in animals were understood, a difference (decrease) between recruitment and derecruitment firing rates was shown (De Luca et al. 1982). In addition to onset-offset hysteresis, firing frequencies of motor units were found to accelerate (De Luca and Contessa 2012) and saturate (Fuglevand et al. 2015) on the ascending portion of the force ramp (De Luca et al. 1982; Mottram et al. 2009; Revill and Fuglevand 2017; Fig. 7e, 1–3). These phenomena all point towards PICs operating in human motoneurons.

To investigate PICs further, Gorassini et al. (2002) took advantage of the phenomenon that low threshold units can be used as a measure of synaptic input to the pool (Bennett et al. 1998; Lee et al. 2003), and developed quantitative methods for assessing PICs and f-I hysteresis in humans. For example, subjects were asked to maintain dorsiflexion at a small percentage of their maximal voluntary contraction (MVC) during recording of single units from the TA muscle, assuming that the same excitatory drive was being delivered to all motoneurons in the pool. Proprioceptive afferents, which project to all homonymous motoneurons, were then stimulated by tendon vibration to provide brief excitatory input to the pool. This led to recruitment and sustained firing in higher threshold TA motor units, presumably due to activation of PICs. On slowly reducing force, there was a reduction in firing rates of both low and higher threshold units. By using the firing frequency of the low threshold unit as a surrogate marker of drive to the pool, this "drive" at recruitment and derecruitment of the higher threshold unit could then be estimated. Indeed, firing rate hysteresis was found (lower drive at derecruitment than at recruitment), consistent with the proprioceptive input activating PICs. This technique was then validated in animal studies (Powers et al. 2008; Gorassini et al. 2002) and in computer simulations (Powers and Heckman 2015), and it is now widely used to assess excitability of human motoneurons in health (Herda et al. 2016) and disease states (Mottram et al. 2009, 2014). In addition, this technique has now been used to assess many motor units in a pool using high density EMG arrays (see beginning of Sect. 7), which have provided an additional, important dimension to our understanding of human motoneuron physiology (Afsharipour et al. 2020). This knowledge will undoubtedly be enhanced by iterative approaches between the studies of animal and human motoneurons to gain mechanistic insights into human motor physiology.

8 Conclusion

Much progress has been made in our understanding of how motoneurons integrate synaptic inputs to produce trains of action potentials that stimulate muscles to contract with precise timing and force. This progress has been facilitated by experiments in both animals and humans, and the transfer of knowledge between the two. The fundamental basis of this knowledge derived from elegant work studying cat motoneuron physiology in vivo, and continues today largely in rodents in vitro and, to a lesser extent, in vivo. Mouse preparations have provided another dimension to our understanding through the capacity for genetic manipulation of motor circuits. Advances have also been made in our ability to record from motoneurons in behaving humans, allowing for the study of voluntary activation of motor pools.

But many questions remain. To understand behaviour, it is important to understand the fundamentals of movement. How are motoneuron properties tuned to the muscle fibre types they innervate? How do motoneurons integrate their many synaptic inputs to produce just the right amount of muscle contraction at a particular time and during a particular state? How do motoneuron pools work as an ensemble to ensure that joint angles change, or joints are stiffened, for the task at hand? How does the task-setting performed by higher centres change the modulatory state of motoneurons for the intended behaviour? These and other questions will be answered by ever expanding tools, interactions between those investigators of human and animal motoneuron function, and new, creative studies of this adaptable sensorimotor system.

References

Adrian ED, Bronk DW (1929) The discharge of impulses in motor nerve fibres: Part II. The frequency of discharge in reflex and voluntary contractions. J Physiol 67:9–151

Afsharipour B, Manzur N, Duchcherer J, Fenrich KF, Thompson CK, Negro F, Quinlan KA, Bennett DJ, Gorassini MA (2020) Estimation of self-sustained activity produced by persistent inward currents using firing rate profiles of multiple motor units in humans. J Neurophysiol 124:63–85

Alaburda A, Perrier JF, Hounsgaard J (2002) An M-like outward current regulates the excitability of spinal motoneurones in the adult turtle. J Physiol 540:875–881

Alvarez FJ, Pearson JC, Harrington D, Dewey D, Torbeck L, Fyffe RE (1998) Distribution of 5-hydroxytryptamine-immunoreactive boutons on α-motoneurons in the lumbar spinal cord of adult cats. J Comp Neurol 393:69–83
Araki T, Otani T (1955) Response of single motoneurons to direct stimulation in toad's spinal cord. J Neurophysiol 18:472–485
Aston-Jones G, Rajkowski J, Cohen J (2000) Locus coeruleus and regulation of behavioral flexibility and attention. Prog Brain Res. Elsevier
Baldissera F, Gustafsson B (1971) Regulation of repetitive firing in motoneurones by the afterhyperpolarization conductance. Brain Res 30:431–434
Baldissera BF, Gustafsson B (1974a) Afterhyperpolarization conductance time course in lumbar motoneurones of the cat. Acta Physiol Scand 91:512–527
Baldissera BF, Gustafsson B (1974b) Firing behaviour of a neurone model based on the afterhyperpolarization conductance time course. First interval firing. Acta Physiologica Scandinavica 91:528–544
Baldissera F, Gustafsson B, Parmiggiani F (1978) Saturating summation of the afterhyperpolarization conductance in spinal motoneurones: a mechanism for 'secondary range' repetitive firing. Brain Res 146:69–82
Baldissera F, Hultborn H, Illert M (1981) Integration in spinal neuronal systems. Compr Physiol
Barrett E, Barret J (1976) Separation of two voltage-sensitive potassium currents, and demonstration of a tetrodotoxin-resistant calcium current in frog motoneurones. J Physiol 255:737–774
Barrett J, Crill W (1980) Voltage clamp of cat motoneurone somata: properties of the fast inward current. J Physiol 304:231–249
Barron DH, Matthews BH (1938) The interpretation of potential changes in the spinal cord. J Physiol 92:276–321
Bayliss DA, Viana F, Bellingham MC, Berger AJ (1994) Characteristics and postnatal development of a hyperpolarization-activated inward current in rat hypoglossal motoneurons in vitro. J Neurophysiol 71:119–128
Bayliss DA, Umemiya M, Berger AJ (1995) Inhibition of N-and P-type calcium currents and the after-hyperpolarization in rat motoneurones by serotonin. J Physiol 485:635–647
Bennett DJ, Hultborn H, Fedirchuk B, Gorassini M (1998) Synaptic activation of plateaus in hindlimb motoneurons of decerebrate cats. J Neurophysiol 80:2023–2037
Berridge MJ (1998) Neuronal calcium signaling. Neuron 21:13–26
Bertrand S, Cazalets J-R (1998) Postinhibitory rebound during locomotor-like activity in neonatal rat motoneurons in vitro. J Neurophysiol 79:342–351
Bhumbra GS, Beato M (2018) Recurrent excitation between motoneurones propagates across segments and is purely glutamatergic. PLoS Biol 16:e2003586
Bigland B, Lippold O (1954) Motor unit activity in the voluntary contraction of human muscle. J Physiol 125:322–335
Bigland-Ritchie B, Johansson R, Lippold O, Woods J (1983a) Contractile speed and EMG changes during fatigue of sustained maximal voluntary contractions. J Neurophysiol 50:313–324
Bigland-Ritchie B, Johansson R, Lippold OCT, Smith S, Woods JJ (1983b) Changes in motoneurone firing rates during sustained maximal voluntary contractions. J Physiol 340:335–346
Binder MD, Powers RK, Heckman C (2020) Nonlinear input-output functions of motoneurons. Physiology 35:31–39
Björklund A, Skagerberg G (1982) Descending monoaminergic projections to the spinal cord. Brain Stem Control Spinal Mech:55–88
Bos R, Harris-Warrick RM, Brocard C, Demianenko LE, Manuel M, Zytnicki D, Korogod SM, Brocard F (2018) Kv1. 2 channels promote nonlinear spiking motoneurons for powering up locomotion. Cell Rep 22:3315–3327
Brock L, Coombs J, Eccles J (1951) Action potentials of motoneurones with intracellular electrode. Proc Univ Otago Med Sch:14–15
Brownstone RM (2006) Beginning at the end: repetitive firing properties in the final common pathway. Prog Neurobiol 78:156–172

Brownstone RM, Jordan L, Kriellaars D, Noga B, Shefchyk S (1992) On the regulation of repetitive firing in lumbar motoneurones during fictive locomotion in the cat. Exp Brain Res 90:441–455
Brownstone R, Gossard J-P, Hultborn H (1994) Voltage-dependent excitation of motoneurones from spinal locomotor centres in the cat. Exp Brain Res 102:34–44
Brownstone RM, Krawitz S, Jordan LM (2011) Reversal of the late phase of spike frequency adaptation in cat spinal motoneurons during fictive locomotion. J Neurophysiol 105:1045–1050
Büdingen HJ, Freund H-J (1976) The relationship between the rate of rise of isometric tension and motor unit recruitment in a human forearm muscle. Pflugers Arch 362:61–67
Bui TV, Grande G, Rose PK (2008) Relative location of inhibitory synapses and persistent inward currents determines the magnitude and mode of synaptic amplification in motoneurons. J Neurophysiol 99:583–594
Burger C, Ribera AB (1996) Xenopus spinal neurons express Kv2 potassium channel transcripts during embryonic development. J Neurosci 16:1412–1421
Burke RE (1967) Composite nature of the monosynaptic excitatory postsynaptic potential. J Neurophysiol 30:1114–1137
Burke R, Glenn LL (1996) Horseradish peroxidase study of the spatial and electrotonic distribution of group Ia synapses on type-identified ankle extensor motoneurons in the cat. J Comp Neurol 372:465–485
Burke R, Rudomin P, Zajac F (1970) Catch property in single mammalian motor units. Science 168:122–124
Burke R, Levine D, Tsairis P, Zajac F III (1973) Physiological types and histochemical profiles in motor units of the cat gastrocnemius. J Physiol 234:723–748
Button DC, Gardiner K, Marqueste T, Gardiner PF (2006) Frequency–current relationships of rat hindlimb α-motoneurones. J Physiol 573:663–677
Canto-Bustos M, Loeza-Alcocer E, González-Ramírez R, Gandini MA, Delgado-Lezama R, Felix R (2014) Functional expression of T-type ca 2+ channels in spinal motoneurons of the adult turtle. PLoS One 9:e108187
Carlin K, Jones K, Jiang Z, Jordan L, Brownstone R (2000) Dendritic L-type calcium currents in mouse spinal motoneurons: implications for bistability. Eur J Neurosci 12:1635–1646
Casanovas A, Salvany S, Lahoz V, Tarabal O, Piedrafita L, Sabater R, Hernández S, Calderó J, Esquerda JE (2017) Neuregulin 1-ErbB module in C-bouton synapses on somatic motor neurons: molecular compartmentation and response to peripheral nerve injury. Sci Rep 7:1–17
Chandler SH, Hsaio C-F, Inoue T, Goldberg LJ (1994) Electrophysiological properties of Guinea pig trigeminal motoneurons recorded in vitro. J Neurophysiol 71:129–145
Chen Y, Frank HY, Surmeier DJ, Scheuer T, Catterall WA (2006) Neuromodulation of Na+ channel slow inactivation via cAMP-dependent protein kinase and protein kinase C. Neuron 49:409–420
Collins D, Burke D, Gandevia S (2001) Large involuntary forces consistent with plateau-like behavior of human motoneurons. J Neurosci 21:4059–4065
Conradi S (1969) Ultrastructure and distribution of neuronal and glial elements on the surface of the proximal part of a motoneuron dendrite, as analyzed by serial sections. Acta Physiol Scand Suppl 332:49
Conradi S, Kellerth JO, Berthold CH, Hammarberg C (1979) Electron microscopic studies of serially sectioned cat spinal α-motoneurons. IV. Motoneurons innervating slow-twitch (type s) units of the soleus muscle. J Comp Neurol 184:769–782
Conway B, Hultborn H, Kiehn O, Mintz I (1988) Plateau potentials in alpha-motoneurones induced by intravenous injection of L-dopa and clonidine in the spinal cat. J Physiol 405:369–384
Coombs J, Curtis D, Eccles J (1957) The interpretation of spike potentials of motoneurones. J Physiol 139:198–231
Cooper S, Eccles J (1930) The isometric responses of mammalian muscles. J Physiol 69:377–385
Crone C, Hultborn H, Kiehn O, Mazieres L, Wigström H (1988) Maintained changes in motoneuronal excitability by short-lasting synaptic inputs in the decerebrate cat. J Physiol 405:321–343

De Luca CJ, Contessa P (2012) Hierarchical control of motor units in voluntary contractions. J Neurophysiol 107:178–195
De Luca C, Lefever R, Mccue M, Xenakis A (1982) Behaviour of human motor units in different muscles during linearly varying contractions. J Physiol 329:113–128
Deardorff AS, Romer SH, Deng Z, Bullinger KL, Nardelli P, Cope TC, Fyffe RE (2013) Expression of postsynaptic Ca2+-activated K+ (SK) channels at C-bouton synapses in mammalian lumbar α-motoneurons. J Physiol 591:875–897
Deardorff AS, Romer SH, Sonner PM, Fyffe RE (2014) Swimming against the tide: investigations of the C-bouton synapse. Front Neural Circ 8:106
Deardorff AS, Romer SH, Fyffe RE (2021) Location, location, location: the organization and roles of potassium channels in mammalian motoneurons. J Physiol 599:1391–1420
Desmedt JE, Godaux E (1977) Ballistic contractions in man: characteristic recruitment pattern of single motor units of the tibialis anterior muscle. J Physiol 264:673–693
Dodt H-U, Zieglgänsberger W (1990) Visualizing unstained neurons in living brain slices by infrared DIC-videomicroscopy. Brain Res 537:333–336
Du J, Haak LL, Phillips-Tansey E, Russell JT, Mcbain CJ (2000) Frequency-dependent regulation of rat hippocampal somato-dendritic excitability by the K+ channel subunit Kv2. 1. J Physiol 522:19–31
Duflocq A, Le Bras B, Bullier E, Couraud F, Davenne M (2008) Nav1. 1 is predominantly expressed in nodes of Ranvier and axon initial segments. Mol Cell Neurosci 39:180–192
Duflocq A, Chareyre F, Giovannini M, Couraud F, Davenne M (2011) Characterization of the axon initial segment (AIS) of motor neurons and identification of a para-AIS and a juxtapara-AIS, organized by protein 4.1 B. BMC Biol 9:1–19
Eccles J (1961) Membrane time constants of cat motoneurons and time courses of synaptic action. Exp Neurol 4:1–22
Eccles JC, Hoff H (1932) The rhythmic discharge of motoneurones. Proc R Soc Lond Ser B Contain Pap Biol Character 110:483–514
Eccles JC, Eccles RM, Lundberg A (1958) The action potentials of the alpha motoneurones supplying fast and slow muscles. J Physiol 142:275–291
Edwards FA, Konnerth A, Sakmann B, Takahashi T (1989) A thin slice preparation for patch clamp recordings from neurones of the mammalian central nervous system. Pflugers Arch 414:600–612
Eken T, Kiehn O (1989) Bistable firing properties of soleus motor units in unrestrained rats. Acta Physiol Scand 136:383–394
Enoka RM (2019) Physiological validation of the decomposition of surface EMG signals. J Electromyogr Kinesiol 46:70–83
Evinger C (1988) Extraocular motor nuclei: location, morphology and afferents. Rev Oculomot Res 2:81–117
Farina D, Negro F (2015) Common synaptic input to motor neurons, motor unit synchronization, and force control. Exerc Sport Sci Rev 43:23–33
Farina D, Yoshida K, Stieglitz T, Koch KP (2008) Multichannel thin-film electrode for intramuscular electromyographic recordings. J Appl Physiol 104:821–827
Fletcher EV, Simon CM, Pagiazitis JG, Chalif JI, Vukojicic A, Drobac E, Wang X, Mentis GZ (2017) Reduced sensory synaptic excitation impairs motor neuron function via Kv2. 1 in spinal muscular atrophy. Nat Neurosci
Florestal J, Mathieu P, Mcgill K (2009) Automatic decomposition of multichannel intramuscular EMG signals. J Electromyogr Kinesiol 19:1–9
Fox PD, Loftus RJ, Tamkun MM (2013) Regulation of Kv2. 1 K+ conductance by cell surface channel density. J Neurosci 33:1259–1270
Fuglevand AJ, Lester RA, Johns RK (2015) Distinguishing intrinsic from extrinsic factors underlying firing rate saturation in human motor units. J Neurophysiol 113:1310–1322

Fukuda S, Maeda H, Sakurai M (2020) Reevaluation of motoneuron morphology: diversity and regularity among motoneurons innervating different arm muscles along a proximal–distal axis. Sci Rep 10:1–10

Fuortes M, Frank K, Becker MC (1957) Steps in the production of motoneuron spikes. J Gen Physiol 40:735–752

Gallart-Palau X, Tarabal O, Casanovas A, Sábado J, Correa FJ, Hereu M, Piedrafita L, Calderó J, Esquerda JE (2014) Neuregulin-1 is concentrated in the postsynaptic subsurface cistern of C-bouton inputs to α-motoneurons and altered during motoneuron diseases. FASEB J 28:3618–3632

Gandevia SC (2001) Spinal and supraspinal factors in human muscle fatigue. Physiol Rev 81:1725–1789

Garland SJ, Griffin L (1999) Motor unit double discharges: statistical anomaly or functional entity? Can J Appl Physiol 24:113–130

Garrido JJ, Giraud P, Carlier E, Fernandes F, Moussif A, Fache M-P, Debanne D, Dargent B (2003) A targeting motif involved in sodium channel clustering at the axonal initial segment. Science 300:2091–2094

Gestreau C, Dutschmann M, Obled S, Bianchi AL (2005) Activation of XII motoneurons and premotor neurons during various oropharyngeal behaviors. Respir Physiol Neurobiol 147:159–176

Giroux N, Rossignol S, Reader TA (1999) Autoradiographic study of α1-and α2-noradrenergic and serotonin1A receptors in the spinal cord of normal and chronically transected cats. J Comp Neurol 406:402–414

Gorassini M, Yang JF, Siu M, Bennett DJ (2002) Intrinsic activation of human motoneurons: possible contribution to motor unit excitation. J Neurophysiol 87:1850–1858

Grande G, Bui TV, Rose PK (2007) Estimates of the location of L-type Ca2+ channels in motoneurons of different sizes: a computational study. J Neurophysiol 97:4023–4035

Granit R (1975) The functional role of the muscle spindles – facts and hypotheses. Brain J Neurol 98:531–556

Granit R, Kernell D, Shortess G (1963a) Quantitative aspects of repetitive firing of mammalian motoneurones, caused by injected currents. J Physiol 168:911–931

Granit R, Kernell D, Smith R (1963b) Delayed depolarization and the repetitive response to intracellular stimulation of mammalian motoneurones. J Physiol 168:890–910

Granit R, Kernell D, Lamarre Y (1966a) Algebraical summation in synaptic activation of motoneurones firing within the 'primary range' to injected currents. J Physiol 187:379–399

Granit R, Kernell D, Lamarre Y (1966b) Synaptic stimulation superimposed on motoneurones firing in the 'secondary range' to injected current. J Physiol 187:401–415

Guan D, Armstrong WE, Foehring RC (2013) Kv2 channels regulate firing rate in pyramidal neurons from rat sensorimotor cortex. J Physiol 591:4807–4825

Heckman C, Lee RH, Brownstone RM (2003) Hyperexcitable dendrites in motoneurons and their neuromodulatory control during motor behavior. Trends Neurosci 26:688–695

Heckman C, Gorassini MA, Bennett DJ (2005) Persistent inward currents in motoneuron dendrites: implications for motor output. Muscle Nerve 31:135–156

Hellström J, Oliveira AL, Meister B, Cullheim S (2003) Large cholinergic nerve terminals on subsets of motoneurons and their relation to muscarinic receptor type 2. J Comp Neurol 460:476–486

Henneman E (1957) Relation between size of neurons and their susceptibility to discharge. Science 126:1345–1347

Herda T, Miller J, Trevino M, Mosier E, Gallagher P, Fry A, Vardiman J (2016) The change in motor unit firing rates at de-recruitment relative to recruitment is correlated with type I myosin heavy chain isoform content of the vastus lateralis in vivo. Acta Physiol 216:454–463

Hochman S, Jordan LM, Schmidt BJ (1994) TTX-resistant NMDA receptor-mediated voltage oscillations in mammalian lumbar motoneurons. J Neurophysiol 72:2559–2562

Hodgkin AL, Huxley AF (1939) Action potentials recorded from inside a nerve fibre. Nature 144:710–711

Holobar A, Farina D, Gazzoni M, Merletti R, Zazula D (2009) Estimating motor unit discharge patterns from high-density surface electromyogram. Clin Neurophysiol 120:551–562
Hounsgaard J, Kiehn O (1985) Ca++ dependent bistability induced by serotonin in spinal motoneurons. Exp Brain Res 57:422–425
Hounsgaard J, Kiehn O (1993) Calcium spikes and calcium plateaux evoked by differential polarization in dendrites of turtle motoneurones in vitro. J Physiol 468:245–259
Hounsgaard J, Mintz I (1988) Calcium conductance and firing properties of spinal motoneurones in the turtle. J Physiol 398:591–603
Hounsgaard J, Hultborn H, Kiehn O (1986) Transmitter-controlled properties of α-motoneurones causing long-lasting motor discharge to brief excitatory inputs. Prog Brain Res. Elsevier
Hounsgaard J, Hultborn H, Jespersen B, Kiehn O (1988a) Bistability of alpha-motoneurones in the decerebrate cat and in the acute spinal cat after intravenous 5-hydroxytryptophan. J Physiol 405:345–367
Hounsgaard J, Kiehn O, Mintz I (1988b) Response properties of motoneurones in a slice preparation of the turtle spinal cord. J Physiol 398:575–589
Hultborn H, Wigström H, Wängberg B (1975) Prolonged activation of soleus motoneurones following a conditioning train in soleus Ia afferents – a case for a reverberating loop? Neurosci Lett 1:147–152
Hultborn H, Denton ME, Wienecke J, Nielsen JB (2003) Variable amplification of synaptic input to cat spinal motoneurones by dendritic persistent inward current. J Physiol 552:945–952
Hwang P, Cunningham A, Peng Y, Snyder S (1993) CDRK and DRK1 K+ channels have contrasting localizations in sensory systems. Neuroscience 55:613–620
Iansek R, Redman S (1973a) The amplitude, time course and charge of unitary excitatory postsynaptic potentials evoked in spinal motoneurone dendrites. J Physiol 234:665–688
Iansek R, Redman S (1973b) An analysis of the cable properties of spinal motoneurones using a brief intracellular current pulse. J Physiol 234:613–636
Iglesias C, Meunier C, Manuel M, Timofeeva Y, Delestrée N, Zytnicki D (2011) Mixed mode oscillations in mouse spinal motoneurons arise from a low excitability state. J Neurosci 31:5829–5840
Issa AN, Zhan WZ, Sieck GC, Mantilla CB (2010) Neuregulin-1 at synapses on phrenic motoneurons. J Comp Neurol 518:4213–4225
Ito M, Oshima T (1962) Temporal summation of after-hyperpolarization following a motoneurone spike. Nature 195:910–911
Jacques, Duchateau Roger M., Enoka (2011) Human motor unit recordings: Origins and insight into the integrated motor system. Brain Research 140942-61 10.1016/j.brainres.2011.06.011
Jacobs BL, Martín-Cora FJ, Fornal CA (2002) Activity of medullary serotonergic neurons in freely moving animals. Brain Res Rev 40:45–52
Jensen DB, Stecina K, Wienecke J, Hedegaard A, Sukiasyan N, Hultborn HR, Meehan CF (2018) The subprimary range of firing is present in both cat and mouse spinal motoneurons and its relationship to force development is similar for the two species. J Neurosci 38:9741–9753
Jensen DB, Kadlecova M, Allodi I, Meehan CF (2020) Spinal motoneurones are intrinsically more responsive in the adult G93A SOD1 mouse model of Amyotrophic Lateral Sclerosis. J Physiol 598:4385–4403
Johnson K, Edwards S, Van Tongeren C, Bawa P (2004) Properties of human motor units after prolonged activity at a constant firing rate. Exp Brain Res 154:479–487
Johnson B, Leek AN, Tamkun MM (2019) Kv2 channels create endoplasmic reticulum/plasma membrane junctions: a brief history of Kv2 channel subcellular localization. Channels 13:88–101
Kanning KC, Kaplan A, Henderson CE (2010) Motor neuron diversity in development and disease. Annu Rev Neurosci 33:409–440
Kellerth JO, Berthold CH, Conradi S (1979) Electron microscopic studies of serially sectioned cat spinal α-motoneurons. III. Motoneurons innervating fast-twitch (type FR) units of the gastrocnemius muscle. J Comp Neurol 184:755–767

Kellerth JO, Conradi S, Berthold CH (1983) Electron microscopic studies of serially sectioned cat spinal α-motoneurons: V. motoneurons innervating fast-twitch (type FF) units of the gastrocnemius muscle. J Comp Neurol 214:451–458
Kernell D (1964) The delayed depolarization in cat and rat motoneurones. Prog Brain Res. Elsevier
Kernell D (1965) Synaptic influence on the repetitive activity elicited in cat lumbosacral motoneurones by long-lasting injected currents. Acta Physiol Scand 63:409–410
Kernell D (1968) The repetitive impulse discharge of a simple neurone model compared to that of spinal motoneurones. Brain Res 11:685–687
Kernell D (1972) The early phase of adaptation in repetitive impulse discharges of cat spinal motoneurones. Brain Res 41:184–186
Kernell D (1983) Functional properties of spinal motoneurons and gradation of muscle force. Adv Neurol 39:213–226
Kernell D (1999) Repetitive impulse firing in motoneurons: facts and perspectives. Prog Brain Res 123:31–37
Kernell D (2006) The motoneurone and its muscle fibres
Kernell D, Monster A (1982) Time course and properties of late adaptation in spinal motoneurones of the cat. Exp Brain Res 46:191–196
Kernell D, Sjöholm H (1973) Repetitive impulse firing: comparisons between neurone models based on 'voltage clamp equations' and spinal motoneurones. Acta Physiol Scand 87:40–56
Kihira Y, Hermanstyne TO, Misonou H (2010) Formation of heteromeric Kv2 channels in mammalian brain neurons. J Biol Chem 285:15048–15055
Kleinfeld D, Moore JD, Wang F, Deschênes M (2014) The brainstem oscillator for whisking and the case for breathing as the master clock for orofacial motor actions. In: Cold Spring Harbor symposia on quantitative biology. Cold Spring Harbor Laboratory Press, pp 29–39
Kobayashi M, Inoue T, Matsuo R, Masuda Y, Hidaka O, Kang Y, Morimoto T (1997) Role of calcium conductances on spike afterpotentials in rat trigeminal motoneurons. J Neurophysiol 77:3273–3283
Kolmodin G, Skoglund C (1958) Slow membrane potential changes accompanying excitaltion and inhibition in spinal noto-and interneurons in the cat during natural activation. Acta Physiol Scand 44:11–54
Konishi S, Otsuka M (1974) Excitatory action of hypothalamic substance P on spinal motoneurones of newborn rats. Nature 252:734
Krawitz S, Fedirchuk B, Dai Y, Jordan L, Mccrea D (2001) State-dependent hyperpolarization of voltage threshold enhances motoneurone excitability during fictive locomotion in the cat. J Physiol 532:271–281
Krnjević K, Lamour Y, Macdonald J, Nistri A (1978) Motoneuronal after-potentials and extracellular divalent cations. Can J Physiol Pharmacol 56:516–520
Krutki P, Mrówczyński W, Bączyk M, Łochyński D, Celichowski J (2017) Adaptations of motoneuron properties after weight-lifting training in rats. J Appl Physiol 123:664–673
Lagerbäck P-Å, Cullheim S, Ulfhake B (1986) Electron microscopic observations on the synaptology of cat sciatic γ-motoneurons after intracellular staining with horseradish peroxidase. Neurosci Lett 70:23–27
Landoni LM, Myles JR, Wells TL, Mayer WP, Akay T (2019) Cholinergic modulation of motor neurons through the C-boutons are necessary for the locomotor compensation for severe motor neuron loss during amyotrophic lateral sclerosis disease progression. Behav Brain Res 369:111914
Larkman P, Kelly J (1992) Ionic mechanisms mediating 5-hydroxytryptamine-and noradrenaline-evoked depolarization of adult rat facial motoneurones. J Physiol 456:473–490
Le Bras B, Fréal A, Czarnecki A, Legendre P, Bullier E, Komada M, Brophy PJ, Davenne M, Couraud F (2014) In vivo assembly of the axon initial segment in motor neurons. Brain Struct Funct 219:1433–1450
Lee R, Heckman C (1999) Paradoxical effect of QX-314 on persistent inward currents and bistable behavior in spinal motoneurons in vivo. J Neurophysiol 82:2518–2527

Lee RH, Heckman CJ (2000) Adjustable amplification of synaptic input in the dendrites of spinal motoneurons in vivo. J Neurosci 20:6734–6740
Lee R, Kuo J, Jiang M, Heckman C (2003) Influence of active dendritic currents on input-output processing in spinal motoneurons in vivo. J Neurophysiol 89:27–39
Leroy F, D'incamps BL, Imhoff-Manuel RD, Zytnicki D (2014) Early intrinsic hyperexcitability does not contribute to motoneuron degeneration in amyotrophic lateral sclerosis. Elife 3:e04046
Li Y, Bennett DJ (2003) Persistent sodium and calcium currents cause plateau potentials in motoneurons of chronic spinal rats. J Neurophysiol 90:857–869
Li X, Bennett DJ (2007) Apamin-sensitive calcium-activated potassium currents (SK) are activated by persistent calcium currents in rat motoneurons. J Neurophysiol 97:3314–3330
Li W-C, Soffe SR, Roberts A (2004a) A direct comparison of whole cell patch and sharp electrodes by simultaneous recording from single spinal neurons in frog tadpoles. J Neurophysiol
Li Y, Gorassini MA, Bennett DJ (2004b) Role of persistent sodium and calcium currents in motoneuron firing and spasticity in chronic spinal rats. J Neurophysiol 91:767–783
Liddell EGT, Sherrington C (1924) Reflexes in response to stretch (myotatic reflexes). Proc R Soc Lond Ser B Contain Pap Biol Character 96:212–242
Liu PW, Bean BP (2014) Kv2 channel regulation of action potential repolarization and firing patterns in superior cervical ganglion neurons and hippocampal CA1 pyramidal neurons. J Neurosci 34:4991–5002
Lorincz A, Nusser Z (2008) Cell-type-dependent molecular composition of the axon initial segment. J Neurosci 28:14329–14340
Mandikian D, Bocksteins E, Parajuli LK, Bishop HI, Cerda O, Shigemoto R, Trimmer JS (2014) Cell type–specific spatial and functional coupling between mammalian brain Kv2. 1 K+ channels and ryanodine receptors. J Comp Neurol 522:3555–3574
Manuel M, Heckman C (2011) Adult mouse motor units develop almost all of their force in the subprimary range: a new all-or-none strategy for force recruitment? J Neurosci 31:15188–15194
Manuel M, Iglesias C, Donnet M, Leroy F, Heckman C, Zytnicki D (2009) Fast kinetics, high-frequency oscillations, and subprimary firing range in adult mouse spinal motoneurons. J Neurosci 29:11246–11256
Marateb HR, Muceli S, Mcgill KC, Merletti R, Farina D (2011) Robust decomposition of single-channel intramuscular EMG signals at low force levels. J Neural Eng 8:066015
Mavlyutov TA, Epstein ML, Andersen KA, Ziskind-Conhaim L, Ruoho AE (2010) The sigma-1 receptor is enriched in postsynaptic sites of C-terminals in mouse motoneurons. An anatomical and behavioral study. Neuroscience 167:247–255
Mavlyutov TA, Epstein ML, Liu P, Verbny YI, Ziskind-Conhaim L, Ruoho AE (2012) Development of the sigma-1 receptor in C-terminals of motoneurons and colocalization with the N, N′-dimethyltryptamine forming enzyme, indole-N-methyl transferase. Neuroscience 206:60–68
Meehan CF, Sukiasyan N, Zhang M, Nielsen JB, Hultborn H (2010) Intrinsic properties of mouse lumbar motoneurons revealed by intracellular recording in vivo. J Neurophysiol 103:2599–2610
Miles GB, Yohn DC, Wichterle H, Jessell TM, Rafuse VF, Brownstone RM (2004) Functional properties of motoneurons derived from mouse embryonic stem cells. J Neurosci 24:7848–7858
Miles G, Dai Y, Brownstone R (2005) Mechanisms underlying the early phase of spike frequency adaptation in mouse spinal motoneurones. J Physiol 566:519–532
Miles GB, Hartley R, Todd AJ, Brownstone RM (2007) Spinal cholinergic interneurons regulate the excitability of motoneurons during locomotion. Proc Natl Acad Sci 104:2448–2453
Milner-Brown H, Stein R, Yemm R (1973) The orderly recruitment of human motor units during voluntary isometric contractions. J Physiol 230:359–370
Misonou H, Mohapatra DP, Menegola M, Trimmer JS (2005a) Calcium-and metabolic state-dependent modulation of the voltage-dependent Kv2. 1 channel regulates neuronal excitability in response to ischemia. J Neurosci 25:11184–11193
Misonou H, Mohapatra DP, Trimmer JS (2005b) Kv2. 1: a voltage-gated k+ channel critical to dynamic control of neuronal excitability. Neurotoxicology 26:743–752

Mitra P, Brownstone RM (2012) An in vitro spinal cord slice preparation for recording from lumbar motoneurons of the adult mouse. J Neurophysiol 107:728–741
Mòdol-Caballero G, Santos D, Navarro X, Herrando-Grabulosa M (2018) Neuregulin 1 reduces motoneuron cell death and promotes neurite growth in an in vitro model of motoneuron degeneration. Front Cell Neurosci 11:431
Moreno CM, Dixon RE, Tajada S, Yuan C, Opitz-Araya X, Binder MD, Santana LF (2016) Ca2+ entry into neurons is facilitated by cooperative gating of clustered CaV1. 3 channels. Elife 5:e15744
Moritz AT, Newkirk G, Powers RK, Binder MD (2007) Facilitation of somatic calcium channels can evoke prolonged tail currents in rat hypoglossal motoneurons. J Neurophysiol 98:1042–1047
Mottram CJ, Suresh NL, Heckman C, Gorassini MA, Rymer WZ (2009) Origins of abnormal excitability in biceps brachii motoneurons of spastic-paretic stroke survivors. J Neurophysiol 102:2026–2038
Mottram CJ, Heckman CJ, Powers RK, Rymer WZ, Suresh NL (2014) Disturbances of motor unit rate modulation are prevalent in muscles of spastic-paretic stroke survivors. J Neurophysiol 111:2017–2028
Muceli S, Bergmeister KD, Hoffmann K-P, Aman M, Vukajlija I, Aszmann OC, Farina D (2018) Decoding motor neuron activity from epimysial thin-film electrode recordings following targeted muscle reinnervation. J Neural Eng 16:016010
Muennich EA, Fyffe RE (2004) Focal aggregation of voltage-gated, Kv2. 1 subunit-containing, potassium channels at synaptic sites in rat spinal motoneurones. J Physiol 554:673–685
Müller D, Cherukuri P, Henningfeld K, Poh CH, Wittler L, Grote P, Schlüter O, Schmidt J, Laborda J, Bauer SR (2014) Dlk1 promotes a fast motor neuron biophysical signature required for peak force execution. Science 343:1264–1266
Murakoshi H, Shi G, Scannevin RH, Trimmer JS (1997) Phosphorylation of the Kv2. 1 K+ channel alters voltage-dependent activation. Mol Pharmacol 52:821–828
Nagy J, Yamamoto T, Jordan L (1993) Evidence for the cholinergic nature of c-terminals associated with subsurface cisterns in α-motoneurons of rat. Synapse 15:17–32
Nascimento F, Broadhead MJ, Tetringa E, Tsape E, Zagoraiou L, Miles G (2020) Synaptic mechanisms underlying modulation of locomotor-related motoneuron output by premotor cholinergic interneurons. eLife 9:e54170
Negro F, Muceli S, Castronovo AM, Holobar A, Farina D (2016) Multi-channel intramuscular and surface EMG decomposition by convolutive blind source separation. J Neural Eng 13:026027
Nelson PG, Burke R (1967) Delayed depolarization in cat spinal motoneurons. Exp Neurol 17:16–26
Nelson P, Lux H (1970) Some electrical measurements of motoneuron parameters. Biophys J 10:55–73
Nordstrom MA, Gorman RB, Laouris Y, Spielmann JM, Stuart DG (2007) Does motoneuron adaptation contribute to muscle fatigue? Muscle Nerve 35:135–158
Pambo-Pambo A, Durand J, Gueritaud J-P (2009) Early excitability changes in lumbar motoneurons of transgenic SOD1G85R and SOD1G93A-low mice. J Neurophysiol 102:3627–3642
Pan Z, Kao T, Horvath Z, Lemos J, Sul J-Y, Cranstoun SD, Bennett V, Scherer SS, Cooper EC (2006) A common ankyrin-G-based mechanism retains KCNQ and NaV channels at electrically active domains of the axon. J Neurosci 26:2599–2613
Parmiggiani F, Stein R (1981) Nonlinear summation of contractions in cat muscles. II. Later facilitation and stiffness changes. J Gen Physiol 78:295–311
Perrier J-F, Hounsgaard J (2003) 5-HT2 receptors promote plateau potentials in turtle spinal motoneurons by facilitating an L-type calcium current. J Neurophysiol 89:954–959
Pinter M, Curtis R, Hosko M (1983) Voltage threshold and excitability among variously sized cat hindlimb motoneurons. J Neurophysiol 50:644–657
Piotrkiewicz M, Wilanowski G (2012) Is spike frequency adaptation an artefact? Insight from human studies. Front Cell Neurosci 6:50

Powers RK, Heckman CJ (2015) Contribution of intrinsic motoneuron properties to discharge hysteresis and its estimation based on paired motor unit recordings: a simulation study. J Neurophysiol 114:184–198
Powers RK, Sawczuk A, Musick JR, Binder MD (1999) Multiple mechanisms of spike-frequency adaptation in motoneurones. J Physiol Paris 93:101–114
Powers RK, Nardelli P, Cope TC (2008) Estimation of the contribution of intrinsic currents to motoneuron firing based on paired motoneuron discharge records in the decerebrate cat. J Neurophysiol 100:292–303
Rall W (1960) Membrane potential transients and membrane time constant of motoneurons. Exp Neurol 2:503–532
Rall W (1967) Distinguishing theoretical synaptic potentials computed for different soma-dendritic distributions of synaptic input. J Neurophysiol 30:1138–1168
Rall W (2011) Core conductor theory and cable properties of neurons. Compr Physiol:39–97
Revill AL, Fuglevand AJ (2017) Inhibition linearizes firing rate responses in human motor units: implications for the role of persistent inward currents. J Physiol 595:179–191
Romer SH, Deardorff AS, Fyffe RE (2019) A molecular rheostat: Kv2. 1 currents maintain or suppress repetitive firing in motoneurons. J Physiol 597:3769–3786
Sah P, Mclachlan EM (1992) Potassium currents contributing to action potential repolarization and the afterhyperpolarization in rat vagal motoneurons. J Neurophysiol 68:1834–1841
Sakmann B, Neher E (1984) Patch clamp techniques for studying ionic channels in excitable membranes. Annu Rev Physiol 46:455–472
Santiago LJ, Abrol R (2019) Understanding G protein selectivity of muscarinic acetylcholine receptors using computational methods. Int J Mol Sci 20:5290
Sawczuk A, Powers RK, Binder MD (1995) Spike frequency adaptation studied in hypoglossal motoneurons of the rat. J Neurophysiol 73:1799–1810
Schwindt PC (1973) Membrane-potential trajectories underlying motoneuron rhythmic firing at high rates. J Neurophysiol 36:434–439
Schwindt P, Crill WE (1977) A persistent negative resistance in cat lumbar motoneurons. Brain Res 120:173–178
Schwindt P, Crill W (1980a) Role of a persistent inward current in motoneuron bursting during spinal seizures. J Neurophysiol 43:1296–1318
Schwindt PC, Crill W (1980b) Effects of barium on cat spinal motoneurons studied by voltage clamp. J Neurophysiol 44:827–846
Schwindt PC, Crill WE (1980c) Properties of a persistent inward current in normal and TEA-injected motoneurons. J Neurophysiol 43:1700–1724
Schwindt PC, Crill WE (1981) Differential effects of TEA and cations on outward ionic currents of cat motoneurons. J Neurophysiol 46:1–16
Schwindt P, Crill W (1982) Factors influencing motoneuron rhythmic firing: results from a voltage-clamp study. J Neurophysiol 48:875–890
Smith CC, Brownstone RM (2020) Spinal motoneuron firing properties mature from rostral to caudal during postnatal development of the mouse. J Physiol 598:5467–5485
Soulard C, Salsac C, Mouzat K, Hilaire C, Roussel J, Mezghrani A, Lumbroso S, Raoul C, Scamps F (2020) Spinal Motoneuron TMEM16F acts at C-boutons to modulate motor resistance and contributes to ALS pathogenesis. Cell Rep 30(2581–2593):e7
Spielmann J, Laouris Y, Nordstrom M, Robinson G, Reinking R, Stuart D (1993) Adaptation of cat motoneurons to sustained and intermittent extracellular activation. J Physiol 464:75–120
Stecko GAJ (1962) A new bipolar electrode for electromyography. J Appl Physiol 17:849
Stein R, Parmiggiani F (1979) Optimal motor patterns for activating mammalian muscle. Brain Res 175:372–376
Stein R, French A, Mannard A, Yemm R (1972) New methods for analysing motor function in man and animals. Brain Res
Stifani N (2014) Motor neurons and the generation of spinal motor neurons diversity. Front Cell Neurosci 8:293

Takahashi T (1978) Intracellular recording from visually identified motoneurons in rat spinal cord slices. Proc R Soc Lond Ser B Biol Sci 202:417–421

Takahashi T, Berger AJ (1990) Direct excitation of rat spinal motoneurones by serotonin. J Physiol 423:63–76

Taylor JL, Amann M, Duchateau J, Meeusen R, Rice CL (2016) Neural contributions to muscle fatigue: from the brain to the muscle and back again. Med Sci Sports Exerc 48:2294

Thomas CK, Ross BH, Calancie B (1987) Human motor-unit recruitment during isometric contractions and repeated dynamic movements. J Neurophysiol 57:311–324

Thompson CK, Negro F, Johnson MD, Holmes MR, Mcpherson LM, Powers RK, Farina D, Heckman CJ (2018) Robust and accurate decoding of motoneuron behaviour and prediction of the resulting force output. J Physiol 596:2643–2659

Tsuchida T, Ensini M, Morton S, Baldassare M, Edlund T, Jessell T, Pfaff S (1994) Topographic organization of embryonic motor neurons defined by expression of LIM homeobox genes. Cell 79:957–970

Tsuzuki S, Yoshida S, Yamamoto T, Oka H (1995) Developmental changes in the electrophysiological properties of neonatal rat oculomotor neurons studied in vitro. Neurosci Res 23:389–397

Turkin VV, O'neilL D, Jung R, Iarkov A, Hamm TM (2010) Characteristics and organization of discharge properties in rat hindlimb motoneurons. J Neurophysiol 104:1549–1565

Ulfhake B, Cullheim S, Franson P (1988) Postnatal development of cat hind limb motoneurons. I: Changes in length, branching structure, and spatial distribution of dendrites of cat triceps surae motoneurons. J Comp Neurol 278:69–87

Umemiya M, Berger AJ (1994) Properties and function of low-and high-voltage-activated Ca2+ channels in hypoglossal motoneurons. J Neurosci 14:5652–5660

Veasey SC, Fornal C, Metzler C, Jacobs BL (1995) Response of serotonergic caudal raphe neurons in relation to specific motor activities in freely moving cats. J Neurosci 15:5346–5359

Viana F, Bayliss DA, Berger AJ (1993) Multiple potassium conductances and their role in action potential repolarization and repetitive firing behavior of neonatal rat hypoglossal motoneurons. J Neurophysiol 69:2150–2163

Vrbova G (1983) Hypothesis: Duchenne dystrophy viewed as a disturbance of nerve-muscle interactions. Muscle Nerve 6:671–675

Wachholder K (1928) Willkürliche Haltung und Bewegung, insbesondere im Lichte elektrophysiologischer Untersuchungen. Willkürliche Haltung und Bewegung, insbesondere im Lichte elektrophysiologischer Untersuchungen. Springer

Wienecke J, Zhang M, Hultborn H (2009) A prolongation of the postspike afterhyperpolarization following spike trains can partly explain the lower firing rates at derecruitment than those at recruitment. J Neurophysiol 102:3698–3710

Wilson JM, Rempel J, Brownstone RM (2004) Postnatal development of cholinergic synapses on mouse spinal motoneurons. J Comp Neurol 474:13–23

Yamada Y, Yamamura K, Inoue M (2005) Coordination of cranial motoneurons during mastication. Respir Physiol Neurobiol 147:177–189

Zagoraiou L, Akay T, Martin JF, Brownstone RM, Jessell TM, Miles GB (2009) A cluster of cholinergic premotor interneurons modulates mouse locomotor activity. Neuron 64:645–662

Zemankovics R, Káli S, Paulsen O, Freund TF, Hájos N (2010) Differences in subthreshold resonance of hippocampal pyramidal cells and interneurons: the role of h-current and passive membrane characteristics. J Physiol 588:2109–2132

Zengel JE, Reid SA, Sypert GW, Munson JB (1985) Membrane electrical properties and prediction of motor-unit type of medial gastrocnemius motoneurons in the cat. J Neurophysiol 53:1323–1344

Zhang L, Krnjević K (1987) Apamin depresses selectively the after-hyperpolarization of cat spinal motoneurons. Neurosci Lett 74:58–62

Zia M, Chung B, Sober SJ, Bakir MS (2018) Fabrication and characterization of 3D multi-electrode array on flexible substrate for in vivo EMG recording from expiratory muscle of songbird. In: 2018 IEEE international electron devices meeting (IEDM). IEEE, 29.4. 1–29.4. 4

The Cellular Basis for the Generation of Firing Patterns in Human Motor Units

Obaid U. Khurram, Gregory E. P. Pearcey, Matthieu K. Chardon, Edward H. Kim, Marta García, and C. J. Heckman

Abstract Motor units, which comprise a motoneuron and the set of muscle fibers it innervates, are the fundamental neuromuscular transducers for all motor commands. The one to one relationship between a motoneuron and its innervated muscle fibers allow motoneuron firing patterns to be readily measured in humans. In this chapter, we summarize the current understanding of the cellular basis for the generation of firing patterns in human motor units. We provide a brief review of land-

Authors Obaid U. Khurram and Gregory E. P. Pearcey have equally contributed to this chapter.

O. U. Khurram · E. H. Kim
Departments of Physiology, Feinberg School of Medicine, Northwestern University, Chicago, IL, USA

G. E. P. Pearcey
Departments of Physiology, Feinberg School of Medicine, Northwestern University, Chicago, IL, USA

Physical Medicine and Rehabilitation, Feinberg School of Medicine, Northwestern University, Chicago, IL, USA

M. K. Chardon
Departments of Physiology, Feinberg School of Medicine, Northwestern University, Chicago, IL, USA

Northwestern-Argonne Institute of Science and Engineering, Evanston, IL, USA

M. García
Northwestern-Argonne Institute of Science and Engineering, Evanston, IL, USA

Computational Science Division, Argonne National Laboratory, Lemont, IL, USA

C. J. Heckman (✉)
Departments of Physiology, Feinberg School of Medicine, Northwestern University, Chicago, IL, USA

Physical Medicine and Rehabilitation, Feinberg School of Medicine, Northwestern University, Chicago, IL, USA

Physical Therapy and Human Movement Sciences, Northwestern University, Chicago, IL, USA
e-mail: c-heckman@northwestern.edu

M. J. O'Donovan, M. Falgairolle (eds.), *Vertebrate Motoneurons*, Advances in Neurobiology 28, https://doi.org/10.1007/978-3-031-07167-6_10

mark insights from classic studies and then proceed to consider the features of motor unit firing patterns that are most likely to be sensitive estimators of motoneuron inputs and properties. In addition, we discuss recent advances in technology for recording human motor unit firing patterns and highly realistic computer simulations of motoneurons. The final section presents our recent efforts to use the power of supercomputers for implementation of the motoneuron models, with a goal of achieving a true "reverse engineering" approach that maximizes the insights from motor unit firing patterns into the synaptic structure of motor commands.

Keywords Motor unit · Motoneuron · Motor control · Persistent inward currents · Neuromodulation

1 Introduction

Motor units are the quantal elements of motor control, serving as the fundamental neuromechanical transducers for all motor commands (Heckman and Enoka 2012). The one to one relation between action potentials in the motoneuron and its innervated muscle fibers allows motoneuron firing patterns to be recorded via electromyography (EMG) in humans and other animals. As a result, motoneurons are the only neurons whose firing patterns can be readily measured in humans. In fact, some of the first firing patterns to be recorded from neurons—at least indirectly—were obtained via electrodes inserted into human muscle, in studies done over 90 years ago (Adrian and Bronk 1929; see Duchateau and Enoka 2011 for historical review). About 70 years ago, both Eccles as well as Woodbury and Patton performed intracellular recordings in the lumbar spinal cord of feline preparations, making motoneurons the first mammalian neurons of any type from which direct intracellular recordings were made (Eccles 1952; Woodbury and Patton 1952; see Stuart and Brownstone 2011 for historical review). These pioneering studies made it clear that investigating motor unit firing patterns in light of the inputs and properties of motoneurons has enormous potential for revealing the cellular basis of motor commands. This potential has dramatically increased in recent years for two main reasons. The first is the rapid improvement in technology for recording motor unit firing patterns in humans, due to the development of new electrodes and new signal decomposition methods, which allows for non-invasive recordings of many motor units simultaneously. The second is the development of highly realistic computer simulations of motoneurons, which provide a quantitative bridge between motor unit firing patterns in humans and cellular data from other animal preparations. In this chapter, we summarize the current understanding of the cellular basis for the generation of firing patterns in human motor units. First, we outline how motor commands can affect the recruitment and rate coding of motor units. In doing so, we provide a brief review of landmark insights from classic studies and then proceed to consider the features of motor unit firing patterns that are most likely to be sensitive estimators of motoneuron inputs and properties. The final section presents our recent efforts to use the power of supercomputers for implementation of the motoneuron models, with a goal of achieving a true "reverse engineering" approach that maximizes the insights from motor unit firing patterns into the synaptic structure of motor commands.

2 Advancing Our Understanding of Human Motor Commands with the Help of Population Motor Unit Recordings

Given the potential complexity of the structure of motor commands, it is reasonable to ask if individual motor unit firing patterns have sufficient information to provide significant insights into that structure. Fortunately, recent technical advances now allow simultaneous recording of many human motor units using non-invasive high-density surface EMG (HD sEMG) arrays. These methods use a flexible array of surface electrodes placed on the skin over a muscle and convolutive blind source separation to decompose this surface EMG into the firing patterns of multiple individual motor units – as many as 20–40 simultaneously. An example of the data obtained with this approach during a linearly increasing and decreasing isometric ankle dorsiflexion is shown in Fig. 1, using the decomposition approach developed by Holobar et al. (2009; Holobar and Zazula 2008; Merletti et al. 2008; Negro et al. 2016). These decomposition methods have been the subject of several reviews (Farina and Holobar 2016; Farina et al. 2016; Merletti et al. 2008) and systematic

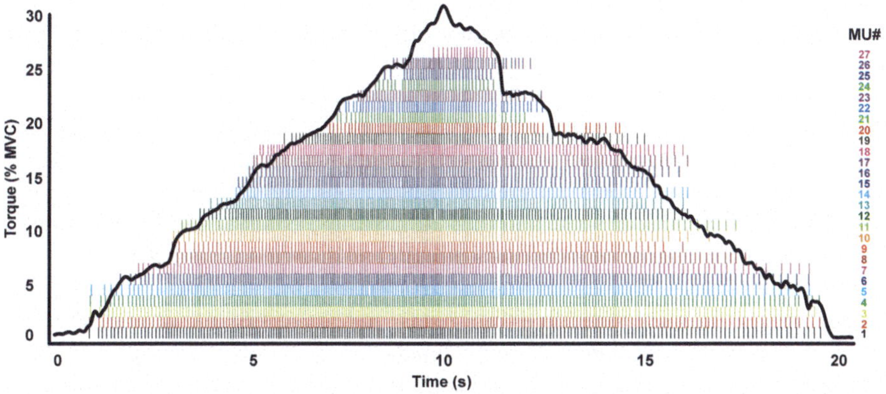

Fig. 1 Decomposition of HD sEMG signals recorded from a single human subject's tibialis anterior yields many individual motor unit spike trains during a slow isometric torque ramp (black trace). Individual vertical lines with unique colors represent the firing instances of each of the 27 motor units discriminated during this trial. Note the tight coupling between the intensity of the torque ramp and the interspike intervals of the discriminated motor units

validation has been implemented by both computer simulations and spike to spike comparisons with the traditional intra-muscular fine wire approach. The correspondence between the two methods in the latter case is typically about 90%. As discussed throughout this chapter, the information provided by this remarkable increase in number of units is essential for command structure estimation.

3 Classic Studies on the Recruitment of Motor Units

A fundamental feature of motor behavior is that muscle force generation is accomplished by recruitment of motor units and/or increases in the firing rates of already recruited motor units, which was first appreciated by Adrian and Bronk (1929). Shortly thereafter, it was noted that recruitment had a clear "size" pattern, i.e. as force increased, motor units with smaller amplitude action potentials were recruited before those with larger amplitude action potentials (Denny-Brown and Pennybacker 1938). With the advent of improved amplifiers and development of oscilloscopes, studies of recruitment of motor axons in filaments isolated from the ventral roots of the feline cord by Henneman (1957) revealed the same small to large pattern. In a series of papers in 1965, Henneman and colleagues achieved major breakthroughs in understanding both the function and the mechanism of this size-based recruitment order (Henneman and Mendell 1981; Henneman and Olson 1965; Henneman et al. 1965a, b). By stimulating single motor axons in thin ventral root filaments, they showed that motor unit action potential size was proportional to motor unit speed, force, and fatigability. This breakthrough defined the functional significance of size-based recruitment, showing that it provided optimal activation of motor units in terms of precision, fatigue resistance, and energy efficiency. A second breakthrough was conceptual: Henneman hypothesized that size was both the phenomenon and its cellular mechanism (Henneman and Mendell 1981). That is, small motor units have small motoneurons with high intrinsic resistance and thus greater responses to a given level of synaptic current – a straightforward consequence of Ohm's law. Thus, the properties of motoneurons themselves define recruitment order, an elegant evolutionary adaptation that greatly simplifies the computational burden for specifying motor commands. Many subsequent studies strongly supported Henneman's initial work and the size principle continues to be the foundation of our understanding of all motor output (Heckman and Enoka 2012).

Systematic studies on mechanisms underlying the size principle have since followed. Early studies suggested that the size principle arose because motoneurons all received approximately equal synaptic inputs but responded according to their threshold differences. For example, Kernell (1966) used intracellular recordings to show that motoneurons indeed had a wide range of threshold currents for generation of action potentials – approximately 10-fold. Mendell and Henneman (1971) subsequently developed the spike triggered average technique to show that individual muscle spindle Ia axons projected similarly to more than 90% of the motoneurons within a single pool. Yet, studies by Burke and colleagues challenged this notion by demonstrating that synaptic inputs were in fact not distributed uniformly within a pool (Burke 1981). Using the differences in mechanical properties of motor units, they divided motor units into three types (slow [S], fast fatigue resistant [FR], and fast fatigable [FF]) and showed that several different synaptic input systems generated EPSPs of differing relative amplitudes across the three types (Burke et al. 1973). Even the monosynaptic input from muscle spindle Ia afferents, which Henneman and Mendell (1981) had shown to project widely across a pool, nonetheless proved to generate much greater EPSPs in smaller motoneurons. Thus, Ia EPSP amplitude was scaled with motor unit type, with S > FR > FF. In contrast, EPSPs from stimulation of the sural cutaneous nerve and the red nucleus showed the opposite pattern (FF > FR > S). These patterns within a single motor nucleus were consistent with previous studies showing differences in EPSPs to motoneurons innervating muscles with predominantly slow (i.e. soleus) and fast (i.e. medial gastrocnemius) muscle fibers (Burke 1981).

A problem with the PSPs approach is that their amplitudes are just as dependent on the resistance of the neuron as on the underlying synaptic current. For example, even if Ia synaptic currents were equal in all motoneurons, Ia EPSPs would be much greater in higher-resistance (i.e. smaller) motoneurons. Switching to prolonged, quasi-steady synaptic inputs instead of transient PSPs allowed separation of these two factors and revealed that Ia synaptic currents had a non-uniform distribution, with a bias towards the smaller motoneurons of Type S motor units (Heckman and Binder 1988). Systematic studies using this steady synaptic current approach by Powers and Binder (2001) showed that all excitatory inputs studied (including vestibulospinal, rubrospinal, corticospinal, and spindle Ia) projected non-uniformly, as illustrated in Fig. 2. All of these descending inputs proved to generate much larger currents in lower-resistance (i.e. larger) motoneurons; thus far only Ia afferents have been shown to have the opposite distribution. Both Ia reciprocal inhibition (Heckman and Binder 1991a) and Renshaw cell recurrent inhibition (Lindsay and Binder 1991) proved to be approximately equally distributed to all motoneurons. Realistic computer simulations showed that none of these distributions on their own were sufficient to reverse recruitment order, though the descending inputs markedly reduced the range of recruitment while Ia input increased this range (Heckman and Binder 1993b). These differences have proved important in understanding recruitment patterns in human data – or to put it another way, make it likely that muscles with narrow recruitment ranges are likely to be primarily driven by descending inputs with at most small contributions from Ia afferents. These simulations did however show that combining rubrospinal excitation with inhibitory currents could result in recruitment reversals. The selective recruitment of type F motor units, therefore, cannot be totally ruled out. Indeed, some supporting data emerged in studies of cat motoneurons (Kanda et al. 1977) and human motor units (Garnett and Stephens 1981) in response to cutaneous stimulation. Since then, however, systematic work on the effects of cutaneous inputs by Cope and Clark (1991) have strongly supported the assumptions underlying the size principle. It appears that motor commands rarely, if ever, utilize reversed recruitment order (Heckman and Enoka 2012).

Overall, Henneman's size principle remains the intellectual bedrock for understanding motor unit recruitment.

4 Effects of Differences in the Distribution of Synaptic Input to Low Versus High Threshold Motoneurons on Their Recruitment Patterns

Although the tendency for many synaptic input systems to generate nonuniform distributions of synaptic currents in motoneurons of type S vs type F motor units does not result in systematic reversals of recruitment, it does impact the overall range of recruitment thresholds. A motoneuron's recruitment threshold is specified by its intrinsic threshold, which is the amount of synaptic current required to reach threshold for spiking, and its relative share of the total synaptic input, which is determined by the input distribution from active input sources (Burke 1981; Powers and Binder 2001). Multiple studies have supported Kernell's classic finding that there is an approximately 10-fold range in the threshold currents from the smallest to the largest motoneurons (Gustafsson and Pinter 1984; Zengel et al. 1985). This difference in recruitment threshold reflects differences partly in cell size, as emphasized by Henneman, and partly in specific membrane resistance, which together generate a 10-fold range in total input resistance (reviewed in Powers and Binder 2001). This 10-fold range confers stability on the recruitment process, hugely favoring the size principle. If an input tends to generate greater synaptic current in motoneurons of type F rather than type S motor units, the range of recruitment will be compressed and the "spacing" between each successive motor unit will be reduced. The recruitment order for each motor unit type and the differential strength of various synaptic inputs to the motoneurons are illustrated in Fig. 2. Briefly, rubrospinal and corticospinal inputs have a strong bias and vestibulospinal inputs have a slight bias towards motoneurons of type F motor units; by contrast, Ia inputs have a slight bias towards motoneurons of type S motor units. This nonuniformity in the distribution of inputs effectively compresses or expands the recruitment range. Therefore, the range of recruitment thresholds also reflects the differences in synaptic current between low and high threshold motor units generated by all excitatory inputs yet studied. In this context, the upper threshold of motor unit recruitment (i.e. the threshold at which the last motor unit is recruited during a maximal contraction) may provide information about the proportion of different synaptic inputs to – and the control strategies favored by – each muscle (Johnson et al. 2017).

The recruitment range varies across different muscles, likely reflecting differences in the synaptic inputs to these muscles. For example, in the human arm, the upper limit of recruitment is around 90% of the maximum voluntary contraction (MVC) for the biceps brachii (Gydikov and Kosarov 1974; Kukulka and Clamann 1981). Indeed, the biceps brachii and deltoid muscles in the upper limb as well as the soleus and tibialis anterior muscles in the lower limb have a relatively high

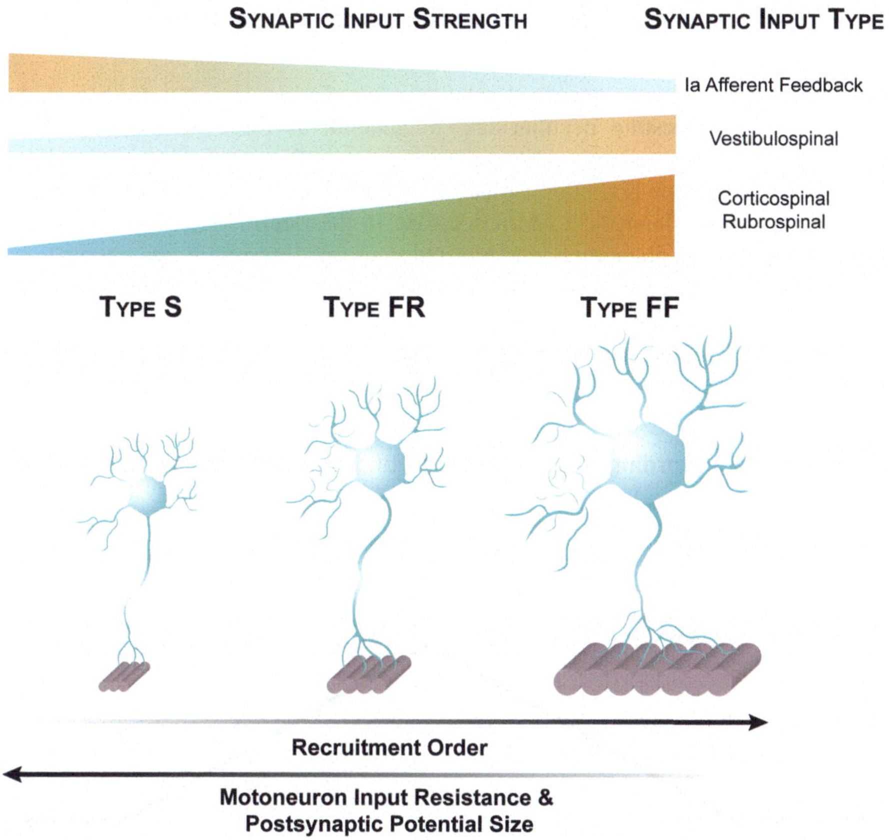

Fig. 2 Motor unit types are shown with Burke et al.' (1973) original classification of slow (Type S), fast-twitch fatigue resistant (Type FR), and fast-twitch fatigable (Type FF). The differential strength of synaptic inputs to motoneurons is represented by both the size and warmth (i.e. blue is weak, whereas orange is strong) of the bar associated with each type of synaptic input. The inverse relationship between the motoneuron size and the recruitment order as well as the amplitude of the postsynaptic potentials (PSPs) is shown at the bottom

upper limit of motor unit recruitment threshold, with some units being recruited within 5% of the MVC (De Luca et al. 1982; Gydikov and Kosarov 1974; Klass et al. 2008; Kukulka and Clamann 1981; Oya et al. 2009). By contrast, intrinsic hand muscles, which are often involved in manipulations necessitating a greater degree of precision and tend to receive much greater corticospinal input, have much lower upper limits of recruitment. Both the adductor pollicis and first dorsal interosseus muscles have upper limits of recruitment around 50% MVC (Kukulka and Clamann 1981) and 75% MVC (Moritz et al. 2005; Thomas et al. 1986), respectively. These compressed recruitment ranges may reflect increased corticospinal input to the motoneuron pools innervating hand muscles.

5 Classic Studies on Rate Modulation in Motor Units

Rate modulation initially appeared relatively simple in form, but further study exposed some interesting peculiarities. In general, as force increases, already recruited motor units undergo subsequent increases in firing rate (i.e. rate modulation). Yet during slowly rising isometric contractions, a striking phenomenon becomes manifest – namely a sharp decrease in the rate of rise of firing of low threshold units. This phenomenon is referred to as rate limiting or rate attenuation. Rate attenuation has been consistently demonstrated in many experiments (De Luca and Contessa 2012; De Luca et al. 1982; Fuglevand et al. 2015; Gydikov and Kosarov 1973; Monster and Chan 1977). Figure 3 shows a clear example of the firing rates of a cat motoneuron (a) and human motor unit (b), during a slow increase

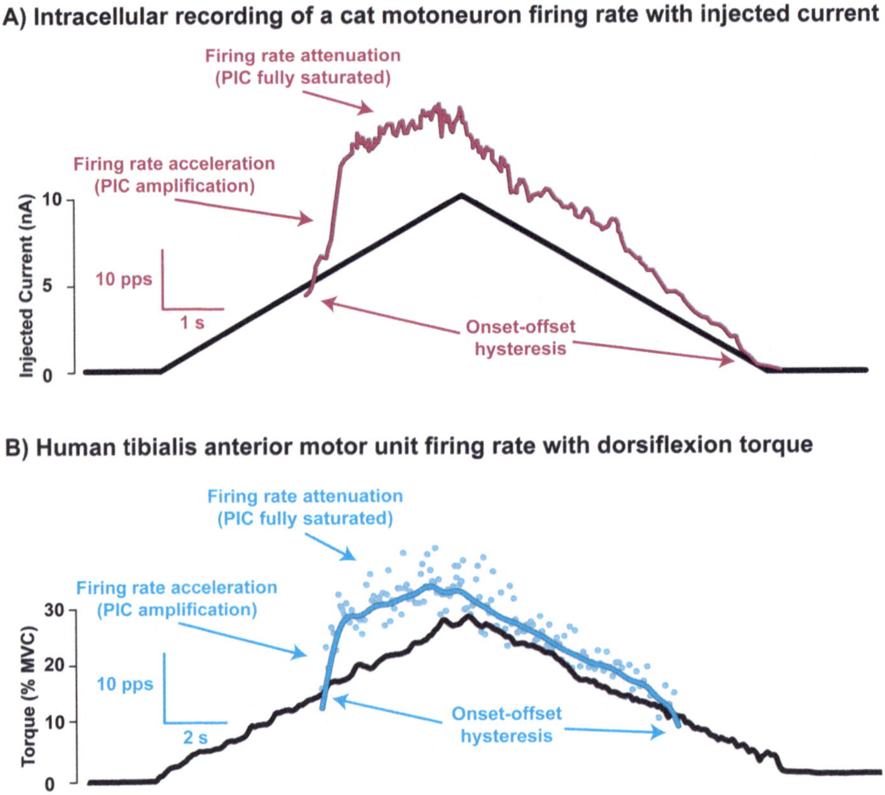

Fig. 3 Firing pattern of a spinal motoneuron (magenta) from a decerebrate cat (**a**) during current injection (black trace) and a human motor unit (cyan) (**b**) during slow isometric volitional torque generation (black trace). The basic effects of the PIC are the initial amplification which results in a steep increase of the firing rate, the PIC attenuation which causes the slope of the firing rate to decrease substantially and firing rate hysteresis due to the prolonged activation of the Ca^{2+} PIC. From these examples, analogous effects of the PIC in both species can be appreciated

and subsequent decrease in injected current or synaptic input. The mechanism of the rate attenuation evident in these types of motor unit firing patterns was perplexing for many years and proved to be a harbinger of unexpectedly complex intrinsic electrical properties of motoneurons.

Initially, from about 1950 through 1980, intracellular studies of motoneurons revealed rather simple input-output behavior: a threshold (corresponding to recruitment) and a slope for conversion of synaptic current to firing rate (i.e. rate modulation) (Powers and Binder 2001). When the measurements of distributions of synaptic currents from various sources became available (i.e. those outlined in the previous section), efforts were made to recreate normal recruitment and rate modulation patterns seen in experiments in humans. These efforts were reasonably successful for recruitment (Heckman and Binder 1993b), but problematic for rate modulation (Heckman and Binder 1993a and unpublished results, Heckman lab). The issue turned out to be that the motoneuron input-output functions were based on data from anesthetized preparations, which suppressed a fundamentally important form of synaptic input to motoneurons – neuromodulatory drive from the brainstem.

This reliance on anesthetized motoneuron properties was a surprisingly myopic approach, given the wealth of studies in invertebrate preparations that had demonstrated the power of neuromodulatory input (Marder 1996; Selverston et al. 1998). Unfortunately, at the time of these initial attempts to model human motor unit firing patterns, the profound effects of neuromodulatory input were still not fully appreciated. Studies of both spinal and brainstem motoneurons had shown that ionophoretic administration of the neurotransmitters serotonin (or 5-hydroxytryptamine; 5-HT) and norepinephrine (NE) induced a marked and long lasting increase in motoneuron excitability (VanderMaelen and Aghajanian 1980, 1982; White et al. 1991). These studies assumed that the primary effects of 5-HT and NE were mediated by actions on leak channels, producing subthreshold depolarization and increased resistance, and on spike afterhyperpolarization (AHP), accelerating firing rates. These actions are substantial but probably the most important role of neuromodulatory input in generating human motor unit firing patterns has proved to be the potent action of 5-HT and NE on persistent inward currents (PICs). PICs are mediated by G-protein coupled metabotropic membrane receptors that facilitate the transfer of ions via persistent Na^+ and L-type Ca^{2+} channels. They were discovered in motoneurons by Schwindt and Crill (1980), who employed blockers of K^+ currents in anesthetized preparations to uncover a PIC. Subsequently, Hounsgaard and colleagues, using the un-anesthetized decerebrate preparation along with drug administration in slice preparations, showed that strong PICs emerged in the presence of 5-HT and were thus a natural consequence of endogenous neurotransmitters (Hounsgaard et al. 1988; Hounsgaard and Kiehn 1985). Subsequently, Lee and Heckman (Lee and Heckman, 1999a) demonstrated that NE had a similar effect.

In these classic studies, the potent effects of PICs on motoneuron firing patterns were evident by a marked acceleration in firing rate in response to injected current and the tendency for firing to continue after the current ceased. The prolonged, self-sustained firing is especially striking (see Fig. 4), continuing for many seconds unless terminated by an inhibitory injected or synaptic current and was thus termed

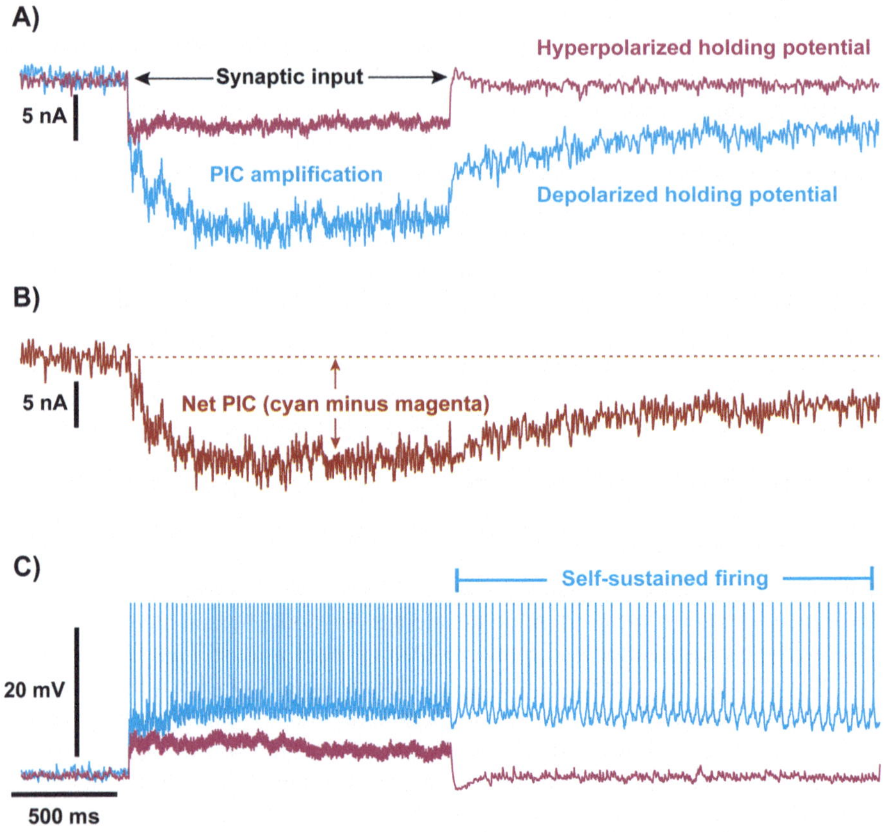

Fig. 4 Amplification and prolongation of synaptic input by the PIC in a cat motoneuron. In (**a**), a low threshold motoneuron is voltage clamped to either −90 mV (magenta trace) or −55 mV (cyan trace). At a hyperpolarized holding potential (i.e. −90 mV) activation of homonymous Ia afferents resulted in a steady current response with a sharp onset and offset that matched temporal parameters of the input. At a depolarized holding potential (i.e. −55 mV), the same input was amplified and prolonged, highlighting the voltage dependence of the PIC. In (**b**), the net PIC (red trace) is the difference between the motoneuron current response at a depolarized and at a hyperpolarized holding potential from (**a**). In (**c**), the motoneuron is unclamped. At a hyperpolarized level (~ −90 mV; magenta trace), synaptic input does not reach the threshold for repetitive firing. By contrast, at a more depolarized level (~ −70 mV; cyan trace), the same input evokes repetitive firing that is sustained well after the input is turned off, albeit at a lower firing rate

bistable behavior. It only slowly became evident, however, that PICs are essential for the generation of repetitive firing in all spinal motoneurons during slowly varying motor behaviors. The PIC mediated by Na^+ channels (NaPIC) is essential for the initiation of repetitive firing, acting to accelerate the relatively slow rate of change of membrane potential during the decay of the AHP into a rate of rise sufficiently fast to allow activation of transient Na^+ channels that generate each action potential (Harvey et al. 2006; Kuo et al. 2006; Lee and Heckman 2001). This action, while necessary, simply allows normal repetitive firing to occur and is thus implicit but not evident in motoneuron firing patterns in humans. The PIC mediated by L-type Ca^{2+} channels (CaPIC) is responsible for the acceleration and prolongation of motoneuron firing (Harvey et al. 2006; Kuo et al. 2006; Lee and Heckman 2001). These long-lasting effects cannot be mediated by the NaPIC, which activates very rapidly but slowly inactivates with a time constant of a few seconds (Lee and Heckman 1999b; Powers and Binder 2001).

6 Fundamental Role of Neuromodulatory Inputs in Shaping Nonlinear Human Motor Unit Firing Patterns

Realization that the CaPIC-induced acceleration and prolongation of motoneuron firing is a fundamental feature of human motor unit firing patterns took a surprisingly long time. Some 10 years after the initial demonstration of these phenomena in cat motoneurons, further studies with injected currents with slow triangular time courses revealed firing patterns (see Fig. 4C in Lee and Heckman (1998) in and Fig. 10A in Bennett et al. (1998)) that mimicked human motor unit firing patterns to a surprising degree (refer back to Fig. 3a). The reason for this delay in understanding was likely that current injection proved to be a very non-physiological means of activating the CaPIC. Intracellular electrodes inject current at or near the soma, whereas the CaPIC is generated primarily in the extensive dendritic trees of motoneurons (Bennett et al. 1998; Hounsgaard and Kiehn 1993; Lee and Heckman 1996). The human-like firing patterns in these intracellular studies resulted from either using an additional agonist for NE to generate a PIC so large that it could be easily activated from the soma (Lee and Heckman 1998) or by combining synaptic current with injected current to depolarize the dendritic PIC to make somatic activation more effective (Bennett et al. 1998).

The basic PIC effects on motoneuron firing during a slow triangle of injected current and the analogous effects in human motor units during slow isometric torque triangles generated volitionally can be appreciated in Fig. 3a, b, respectively. The PIC activates with a surge in firing rate that lasts about 1 s, representing the PIC amplification phase. The slope of firing rate then sharply decreases, to produce firing rate attenuation. Voltage clamp data and computer simulations show that this phase is due to the dendritic location of $Ca_V1.3$ channels, where PIC activation results in very depolarized levels approaching the reversal potential of the EPSP (Elbasiouny et al. 2006; Hyngstrom et al. 2008; Lee and Heckman 2000; Powers et al. 2012). This mechanism predicts that, once the PIC is activated and firing rate saturates, the motoneuron should become much less sensitive to additional synaptic input. This reduced sensitivity has been clearly demonstrated in intracellular studies (Hyngstrom et al. 2008; Lee and Heckman 2000). It is also the case in human motor units as was demonstrated by Fuglevand et al. (2015). The additional input does of course increase net torque, but primarily by recruitment of more motor units that then go through the same PIC acceleration and attenuation effects. By the same token, the voltage clamp and modeling studies show the motoneuron becomes much more sensitive to inhibition during the attenuation phase, as the dendritic depolarization greatly increases inhibitory driving force (Bui et al. 2008; Hyngstrom et al. 2008; Kuo et al. 2003). This high sensitivity has yet to be systematically evaluated in human subjects. Last, but certainly not least, the PIC tends to prolong firing because the $Ca_V1.3$ channels show little or no inactivation (Binder et al. 2020; Moritz et al. 2007). These results show motor unit de-recruitment at a lower input level than recruitment, illustrating a marked hysteresis in the firing rate. As discussed in the following section, this hysteresis provides the best estimation of the contribution of PICs in human motor unit firing pattern.

The remarkable similarity between intracellularly recorded motoneuron firing patterns when PICs are present and human motor unit firing patterns during volitional movements (Fig. 3) is perhaps the strongest evidence that neuromodulatory input from the brainstem is a fundamental component of normal motor behavior (Johnson et al. 2017). This conclusion is also strongly supported by several other lines of evidence. The intracellular studies clearly show that PIC amplitudes are proportional to the level of neuromodulatory drive from the brainstem – that is, to the levels of 5-HT and NE (Lee and Heckman 1999a, 2000). In reduced preparations and computer simulations, the maximal amplitudes of excitatory synaptic currents from both descending and sensory systems have been shown to be too small to have much of an impact on motoneuron firing, unless amplified by PICs (Powers and Binder 2001). Brainstem neurons in the caudal raphe nucleus that are the source of the 5-HT input to the cord, are tonically active in the awake cat, and increase their firing rates with increasing motor output (Jacobs et al. 2002). In humans, drugs that alter levels of 5-HT alter the gain of reflexes and of volitional commands (Wei et al. 2014) and increasing NE levels increases the PIC-like behaviors in motor unit firing patterns (Udina et al. 2010; Walton et al. 2002).

7 Characteristics of Motor Unit Firing Patterns That Provide Estimates of Cellular Mechanisms: *Estimation of Motoneuron PICs from Motor Unit Firing Patterns*

The PIC-induced bistable firing behavior became the initial focus for detection of PICs in human motor unit firing patterns. Bistable behavior could sometimes be identified (Kiehn and Eken 1997), but the hysteresis in motoneuron firing rates has proven to be the most consistent hallmark for non-invasive PIC estimation. This important point was first clearly realized by Gorassini et al. (2002), who devised a now standard technique for measuring firing rate hysteresis by quantifying the difference in the firing frequency of a lower-threshold motor unit at the onset and offset of firing of a higher threshold motor unit ($\Delta F = F_{Recruitment} - F_{Derecruitment}$). This paired motor unit analysis technique uses the lower-threshold motor unit as a marker of the

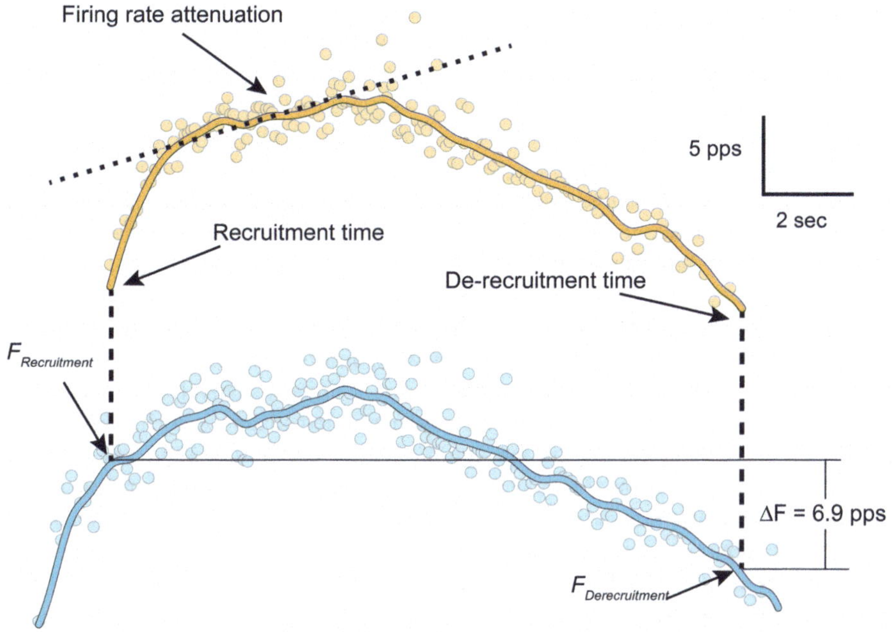

Fig. 5 Motor unit firing patterns from two tibialis anterior motor units during a slow isometric torque ramp recorded from a single human subject. The paired motor unit analysis technique quantifies the difference between the firing rate of a lower-threshold (blue trace) motor unit at the recruitment and de-recruitment of a higher-threshold (orange trace) motor unit. The lower-threshold motor unit's firing rate is used to estimate the synaptic drive to the motor unit pool. This technique allows for quantification of firing rate hysteresis, an estimate of the magnitude of the PIC

synaptic drive to the motoneuron pool, as can be appreciated in Fig. 5. The ΔF technique has been subjected to rigorous evaluations of its effectiveness as a means of estimating PICs and of the factors affecting the accuracy of these estimations (Afsharipour et al. 2020; Bennett et al. 2001a, b; Powers and Heckman 2015; Powers et al. 2008; Revill and Fuglevand 2011).

Bennett et al. (2001a, b) clearly demonstrated, with parallel motor unit and intracellular recordings in rat motoneurons, that ΔF reflects features of PICs. Results from human experiments and from realistic simulations of motoneurons also suggest that changes in ΔF primarily correspond to changes in the amplitude of PICs, especially in the range of values commonly seen in human motor units (3–6 imp/s) (Powers and Heckman 2015; Vandenberk and Kalmar 2014). Additionally, amphetamines (primarily affecting levels of NE) amplify ΔF (Udina et al. 2010) whereas inhibition tends to linearize motor unit firing patterns (Revill and Fuglevand 2017). Thus, ΔF is proportional to the PIC amplitude, which is itself an indicator of the level of brainstem monoaminergic drive. The self-sustained firing that is a hallmark of the PIC is particularly important from functional significance standpoint. Indeed, even in the initial studies of PICs in cat motoneurons, it was appreciated that motoneuron self-sustained firing behavior may be important for accomplishing sustained motor output for control of posture (Hounsgaard et al. 1988).

In the past 20 years, a plethora of studies have employed the paired motor unit analysis technique to estimate the level of neuromodulation across a variety of different muscles in humans. Some of the earlier studies using this method were centered on quantifying ΔF during long-lasting spasms in spinal cord injury subjects (Gorassini et al. 2004). Subsequently, numerous investigators have used the paired motor unit analysis technique to obtain estimates of PICs in human arm (Hassan et al. 2019, 2020; Mottram et al. 2009; Wilson et al. 2015), trunk (Stephenson and Maluf 2010), and leg muscles (Foley and Kalmar 2019; Hassan et al. 2019; Kim et al. 2020; Oya et al. 2009). Until recently, most studies investigating the influence of PICs on human motor unit firing patterns were done using wires inserted into the muscle to record the activity of a handful of motor units (usually 1–5) during relatively low-force contractions. The previously mentioned advances in HD sEMG and associated decomposition algorithms signify a major step in the field – one that has allowed the discrimination of tens of motor units from muscles throughout the body. As such, our lab and others have used these state-of-the-art techniques to measure ΔF and other parameters of motor unit firing in upper and lower limb muscles (Afsharipour et al. 2020; Hassan et al. 2020; Kim et al. 2020; Taylor et al. 2020). These and other studies in humans have revealed striking differences in ΔF in muscles throughout the body that likely reflect both the functional role and evolutionary precursors of the muscles in question. Nevertheless, our understanding of the relations between the structure of motor commands and the demands of motor tasks remains far from sufficient. Systematic analyses of muscles throughout the body will likely reveal the underlying structure of these motor commands and remains an area of active research (Johnson et al. 2017).

8 Characteristics of Motor Unit Firing Patterns That Provide Estimates of Cellular Mechanisms: *Estimation of the Spike Afterhyperpolarization*

Temporal parameters of motor unit spike times during sustained activation can provide considerable information about intrinsic properties of motoneurons because repetitive firing of motoneurons is closely related to the post-spike AHP (Kernell 1983). The post-spike AHP reflects Ca^{2+}-mediated K^+ currents, but since measuring directly from human motoneurons is not possible, we are reliant on methods that can estimate its duration from the variability of interspike intervals (ISI) during low frequency repetitive firing.

The relative variability in motor unit firing rates depends on the amplitude and duration of AHP, in addition to the amplitude and frequency content of concurrent synaptic noise (Powers and Binder 2000). Because of this, two methods have been developed to estimate the AHP duration in humans, both of which have been validated in motoneurons of the cat: (1) variability analysis, and (2) interval death rate analysis. The variability of the ISI in relation to the mean ISI has an upward bend; the onset of the bend is related to the AHP duration, which can be appreciated with break point analysis of this relationship (Person and Kudina 1972; Piotrkiewicz 1999). These break points were confirmed to be similar to the measured AHP duration, but are often difficult to measure and are therefore considered a relatively crude estimate of AHP duration (Powers and Binder 2000). The interval death rate analysis, on the other hand, uses the instantaneous probability of spike occurrence as a function of time since the last spike to estimate the AHP duration. In brief, ISI histories are plotted as histograms with small (i.e. 5 ms) bins, and the probability of that ISI being terminated (or "dying") as a result of a subsequent spike is determined as a function of time since the last spike. The likelihood of a subsequent spike increases as the ISI becomes longer, until the AHP duration has been exceeded, in which case there is a plateau (Matthews 1996, 1999, 2002). The initiation of the plateau has been confirmed to reflect the AHP duration in cat motoneurons (Powers and Binder 2000).

These analysis techniques, and in particular the interval death rate analysis, have proven reliable for estimating the time course of AHP (MacDonell et al. 2007), provided fruitful information on the intrinsic properties of motoneurons, and how these mechanisms contribute to normal motor control. The AHP duration estimated from the interval death rate analysis has an inverse correlation with the minimal firing rate of low threshold motor units. However, synaptic noise contributes to ISIs because the AHP is typically shorter than the duration of the ISI (MacDonell et al. 2008). Post-spike AHP is also modulated by sensory input, such as antagonist muscle vibration to induce Ia reciprocal inhibition or agonist muscle vibration to provide additional Ia excitatory synaptic input (MacDonell et al. 2010). Reciprocal inhibition prolongs estimates of AHP duration, whereas excitatory input reduces estimates of AHP duration. This is probably due to the bombardment of excitatory synaptic noise, even when the firing rate of motor units is tightly controlled. Although the role that AHP duration plays in the minimal firing rate of low-threshold motor units seems quite clear, much work remains to be done if we are to understand how AHP duration influences motor unit output and systematically differs across the human body.

9 Characteristics of Motor Unit Firing Patterns That Provide Estimates of Cellular Mechanisms: *Estimation of Motoneuron Post-synaptic Potentials from Motor Unit Firing Patterns*

Two related methods have been developed to estimate the time course of EPSPs and IPSPs evoked by electrical stimulation of sensory inputs in humans, the post-stimulus time histogram (PSTH) and the post-stimulus frequencygram (PSF). The

PSTH approach was first introduced by Gerstein and Kiang (1960) and measures the stimulus-induced change in the probability of a spike as a function of the time since the stimulus and has been used widely to infer PSPs caused by afferent and descending inputs. The limitations of this technique were acknowledged from the inception (Moore et al. 1970), but, with due consideration, this method remains useful. The elegant work of Turker and Powers (1999, 2005, 2010) suggested that the post-stimulus frequencygram (PSF) would complement the PSTH approach. By comparing the estimates of PSPs with the PSTH and PSF approaches to measured PSPs from hypoglossal motoneurons recorded in slices of rat brain stem, Turker and Powers (1999) were able to highlight some key issues with the PSTH approach. Although the probability-based PSTH approach is useful for identifying the early portions of the PSP, later fluctuations in the probability of firing, which are increasingly influenced by the initial advancement or delay of spikes, are often misinterpreted as either excitatory or inhibitory events (Turker and Powers 2005). For example, the slow rising portion of an IPSP is often interpreted as an EPSP with the PTSH approach.

Multiple studies have since implemented this combined PSTH and PSF approaches with great success (Kahya et al. 2010; Norton et al. 2008; Yavuz et al. 2014). For instance, the duration of the IPSP caused by painful stimulation to cutaneous afferents of the hand (i.e. cutaneous silent period) was shown to be significantly longer with the PSF approach than was once believed with the PSTH approach (Kahya et al. 2010). A re-examination of the stretch reflex with a combined PSTH and PSF approach revealed that there is a prolonged excitatory (or multiple EPSPs) rather than a mixture of excitatory and inhibitory PSPs associated with a tendon tap, as was once believed with the exclusive probabilistic approach (Yavuz et al. 2014). Firing rate approaches have also improved our understanding of some underlying mechanisms of neurological impairment. For instance, Norton et al. (2008) showed that brief (~20 ms) cutaneomuscular stimulation causes a short (i.e. 300 ms) EPSP that is immediately followed by an intervening IPSP, with the latter component suppressed following spinal injury. The recent developments described for simultaneous recording of many motor units have further advanced efforts to identify PSP time courses. For example, Yavuz et al. (2015, 2018) recently proposed a technique for estimating reflexes in large populations of motor units using HD sEMG. Combined with the PSF approach, they were able to non-invasively quantify both inhibitory and excitatory reflexes in large populations of individual motor units and to identify asymmetries in the distribution of reciprocal inhibition between antagonists at the ankle. These advances in technology in combination with improved methods of estimating PSPs will continue to provide information about the organization of synaptic inputs involved in human motor commands.

10 Computer-Based Simulations to Study the Synaptic Organization of Motor Commands

The use of computer simulations to understand genesis of spinal motor output in humans has been underway for over 30 years. For example, the estimates of AHP and synaptic potentials in the work summarized above often use simulations to further understand the results of their experiments (Matthews 1996; Norton et al. 2008; Powers et al. 2005). Our goal has been to understand the structure of motor commands in terms of identifying the overall patterns of all three components of motor commands – excitation, inhibition and neuromodulation. As noted above, the initial efforts in this direction (Heckman 1994; Heckman and Binder 1991b, 1993a, b) failed to take into account the neuromodulatory component, which was found later to be crucial in understanding the genesis of human motor unit firing patterns. The Fuglevand model (Fuglevand et al. 1993) of the motoneuron pool and its muscle has been transformative for understanding the relations between basic recruitment and rate modulation patterns with EMG and force, but is based on incomplete motoneuron models due to the omission of the profound influence of neuromodulation on motor output. The development of motoneuron models that include neuromodulatory inputs in a realistic fashion now provides the potential to advance the goal of understanding the interactions of all three components of motor commands (i.e. excitation, inhibition and neuromodulation).

Although the highly nonlinear effects of PICs on motoneuron input-output processing remain difficult to understand in terms of their full functional implications (Binder et al. 2020), these nonlinearities have proven highly advantageous for estimating PIC amplitudes from motor unit firing patterns. In theory, information about these inhibitory and neuromodulatory components could be sufficient to identify the temporal pattern of excitation. Therefore, estimating all 3 components of motor commands is the goal of our supercomputer-based simulations of motoneurons.

11 Implementing Supercomputer-Based Simulations to Estimate Excitation, Inhibition and Neuromodulation

The supercomputing landscape has changed drastically over the last decade. In preparation for the end of Moore's law (Moore 1965), it has increased in complexity with new architectures and programming models, hybrid machines, heterogeneous memory hierarchies and codes in constant refactoring to survive longer than the normal lifetime of a supercomputer. Supercomputers are currently so powerful that the horizons for some scientific fields had expanded considerably and it is just now that they can afford the resolution of problems at the level required by their challenging science. Upcoming U.S. exascale supercomputers to be delivered in 2021 will be capable of performing one quadrillion calculations per second. Carefully designed to address the needs of projects spanning the three pillars: simulation, data and learning, they are ready to accelerate research at unprecedented pace and shape the future of how neuroscientists are learning to decode the dialogue among neurons.

The exploding demand of artificial intelligence (AI) and big data projects was one of the motivations for a series of AI for Science town halls hosted by Argonne, Oak Ridge and Berkeley national laboratories to capture the big ideas, major challenges, advances in the next decade and next steps to realizing these opportunities. The result is a 224-page report (Stevens et al. 2020) which highlights the growing opportunities for sixteen main areas that emerged for AI applications in science and outlines the research and infrastructure needed to accelerate existing AI methods and techniques for science applications. The dawn of AI-enabled discovery in biology has occurred and computational neuroscience has a merited place. Neuroscience challenges are mentioned in Sect. 3 (*Biology and Life Sciences*) of the report (Stevens et al. 2020) and here below we present our attempt to use supercomputers and AI to shed light into the basic elements for motor control.

Figure 6 illustrates our approach for taking advantage of supercomputers to advance our understanding of the cellular basis of motor commands. We focused on a simple motor behavior, a linearly rising and falling motor output (i.e., a triangle). For these simulations, a pool of 20 multi-compartment motoneuron models were derived from previously-developed models of the cat medial gastrocnemius motoneuron pool (Kim and Jones 2011, 2012; Kim et al. 2009, 2014; Powers and Heckman 2017). These models accurately recreate our extensive current and voltage clamp data on how PICs affect motoneuron input-output functions (reviewed in Binder et al. 2020), but were modified to recreate the lower firing rates and more restricted recruitment range characteristic of human motor unit firing patterns during moderate isometric contractions (Johnson et al. 2017). In each individual, the target was a linear increase and decrease in the summed firing patterns of all 20 motoneuron models (this is essentially analogous to EMG) (bottom traces). In both simulations, inhibition is configured in a reciprocal pattern, i.e. starting at a modest background level and then decreasing as excitation increases. But the background level of neuromodulation is different, with PICs being 20% stronger in the simulations on the right-hand side in Fig. 6. Even with moderate PIC amplitudes (Fig. 6, left) nonlinear patterns of both excitation and inhibition are required to achieve the target linear EMG output. The 20% increase in PIC amplitude sharply increases the nonlinearity of the input patterns that produce the linear EMG output, which is consistent with our previous simulations showing how PICs distort the input-output functions of motor pools (Powers and Heckman 2017). Although the average of the firing rate outputs for all 20 model motoneurons (bottom traces) are indeed similar for these two neuromodulatory levels, as required by our simulation process, the motoneuron firing patterns are highly distinctive. These distinctive patterns clearly demonstrate the potential of extracting information about input structure from population recordings of motor units via array electrodes. We have run hundreds of thousands of these simulations so far, with the goal of determining which characteristics of the firing patterns in the population output (e.g. average values of ΔF, post-acceleration rate modulation slope, and spacing of recruitment between units – highlighted in Fig. 5) that best predict the input pattern of excitation, inhibition and neuromodulation. Our analyses thus far suggest that nonlinear regression

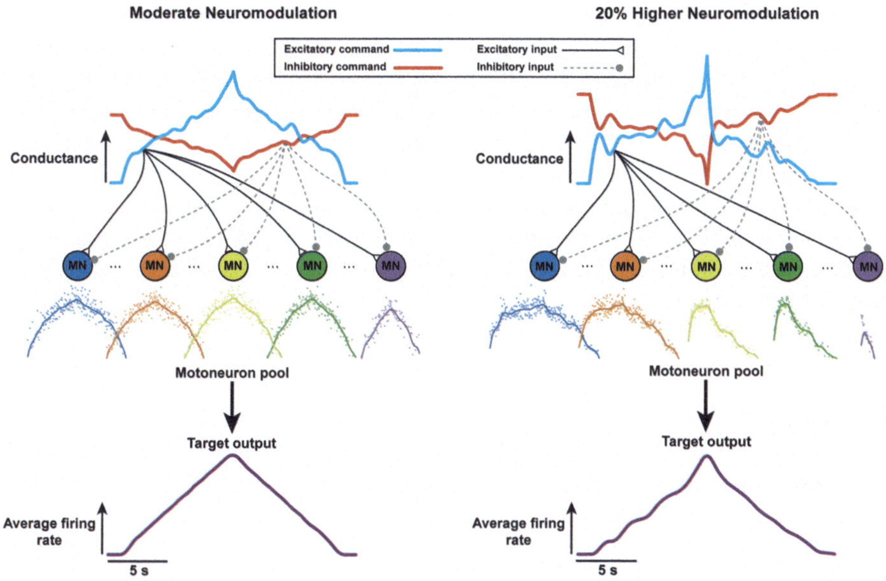

Fig. 6 Two examples of the input-output behavior of a model motoneuron pool (n = 20) run on a supercomputer at Argonne National Laboratory. Each simulation run was generated from a distinct pattern of the three components of motor commands: excitation (cyan trace, top), inhibition (purple trace, top), and neuromodulation (not shown but assumed to be constant at a specified level throughout each run). In all simulations, the output of the pool was defined as the average firing rate produced as the 20 model motoneurons were progressively activated and deactivated (purple trace, bottom). In each run, the temporal patterns of the excitatory and inhibitory synaptic conductances (cyan, purple traces, top) were varied until that particular combination of excitation, inhibition and neuromodulation successfully generated a near linear average firing pattern. In both examples, inhibitory command was set to follow a precisely reciprocal pattern compared to excitation. On the left, neuromodulation was set to produce moderate amplitude PICs in motoneurons. On the right, PICs were increased by 20%. Thus, modest differences require large nonlinearities in the excitatory and inhibitory motor commands to achieve matching outputs. Despite the good output match, the firing patterns of the constituent model motoneurons show clear differences. These examples illustrate the potential power of using motoneuron firing patterns to identify the structure of motor commands

methods applied to combinations of these characteristics may consistently predict input patterns with r^2 values of 0.75–0.80 and above. Thus, the approach is promising, and we plan to soon attempt analyses of human motor unit firing patterns, in both healthy and disease states.

12 Conclusion

The past 10 years have been an exciting era for study of cellular mechanisms of motor output in the human. The availability of techniques to simultaneously record firing patterns of many motor units coupled with analyses methods based on realistic computer simulations have great potential for understanding the cellular basis of motor output. One of the biggest surprises is that, at least during slow movements, persistent inward currents (PICs) are a dominant feature of normal motor output. Implementation of motoneuron simulations with realistic implementations of PICs on supercomputers now allows systematic exploration of the set of synaptic inputs that produce a specified trajectory of motor output. Concerns that a huge range of excitation and inhibitory patterns could all produce nearly the same output are ameliorated by the highly nonlinear effects of PICs. Much further work needs to be done, but the hundreds of thousands of simulations we have performed so far have highly promising results.

References

Adrian ED, Bronk DW (1929) The discharge of impulses in motor nerve fibres: part II. The frequency of discharge in reflex and voluntary contractions. J Physiol 67(2):i3–i151

Afsharipour B, Manzur N, Duchcherer J, Fenrich KF, Thompson CK, Negro F, Quinlan KA, Bennett DJ, Gorassini MA (2020) Estimation of self-sustained activity produced by persistent inward currents using firing rate profiles of multiple motor units in humans. J Neurophysiol 124(1):63–85. https://doi.org/10.1152/jn.00194.2020

Bennett DJ, Hultborn H, Fedirchuk B, Gorassini MA (1998) Synaptic activation of plateaus in hindlimb motoneurons of decerebrate cats. J Neurophysiol 80(4):2023–2037. https://doi.org/10.1152/jn.1998.80.4.2023

Bennett DJ, Li Y, Harvey PJ, Gorassini MA (2001a) Evidence for plateau potentials in tail motoneurons of awake chronic spinal rats with spasticity. J Neurophysiol 86(4):1972–1982. https://doi.org/10.1152/jn.2001.86.4.1972

Bennett DJ, Li Y, Siu M (2001b) Plateau potentials in sacrocaudal motoneurons of chronic spinal rats, recorded in vitro. J Neurophysiol 86(4):1955–1971. https://doi.org/10.1152/jn.2001.86.4.1955

Binder MD, Powers RK, Heckman CJ (2020) Nonlinear input-output functions of motoneurons. Physiology (Bethesda) 35(1):31–39. https://doi.org/10.1152/physiol.00026.2019

Bui TV, Grande G, Rose PK (2008) Multiple modes of amplification of synaptic inhibition to motoneurons by persistent inward currents. J Neurophysiol 99(2):571–582. https://doi.org/10.1152/jn.00717.2007

Burke RE (1981) Motor units: anatomy, physiology, and functional organization. In: Brooks VB (ed) Handbook of physiology, the nervous system, motor control. American Physiological Society, pp 345–422

Burke RE, Levine DN, Tsairis P, Zajac FE 3rd. (1973) Physiological types and histochemical profiles in motor units of the cat gastrocnemius. J Physiol 234(3):723–748. https://doi.org/10.1113/jphysiol.1973.sp010369

Cope TC, Clark BD (1991) Motor-unit recruitment in the decerebrate cat: several unit properties are equally good predictors of order. J Neurophysiol 66(4):1127–1138. https://doi.org/10.1152/jn.1991.66.4.1127

De Luca CJ, Contessa P (2012) Hierarchical control of motor units in voluntary contractions. J Neurophysiol 107(1):178–195. https://doi.org/10.1152/jn.00961.2010

De Luca CJ, LeFever RS, McCue MP, Xenakis AP (1982) Behaviour of human motor units in different muscles during linearly varying contractions. J Physiol 329:113–128. https://doi.org/10.1113/jphysiol.1982.sp014293

Denny-Brown D, Pennybacker J (1938) Fibrillation and fasciculation in voluntary muscle. Brain 61(3):311–312

Duchateau J, Enoka RM (2011) Human motor unit recordings: origins and insight into the integrated motor system. Brain Res 1409:42–61. https://doi.org/10.1016/j.brainres.2011.06.011

Eccles JC (1952) The electrophysiological properties of the motoneurone. Cold Spring Harb Symp Quant Biol 17:175–183. https://doi.org/10.1101/sqb.1952.017.01.017

Elbasiouny SM, Bennett DJ, Mushahwar VK (2006) Simulation of Ca^{2+} persistent inward currents in spinal motoneurones: mode of activation and integration of synaptic inputs. J Physiol 570(Pt 2):355–374. https://doi.org/10.1113/jphysiol.2005.099119

Farina D, Holobar A (2016) Characterization of human motor units from surface EMG decomposition. Proc IEEE 104(2):353–373

Farina D, Negro F, Muceli S, Enoka RM (2016) Principles of motor unit physiology evolve with advances in technology. Physiology (Bethesda) 31(2):83–94. https://doi.org/10.1152/physiol.00040.2015

Foley RCA, Kalmar JM (2019) Estimates of persistent inward current in human motor neurons during postural sway. J Neurophysiol 122(5):2095–2110. https://doi.org/10.1152/jn.00254.2019

Fuglevand AJ, Winter DA, Patla AE (1993) Models of recruitment and rate coding organization in motor-unit pools. J Neurophysiol 70(6):2470–2488. https://doi.org/10.1152/jn.1993.70.6.2470

Fuglevand AJ, Lester RA, Johns RK (2015) Distinguishing intrinsic from extrinsic factors underlying firing rate saturation in human motor units. J Neurophysiol 113(5):1310–1322. https://doi.org/10.1152/jn.00777.2014

Garnett R, Stephens JA (1981) Changes in the recruitment threshold of motor units produced by cutaneous stimulation in man. J Physiol 311:463–473. https://doi.org/10.1113/jphysiol.1981.sp013598

Gerstein GL, Kiang NY (1960) An approach to the quantitative analysis of electrophysiological data from single neurons. Biophys J 1:15–28. https://doi.org/10.1016/s0006-3495(60)86872-5

Gorassini MA, Yang JF, Siu M, Bennett DJ (2002) Intrinsic activation of human motoneurons: possible contribution to motor unit excitation. J Neurophysiol 87(4):1850–1858. https://doi.org/10.1152/jn.00024.2001

Gorassini MA, Knash ME, Harvey PJ, Bennett DJ, Yang JF (2004) Role of motoneurons in the generation of muscle spasms after spinal cord injury. Brain 127(Pt 10):2247–2258. https://doi.org/10.1093/brain/awh243

Gustafsson B, Pinter MJ (1984) An investigation of threshold properties among cat spinal alpha-motoneurones. J Physiol 357:453–483. https://doi.org/10.1113/jphysiol.1984.sp015511

Gydikov A, Kosarov D (1973) Physiological characteristics of the tonic and phasic motor units in human muscles. In: Motor control. Springer, pp 75–94

Gydikov A, Kosarov D (1974) Some features of different motor units in human biceps brachii. Pflugers Arch 347(1):75–88. https://doi.org/10.1007/BF00587056

Harvey PJ, Li Y, Li X, Bennett DJ (2006) Persistent sodium currents and repetitive firing in motoneurons of the sacrocaudal spinal cord of adult rats. J Neurophysiol 96(3):1141–1157. https://doi.org/10.1152/jn.00335.2005

Hassan AS, Kim EH, Khurram OU, Cummings M, Thompson CK, Miller McPherson L, Heckman CJ, Dewald JPA, Negro F (2019) Properties of motor units of elbow and ankle muscles decomposed using high-density surface EMG. Conf Proc IEEE Eng Med Biol Soc 2019:3874–3878. https://doi.org/10.1109/EMBC.2019.8857475

Hassan AS, Thompson CK, Negro F, Cummings M, Powers RK, Heckman CJ, Dewald JPA, McPherson LM (2020) Impact of parameter selection on estimates of motoneuron excitability using paired motor unit analysis. J Neural Eng 17(1):016063. https://doi.org/10.1088/1741-2552/ab5eda

Heckman CJ (1994) Computer simulations of the effects of different synaptic input systems on the steady-state input-output structure of the motoneuron pool. J Neurophysiol 71(5):1727–1739. https://doi.org/10.1152/jn.1994.71.5.1727

Heckman CJ, Binder MD (1988) Analysis of effective synaptic currents generated by homonymous Ia afferent fibers in motoneurons of the cat. J Neurophysiol 60(6):1946–1966. https://doi.org/10.1152/jn.1988.60.6.1946

Heckman CJ, Binder MD (1991a) Analysis of Ia-inhibitory synaptic input to cat spinal motoneurons evoked by vibration of antagonist muscles. J Neurophysiol 66(6):1888–1893. https://doi.org/10.1152/jn.1991.66.6.1888

Heckman CJ, Binder MD (1991b) Computer simulation of the steady-state input-output function of the cat medial gastrocnemius motoneuron pool. J Neurophysiol 65(4):952–967. https://doi.org/10.1152/jn.1991.65.4.952

Heckman CJ, Binder MD (1993a) Computer simulations of motoneuron firing rate modulation. J Neurophysiol 69(4):1005–1008. https://doi.org/10.1152/jn.1993.69.4.1005

Heckman CJ, Binder MD (1993b) Computer simulations of the effects of different synaptic input systems on motor unit recruitment. J Neurophysiol 70(5):1827–1840. https://doi.org/10.1152/jn.1993.70.5.1827

Heckman CJ, Enoka RM (2012) Motor unit. Compr Physiol 2(4):2629–2682. https://doi.org/10.1002/cphy.c100087

Henneman E (1957) Relation between size of neurons and their susceptibility to discharge. Science 126(3287):1345–1347. https://doi.org/10.1126/science.126.3287.1345

Henneman E, Mendell LM (1981) Functional organization of motoneuron pool and its inputs. In: Brooks VB (ed) Handbook of physiology, the nervous system, motor control. American Physiological Society, pp 423–507

Henneman E, Olson CB (1965) Relations between structure and function in the design of skeletal muscles. J Neurophysiol 28:581–598. https://doi.org/10.1152/jn.1965.28.3.581

Henneman E, Somjen G, Carpenter DO (1965a) Excitability and inhibitability of motoneurons of different sizes. J Neurophysiol 28(3):599–620. https://doi.org/10.1152/jn.1965.28.3.599

Henneman E, Somjen G, Carpenter DO (1965b) Functional significance of cell size in spinal motoneurons. J Neurophysiol 28:560–580. https://doi.org/10.1152/jn.1965.28.3.560

Holobar A, Zazula D (2008) On the selection of the cost function for gradient-based decomposition of surface electromyograms. Conf Proc IEEE Eng Med Biol Soc 2008:4668–4671. https://doi.org/10.1109/IEMBS.2008.4650254

Holobar A, Farina D, Gazzoni M, Merletti R, Zazula D (2009) Estimating motor unit discharge patterns from high-density surface electromyogram. Clin Neurophysiol 120(3):551–562. https://doi.org/10.1016/j.clinph.2008.10.160

Hounsgaard J, Kiehn O (1985) Ca++ dependent bistability induced by serotonin in spinal motoneurons. Exp Brain Res 57(2):422–425. https://doi.org/10.1007/BF00236551

Hounsgaard J, Kiehn O (1993) Calcium spikes and calcium plateaux evoked by differential polarization in dendrites of turtle motoneurones in vitro. J Physiol 468:245–259. https://doi.org/10.1113/jphysiol.1993.sp019769

Hounsgaard J, Hultborn H, Jespersen B, Kiehn O (1988) Bistability of alpha-motoneurones in the decerebrate cat and in the acute spinal cat after intravenous 5-hydroxytryptophan. J Physiol 405:345–367. https://doi.org/10.1113/jphysiol.1988.sp017336

Hyngstrom AS, Johnson MD, Heckman CJ (2008) Summation of excitatory and inhibitory synaptic inputs by motoneurons with highly active dendrites. J Neurophysiol 99(4):1643–1652. https://doi.org/10.1152/jn.01253.2007

Jacobs BL, Martin-Cora FJ, Fornal CA (2002) Activity of medullary serotonergic neurons in freely moving animals. Brain Res Brain Res Rev 40(1–3):45–52. https://doi.org/10.1016/s0165-0173(02)00187-x

Johnson MD, Thompson CK, Tysseling VM, Powers RK, Heckman CJ (2017) The potential for understanding the synaptic organization of human motor commands via the firing patterns of motoneurons. J Neurophysiol 118(1):520–531. https://doi.org/10.1152/jn.00018.2017

Kahya MC, Yavuz SU, Turker KS (2010) Cutaneous silent period in human FDI motor units. Exp Brain Res 205(4):455–463. https://doi.org/10.1007/s00221-010-2380-6

Kanda K, Burke RE, Walmsley B (1977) Differential control of fast and slow twitch motor units in the decerebrate cat. Exp Brain Res 29(1):57–74. https://doi.org/10.1007/BF00236875

Kernell D (1966) Input resistance, electrical excitability, and size of ventral horn cells in cat spinal cord. Science 152(3729):1637–1640. https://doi.org/10.1126/science.152.3729.1637

Kernell D (1983) Functional properties of spinal motoneurons and gradation of muscle force. Adv Neurol 39:213–226

Kiehn O, Eken T (1997) Prolonged firing in motor units: evidence of plateau potentials in human motoneurons? J Neurophysiol 78(6):3061–3068. https://doi.org/10.1152/jn.1997.78.6.3061

Kim H, Jones KE (2011) Asymmetric electrotonic coupling between the soma and dendrites alters the bistable firing behaviour of reduced models. J Comput Neurosci 30(3):659–674. https://doi.org/10.1007/s10827-010-0284-x

Kim H, Jones KE (2012) The retrograde frequency response of passive dendritic trees constrains the nonlinear firing behaviour of a reduced neuron model. PLoS One 7(8):e43654. https://doi.org/10.1371/journal.pone.0043654

Kim H, Major LA, Jones KE (2009) Derivation of cable parameters for a reduced model that retains asymmetric voltage attenuation of reconstructed spinal motor neuron dendrites. J Comput Neurosci 27(3):321–336. https://doi.org/10.1007/s10827-009-0145-7

Kim H, Jones KE, Heckman CJ (2014) Asymmetry in signal propagation between the soma and dendrites plays a key role in determining dendritic excitability in motoneurons. PLoS One 9(8):e95454. https://doi.org/10.1371/journal.pone.0095454

Kim EH, Wilson JM, Thompson CK, Heckman CJ (2020) Differences in estimated persistent inward currents between ankle flexors and extensors in humans. J Neurophysiol 124(2):525–535. https://doi.org/10.1152/jn.00746.2019

Klass M, Baudry S, Duchateau J (2008) Age-related decline in rate of torque development is accompanied by lower maximal motor unit discharge frequency during fast contractions. J Appl Physiol (1985) 104(3):739–746. https://doi.org/10.1152/japplphysiol.00550.2007

Kukulka CG, Clamann HP (1981) Comparison of the recruitment and discharge properties of motor units in human brachial biceps and adductor pollicis during isometric contractions. Brain Res 219(1):45–55. https://doi.org/10.1016/0006-8993(81)90266-3

Kuo JJ, Lee RH, Johnson MD, Heckman HM, Heckman CJ (2003) Active dendritic integration of inhibitory synaptic inputs in vivo. J Neurophysiol 90(6):3617–3624. https://doi.org/10.1152/jn.00521.2003

Kuo JJ, Lee RH, Zhang L, Heckman CJ (2006) Essential role of the persistent sodium current in spike initiation during slowly rising inputs in mouse spinal neurones. J Physiol 574(Pt 3):819–834. https://doi.org/10.1113/jphysiol.2006.107094

Lee RH, Heckman CJ (1996) Influence of voltage-sensitive dendritic conductances on bistable firing and effective synaptic current in cat spinal motoneurons in vivo. J Neurophysiol 76(3):2107–2110. https://doi.org/10.1152/jn.1996.76.3.2107

Lee RH, Heckman CJ (1998) Bistability in spinal motoneurons in vivo: systematic variations in rhythmic firing patterns. J Neurophysiol 80(2):572–582. https://doi.org/10.1152/jn.1998.80.2.572

Lee RH, Heckman CJ (1999a) Enhancement of bistability in spinal motoneurons in vivo by the noradrenergic alpha1 agonist methoxamine. J Neurophysiol 81(5):2164–2174. https://doi.org/10.1152/jn.1999.81.5.2164

Lee RH, Heckman CJ (1999b) Paradoxical effect of QX-314 on persistent inward currents and bistable behavior in spinal motoneurons in vivo. J Neurophysiol 82(5):2518–2527. https://doi.org/10.1152/jn.1999.82.5.2518

Lee RH, Heckman CJ (2000) Adjustable amplification of synaptic input in the dendrites of spinal motoneurons in vivo. J Neurosci 20(17):6734–6740

Lee RH, Heckman CJ (2001) Essential role of a fast persistent inward current in action potential initiation and control of rhythmic firing. J Neurophysiol 85(1):472–475. https://doi.org/10.1152/jn.2001.85.1.472

Lindsay AD, Binder MD (1991) Distribution of effective synaptic currents underlying recurrent inhibition in cat triceps surae motoneurons. J Neurophysiol 65(2):168–177. https://doi.org/10.1152/jn.1991.65.2.168

MacDonell CW, Ivanova TD, Garland SJ (2007) Reliability of the interval death rate analysis for estimating the time course of the motoneurone afterhyperpolarization in humans. J Neurosci Methods 162(1–2):314–319. https://doi.org/10.1016/j.jneumeth.2007.01.020
MacDonell CW, Ivanova TD, Garland SJ (2008) Afterhyperpolarization time-course and minimal discharge rate in low threshold motor units in humans. Exp Brain Res 189(1):23–33. https://doi.org/10.1007/s00221-008-1400-2
MacDonell CW, Ivanova TD, Garland SJ (2010) Changes in the estimated time course of the motoneuron afterhyperpolarization induced by tendon vibration. J Neurophysiol 104(6):3240–3249. https://doi.org/10.1152/jn.00941.2009
Marder E (1996) Neural modulation: following your own rhythm. Curr Biol 6(2):119–121. https://doi.org/10.1016/s0960-9822(02)00438-4
Matthews PB (1996) Relationship of firing intervals of human motor units to the trajectory of post-spike after-hyperpolarization and synaptic noise. J Physiol 492(Pt 2):597–628. https://doi.org/10.1113/jphysiol.1996.sp021332
Matthews PB (1999) Properties of human motoneurones and their synaptic noise deduced from motor unit recordings with the aid of computer modelling. J Physiol Paris 93(1–2):135–145. https://doi.org/10.1016/s0928-4257(99)80144-2
Matthews PB (2002) Measurement of excitability of tonically firing neurones tested in a variable-threshold model motoneurone. J Physiol 544(Pt 1):315–332. https://doi.org/10.1113/jphysiol.2002.024984
Mendell LM, Henneman E (1971) Terminals of single Ia fibers – location, density, and distribution within a pool of 300 homonymous motoneurons. J Neurophysiol 34(1):171
Merletti R, Holobar A, Farina D (2008) Analysis of motor units with high-density surface electromyography. J Electromyogr Kinesiol 18(6):879–890. https://doi.org/10.1016/j.jelekin.2008.09.002
[Record #159 is using a reference type undefined in this output style]
Monster AW, Chan H (1977) Isometric force production by motor units of extensor digitorum communis muscle in man. J Neurophysiol 40(6):1432–1443. https://doi.org/10.1152/jn.1977.40.6.1432
Moore GE (1965) Cramming more components onto integrated circuits, Electronics, 38(8):114–117
Moore GP, Segundo JP, Perkel DH, Levitan H (1970) Statistical signs of synaptic interaction in neurons. Biophys J 10(9):876–900. https://doi.org/10.1016/S0006-3495(70)86341-X
Moritz CT, Barry BK, Pascoe MA, Enoka RM (2005) Discharge rate variability influences the variation in force fluctuations across the working range of a hand muscle. J Neurophysiol 93(5):2449–2459. https://doi.org/10.1152/jn.01122.2004
Moritz AT, Newkirk G, Powers RK, Binder MD (2007) Facilitation of somatic calcium channels can evoke prolonged tail currents in rat hypoglossal motoneurons. J Neurophysiol 98(2):1042–1047. https://doi.org/10.1152/jn.01294.2006
Mottram CJ, Suresh NL, Heckman CJ, Gorassini MA, Rymer WZ (2009) Origins of abnormal excitability in biceps brachii motoneurons of spastic-paretic stroke survivors. J Neurophysiol 102(4):2026–2038. https://doi.org/10.1152/jn.00151.2009
Negro F, Muceli S, Castronovo AM, Holobar A, Farina D (2016) Multi-channel intramuscular and surface EMG decomposition by convolutive blind source separation. J Neural Eng 13(2):026027. https://doi.org/10.1088/1741-2560/13/2/026027
Norton JA, Bennett DJ, Knash ME, Murray KC, Gorassini MA (2008) Changes in sensory-evoked synaptic activation of motoneurons after spinal cord injury in man. Brain 131(Pt 6):1478–1491. https://doi.org/10.1093/brain/awn050
Oya T, Riek S, Cresswell AG (2009) Recruitment and rate coding organisation for soleus motor units across entire range of voluntary isometric plantar flexions. J Physiol 587(Pt 19):4737–4748. https://doi.org/10.1113/jphysiol.2009.175695
Person RS, Kudina LP (1972) Discharge frequency and discharge pattern of human motor units during voluntary contraction of muscle. Electroencephalogr Clin Neurophysiol 32(5):471–483. https://doi.org/10.1016/0013-4694(72)90058-2

Piotrkiewicz M (1999) An influence of afterhyperpolarization on the pattern of motoneuronal rhythmic activity. J Physiol Paris 93(1–2):125–133. https://doi.org/10.1016/s0928-4257(99)80143-0
Powers RK, Binder MD (2000) Relationship between the time course of the afterhyperpolarization and discharge variability in cat spinal motoneurones. J Physiol 528(Pt 1):131–150. https://doi.org/10.1111/j.1469-7793.2000.t01-1-00131.x
Powers RK, Binder MD (2001) Input-output functions of mammalian motoneurons. Rev Physiol Biochem Pharmacol 143:137–263. https://doi.org/10.1007/BFb0115594
Powers RK, Heckman CJ (2015) Contribution of intrinsic motoneuron properties to discharge hysteresis and its estimation based on paired motor unit recordings: a simulation study. J Neurophysiol 114(1):184–198. https://doi.org/10.1152/jn.00019.2015
Powers RK, Heckman CJ (2017) Synaptic control of the shape of the motoneuron pool input-output function. J Neurophysiol 117(3):1171–1184. https://doi.org/10.1152/jn.00850.2016
Powers RK, Turker KS (2010) Estimates of EPSP amplitude based on changes in motoneuron discharge rate and probability. Exp Brain Res 206(4):427–440. https://doi.org/10.1007/s00221-010-2423-z
Powers RK, Dai Y, Bell BM, Percival DB, Binder MD (2005) Contributions of the input signal and prior activation history to the discharge behaviour of rat motoneurones. J Physiol 562(Pt 3):707–724. https://doi.org/10.1113/jphysiol.2004.069039
Powers RK, Nardelli P, Cope TC (2008) Estimation of the contribution of intrinsic currents to motoneuron firing based on paired motoneuron discharge records in the decerebrate cat. J Neurophysiol 100(1):292–303. https://doi.org/10.1152/jn.90296.2008
Powers RK, Elbasiouny SM, Rymer WZ, Heckman CJ (2012) Contribution of intrinsic properties and synaptic inputs to motoneuron discharge patterns: a simulation study. J Neurophysiol 107(3):808–823. https://doi.org/10.1152/jn.00510.2011
Revill AL, Fuglevand AJ (2011) Effects of persistent inward currents, accommodation, and adaptation on motor unit behavior: a simulation study. J Neurophysiol 106(3):1467–1479. https://doi.org/10.1152/jn.00419.2011
Revill AL, Fuglevand AJ (2017) Inhibition linearizes firing rate responses in human motor units: implications for the role of persistent inward currents. J Physiol 595(1):179–191. https://doi.org/10.1113/JP272823
Schwindt PC, Crill WE (1980) Properties of a persistent inward current in normal and TEA-injected motoneurons. J Neurophysiol 43(6):1700–1724. https://doi.org/10.1152/jn.1980.43.6.1700
Selverston A, Elson R, Rabinovich M, Huerta R, Abarbanel H (1998) Basic principles for generating motor output in the stomatogastric ganglion. Ann N Y Acad Sci 860:35–50. https://doi.org/10.1111/j.1749-6632.1998.tb09037.x
Stephenson JL, Maluf KS (2010) Discharge behaviors of trapezius motor units during exposure to low and high levels of acute psychosocial stress. J Clin Neurophysiol 27(1):52–61. https://doi.org/10.1097/WNP.0b013e3181cb81d3
Stevens R, Taylor V, Nichols J, Maccabe AB, Yelick K, Brown D (2020) AI for Science (ANL-20/17). https://anl.app.box.com/s/f7m53y8beml6hs270h4yzh9l6cnmukph
Stuart DG, Brownstone RM (2011) The beginning of intracellular recording in spinal neurons: facts, reflections, and speculations. Brain Res 1409:62–92. https://doi.org/10.1016/j.brainres.2011.06.007
Taylor C, Kmiec T, Thompson C (2020) Differences in human motoneuron excitability between functionally diverse muscles. CommonHealth 1(1):12–23
Thomas CK, Ross BH, Stein RB (1986) Motor-unit recruitment in human first dorsal interosseous muscle for static contractions in three different directions. J Neurophysiol 55(5):1017–1029. https://doi.org/10.1152/jn.1986.55.5.1017
Turker KS, Powers RK (1999) Effects of large excitatory and inhibitory inputs on motoneuron discharge rate and probability. J Neurophysiol 82(2):829–840. https://doi.org/10.1152/jn.1999.82.2.829
Turker KS, Powers RK (2005) Black box revisited: a technique for estimating postsynaptic potentials in neurons. Trends Neurosci 28(7):379–386. https://doi.org/10.1016/j.tins.2005.05.007

Udina E, D'Amico J, Bergquist AJ, Gorassini MA (2010) Amphetamine increases persistent inward currents in human motoneurons estimated from paired motor-unit activity. J Neurophysiol 103(3):1295–1303. https://doi.org/10.1152/jn.00734.2009

Vandenberk MS, Kalmar JM (2014) An evaluation of paired motor unit estimates of persistent inward current in human motoneurons. J Neurophysiol 111(9):1877–1884. https://doi.org/10.1152/jn.00469.2013

VanderMaelen CP, Aghajanian GK (1980) Intracellular studies showing modulation of facial motoneurone excitability by serotonin. Nature 287(5780):346–347. https://doi.org/10.1038/287346a0

Vandermaelen CP, Aghajanian GK (1982) Intracellular studies on the effects of systemic administration of serotonin agonists on rat facial motoneurons. Eur J Pharmacol 78(2):233–236. https://doi.org/10.1016/0014-2999(82)90242-4

Walton C, Kalmar JM, Cafarelli E (2002) Effect of caffeine on self-sustained firing in human motor units. J Physiol 545(2):671–679. https://doi.org/10.1113/jphysiol.2002.025064

Wei K, Glaser JI, Deng L, Thompson CK, Stevenson IH, Wang Q, Hornby TG, Heckman CJ, Kording KP (2014) Serotonin affects movement gain control in the spinal cord. J Neurosci 34(38):12690–12700. https://doi.org/10.1523/JNEUROSCI.1855-14.2014

White SR, Fung SJ, Barnes CD (1991) Norepinephrine effects on spinal motoneurons. Prog Brain Res 88:343–350. https://doi.org/10.1016/s0079-6123(08)63821-2

Wilson JM, Thompson CK, Miller LC, Heckman CJ (2015) Intrinsic excitability of human motoneurons in biceps brachii versus triceps brachii. J Neurophysiol 113(10):3692–3699. https://doi.org/10.1152/jn.00960.2014

Woodbury JW, Patton HD (1952) Electrical activity of single spinal cord elements. Cold Spring Harb Symp Quant Biol 17:185–188. https://doi.org/10.1101/sqb.1952.017.01.018

Yavuz SU, Mrachacz-Kersting N, Sebik O, Berna Unver M, Farina D, Turker KS (2014) Human stretch reflex pathways reexamined. J Neurophysiol 111(3):602–612. https://doi.org/10.1152/jn.00295.2013

Yavuz US, Negro F, Sebik O, Holobar A, Frommel C, Turker KS, Farina D (2015) Estimating reflex responses in large populations of motor units by decomposition of the high-density surface electromyogram. J Physiol 593(19):4305–4318. https://doi.org/10.1113/JP270635

Yavuz US, Negro F, Diedrichs R, Farina D (2018) Reciprocal inhibition between motor neurons of the tibialis anterior and triceps surae in humans. J Neurophysiol 119(5):1699–1706. https://doi.org/10.1152/jn.00424.2017

Zengel JE, Reid SA, Sypert GW, Munson JB (1985) Membrane electrical-properties and prediction of motor-unit type of medial gastrocnemius motoneurons in the cat. J Neurophysiol 53(5):1323–1344

Motoneuronal Regulation of Central Pattern Generator and Network Function

Mélanie Falgairolle and Michael J. O'Donovan

Abstract This chapter reviews recent work showing that vertebrate motoneurons can trigger spontaneous rhythmic activity in the developing spinal cord and can modulate the function of several different central pattern generators later in development. In both the embryonic chick and the fetal mouse spinal cords, antidromic activation of motoneurons can trigger bouts of rhythmic activity. In the neonatal mouse, optogenetic manipulation of motoneuron firing can modulate the frequency of fictive locomotion activated by a drug cocktail. In adult animals, motoneurons have been shown to regulate swimming in the zebrafish, and vocalization in fish and frogs. We discuss the significance of these findings and the degree to which motoneurons may be considered a part of these central pattern generators.

Keywords Motoneuron · Locomotion · Central pattern generator · Calcium imaging · Optogenetics

1 Introduction

Motoneurons have been exhaustively investigated with intracellular recording methods since the early 1950s (Brock et al. 1952). Despite such intense study, work in the in the twenty-first century is providing new and exciting insights into their properties and functions. In this chapter, we will discuss these novel findings and consider their significance for motor function. We will first describe experiments that implicate motoneurons in the genesis of spontaneous activity in the developing spinal cord. We will then report the results of experiments in the mammalian spinal cord showing that motoneurons release an excitatory amino acid in addition to

M. Falgairolle
NCCIH, NINDS, NIH, Bethesda, MD, USA
e-mail: melanie.falgairolle@nih.gov

M. J. O'Donovan (✉)
NINDS, NIH, Bethesda, MD, USA
e-mail: odonovm@ninds.nih.gov

M. J. O'Donovan, M. Falgairolle (eds.), *Vertebrate Motoneurons*, Advances in Neurobiology 28, https://doi.org/10.1007/978-3-031-07167-6_11

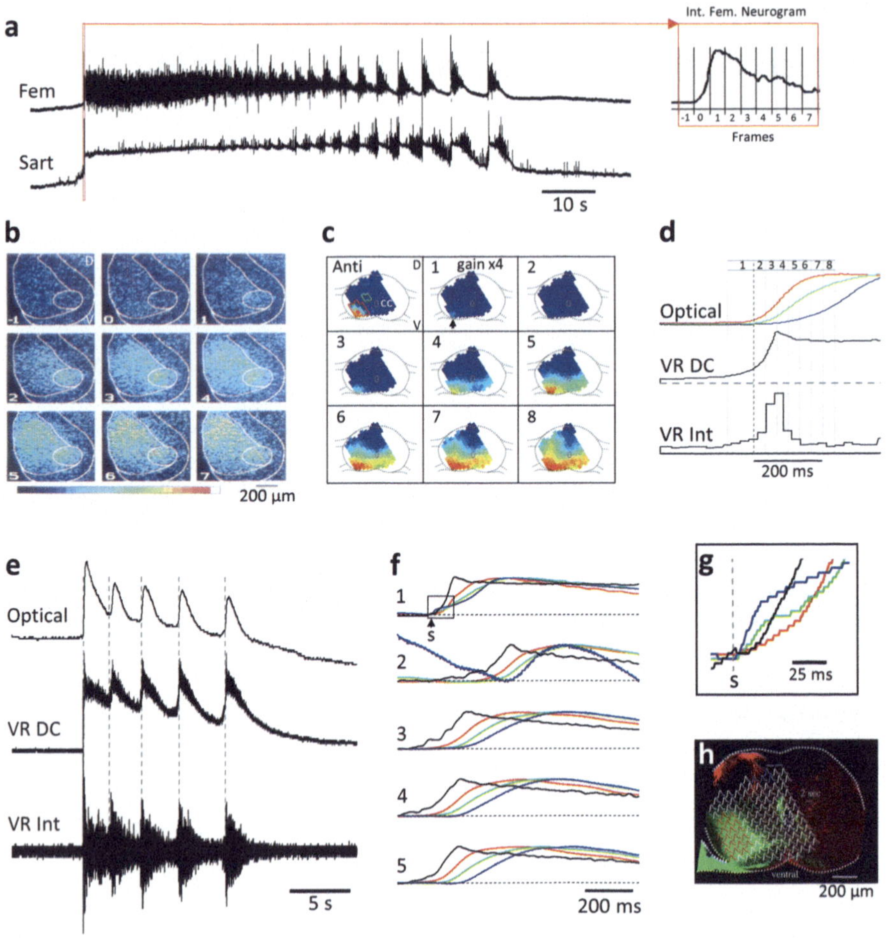

Fig. 1 Calcium and voltage imaging shows that activity begins ventro-laterally in the cut transverse face of the lumbosacral cord of the chick embryo at the onset of an episode of spontaneous bursting. (**a**) Spontaneous burst recorded from the femorotibialis (Fem) and the sartorius (Sart) muscle nerves in an E11 chick embryo. The red box at the beginning of the episode shows the approximate timing of the images shown in panel **b**. The expanded region indicated by the arrow shows the integrated femorotibialis neurogram corresponding to the frames shown in panel **b**. The first frame (−1) is the frame before the onset of the electrical activity and this and the subsequent frames are demarcated by the bars over the simultaneously recorded femorotibialis neurogram. Note that the data in the inset and the images of panel **b** are from a different embryo than that shown in **a**. (**b**) Calcium imaging of the onset of a spontaneous episode like the one shown in panel **a**. The panels are single frame difference images (active-control) averaged from 3 separate episodes all synchronized to the onset of the electrical activity recorded from the femorotibialis nerve. The color map below the image shows the color mapping of the fully expanded 8-bit images. (**c**) Voltage-sensitive dye imaging of a spontaneous episode of activity imaged with a 128-photodiode array. Each small square in the image corresponds to the output of one of the photodiodes. The top left-hand panel (Anti) shows the signals accompanying antidromic stimulation of the homonymous ventral root. The colored outlines identify the diodes whose signals were averaged together to

acetylcholine at their terminals with Renshaw cells and provide feedback to regulate the locomotor central pattern generator. This will be followed by a discussion of motoneuronal feedback to rhythmic circuits in several systems and we will conclude by considering the relevance of the new findings for motor network function.

2 The Role of Motoneurons in Initiating Spontaneous Bursting in the Developing Spinal Cord

Early in development, vertebrate embryos exhibit spontaneous movements that are crucial for the proper functioning of joints, muscles, and neural networks (Toutant et al. 1979; Hall and Herring 1990; Borodinsky et al. 2004; Hanson and Landmesser 2004; Wenner 2014). These movements are generated by spontaneously active neural networks in the spinal cord (Provine et al. 1970; Landmesser and O'Donovan 1984). At embryonic day 4 (E4) in the chick embryo, spontaneous bursting can be recorded from lumbosacral motoneurons before their axons have innervated limb muscles (Milner and Landmesser 1999). At this stage, the neurotransmitter acetylcholine drives the spontaneous episodes and is thought to originate from motoneurons (Milner and Landmesser 1999) which are among the earliest spinal neurons to differentiate (Hamburger 1990). More direct evidence for the role of motoneurons in the initiation of spontaneous activity emerges later in development when motoneurons have innervated their respective limb muscles. At E10–11, calcium (Fig. 1b) and voltage sensitive dye (Fig. 1c, d) imaging reveals that an episode of spontaneous

Fig. 1 (continued) provide the colored traces in panel **d**. The numbered panels correspond to the signals averaged from frames acquired during the temporal windows indicated in panel **d**. Panel 1 is the average of 105 fames (67 ms) before the onset of the main ventral root discharge indicated by the grey dotted line in **d** and amplified 4-fold compared to the remaining panels. Panels 2–8 were averaged from 45 frames (29 ms) and the data were acquired at 636 frames/s. The arrow shows that the earliest detectable activity begins over the motor nucleus from where it spreads contralaterally and ventro-dorsally. The data were combined from 2 E10 embryos and synchronized to the onset of the electrical activity. (**d**) Comparison between the timing of the optical signals and the low-passed ventral root activity (VR DC) and the integrated ventral root discharge (VR int.). (Modified from Arai et al. 2007). (**e–h**) Voltage-sensitive dye imaging reveals that the pattern of activity seen at the onset of a spontaneous burst recurs during each subsequent cycle of activity. (**e**) An episode of rhythmic activity initiated by a single stimulus to a dorsal root in an E10 embryo. The optical recording is from a single diode located over the motor nucleus ipsilateral to the ventral root recording. The dotted lines indicate the initiation of ventral root discharge. VR DC – ventral root activity filtered from DC-20Hz. VR Int – ventral root spiking. (**f**) Comparison of the timing of the ventral root electrical activity with the optical signals averaged from multiple diodes over three regions of the cord as shown in panel H. Note that with the exception of the first cycle, optical activity arises over the motor nucleus and propagates dorsally in each cycle. (**g**) Expansion of the rectangular region shown at the start of first cycle in **f**, showing that the dorsal root stimulus (s) initially triggers dorsal activity that subsequently propagates dorso-ventrally. (**h**) Transverse section of the cord showing the superimposed diode signals. The motoneurons (green) were labeled with DiO and the dorsal roots (red) with DiI. The diodes signals indicated in the different colors were averaged to generate the optical signals shown in (**f**). (Modified from Arai et al. 2007) activity begins in the region of the motor nucleus and spreads to encompass the rest of the cord from there (O'Donovan et al. 1994; Arai et al. 2007) (Fig. 1). Voltage-sensitive dye imaging also demonstrated that the initial optical activity over the motor nucleus occurred at the onset of each cycle of activity in the episode, suggesting that motoneurons were critical for rhythmogenesis in this preparation (Fig. 1e–g). Consistent with motoneurons initiating a spontaneous episode, it was found that an episode could be triggered by a brief train of stimuli applied to motor axons in a deafferented muscle nerve or ventral root. Further work revealed that the connection between motoneurons and the spinal motor network was mediated by the projection of motoneurons to the avian equivalent of mammalian Renshaw cells – R-interneurons (Wenner and O'Donovan 1999).

Although R-interneurons release the inhibitory neurotransmitter GABA (Wenner and O'Donovan 1999), their synaptic actions are depolarizing because the chloride equilibrium potential is elevated above rest potential in developing spinal neurons (Chub and O'Donovan 2001). To show that R-interneurons were the likely mediators communicating the motoneuronal activity to the rest of the network, we applied a calcium-sensitive dye to the ventrolateral funiculus to back-label the interneurons whose axons projected therein. The cut transverse face of the cord was then imaged during stimulation of motor axons sub-threshold for initiating an episode of bursting (Fig. 2a, panel labeled VR-stim). Intracellular recording from individual R-interneurons confirmed that the labeled region contained the neurons activated monosynaptically by ventral root stimulation of motoneurons (Wenner and O'Donovan 2001). Once this location was established, we then identified the first region to become active during a spontaneous episode and found it overlapped with the R-interneuron region (Fig. 2 panels 2 and 3). One limitation of these experiments is that only a subset of ventral interneurons is labeled with the calcium-sensitive dye applied to the ventro-lateral funiculus. Nevertheless, the observation that the first region to become active at the onset of a spontaneous episode contains neurons monosynaptically activated by motoneurons is consistent with their role in communicating the initial motoneuron activity to the rest of the network. However, we also found that the motoneuron-R-interneuron pathway was not obligatory for transmitting motoneuron activity to the rest of the network because when the recurrent connection between motoneurons and R-interneurons was depressed by cholinergic antagonists, spontaneous activity still occurred (Fig. 2b) but under this condition the interneuronal activity following motoneuronal bursting began medial to the motor column and not in the R-interneuron area (Wenner and O'Donovan 2001).

We were concerned that activation of afferent fibers in the ventral roots (Coggeshall 1979) might contribute to the optical signals, so we performed experiments in the presence of CNQX and APV to block glutamatergic transmission. We found that neither the amplitude nor the location of the optical signals changed indicating that any contribution from afferent excitation was negligible (Wenner and O'Donovan 2001).

Similar findings have been made in the developing mouse spinal cord where a single stimulus to a ventral root can trigger a network burst (Hanson and Landmesser 2003). This is illustrated in Fig. 3b, which shows electrical recordings from the left

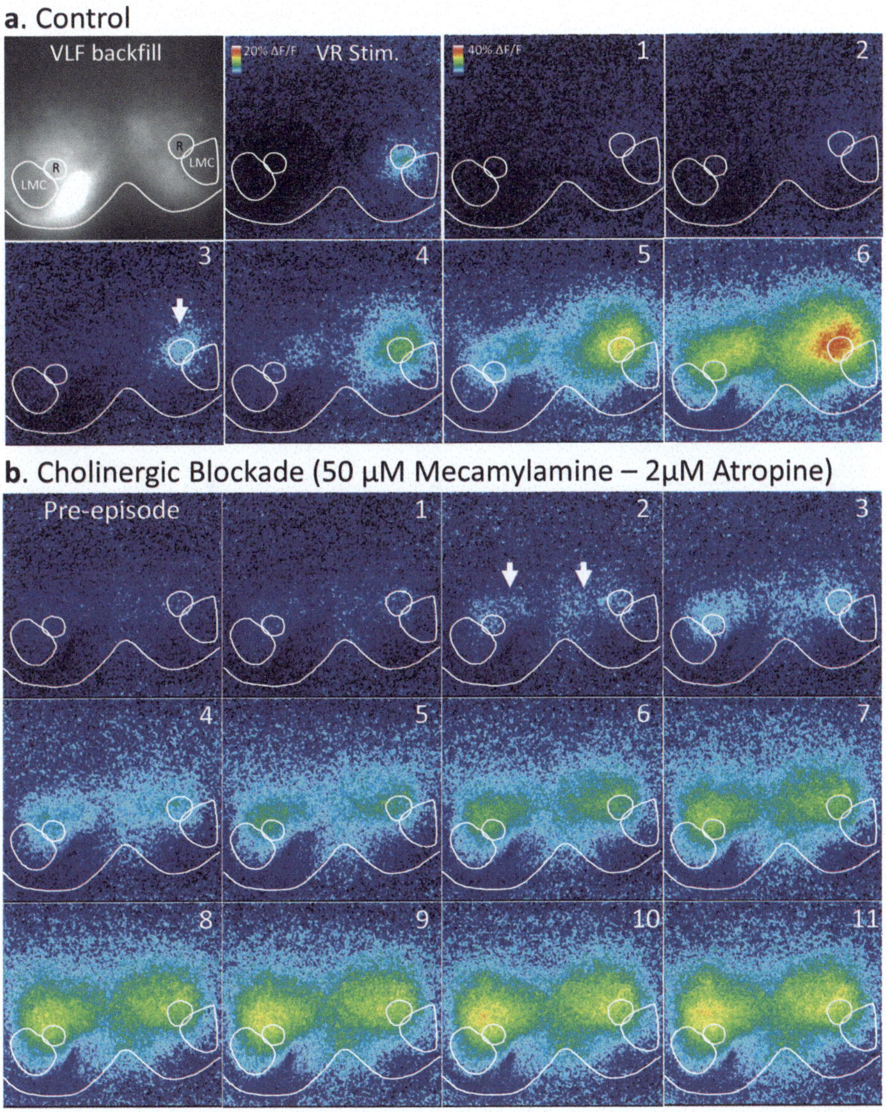

Fig. 2 Recruitment of interneurons retrogradely labeled with calcium green dextran at the onset of a spontaneous episode under control conditions and in the presence of cholinergic blockade. (**a**) The left most panel shows the cut transverse face of a single segment of the cord in which calcium green dextran was backfilled bilaterally from the ventrolateral funiculus (VLF backfill). The lateral motor columns (LMC) and the R-interneuron regions (R) are outlined. The next panel shows the optical signal averaged from 10 successive frames generated in response to a train of stimuli applied to the ventral root on that side of the cord. The other R-interneuron region was defined similarly. The remaining frames (1–6) show the interneuronal activity at the onset of a spontaneous episode. The activity commences in the R-interneuron region on that side of the cord (arrow) and propagates contralaterally from there. (**b**) The same cord section is shown during a spontaneous episode occurring in the presence of cholinergic antagonists. Under this condition the optical activity emerges bilaterally and medial to the lateral motor column (arrows) and intensifies to occupy most of the ventral cord. The frame rate was 30 Hz. The images in A were averaged from 4 episodes and those in B from 3 episodes. (Modified from Wenner and O'Donovan 2001)

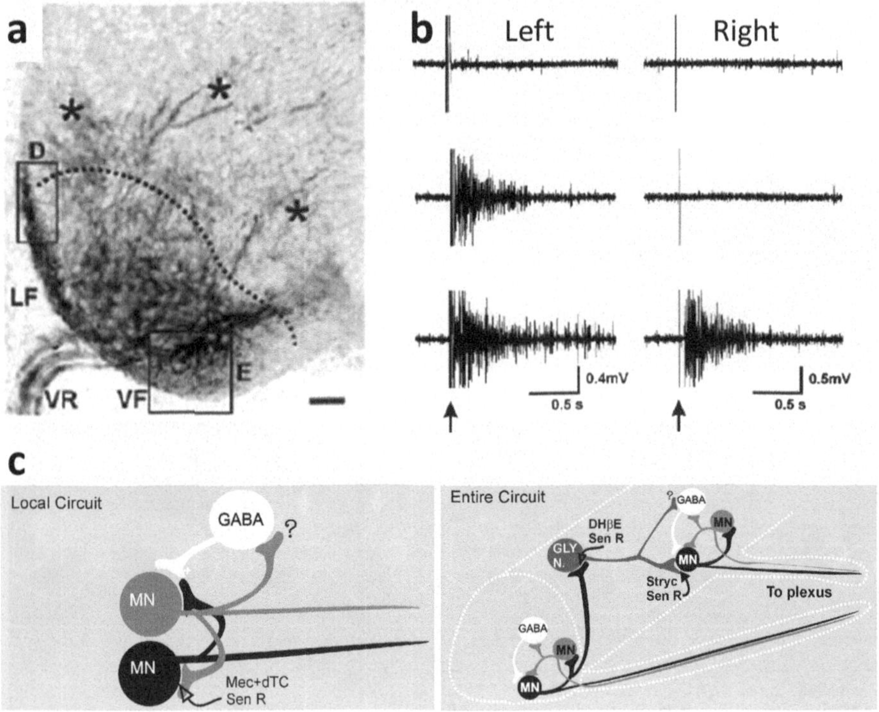

Fig. 3 Stimulation of motoneuron axons in the sciatic nerve triggers network bursts in the embryonic mouse spinal cord. (**a**) Immunocytochemistry for the vesicular acetylcholine transporter (VAChT) in an E12.5 mouse embryo. Notice the limited extent of the VAChT positive neurites (asterisks) and the projections into the lateral (LF) and ventral (VF) funiculi. (**b**) Single stimuli applied at increasing current intensities to the left sciatic nerve. At the lowest intensity (top traces) no bursts are evoked in the left and right ventral roots. As the stimulus intensity is increased an ipsilateral burst is evoked and at the highest intensity bursting propagates and can be recorded contralaterally. (**c**) Schematic showing the hypothesized circuitry responsible for the local (Local Circuit) and the propagating (Entire Circuit) bursts. *MN* motoneuron, *GLY N* glycinergic neuron, *Sen R* sensitive receptor, *Mec + dTC* mecamylamine + d-tubocurarine, *DHβE* dihydro-β-erythroidine hydrobromide, *Stryc* strychnine. (Modified from Hanson and Landmesser 2003. Copyright (2003) Society for Neuroscience, U.S.A)

and right sciatic nerve while stimulating the left sciatic nerve at increasing intensity in an E12.5 mouse embryo. At the lowest intensities (top traces) no bursts are evoked; with higher intensity a burst is evoked ipsilateral to the stimulated root (local burst) and at the highest intensity (lowest traces) bursts are evoked on both sides of the cord (propagated burst). The hypothesized circuitry responsible for the local and propagated bursts is shown in Fig. 3c. The local burst is presumed to be generated by reciprocal cholinergic connections between motoneurons and depolarizing responses from GABAergic neurons activated by motoneuron collaterals – presumably Renshaw cells. The propagated burst is hypothesized to be generated by motoneuronal connections with a glycinergic interneuronal population that projects to motoneurons and Renshaw cells in other segments and also contralaterally (Hanson and Landmesser 2003). It is not clear whether this glycinergic neuron is a novel cell class, because although Renshaw cells are known to project to each other, contralateral projections have not been demonstrated in the adult.

3 Excitatory Effects of Ventral Root Stimulation on Neonatal Mammalian Spinal Networks

The excitatory effects of ventral root stimulation on spinal networks persist into the neonatal period in both the mouse and rat. Stimulation of a ventral root can trigger bursting in an adjacent root (Nishimaru et al. 2005; Machacek and Hochman 2006) and entrain disinhibited bursting in both the neonatal rat (Machacek and Hochman 2006) and the mouse (Bonnot et al. 2009) spinal cords. However, unlike the situation earlier in development, these excitatory effects are labile, and they are not observed in every preparation. For example, in the neonatal mouse spinal cord, the ability to entrain bursting depends on the stimulated ventral root and varied from a minimum of ~15% for L2 stimulation to a maximum of ~50% for stimulation of the L6 root (Bonnot et al. 2009). In the neonatal rat spinal cord in vitro, entrainment of disinhibited bursting was observed in 4/11 experiments in P11–P14 rats (Machacek and Hochman 2006). In this study, other excitatory effects of ventral root stimulation, including ventral root evoked bursting in motoneurons and modulation of the frequency of the locomotor rhythm were seen in younger animals. Excitatory effects were observed in ~18% of the preparations although their frequency increased in the presence of bath-applied noradrenaline. The reason for this variability is unknown. In the neonatal rat, the latency of the VR-evoked bursts in the ventral roots was compatible with a disynaptic pathway, suggesting that motoneurons project to an excitatory interneuron that in turn projects back to motoneurons. Evidence for such a recurrent excitatory pathway in the neonatal mouse was found when it was demonstrated that stimulation of a ventral root produces monosynaptic EPSPs in glutamatergic V3 interneurons that in turn, project monosynaptically back to motoneurons (Chopek et al. 2018), and that motoneurons also make recurrent projections to glutamatergic spinocerebellar neurons (Chalif et al. 2022). However, the extent to which these connections mediate the excitatory effects of motoneurons is not clear, because there is no evidence that they are variable or labile.

4 Motoneuronal Regulation of Locomotion

The study of locomotion has been greatly facilitated using preparations in which the movements accompanying locomotion have been abrogated. In adult animals, this can be achieved by paralyzing muscles (Viala and Buser 1969; Grillner and Zangger 1979; Iles and Nicolopoulos-Stournaras 1996; Meehan et al. 2012) and in neonatal

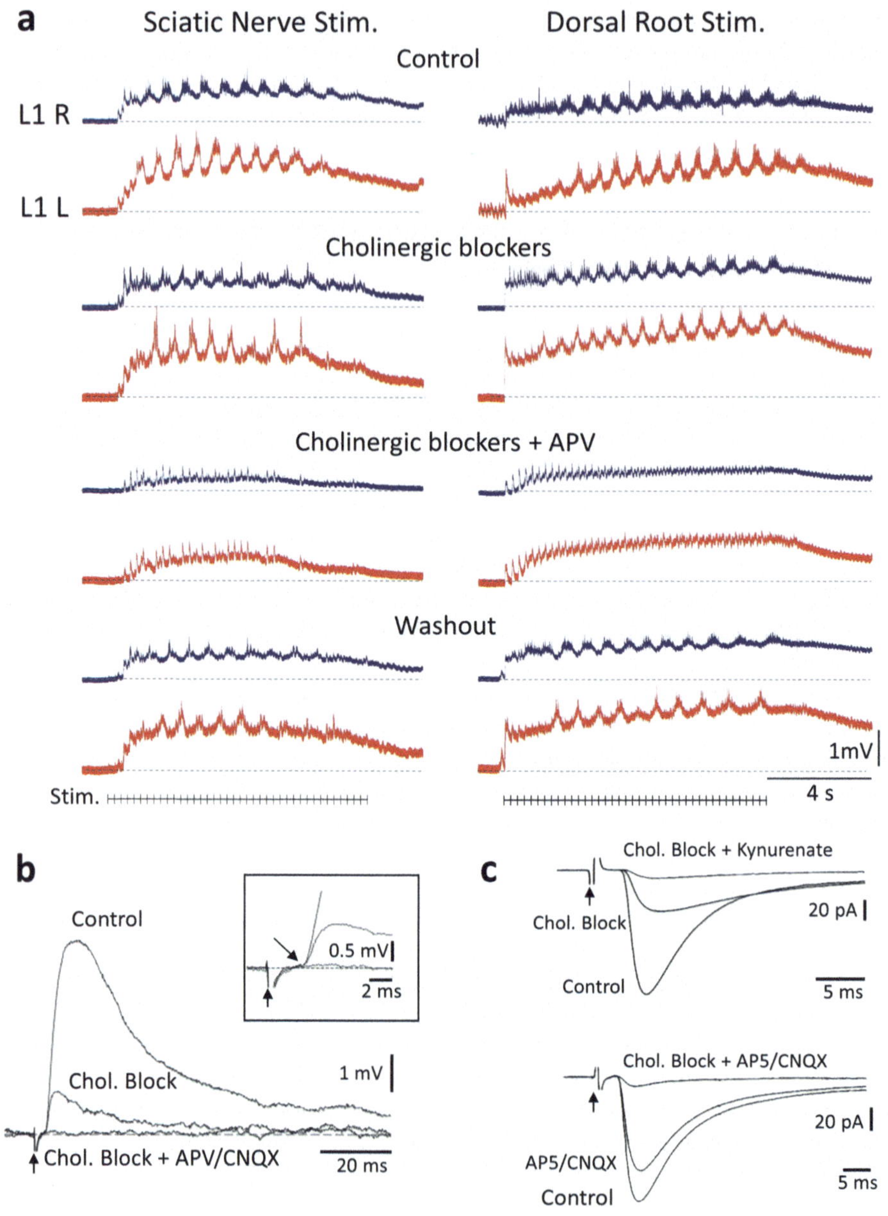

Fig. 4 Stimulation of the sciatic nerve with cut dorsal roots can evoke an episode of fictive locomotion in the isolated cord of the neonatal mouse. (**a**) Comparison between an episode of locomotor-like activity triggered by a train of stimuli (4 Hz for 10 s) applied to either the deafferented sciatic nerve (left hand panels) or a dorsal root (right hand panels), under control conditions, in the presence of cholinergic blockers (50 μM mecamylamine, 50 μM dihydro-β-erythroidine, and 5 μM atropine), cholinergic blockers plus APV (100 μM) followed by washout of the drugs. The recordings are DC coupled from the left (L) and right (R) Lumbar(L) 1 ventral roots. (**b** and **c**)

animals where the spinal cord is isolated and detached from the musculature (Grillner and Wallén 1980; Kudo and Yamada 1987; Smith and Feldman 1987; Whelan et al. 2000). Because no movements accompany the activity, it is referred to as fictive locomotion or locomotor-like activity. In such preparations, fictive locomotion is characterized by a rhythmic alternation between flexor and extensor muscle nerves and between bilateral flexor or extensor muscle nerves. A further simplification occurs in the isolated spinal cord preparation because the rostral lumbar ventral roots (L1 and L2) comprise predominantly flexor motoneurons and the caudal lumbar roots (L5 and L6) mostly extensor motoneurons (Cazalets et al. 1992; Kiehn et al. 1992). In the isolated spinal cord of the mouse, locomotor-like activity can be induced by a cocktail of drugs including NMDA (n-methyl-D-aspartate) and serotonin which can also be supplemented with dopamine (Jiang et al. 1999; Whelan et al. 2000). In addition, tonic low frequency (1–4 Hz) stimulation of sensory afferents in the dorsal roots or the locomotor centers in the brainstem can trigger locomotion in the neonatal mouse and rat spinal cords (Whelan et al. 2000; Zaporozhets et al. 2004)

The role of motoneurons in locomotion was first examined in the isolated lamprey spinal cord. This was done by antidromically stimulating a ventral root during fictive swimming induced by drugs (Wallen 1984). It was found that antidromic stimulation of 1–3 ventral roots failed to modify the ongoing locomotor pattern or frequency (Wallen 1984), suggesting that motoneuronal activity does not modulate the swimming central pattern generator. However, subsequent work in the fictively swimming *Xenopus* tadpole, challenged this idea by showing that the amplitude of the rhythmic drive potentials recorded from spinal interneurons was reduced by ~20% in the presence of bath-applied nicotinic antagonists (Perrins and Roberts 1995). Initially it was assumed that the source of acetylcholine was motoneurons, but later work revealed that spinal glutamatergic interneurons could also release acetylcholine (Li et al. 2004), raising doubts about this assumption.

In mammals, studies of the role of motoneurons in locomotion have been restricted to isolated preparations of neonatal mice (Mentis et al. 2005; Humphreys and Whelan 2012; Falgairolle et al. 2017) and rats (Machacek and Hochman 2006). In the de-afferented isolated cord of the neonatal mouse, low frequency (1–4 Hz) tonic stimulation of a ventral root or the deafferented sciatic nerve can trigger an episode of locomotor-like activity (Fig. 4) (Mentis et al. 2005).

Fig. 4 (continued) **Motoneurons release acetylcholine and an excitatory amino acid at their terminals with Renshaw cells.** (**b**) Synaptic potentials recorded from a Renshaw cell in response to a single stimulus (arrow) applied to the ipsilateral ventral root. In the presence of cholinergic blockade (Chol. Block), a small potential persisted that had the same latency as the pre-drug potential (see inset). The smaller potential was abolished by glutamatergic antagonists (100 μM APV and 10 μM CNQX). (**c**) Voltage clamp recordings of a Renshaw cell following a single stimulus to the ipsilateral ventral root. After application of the nicotinic antagonist mecamylamine (50 μM Chol. Block) the evoked current was reduced and subsequently almost abolished when the broad-spectrum glutamatergic antagonist kynurenate (2 mM) was added. In the lower traces, the order of antagonists was reversed so that the glutamatergic antagonists (20 μM CNQX and 20 μM AP5) were applied first followed by the addition of 50 μM mecamylamine. (Panels **a** and **b** modified from (Mentis et al. 2005) and **c** from (Nishimaru et al. 2005) Copyright (2005) National Academy of Sciences, U.S.A)

The locomotor-like activity evoked by ventral root stimulation was blocked by ionotropic glutamate antagonists but not by cholinergic antagonists or a gap junction blocker (carbenoxolone). This was a very surprising result given that acetylcholine was assumed to be the only fast neurotransmitter released from motoneurons. However, it was shown that motoneurons release from their synaptic connections with Renshaw cells, an excitatory amino acid that binds to glutamatergic receptors (Fig. 4b, c) (Mentis et al. 2005; Nishimaru et al. 2005). Whether glutamate or aspartate is activating glutamatergic receptors has not yet be resolved (Richards et al. 2014). Nevertheless, if the excitatory effects of motoneurons were mediated by an excitatory interneuron then presumably both glutamatergic and cholinergic receptors would be activated when motoneurons were stimulated. The discovery that the excitatory projections of motoneurons to V3 interneurons were exclusively glutamatergic in the neonatal mouse spinal cord (Chopek et al. 2018) provided a possible explanation for this effect. Unfortunately, it is not clear that V3 interneurons mediate the locomotor actions of ventral root stimulation because optogenetic activation of V3 interneurons slows the locomotor rhythm (Danner et al. 2019) and silencing them leads to increased variability of the cycle length and flexor bursting (Zhang et al. 2008). A more likely candidate for mediating the effects of motoneuronal activity on the CPG are ventral spinocerebellar neurons that receive glutamatergic and cholinergic input from motoneurons in the neonatal mouse cord (Chalif et al. 2022). This is because optogenetic hyperpolarization of these neurons blocked the ability of motoneuron stimulation to trigger the locomotor rhythm in the neonatal mouse spinal cord, and these neurons were shown to be both necessary and sufficient for the generation of the locomotor rhythm in the neonatal mouse spinal cord (Chalif et al. 2022).

5 Ventral Root Afferents

One complicating factor in attributing the excitatory effects of ventral root stimulation to motor axons in the ventral roots is the possibility that they are mediated or complimented by the activation of sensory afferents that enter the cord through the ventral roots (Coggeshall 1979), or by excitation of sensory neurons in the ventral root itself (Windle 1931; Yamamoto et al. 1977). The existing data on this issue is somewhat contradictory. For example, direct injection of horseradish peroxidase into the spinal cord, labels cell bodies in the dorsal root ganglion when the dorsal roots have been cut (Maynard et al. 1977). Furthermore, a few studies using horseradish peroxidase applied to the ventral roots have revealed the existence of the occasional axon projecting to the dorsal horn and to preganglionic sympathetic neurons (Light and Metz 1978; Mawe et al. 1984; Beattie et al. 1987). Whether these projections are functional is unclear because activation of ventral root afferents could excite spinal neurons only if the dorsal roots remained intact (Clifton et al. 1976; Chung et al. 1983, 1985). Furthermore, more recent work has suggested that ventral root afferents end blindly, innervate the meninges or loop into the dorsal root (Shin et al. 1986; Hildebrand et al. 1997) contradicting the work showing dorsal root ganglion cells labeled from within the de-afferented cord. If ventral root afferents exist in the neonatal mouse is not known. However, it seems unlikely that such projections, even if they exist in the neonatal mouse, can account for the ability of ventral root stimulation to activate the locomotor CPG, because electrical stimulation of the dorsal root ganglion, with the dorsal roots cut, does not trigger locomotor activity (Pujala et al. 2016), although such stimulation would probably not activate neurons in the ventral root (Windle 1931; Yamamoto et al. 1977).

6 Optogenetic Manipulation of Motoneuron Activity During Locomotion

One approach to circumvent the potential activation of ventral root afferents, is to manipulate the firing of motoneurons directly and to establish if this affects the function of the locomotor CPG. In the neonatal mouse, this was accomplished in a series of optogenetic experiments in which the light-sensitive opsin archaerhodopsin was introduced into cholinergic neurons expressing the enzyme responsible the synthesis of the neurotransmitter acetylcholine (choline acetyltransferase -ChAT) or neurons expressing the transcription factor islet-1, of which motoneurons are a subset (Falgairolle et al. 2017). Archaerhodopsin is a light-gated outward proton pump that hyperpolarizes neurons (Chow et al. 2010) and independently reduces synaptic transmission (El-Gaby et al. 2016) when illuminated with green light. During drug-induced locomotor-like activity, illumination of cords expressing archaerhodopsin in either ChAT-positive or Islet1 expressing neurons transiently abolished or slowed the locomotor rhythm and often made it less regular (Fig. 5).

To control for the intra- and extracellular changes in pH that accompany activation of archaerhodopsin (Chow et al. 2010), similar experiments were performed in $ChAT^+$ neurons expressing another hyperpolarizing opsin – halorhodopsin. In contrast to archaerhodopsin, halorhodopsin is a light-gated chloride channel that hyperpolarizes the cell membrane by moving chloride ions into the cell (Zhang et al. 2007). Illumination of halorhodopsin during locomotor-like activity had similar results to the archaerhodopsin experiments indicating that changes in pH were not responsible for the slowing and disruption of the locomotor rhythm (Falgairolle et al. 2017). When the light was turned off in either experiment, the bursting of motoneurons was transiently enhanced and this was accompanied by a corresponding increase in the locomotor-like frequency. Confirmation that increased motoneuron firing could accelerate the locomotor-like frequency came from experiments in which the excitatory opsin channelrhodopsin was introduced into $ChAT^+$ neurons. The pharmacology of the modulatory effects of motoneuron firing on the locomotor-like rhythm were similar to that of the locomotor-like activity evoked by ventral root stimulation. Specifically, it was not blocked by cholinergic antagonists but was

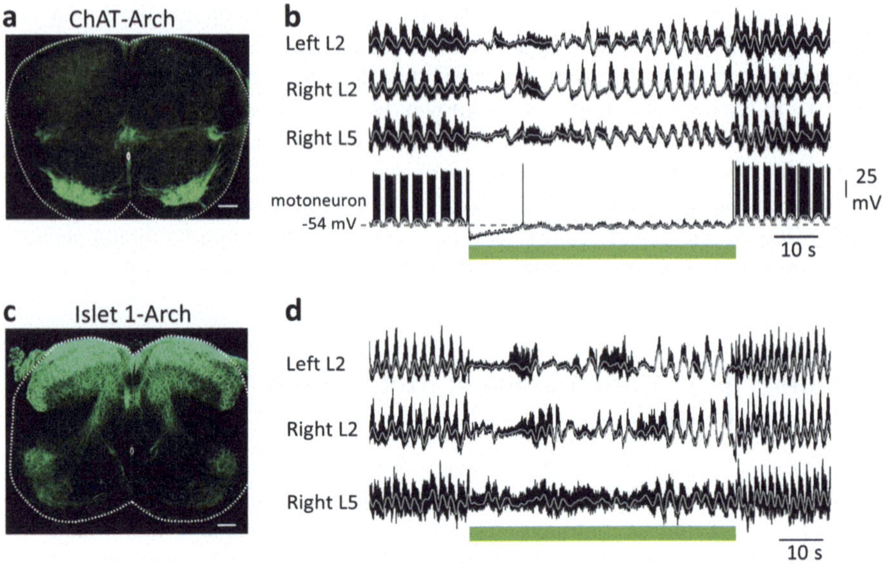

Fig. 5 Optogenetic hyperpolarization of motoneurons transiently abolishes and disrupts the locomotor rhythm during drug-induced fictive locomotion. (**a**) Z-stack projection of a 60µm transverse section of the second Lumbar (L2) segment of a P3 mouse spinal cord in which cholinergic (ChAT) neurons expressed the inhibitory opsin archaerhodopsin (Arch) coupled to enhanced green fluorescent protein (eGFP). (**b**) Green light (timing indicated by green bar) transiently inhibits and disrupts the locomotor rhythm induced by NMDA and serotonin in a ChAT/Arch animal. The lowest trace is an intracellular recording from an antidromically identified motoneuron. (**c**) Z-stack projection of a 60 µm transverse section of the L5 segment of a P2 mouse spinal cord in which Islet-1 positive neurons express archaerhodopsin coupled to eGFP. (**d**) Green light transiently suppresses and disrupts the locomotor rhythm induced by NMDA and serotonin in the Islet-1/Arch animal. The light grey lines superimposed on the neurograms in **b** and **d** are the slow potentials obtained by low pass filtering the raw ventral root signals (black traces). (Modified from Falgairolle et al. 2017)

abrogated by the AMPA receptor antagonist CNQX and persisted in the presence of carbenoxolone to block gap junctions. Ventral spinocerebellar neurons, which have reciprocal excitatory and electrical connections with motoneurons, are an obvious candidate to mediate the modulation of the CPG by motoneurons because their optogenetic excitation activates the locomotor rhythm and their hyperpolarization abolishes it (Chalif et al. 2022).

7 Motoneuronal Regulation of Central Pattern Generating Circuitry in Non-mammalian Vertebrates

Motoneurons have been shown to regulate central pattern generator function for swimming and vocalization in fish and frogs. We will first consider the role of motoneurons in the regulation of fictive swimming in the adult zebrafish and then discuss the motoneuronal control of vocalization in toadfish and the frog *Xenopus Laevis*. The adult zebrafish can generate fictive swimming in response to tonic stimulation applied between the brainstem and the spinal cord (Gabriel et al. 2008). Research in this preparation had revealed that the rhythmic drive to motoneurons during fictive swimming is derived from a class of excitatory glutamatergic interneurons (V2a) that express the Chox-10 transcription factor (Eklof-Ljunggren et al. 2012). These interneurons express pacemaker properties and are organized into modules that are sequentially recruited as swimming speed increases (Ampatzis et al. 2014). They form hybrid chemical bidirectional electrical synapses with motoneurons (Song et al. 2016) so that motoneuron depolarization or hyperpolarization is transmitted electrotonically to the V2a synapses on motoneurons which increases or attenuates transmitter release from the V2a synapse (Fig. 6b). Furthermore, changes in motoneuronal membrane potential can regulate the membrane potential of the V2a interneurons and thereby affect their firing (Fig. 6c). When motoneurons expressing the inhibitory opsin halorhodopsin were illuminated by yellow light, swim episodes were shortened and occurred at a reduced frequency (Fig. 6d). This was accompanied by hyperpolarization of the V2a interneurons sufficient to block firing, thereby reducing the excitatory drive to motoneurons (Song et al. 2016). It seems unlikely that a similar mechanism accounts for the modulation of the locomotor rhythm in the neonatal mouse spinal cord because the phenomenon persists in the presence of the gap junction blocker carbenoxolone (Falgairolle et al. 2017) and motoneurons are not electrically coupled to V2a interneurons in the mouse spinal cord (Bhumbra and Beato 2018).

Another system in which gap junctions between motoneurons and interneurons may be involved in the regulation of a CPG, is the vocalization network of toadfish. Toadfish make two types of vocalization – grunts and boatwhistles (advertisement calls) – that are produced by superfast muscles attached to the swim bladder (Chagnaud and Bass 2014). The frequency of the rhythm is generated by the vocal pacemaker nucleus and the duration of the vocalization by the vocal pre-pacemaker nucleus located in the hindbrain (Chagnaud et al. 2011) (Fig. 7a). Vocal motoneurons are extensively connected through gap junctions to the premotor interneurons generating the vocal input to motoneurons including the neurons of the pacemaker nucleus (Bass et al. 1994), suggesting that they can directly influence pacemaker function. Moreover, antidromic stimulation of the vocal nerve reveals a short latency depolarization in intracellularly recorded motoneurons indicative of electrical coupling between motoneurons. As the stimulus intensity is increased a short latency hyperpolarization is also recorded. This is unlikely to be due to classic recurrent inhibition mediated by Renshaw cells because the vocal motoneurons lack recurrent collaterals (Chagnaud and Bass 2014), and appears to be the result of electrical coupling between motoneurons and a glycinergic interneuron. It is hypothesized that this antidromically driven inhibition is essential for repetitive firing of the vocal motoneurons because intracellular depolarization of individual motoneurons does not result in repetitive firing possibly because of weak repolarization after the action potential (Chagnaud et al. 2021). Thus, motoneuronal activity may contribute directly to the depolarization of the pacemaker neurons and to the synchronization

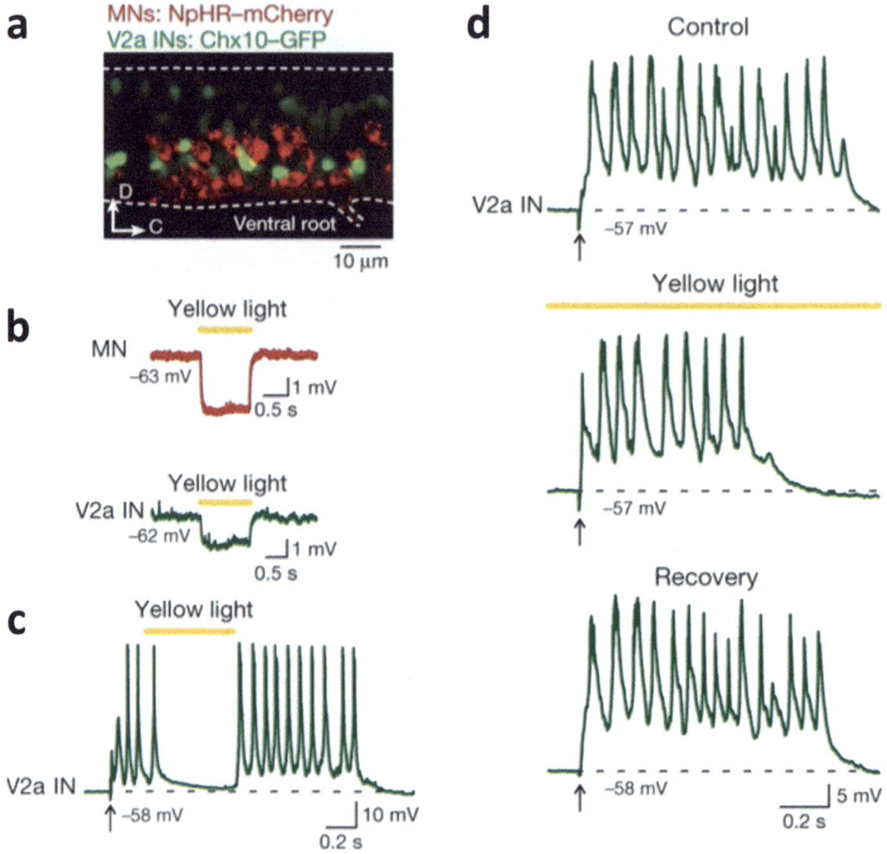

Fig. 6 Hyperpolarizing motoneurons expressing halorhodopsin lowers the swim frequency and shortens the duration of the swim bout in the adult zebrafish. (**a**) Micrograph of the spinal cord showing motoneurons (MNs) expressing halorhodopsin coupled with mCherry (NpHR-mCherry, red) and V2a interneurons (V2a Ins) expressing Chox10 coupled with GFP (Chx10-GFP, green). (**b**) Intracellular recordings showing that yellow light on the cord hyperpolarizes the motoneuron (MN, upper red trace) and also the V2a interneuron (V2a IN, lower green trace). (**c**) Intracellular recording from a V2a interneuron showing that light-induced hyperpolarization of motoneurons can block firing in the interneuron during a swim episode. (**d**) More frequently, however, motoneuron hyperpolarization reversibly slows the swim frequency and shortens the swim duration recorded in a V2a interneuron. Top trace control, middle trace light-induced hyperpolarization of motoneurons, lowest trace recovery. In all panels the duration of the light stimulus is shown by the yellow bar. (Modified from Song et al. 2016 with permission)

of the motoneuronal bursts through gap junctional coupling (Chagnaud et al. 2021). Whether these gap junction connections are hybrid chemical/electrical synapses, as in the connection of spinal motoneurons to V2a neurons in the zebrafish, is not known.

Xenopus Laevis is another species in which motoneuronal activity can modulate a vocal CPG (Lawton et al. 2017). In male frogs this CPG produces mating songs

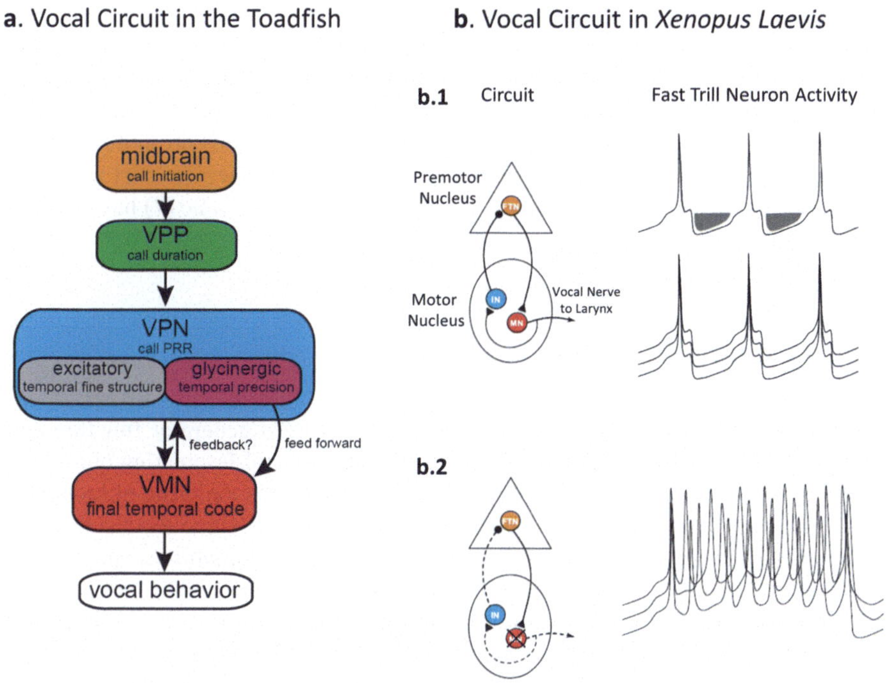

Fig. 7 Motoneurons regulate the vocalization circuitry in the toadfish and in *Xenopus Laevis*. (**a**) Schematic of the nuclei and connectivity in the circuit controlling vocalization in the toadfish. Motoneurons in the vocal motor nucleus (VMN) are gap-junction coupled to excitatory interneurons in the vocal pacemaker nucleus (VPN) and to a glycinergic inhibitory population. The VPN sets the frequency of the calls and is controlled by the vocal pre-pacemaker nucleus (VPP) that determines the call duration. It is hypothesized that motoneuronal activity feeds back to the VPN and glycinergic populations through gap junctions to regulate its function. (Modified from Fig. 9 in Chagnaud et al. 2021). (**b**) The organization of the vocal circuitry in *Xenopus Laevis*. **b1**. The vocal circuit comprises a premotor nucleus containing neurons that generate fast trills that project excitatory connections to motoneurons innervating the larynx. The hypothetical intracellular recordings from these premotor interneurons (shown on the right) illustrate how the proposed inhibitory feedback (grayed area) slows the frequency and synchronizes the population activity (lower panel). **b2**. The effect of interrupting the feedback from the motor nucleus to the premotor interneurons either by silencing motoneuron firing with QX-314 (an intracellular sodium channel blocker) or by sectioning the preparation between the premotor nucleus and the motor nucleus abolishes the feedback inhibition leading to an increased frequency of firing and population asynchrony of the fast trill neurons. (Modified from Lawton et al. 2017)

which can be activated in vitro by bath applying serotonin to the isolated brain (Zornik and Yamaguchi 2012). The evoked song comprises a brief 'trill' at 50–60 Hz and lasting ~500 ms which is similar to the naturally occurring vocalizations. The laryngeal motoneurons are driven by premotor neurons termed fast trill neurons that are located in the dorsal tegmental area of the medulla (DTAM). There is also an ascending inhibitory projection from the laryngeal motor nucleus to the DTAM (Fig. 7b1). When laryngeal motoneuron activity is silenced by retrograde loading of motoneuron axons with the sodium channel blocker QX-314, the fast trill neurons fire faster and asynchronously instead of synchronously (Fig. 7b2). This modulation appeared to be mediated by recurrently activated inhibitory neurons that synapse on the fast trill neurons. Whether or not these inhibitory neurons are the equivalent of mammalian Renshaw cells is not known. Spinal Renshaw cells are not known to project to locomotor CPG neurons and silencing them pharmacologically in the cat (Pratt and Jordan 1987) or genetically in the mouse (Enjin et al. 2017) has minimal effect on the locomotor pattern.

8 Concluding Remarks

The finding that motoneurons can regulate central pattern generator function and project to several different interneuronal types raises several fascinating questions. First, because motoneuronal excitation of the CPG increases the firing of motoneurons, this could lead to runaway excitation of the CPG by positive feedback. Since such runaway excitation does not occur, it is likely that the excitation is balanced by inhibition. Currently only inhibitory Renshaw cells receive input from motoneurons, and their activity is not believed to significantly affect locomotion (Pratt and Jordan 1987; Enjin et al. 2017). However, in the Vglut2 knockout mouse, stimulation of a ventral root can inhibit and slow rhythmogenesis (Talpalar et al. 2011), potentially providing inhibitory feedback from motoneurons to the CPG. Whether this pathway can inhibit the CPG in the intact cord is not clear because electrical stimulation of the ventral roots accelerates the drug-induced locomotor rhythm (Machacek and Hochman 2006). It is possible, therefore, that motoneurons synapse with another class of inhibitory interneuron in addition to Renshaw cells as hypothesized by Hanson and Landmesser in their work on the developing mouse lumbar cord (Hanson and Landmesser 2003) (Fig. 3c).

The presence of motoneuronal projections to spinal interneurons including V3 (Chopek et al. 2018) and spinocerebellar interneurons (Chalif et al. 2022) raises the question of additional functions of recurrent excitation within spinal and ascending circuits. One potential function of such projections could be efference copy. It is well established that sensory feedback from muscle proprioceptors and joint receptors provides information about the movements produced by muscle contraction. The existence of direct feedback from motoneurons provides the efferent signal driving muscle contraction. Thus, the intended action (motoneuronal activity) and the actual action (sensory feedback from muscles) can be compared both within spinal circuits and/or remotely in the cerebellum (Wolpert and Miall 1996; Popa and Ebner 2018). It might be argued that central command coming from the CPG or from descending control systems might be sufficient for efference copy. However, given the multiplicity of inputs to motoneurons and the non-linear behavior of the motoneuron membrane, such signals would provide a poor representation of the firing behavior of the motoneuron and the command signal to muscle.

It is not known if motoneurons regulate locomotion in the adult rodent. This has not been addressed experimentally because of the challenges of studying the issue in the mature spinal cord. However, with the development of isolated sacral cord (Manuel et al. 2012) and decerebrate preparations of the adult mouse that can generate fictive locomotion (Nakanishi and Whelan 2012), it should be possible to answer the questions using opto- or chemo-genetics. In addition, we do not know if all classes of motoneuron are capable of modulating locomotor activity or whether it is restricted to a subset. Previous work has suggested that type S motoneurons or even gamma motoneurons might mediate the effects (Pujala et al. 2016), but definitive resolution of this issue must await the ability to selectively activate the different classes of motoneuron.

Finally, we must ask if motoneurons could be part of the central pattern generator for locomotion. If the CPG is a dedicated class of interneuron, such as ventral spinocerebellar neurons, obviously the answer is no. Current evidence suggests that several different neuronal classes, in addition to spinocerebellar neurons, contribute to locomotor rhythmogenesis in the neonatal mouse cord, including HB9 interneurons (Hinckley et al. 2005; Wilson et al. 2005; Kwan et al. 2009), Shox2 glutamatergic interneurons (Dougherty et al. 2013) and motoneurons (Falgairolle et al. 2017). Here, we hypothesize that many neuron classes participate in rhythmogenesis and the membership varies according to motor task and the state of the spinal networks. This is most likely to be true for drug-induced fictive locomotion because the drug cocktail used to induce locomotion contains NMDA which induces membrane potential oscillations in many spinal neurons including motoneurons (MacLean et al. 1997; Wilson et al. 2005). Consistent with this thinking, when the vesicular glutamate transporter VGluT2 is knocked out, rendering the great majority of glutamatergic neurons non-functional, a drug cocktail containing NMDA can still generate fictive locomotion. It has been suggested that NMDA-induced oscillations in 1a inhibitory interneurons, Renshaw cell and motoneurons may support the rhythm under these conditions – a very different neuronal cohort than when glutamatergic neurons are also present (Talpalar et al. 2011). Another piece of evidence in support of the idea of a fluid CPG comes from calcium imaging of neurons in the cut transverse face of the cord during fictive locomotion induced by NMDA and serotonin at different frequencies. As locomotor speed increases interneuronal recruitment in the ventral part of the cord shifted from lateral to medial with few neurons co-active at the different speeds (Rancic et al. 2020). Variability in the recruitment of interneurons during fictive locomotion is not restricted to the neonatal mouse and has also been observed in the adult mouse during successive bouts of treadmill locomotion (Pham et al. 2020). Using the FosTRAP mouse that allows a comparison between the neurons activated on two separate occasions it was shown that only 20% of the spinal neurons active in an episode of treadmill locomotion are also active in a second bout 2 weeks later (Pham et al. 2020). Whether this reflects a change in the membership of the CPG or is an artefact of the indirect measure of neuronal activity provided by Fos signaling is not clear at present and must await chronic single unit recordings or 2-photon imaging of single neurons over extended periods. Nevertheless, if the neonatal mouse CPG is organized with a varying

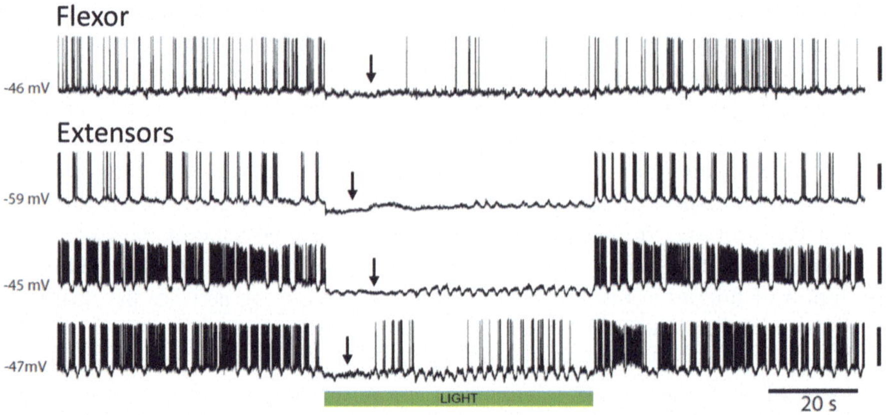

Fig. 8 Intracellular recording from motoneurons in 4 different cords in which ChAT⁺ neurons expressed archaerhodopsin, showing a transient cessation of rhythmic drive potentials during illumination with green light. The motoneurons are separated into flexor and extensor motoneurons according to the phase relation of their bursting with simultaneous ventral root recordings (not shown). The arrows indicate periods when the rhythmic drive potentials could not be detected. The calibration bars to the right of the records are 40 mV

cellular membership it may account for the apparent variability in the participation of motoneurons in rhythmogenesis.

The idea that motoneurons might participate directly in locomotor rhythmogenesis is suggested by a number of observations. First, optical hyperpolarization of motoneurons in either islet-1 or Chat mice expressing archaerhodopsin slows the locomotor rhythm and in some cases abolishes the rhythmic locomotor drive potentials recorded intracellularly from motoneurons (Fig. 8). Second, neonatal lumbar and adult sacral motoneurons exhibit TTX-resistant NMDA-induced membrane potential oscillations that are likely activated during locomotion. In the adult mouse sacral cord preparation, where motoneurons exhibit NMDA induced oscillations (Manuel et al. 2012), it was argued that these are unlikely to contribute to locomotion because motoneurons are not coupled by electrical or chemical synapses in the adult cord. As a result, in the absence of interneuronal drive, the activity of different motoneurons would not be synchronized. However, later work in both the neonate and the adult mouse spinal cord revealed that motoneurons are coupled by chemical excitatory synapses (Bhumbra and Beato 2018), rendering this objection moot.

References

Ampatzis K, Song J, Ausborn J, El Manira A (2014) Separate microcircuit modules of distinct v2a interneurons and motoneurons control the speed of locomotion. Neuron 83(4):934–943

Arai Y, Mentis GZ, Wu JY, O'Donovan MJ (2007) Ventrolateral origin of each cycle of rhythmic activity generated by the spinal cord of the chick embryo. PLoS One 2(5):e417

Bass AH, Marchaterre MA, Baker R (1994) Vocal-acoustic pathways in a teleost fish. J Neurosci 14(7):4025–4039

Beattie MS, Bresnahan JC, Mawe GM, Finn S (1987) Distribution and ultrastructure of ventral root afferents to lamina I of the cat sacral spinal cord. Neurosci Lett 76(1):1–6

Bhumbra GS, Beato M (2018) Recurrent excitation between motoneurones propagates across segments and is purely glutamatergic. PLoS Biol 16(3):e2003586

Bonnot A, Chub N, Pujala A, O'Donovan MJ (2009) Excitatory actions of ventral root stimulation during network activity generated by the disinhibited neonatal mouse spinal cord. J Neurophysiol 101(6):2995–3011

Borodinsky LN, Root CM, Cronin JA, Sann SB, Gu X, Spitzer NC (2004) Activity-dependent homeostatic specification of transmitter expression in embryonic neurons. Nature 429(6991):523–530

Brock LG, Coombs JS, Eccles JC (1952) The recording of potentials from motoneurones with an intracellular electrode. J Physiol 117(4):431–460

Cazalets JR, Sqalli-Houssaini Y, Clarac F (1992) Activation of the central pattern generators for locomotion by serotonin and excitatory amino acids in neonatal rat. J Physiol 455:187–204

Chagnaud BP, Baker R, Bass AH (2011) Vocalization frequency and duration are coded in separate hindbrain nuclei. Nat Commun 2(1):346

Chagnaud BP, Bass AH (2014) Vocal behavior and vocal central pattern generator organization diverge among toadfishes. Brain Behav Evol 84(1):51–65

Chagnaud BP, Perelmuter JT, Forlano PM, Bass AH (2021) Gap junction-mediated glycinergic inhibition ensures precise temporal patterning in vocal behavior. elife 10

Chalif JI, Martínez-Silva MDL, Pagiazitis JG, Murray AJ, Mentis GZ (2022) Control of mammalian locomotion by ventral spinocerebellar tract neurons. Cell 185(2):328–344.e326

Chopek JW, Nascimento F, Beato M, Brownstone RM, Zhang Y (2018) Sub-populations of spinal V3 interneurons form focal modules of layered pre-motor microcircuits. Cell Rep 25(1):146–156 e143

Chow BY, Han X, Dobry AS, Qian X, Chuong AS, Li M, Henninger MA, Belfort GM, Lin Y, Monahan PE, Boyden ES (2010) High-performance genetically targetable optical neural silencing by light-driven proton pumps. Nature 463(7277):98–102

Chub N, O'Donovan MJ (2001) Post-episode depression of GABAergic transmission in spinal neurons of the chick embryo. J Neurophysiol 85(5):2166–2176

Chung JM, Lee KH, Endo K, Coggeshall RE (1983) Activation of central neurons by ventral root afferents. Science 222(4626):934–935

Chung JM, Lee KH, Kim J, Coggeshall RE (1985) Activation of dorsal horn cells by ventral root stimulation in the cat. J Neurophysiol 54(2):261–272

Clifton GL, Coggeshall RE, Vance WH, Willis WD (1976) Receptive fields of unmyelinated ventral root afferent fibres in the cat. J Physiol 256(3):573–600

Coggeshall R (1979) Afferent fibers in the ventral root. Neurosurgery 4(5):443–448

Danner SM, Zhang H, Shevtsova NA, Borowska-Fielding J, Deska-Gauthier D, Rybak IA, Zhang Y (2019) Spinal V3 interneurons and left-right coordination in mammalian locomotion. Front Cell Neurosci 13:516

Dougherty KJ, Zagoraiou L, Satoh D, Rozani I, Doobar S, Arber S, Jessell TM, Kiehn O (2013) Locomotor rhythm generation linked to the output of spinal shox2 excitatory interneurons. Neuron 80(4):920–933

Eklof-Ljunggren E, Haupt S, Ausborn J, Dehnisch I, Uhlen P, Higashijima S, El Manira A (2012) Origin of excitation underlying locomotion in the spinal circuit of zebrafish. Proc Natl Acad Sci U S A 109(14):5511–5516

El-Gaby M, Zhang Y, Wolf K, Schwiening J, Christof OP, Shipton A, Olivia (2016) Archaerhodopsin selectively and reversibly silences synaptic transmission through altered pH. Cell Rep 16(8):2259–2268

Enjin A, Perry S, Hilscher MM, Nagaraja C, Larhammar M, Gezelius H, Eriksson A, Leao KE, Kullander K (2017) Developmental disruption of recurrent inhibitory feedback results in compensatory adaptation in the Renshaw cell-motor neuron circuit. J Neurosci 37(23):5634–5647
Falgairolle M, Puhl JG, Pujala A, Liu W, O'Donovan MJ (2017) Motoneurons regulate the central pattern generator during drug-induced locomotor-like activity in the neonatal mouse. elife 6
Gabriel JP, Mahmood R, Walter AM, Kyriakatos A, Hauptmann G, Calabrese RL, El Manira A (2008) Locomotor pattern in the adult zebrafish spinal cord in vitro. J Neurophysiol 99(1):37–48
Grillner S, Wallén P (1980) Does the central pattern generation for locomotion in lamprey depend on glycine inhibition? Acta Physiol Scand 110(1):103–105
Grillner S, Zangger P (1979) On the central generation of locomotion in the low spinal cat. Exp Brain Res 34(2)
Hall BK, Herring SW (1990) Paralysis and growth of the musculoskeletal system in the embryonic chick. J Morphol 206(1):45–56
Hamburger V (1990) The developmental history of the motor neuron. Neuroembryology 15:1–37
Hanson MG, Landmesser LT (2003) Characterization of the circuits that generate spontaneous episodes of activity in the early embryonic mouse spinal cord. J Neurosci 23(2):587–600
Hanson MG, Landmesser LT (2004) Normal patterns of spontaneous activity are required for correct motor axon guidance and the expression of specific guidance molecules. Neuron 43(5):687–701
Hildebrand C, Karlsson M, Risling M (1997) Ganglionic axons in motor roots and PIA mater. Prog Neurobiol 51(2):89–128
Hinckley CA, Hartley R, Wu L, Todd A, Ziskind-Conhaim, L (2005) Locomotor-Like Rhythms in a Genetically Distinct Cluster of Interneurons in the Mammalian Spinal Cord. J Neurophysiol 93(3):1439–1449. https://doi.org/10.1152/jn.00647.2004
Humphreys JM, Whelan PJ (2012) Dopamine exerts activation-dependent modulation of spinal locomotor circuits in the neonatal mouse. J Neurophysiol 108(12):3370–3381
Iles JF, Nicolopoulos-Stournaras S (1996) Fictive locomotion in the adult decerebrate rat. Exp Brain Res 109(3):393–398
Jiang Z, Carlin KP, Brownstone RM (1999) An in vitro functionally mature mouse spinal cord preparation for the study of spinal motor networks. Brain Res 816(2):493–499
Kiehn O, Iizuka M, Kudo N (1992) Resetting from low threshold afferents of N-methyl-D-aspartate-induced locomotor rhythm in the isolated spinal cord-hindlimb preparation from newborn rats. Neurosci Lett 148(1-2):43–46
Kudo N, Yamada T (1987) locomotor activity in a spinal cord-indlimb muscles preparation of the newborn rat studied in vitro. Neurosci Lett 75(1):43–48
Kwan AC, Dietz SB, Webb WW, Harris-Warrick RM (2009) Activity of Hb9 interneurons during fictive locomotion in mouse spinal cord. J Neurosci 29(37):11601–11613
Landmesser LT, O'Donovan MJ (1984) Activation patterns of embryonic chick hind limb muscles recorded in ovo and in an isolated spinal cord preparation. J Physiol 347(1):189–204
Lawton KJ, Perry WM, Yamaguchi A, Zornik E (2017) Motor neurons tune premotor activity in a vertebrate central pattern generator. J Neurosci 37(12):3264–3275
Li WC, Soffe SR, Roberts A (2004) Glutamate and acetylcholine corelease at developing synapses. Proc Natl Acad Sci U S A 101(43):15488–15493
Light AR, Metz CB (1978) The morphology of the spinal cord efferent and afferent neurons contributing to the ventral roots of the cat. J Comp Neurol 179(3):501–515
Machacek DW, Hochman S (2006) Noradrenaline unmasks novel self-reinforcing motor circuits within the mammalian spinal cord. J Neurosci 26(22):5920–5928
MacLean JN, Schmidt BJ, Hochman S (1997) NMDA receptor activation triggers voltage oscillations, plateau potentials and bursting in neonatal rat lumbar motoneurons in vitro. Eur J Neurosci 9(12):2702–2711
Manuel M, Li Y, Elbasiouny SM, Murray K, Griener A, Heckman CJ, Bennett DJ (2012) NMDA induces persistent inward and outward currents that cause rhythmic bursting in adult rodent motoneurons. J Neurophysiol 108(11):2991–2998

Mawe GM, Bresnahan JC, Beattie MS (1984) Primary afferent projections from dorsal and ventral roots to autonomic preganglionic neurons in the cat sacral spinal cord: light and electron microscopic observations. Brain Res 290(1):152–157

Maynard CW, Leonard RB, Dan Coulter J, Coggeshall RE (1977) Central connections of ventral root afferents as demonstrated by the HRP method. J Comp Neurol 172(4):601–608

Meehan CF, Grondahl L, Nielsen JB, Hultborn H (2012) Fictive locomotion in the adult decerebrate and spinal mouse in vivo. J Physiol 590(2):289–300

Mentis GZ, Alvarez FJ, Bonnot A, Richards DS, Gonzalez-Forero D, Zerda R, O'Donovan MJ (2005) Noncholinergic excitatory actions of motoneurons in the neonatal mammalian spinal cord. Proc Natl Acad Sci U S A 102(20):7344–7349

Milner LD, Landmesser LT (1999) Cholinergic and GABAergic inputs drive patterned spontaneous motoneuron activity before target contact. J Neurosci 19(8):3007–3022

Nakanishi ST, Whelan PJ (2012) A decerebrate adult mouse model for examining the sensorimotor control of locomotion. J Neurophysiol 107(1):500–515

Nishimaru H, Restrepo CE, Ryge J, Yanagawa Y, Kiehn O (2005) Mammalian motor neurons corelease glutamate and acetylcholine at central synapses. Proc Natl Acad Sci U S A 102(14):5245–5249

O'Donovan M, Ho S, Yee W (1994) Calcium imaging of rhythmic network activity in the developing spinal cord of the chick embryo. J Neurosci 14(11 Pt 1):6354–6369

Perrins R, Roberts A (1995) Cholinergic contribution to excitation in a spinal locomotor central pattern generator in Xenopus embryos. J Neurophysiol 73(3):1013–1019

Pham BN, Luo J, Anand H, Kola O, Salcedo P, Nguyen C, Gaunt S, Zhong H, Garfinkel A, Tillakaratne N, Edgerton VR (2020) Redundancy and multifunctionality among spinal locomotor networks. J Neurophysiol 124(5):1469–1479

Popa LS, Ebner TJ (2018) Cerebellum, predictions and errors. Front Cell Neurosci 12:524

Pratt CA, Jordan LM (1987) Ia inhibitory interneurons and Renshaw cells as contributors to the spinal mechanisms of fictive locomotion. J Neurophysiol 57(1):56–71

Provine RR, Sharma SC, Sandel TT, Hamburger V (1970) Electrical activity in the spinal cord of the chick embryo, in situ. Proc Natl Acad Sci 65(3):508–515

Pujala A, Blivis D, O'Donovan MJ (2016) Interactions between dorsal and ventral root stimulation on the generation of locomotor-like activity in the neonatal mouse spinal cord. eneuro 3(3):ENEURO.0101-0116

Rancic V, Ballanyi K, Gosgnach S (2020) Mapping the dynamic recruitment of spinal neurons during fictive locomotion. J Neurosci 40(50):9692–9700

Richards DS, Griffith RW, Romer SH, Alvarez FJ (2014) Motor axon synapses on Renshaw cells contain higher levels of aspartate than glutamate. PLoS One 9(5):e97240

Shin HK, Kim J, Nam SC, Paik KS, Chung JM (1986) Spinal entry route for ventral root afferent fibers in the cat. Exp Neurol 94(3):714–725

Smith JC, Feldman JL (1987) In vitro brainstem-spinal cord preparations for study of motor systems for mammalian respiration and locomotion. J Neurosci Methods 21(2-4):321–333

Song J, Ampatzis K, Bjornfors ER, El Manira A (2016) Motor neurons control locomotor circuit function retrogradely via gap junctions. Nature 529(7586):399–402

Talpalar AE, Endo T, Low P, Borgius L, Hagglund M, Dougherty KJ, Ryge J, Hnasko TS, Kiehn O (2011) Identification of minimal neuronal networks involved in flexor-extensor alternation in the mammalian spinal cord. Neuron 71(6):1071–1084

Toutant JP, Toutant MN, Renaud D, Le Douarin GH (1979) Enzymatic differentiation of muscle fibre types in embryonic Latissimus dorsii of the chick: effects of spinal cord stimulation. Cell Diff 8(5):375–382

Viala D, Buser P (1969) The effects of DOPA and 5-HTP on rhythmic efferent discharges in hind limb nerves in the rabbit. Brain Res 12(2):437–443

Wallen PLA (1984) Do the motoneurones constitute a part of the spinal network generating the Swimming rhythm in the lamprey? J Exp Biol 113(November):493–497

Wenner P (2014) Homeostatic synaptic plasticity in developing spinal networks driven by excitatory GABAergic currents. Neuropharmacology 78:55–62

Wenner P, O'Donovan MJ (1999) Identification of an interneuronal population that mediates recurrent inhibition of motoneurons in the developing chick spinal cord. J Neurosci 19(17):7557–7567

Wenner P, O'Donovan MJ (2001) Mechanisms that initiate spontaneous network activity in the developing chick spinal cord. J Neurophysiol 86(3):1481–1498

Whelan P, Bonnot A, O'Donovan MJ (2000) Properties of rhythmic activity generated by the isolated spinal cord of the neonatal mouse. J Neurophysiol 84(6):2821–2833

Wilson JM, Hartley R, Maxwell DJ, Todd AJ, Lieberam I, Kaltschmidt JA, Yoshida Y, Jessell TM, Brownstone RM (2005) Conditional rhythmicity of ventral spinal interneurons defined by expression of the Hb9 homeodomain protein. J Neurosci 25(24):5710–5719

Windle WF (1931) Neurons of the sensory type in the ventral roots of man and of other mammals. Arch Neurol Psychiatr 26(4):791

Wolpert DM, Miall RC (1996) Forward models for physiological motor control. Neural Netw 9(8):1265–1279

Yamamoto T, Takahashi K, Satomi H, Ise H (1977) Origins of primary afferent fibers in the spinal ventral roots in the cat as demonstrated by the horseradish peroxidase method. Brain Res 126(2):350–354

Zaporozhets E, Cowley KC, Schmidt BJ (2004) A reliable technique for the induction of locomotor-like activity in the in vitro neonatal rat spinal cord using brainstem electrical stimulation. J Neurosci Methods 139(1):33–41

Zhang F, Wang L-P, Brauner M, Liewald JF, Kay K, Watzke N, Wood PG, Bamberg E, Nagel G, Gottschalk A, Deisseroth K (2007) Multimodal fast optical interrogation of neural circuitry. Nature 446(7136):633–639

Zhang Y, Narayan S, Geiman E, Lanuza GM, Velasquez T, Shanks B, Akay T, Dyck J, Pearson K, Gosgnach S, Fan CM, Goulding M (2008) V3 spinal neurons establish a robust and balanced locomotor rhythm during walking. Neuron 60(1):84–96

Zornik E, Yamaguchi A (2012) Coding rate and duration of vocalizations of the frog, Xenopus laevis. J Neurosci 32(35):12102–12114

Extraocular Motoneurons and Neurotrophism

Angel M. Pastor, Roland Blumer, and Rosa R. de la Cruz

Abstract Extraocular motoneurons are located in three brainstem nuclei: the abducens, trochlear and oculomotor. They control all types of eye movements by innervating three pairs of agonistic/antagonistic extraocular muscles. They exhibit a tonic-phasic discharge pattern, demonstrating sensitivity to eye position and sensitivity to eye velocity. According to their innervation pattern, extraocular muscle fibers can be classified as singly innervated muscle fiber (SIF), or the peculiar multiply innervated muscle fiber (MIF). SIF motoneurons show anatomical and physiological differences with MIF motoneurons. The latter are smaller and display lower eye position and velocity sensitivities as compared with SIF motoneurons.

Keywords Oculomotor system · Vestibular system · Eye movements · Synaptic stripping · Trophic factors · Multiply-innervated fibers · Singly-innervated fibers

These motoneurons have been extensively used as a model for studies of lesion-induced plasticity and the effects of neurotrophic factors, demonstrating that different neurotrophic factors can regulate differentially their discharge mode and synaptic inputs. A link between neurotrophic factors and MIF and SIF motoneuronal types is proposed.

1 Introduction

The eye of vertebrates and many invertebrates can be mechanically described as a ball joint rotating around three perpendicular cartesian axis, the yaw, pitch and roll axes of rotation in space of, for instance, an aircraft. From the cyclopean crustacean

A. M. Pastor (✉) · R. R. de la Cruz
Departamento de Fisiología, Universidad de Sevilla, Seville, Spain
e-mail: ampastor@us.es

R. Blumer
Center of Anatomy and Cell Biology, Medical University of Vienna, Vienna, Austria

M. J. O'Donovan, M. Falgairolle (eds.), *Vertebrate Motoneurons*, Advances in Neurobiology 28, https://doi.org/10.1007/978-3-031-07167-6_12

Daphnia magna (Consi et al. 1987), to jumping spiders (Land 1969), and to all vertebrates (Walls 1962), the eyeball is rotated using six muscles arranged in push-pull antagonistic pairs. The yoked pairs are medial and lateral rectus for rotations around the vertical axis of the eye globe, the yaw axis. The medial and lateral rectus muscles produce adduction and abduction, respectively. The superior and inferior rectus muscles act for rotations around a transverse or pitch axis, and produce elevation and depression of the angle of sight, respectively. The last pair, superior and inferior oblique muscles produce rotations around the longitudinal or roll axis, the axis of sight, and thus, the movements are called intorsions and extorsions, respectively. Except for the lateral and the medial rectus muscles, which generate purely horizontal eye movements, the other four extraocular muscles (vertical and oblique) produce, in addition to their primary action, secondary actions which differ between frontal-eyed and lateral-eyed animals (Graft and Simpson 1981).

This mechanical scheme is the practical solution of convergent evolution across species despite certain variations in insertions and innervation (Isomura 1981). There are five main types of eye movements in vertebrates. Three types of eye movements are gaze shifting and defined as saccadic, smooth pursuit and vergence eye movements that either point, follow or maintain binocularity, respectively. In turn, two types of gaze holding movements, the optokinetic and vestibular reflexes, compensate for self or externally imposed movements (Walls 1962). Extraocular eye muscles participate in all types of eye movements with slight differences in the kinematics across vertebrates. Perhaps the most fundamental difference with the skeletal muscle activation patterns is that the oculomotor muscles pull against a constant inertial load as opposed to the always changing moment of inertia of skeletal muscles.

Extraocular muscles are innervated by motoneurons located in the brainstem, i.e., in the oculomotor, the trochlear and the abducens nuclei. The midbrain oculomotor nucleus comprises the motoneurons that innervate ipsilaterally the medial rectus, inferior rectus, and inferior oblique muscles, and contralaterally the superior rectus muscle. The axons of oculomotor motoneurons form the IIIrd cranial nerve. The trochlear motoneurons (in the caudal midbrain) innervate contralaterally the superior oblique muscle and their axons constitute the IVth cranial nerve. The abducens nucleus, located in the pons, contains the motoneurons that innervate ipsilaterally the lateral rectus muscle. Their axons form the VIth cranial nerve (Spencer and Porter 2006).

2 Discharge Characteristics of Extraocular Motoneurons

The neurophysiological study of oculomotor nuclei was initially addressed to the quantitative analysis of motoneuronal behavior during the vestibulo-ocular reflex. Such work was carried out in anesthetized animals, therefore limiting the repertoire, amplitude and velocity of eye movements in relation to the dynamics, frequency response and patterns of motoneuronal behavior (Lorente de No 1933; Precht et al.

1967; Yamanaka and Bach-Y-Rita 1970). The development of extracellular single-unit recordings in awake animals allowed rapid progress in the knowledge of extraocular motoneuron discharge characteristics during the performance of spontaneous or induced eye movements. In parallel, the development of a precise method for the measurement of eye movements was also crucial for a better understanding of the correlations between firing rate and eye movements. Eye movements were initially recorded by electrooculography, and later on by the most sensitive magnetic-search coil technique (Fuchs and Robinson 1966).

Since eye movements are quite stereotyped in nature, likely because, unlike most skeletal muscles, they deal with the same mechanical load throughout life, and since motoneuron discharge rate is to some extent translated into muscle force, (Goldberg et al. 1998; Miller et al. 2002) and then translated into eye movement, (Ghasia and Angelaki, 1995), it could be anticipated that the discharge pattern of extraocular motoneurons would reflect eye movements (Robinson 1970). Indeed, extraocular motoneurons exhibit a discharge combined of static and dynamic components of different proportions and forming a continuum between groups (Henn and Cohen 1972). The tonic and phasic components of motoneuron discharge translate directly into eye position and eye velocity (Evinger and Baker 1991). The tonic-phasic discharge pattern of extraocular motoneurons has been demonstrated by single-unit extracellular recordings of these motoneurons under alert conditions, carried out initially in monkeys in the abducens (Schiller 1970; Keller and Robinson 1972; Skavenski and Robinson 1973; Goldstein and Robinson 1986; Fuchs et al. 1988), oculomotor (Robinson 1970; Schiller 1970) and trochlear (Fuchs and Luschei 1971) nuclei. These experiments were later carried out in other vertebrates, such as cat abducens (Goldberg 1980; Delgado-García et al. 1986; Davis-López de Carrizosa et al. 2011) and medial rectus (de la Cruz et al. 1990; Hernández et al. 2017a) motoneurons, frog abducens (Dieringer and Precht 1986; Dietrich et al. 2017), goldfish abducens and medial rectus motoneurons (Pastor et al. 1991) and rabbit abducens motoneurons (Stahl and Simpson 1995). In all cases, the discharge of extraocular motoneurons was characterized by a tonic-phasic firing pattern (Fig. 1a). That is, firing rate contained a sustained (tonic, proportional to eye position) and a dynamic component (phasic, related to eye velocity), during the different types of eye movements. This fact led to the early conclusion that only one functional type of extraocular motoneuron subserves the whole repertoire of eye movements, including fixations, saccades and smooth pursuit (Fuchs and Luschei 1971). Later studies showed however that motoneurons could be classified in clusters (Dieringer and Precht 1986; Davis-López de Carrizosa et al. 2011; Dietrich et al. 2017; Hernández et al. 2019). In the absence of genetic and embryological separations of motoneurons into classes, all these sorting into functional groups are just the phenomenological and semantic classification into groups of high, medium and low sensitivities and/or thresholds of motoneuronal pools, that actually are part of a continuous gradient. This has been the recurrent topic at the time of describing, and thus classifying, motoneurons into different functional groups ever since the development of this field (Henn and Cohen 1972).

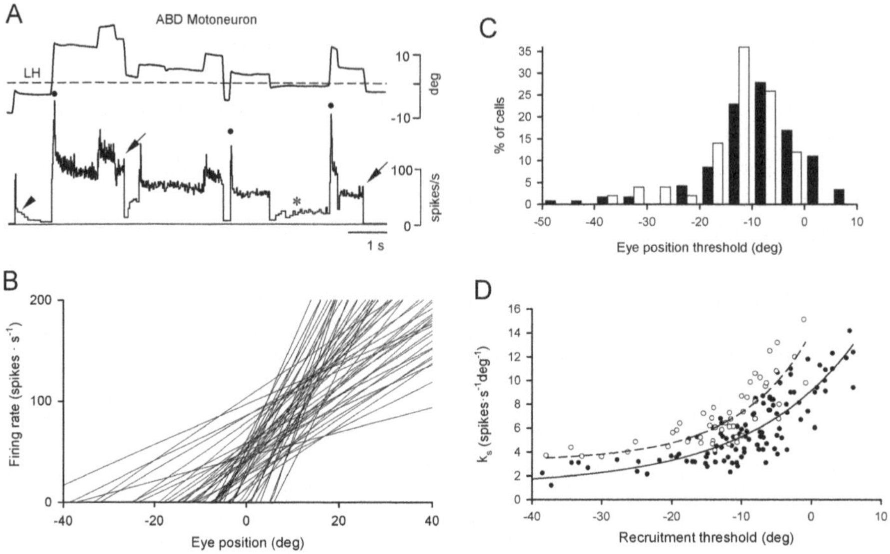

Fig. 1 Firing and pool characteristics of abducens neurons. (**a**) Control abducens motoneurons discharge tonically (firing rate, FR, in spikes/s) in relation to eye position (LH, left horizontal eye position, in degrees; positive values are movements to the left) during spontaneous fixations. Bursts of action potentials were present during saccades in the on direction (•) and pauses (→) occurred during off-directed saccades. Abducens neurons showed minimal firing (10–15 spikes/s) near recruitment eye positions reached after on- (►) or off-directed (*) saccades. Dashed line indicates mid eye position. (**b**) Linear regression lines of firing rate versus eye position for 50 control motoneurons selected throughout the oculomotor range. Scatterplots omitted for the sake of clarity but see examples in Fig. 5. The slope of each line represents the eye position sensitivity (k_s) during spontaneous eye movements. (**c**) Histograms of recruitment threshold for 118 motoneurons (filled bars) and 50 internuclear neurons (empty bars). (**d**) Plots of the eye position sensitivity k_s (in spikes$\cdot$s$^{-1}\cdot$deg.$^{-1}$) versus the recruitment threshold for 118 abducens motoneurons (•) and 50 internuclear neurons (○). Lines are the exponential regression y = 1.19 + 7.94 * exp. (0.06 x) (r = 0.84; $P < 0.001$) for motoneurons (——) and y = 3.23 + 10.8 * exp. (0.09 x) (r = 0.82; $P < 0.001$) for internuclear neurons (- - -). Note in both curves that sensitivities for a given threshold are always larger for internuclear neurons. (Modified with permission from Pastor and González-Forero 2003)

3 Quantitative Analysis of Extraocular Motoneuron Firing Rate

The pioneering studies of extraocular motoneurons in awake animals allowed the mathematical correlation between discharge frequency and eye dynamics. Robinson (1970, 1981) established that motoneuronal firing rate (FR, in spikes/s) depends linearly on both eye position (EP, in degrees) and eye velocity (EV, in degrees/s), and therefore a first-order approximation of the FR, with a single time constant, to eye parameters represent a good fit and it has been universally used since that time. Thus, the following equation was proposed to adjust FR to EP and EV:

$$FR = F_0 + k \cdot EP + r \cdot EV,$$

where F_0 represents the firing rate at straight ahead gaze (i.e., at zero eye position), k represents the neuronal eye position sensitivity (in spikes/s/degree) and r the neuronal eye velocity sensitivity (in spikes/s/degree/s). During eye fixations, since EV = 0, then the equation can be expressed as FR = F_0 + k · EP (Fig. 1b). From this equation, another important parameter can be calculated, the neuronal recruitment threshold, i.e., the eye position at which the motoneuron is recruited into activity (Fig. 1b). The estimated threshold is resolved as the abscissa intercept by making FR = 0 and then the threshold eye position is obtained, according to the eq. FR = F_0 + k · EP, as $-F_0/k$.

Abducens internuclear neurons, that convey information to the contralateral medial rectus motoneurons to keep conjugacy of eye movements also fit this scheme of firing pattern but will not be the main object of this chapter. Most of the abducens neurons are recruited well before the eye reaches its center position (Fig. 1c).The first-order model, described in the equation above, is simple and straightforward to quantitate the transfer function of the firing to the muscle dynamics, considered as a single time constant of a single Voight element (a spring in parallel to a dashpot to account for the elasticity and viscosity of the muscle, respectively). However, it was later recognized that higher order dynamics explain better the orbital mechanics that originally was proposed as the sum of at least two exponentials, with slow and fast time constants (Robinson, 1981). The newer models made use of higher order derivatives of firing rate to account for more than one time constant in response to a step-change in eye position (Fuchs et al. 1988; de la Cruz et al. 1990; Stahl and Simpson 1995).

Correlations between these parameters reflect an important and well known motoneuron pool property, the recruitment order. Thus, an interesting relationship described in the literature for extraocular motoneurons (mainly in abducens motoneurons) is that motoneurons with higher k values tend to have higher eye position thresholds. Notice the progressive increase in slope of the lines represented in Fig. 1b. These data, which is plotted in Fig. 1d for both abducens motoneurons and abducens internuclear neurons indicate that the higher k value neurons recruited at more eccentric eye positions in the on-direction (Robinson 1970; Schiller 1970; Delgado-García et al. 1986; Fuchs et al. 1988; Pastor et al. 1991). The significance of this universal recruitment rule of oculomotor motoneurons is that as the eye is deviated more eccentrically in the orbit, the force required to hold the eye increases exponentially as does the motor output of the nerve, i.e., the total sum of action potentials orthodromically sent through the axons (Pastor and González-Forero 2003; Davis-López de Carrizosa et al. 2011).

Similarly, a positive relationship between neuronal eye velocity sensitivity (r) and recruitment threshold has also been described. Indeed, motoneurons with high eye position sensitivities generally present higher eye velocity sensitivities, which means that those units with a marked tonic component also have a prominent phasic discharge. This rule can be also generalized to other sensitivities obtained during other type of eye movements such as the vestibulo-ocular reflex, so that motoneurons can be ranked according to sensitivity values across different modalities (Pastor and González-Forero 2003; Davis-López de Carrizosa et al. 2011).

In experiments on antidromically identified motoneurons, a significant relationship has been found between antidromic latency and eye position sensitivity, both in cat (Delgado-García et al. 1986) and fish (Pastor et al. 1991) but not monkey (Fuchs et al. 1988). In the cat, this relationship is lost under tetanus neurotoxin treatment indicating that the synaptic inputs that are responsible for the eye position signal are arranged so the size principle is attained (Pastor and González-Forero 2003). Antidromic latency is negatively related to conduction velocity (i.e., the higher the latency the lower the conduction velocity), and conduction velocity provides a basis for inferring cell size, as motoneurons with greater soma size usually have also axons with larger diameter and, consequently, higher axonal conduction velocity. Altogether, it seems that larger motoneurons would have lower antidromic latencies and these would correspond to motoneurons with higher sensitivities. If so, these relationships would imply that extraocular motoneurons follow the "size principle" proposed by Henneman et al. (1965a, b, 1981), contending that hierarchical organization of spinal motoneuronal pools are based on soma size with a rank order that would fit the threshold of each cell in relation to the others. This rank order would allow a fine graduation in muscle tension by successive recruitment of larger motoneurons as muscle force increases.

4 Firing Pattern of Abducens Motoneurons

The discharge activity of extraocular motoneurons reflects the eye position and eye velocity evoked by the activation of the corresponding innervated muscle. Experiments of single-unit recordings in awake animals have demonstrated that trochlear (Fuchs and Luschei 1971), medial rectus (de la Cruz et al. 1990; Pastor et al. 1991; Hernández et al. 2017) and abducens motoneurons (Fuchs and Luschei 1970; Schiller 1970; Delgado-García et al. 1986; Fuchs et al. 1988; Pastor et al. 1991; Stahl and Simpson 1995; Davis-López de Carrizosa et al. 2011) exhibit a tonic-phasic firing pattern in relation to oculomotor performance. However, it should be emphasized that abducens motoneurons have been studied more extensively than other motoneurons of the oculomotor system and, therefore, constitute the extraocular motoneurons in which correlations of discharge with eye movements have been analyzed in more detail and in a more numerous number of vertebrate species. Therefore, we will describe the discharge pattern of these particular motoneurons as a general model for extraocular motoneurons.

Discharge of Abducens Motoneurons During Fixations

Fixations consist of stable eye positions when gaze is stationary on a particular visual target. Abducens motoneurons innervate ipsilaterally the lateral rectus muscle, whose contraction induces a purely horizontal temporally-directed eye movement. Therefore, abducens motoneurons will have, as their on-direction, the horizontal eye movement in the ipsilateral direction with respect to the side in which they are located, for instance, leftwards horizontal eye movements will be the on-direction of left abducens motoneurons.

Abducens motoneurons, and in general abducens internuclear neurons too, discharge a sustained firing rate during eye fixations which increases as the eye deviate successively in the on-direction, i.e., for more eccentric eye positions. Firing rate increases proportionally to horizontal eye position, so that a linear relationship can be established between firing rate and eye position, as stated above ($FR = F_0 + k \cdot EP$). For each motoneuron, firing frequency and eye position data can be fitted by a linear regression line whose slope represents the eye position sensitivity for that particular neuron (Fig. 1b). Abducens motoneurons, and in general all extraocular motoneurons, discharge high tonic firing frequencies (maximum 200–300 Hz) during fixations as compared to spinal motoneurons (Robinson 1970; Delgado-García et al. 1986; Fuchs et al. 1988).

However, the tonic discharge of motoneurons exhibit a certain degree of irregularity (Powers and Binder 2000; González-Forero et al. 2002). Discharge variability has been calculated by measuring the mean and standard deviation of instantaneous firing frequency (reciprocal of interspike intervals) during a set of multiple fixations across the oculomotor field. Then, the coefficient of variation is obtained as the ratio of standard deviation to the mean. The coefficient of variation is higher at lower frequencies but decreases at higher frequencies in a gradual exponential-like fashion. In general, the tonic discharge of abducens motoneurons show low coefficients of variations (i.e., small irregularity) that ranged from 3.5% to 14% (Delgado-García et al. 1986) with an average of around 8% (Davis-López de Carrizosa et al. 2010).

The major input to abducens motoneurons responsible of the generation of an eye position signal arises in the prepositus hypoglossi nucleus, which is located in the pons caudal to the abducens nucleus. The projection from the ipsilateral prepositus hypoglossi neurons on abducens motoneurons is excitatory, whereas the prepositus hypoglossi afferent is inhibitory on the contralateral abducens motoneurons (McCrea and Baker 1985; Escudero et al. 1992).

Discharge of Abducens Motoneurons During Saccades

Saccades are rapid excursions of the eyes across the visual field normally with the purpose to foveate an object of interest. Abducens motoneurons increase their activity during saccades in the on-direction (ipsilateral to the recording site) and decrease abruptly or even ceased their firing for off-directed saccades. This behavior corresponds to the phasic or dynamic component of extraocular motoneuron discharge pattern. The burst precedes eye movement a few milliseconds, most of this time is involved in muscle contraction (approximate range 5–25 ms; Delgado-García et al. 1986). The firing frequency during on-directed saccades increases with the velocity of the saccade and can reach a peak of approximately 250–300 Hz in the cat (Davis-López de Carrizosa et al. 2011) or even more than 400 Hz in the monkey (Robinson 1970).

For on-directed saccades, firing rate increases quickly up to a maximum that occurs coincident or slightly before the peak eye velocity, in the form of a high-frequency burst of action potentials. This peak in firing frequency is followed by an exponential firing decrease, reaching a new tonic discharge proportional to the new eye position. The transition between peak firing frequency and the later tonic discharge is known as the post-saccadic slide (Fuchs and Luschei 1970; Delgado-García et al. 1986). However, the position of the eye changes during the saccade in a pulse-step manner (i.e., without any exponential eye position drift). So the post-saccadic decay is present in motoneuron firing but not in eye position. Experiments in which the tension of the lateral rectus muscle has been recorded simultaneously with neuronal activity and eye position, have demonstrated that the tension profile does reflect the post-saccadic decay (Davis López de Carrizosa et al. 2011). Therefore, although the meaning of the post-saccadic decay is still unclear, it reflects the transition in force from the pulse to the step that is not seen in the eye position trace. That excess of force may be necessary to overcome the viscoelastic properties of the oculomotor plant.

To correlate, for a particular motoneuron, peak firing rate with peak eye velocity during saccades, the firing rate corresponding to the position previous to the saccade has to be subtracted from the peak firing rate during each saccade, once the eye position sensitivity (k) has been obtained for that particular neuron. According to the equation indicated above ($FR = F_0 + k \cdot EP + r \cdot EV$), the fit would be $FR - k \cdot EP = F_0 + r \cdot EV$. In this way, the slope of the rate-velocity regression line represents the neuronal eye velocity sensitivity (r). Alternatively, firing rate may be correlated to eye position and velocity using the multiple regression approach ($FR = F_0 + k \cdot EP + r \cdot EV$), and thus both sensitivities (k and r) are obtained simultaneously in the fit.

The main input to abducens motoneurons responsible for the phasic response of these cells during saccades arises in the ponto-medullary reticular formation. In particular, the so-called excitatory burst neurons (known as EBNs) discharge in a burst-like manner preceding on-directed saccades and contact ipsilaterally with abducens motoneurons by means of excitatory synapses. On the other hand, the inhibitory burst neurons (IBNs) project contralaterally to abducens motoneurons exerting an inhibitory synaptic influence. IBNs discharge also a burst of spikes but in this case preceding off-directed saccades (Hikosaka et al. 1978; Igusa et al. 1980; Strassman et al. 1986a, 1986b) (Fig. 2a).

Discharge of Abducens Motoneurons During Vestibularly-Induced Eye Movements

The vestibulo-ocular reflex is performed by the three neuron arc that includes: (i) the semicircular canal hair cells which senses head acceleration during rotation and constitute the mechanoreceptive element of the reflex, transducing head movement into a local membrane potential that leads to neurotransmitter release; (ii) the first-order vestibular neurons or primary vestibular afferents are bipolar cells whose cell bodies are located in Scarpa's ganglion, are excited by the hair cells in their peripheral axonal projection and terminate centrally in the medial and lateral vestibular nuclei of the brainstem; and (iii) the third element of the reflex are the efferent neurons, that is, the extraocular motoneurons mediating the horizontal vestibulo-ocular reflex (Fig. 2a). These are abducens and medial rectus motoneurons. Abducens motoneurons and abducens internuclear neurons, that can be identified antidromically from the abducens nerve and medial longitudinal fascicle, respectively, (Fig. 2b, c) are innervated by the second-order medial vestibular neurons, whereas medial rectus motoneurons of the oculomotor complex receive their vestibular input through the ascending tract of Deiters' arising in the ipsilateral lateral vestibular nucleus. Abducens motoneurons are innervated ipsilaterally by inhibitory vestibular neurons and contralaterally by excitatory vestibular neurons (Highstein and Holstein 2006; Fig. 2c). However, this "apparently simple" three-element circuit belongs to an intricate network of (mainly) other brainstem and cerebellar neurons that modulate the reflex in a complex processing of the vestibulo-oculomotor information (Pastor et al. 2019).

The vestibulo-ocular reflex in the horizontal plane is induced by the rotation around the vertical axis (generally by sinusoidal waves) of the rotating table or the primate chair where the animal is seated. Table rotation induces, in head-fixed animals, the simultaneous rotation of the head which, in turn, leads to compensatory eye movements approximately at the same velocity but in opposite direction to head velocity. These compensatory eye movements prevent visual images from sweeping across the retina too quickly thus avoiding blurred vision. These so-called slow phases of the nystagmus are interrupted by short-duration fast phases that reset the eye to its central position. Altogether, this compensatory eye response to head rotation is known as the vestibulo-ocular reflex. Control abducens neurons modulate their activity in relation to slow and fast phases of the vestibulo-ocular reflex (Fig. 3a). Thus, the firing rate profile showed a sinusoidal modulation during the slow phases of the nistagmus (Fig. 3a), and bursts or pauses during the on- or

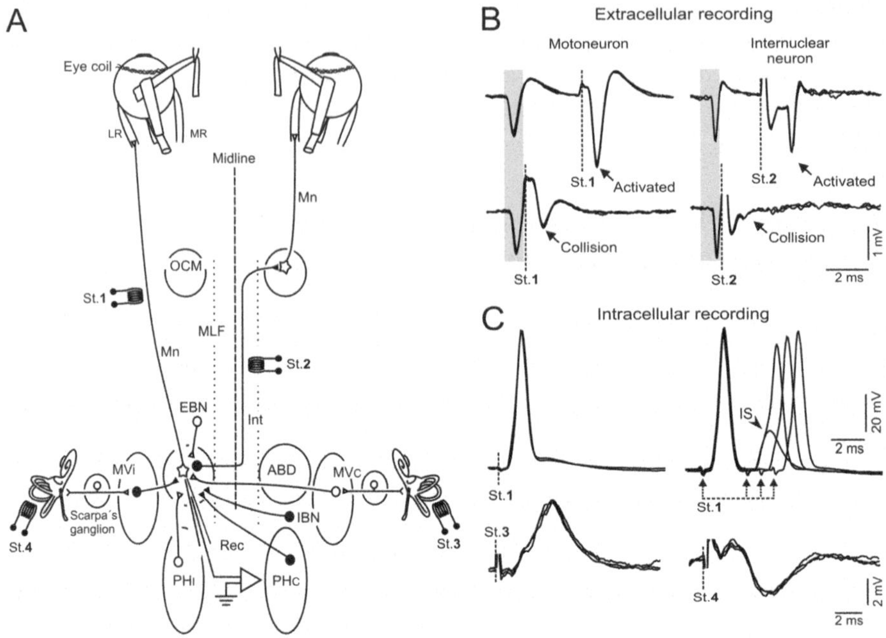

Fig. 2 Recording and identification of abducens neurons. (**a**) Diagram of the experimental design for chronic and acute recordings (Rec). The abducens nucleus (ABD) contains motoneurons (Mn) that innervate the lateral rectus muscle (LR) and internuclear neurons (Int) whose axons cross the midline and contact with the medial rectus muscle (MR) innervating motoneurons (Mn) of the oculomotor nucleus (OCM). Eye movements were recorded using eye coils implanted bilaterally. Bipolar stimulating electrodes were implanted on the VIth nerve (St. 1) and on the medial longitudinal fascicle (MLF; St. 2) for the identification of abducens motoneurons and internuclear neurons, respectively. For acute recordings, two additional electrodes were implanted near the ampulla of the horizontal contralateral (St. 3) and ipsilateral (St. 4) semicircular canals that activate or inhibit disynaptically abducens neurons through second order vestibular neurons (MVc and Mvi, respectively). (**b**) Antidromic activation and collision test of a control motoneuron (left) and internuclear neuron (right) during chronic extracellular recording. Upper traces show the spontaneous orthodromic spike (gray band) followed by the antidromic activation evoked by single shock application to the VIth nerve or MLF (---; St. 1 or St. 2, respectively). By shortening the interval between the orthodromic spike and the antidromic stimulus, antidromic spike occluded (lower traces; collision). (**c**) Antidromic activation (upper traces) and vestibular synaptic potentials (lower traces) in abducens neurons recorded intracellularly. Single (left) or double shock (right) stimulation of the VIth nerve (St. 1, ---) evoked the generation of antidromic spikes in control motoneurons. When minimal intervals were used during double stimulation, a clear dissociation of the initial-segment (IS) and somatodendritic components of the second antidromic spike was present. EPSP (left) and IPSP (right) recorded intracellularly in a control abducens motoneuron following electrical stimulation of the contralateral (St. 3) and ipsilateral (St. 4) vestibular labyrinths, respectively. (Modified with permission (**a** and **b**) from González-Forero et al. 2002; (**c**) unpublished data)

off-directed fast phases of the reflex, respectively. Analysis of the neuronal activity during vestibular stimulation can be correlated to eye movements by selecting regions of slow phases (Fig. 3b–e). Firing rate is proportional to both eye position

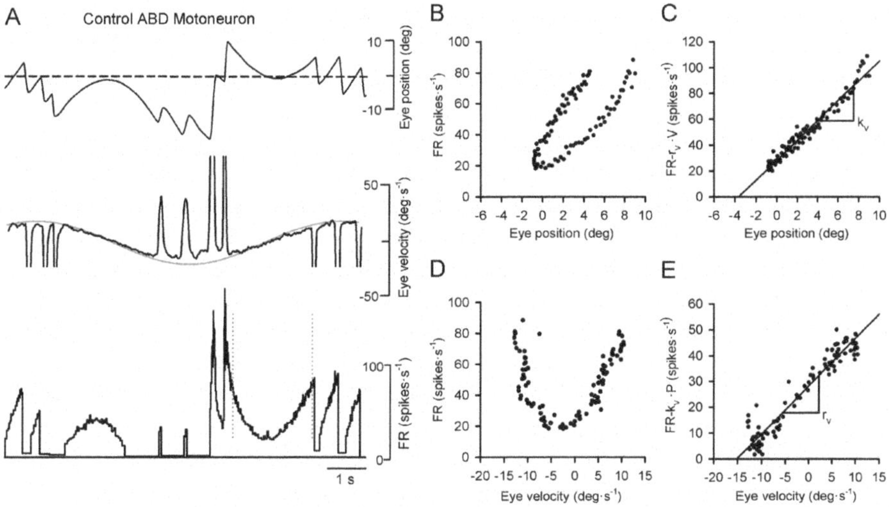

Fig. 3 **Calculation of dynamic sensitivities during vestibular stimulation**. (**a**) Firing of a control motoneuron during a complete cycle of sinusoidal head rotation in the dark at 0.1 Hz. From top to bottom, traces are position of the left eye, eye velocity and the inverted head velocity superimposed, and histogram of neuronal instantaneous firing rate (FR). Dashed lines indicate the epoch used for analysis. (**b**) Scatterplot of FR versus eye position. (**c**) Partial regression plot of the data shown in **b** after subtraction of the firing component due to eye velocity, the scatterplot in **b** collapsed to a line. The slope of the linear regression line corresponds to the neuronal sensitivity to eye position during vestibular stimulation (k_v = 7.69 spikes.s^{-1}.deg.$^{-1}$, r = 0.98; $P < 0.001$). (**d**) Scatterplot of FR versus eye velocity for the same data set as in **b**. (**e**) After subtraction of the component of firing due to eye position, the scatterplot shown in **d** collapsed closer to a line. The slope of the linear regression line represents the neuronal sensitivity to eye velocity during the slow phases of the vestibulo-ocular reflex (r_v = 1.85 spikes.s^{-1}.deg.$^{-1}$.s^{-1}, r = 0.96; $P < 0.001$)

and velocity, so firing-eye position and firing-eye velocity plots do not render true sensitivities but circularization due to the multiple dependence (Fig. 3b, d). It is necessary then to perform, multiple regression analysis to extract neuronal position (k_v) or velocity (r_v) sensitivities, as can be seen in the partial regression plots where data collapse approximately into a straight line (Fig. 3c, e).

5 MIF and SIF Extraocular Motoneurons: Is There a Functional Segregation?

Singly and Multiply Innervated Muscle Fibers

Mammalian extraocular muscles have the peculiarity of containing two main different morphological and functional types of muscle fibers. According to their innervation pattern, extraocular muscle fibers can be singly innervated (SIF, the most frequent type) or multiply innervated (MIF). In contrast, muscle cells in mammalian skeletal muscles comprises only SIFs. The pattern of motoneuronal axonal innervation of these two types of muscle fiber differs. SIFs are innervated by the typical single end-plate (*en plaque* terminal) located at the middle third of the muscle (Fig. 4a, c), and are fibers of large size innervated by axons of large diameter, as happens in general in the skeletal musculature (Browne 1976; Bondi and Chiarandini 1979; Chiarandini and Stefani 1979). However, MIFs are small muscle fibers, less numerous, and are innervated by thinner axons, which establish multiple synaptic boutons *en grappe* extending along the whole length of the fiber (Fig. 4b, d) (Hess and Pilar 1963; Namba et al. 1968). SIFs and MIFs also operate differently. Thus, the electrical stimulation of SIFs leads to propagated impulse activity with non-graded contractile activation and a twitch-like response in tension. However, MIFs usually do not display impulse activity and undergo a graded contraction dependent upon the extent of membrane depolarization. The change in tension produced by MIFs stimulation is of low amplitude and slow time course, so that MIFs are also called non-twitch or slow muscle fibers (Hess and Pilar 1963; Browne 1976). Functional differences correlate with ultrastructural dissimilarities. Thus, SIFs have well-defined myofibrils and a highly developed sarcoplasmic reticulum, as opposed to MIFs (Davidowitz et al. 1996). Moreover, the scenario is much more complex since MIFs and SIFs can be present either in the global (near the eyeball) or in the orbital (next to the periorbital tissue) layers of mammalian extraocular muscles, in which at least six types of muscle fibers have been described (for review see Spencer and Porter 2006; Hoh 2021).

Amphibians are the group in which slow muscle was first recognized and from which there is more detail information available. Nonetheless, teleosts, reptiles and birds have some muscles that have many of the features of frog slow muscle. In contrast, in mammals slow fibers (MIF) are confined to extraocular muscles, intrinsic muscles of the ear, and some muscles of the esophagus (Morgan and Proske 1984). It should be emphasized that MIF fibers are quite different from the S-type fibers of the limb muscles.

It has been proposed that the differences in structural and physiological characteristics between MIFs and SIFs in mammalian extraocular muscles lead to different contributions of these two types of muscle fiber during the various types of eye movements (Büttner-Ennever et al. 2001; Ugolini et al. 2006). MIFs of mammalian extraocular muscles are, in addition, very resistant to fatigue and during repetitive stimulation of their innervating axons, virtually do not exhibit any diminution in tension. Therefore, MIF features has led to propose that these fibers might contribute to the holding system required during eye fixations, and to smooth and damp the action of antagonistic muscles during fixation (Browne 1976). Along these lines, twitch fibers (SIFs) might give rise mainly to the phasic responses of the oculomotor behaviour, whereas non-twitch fibers (MIFs) would generate tonic tension (Chiarandini 1976).

At the muscle-tendon junction of multiply innervated muscle fibers, a unique nerve specialization, the palisade ending, has been found in the eye muscles of all mammalian species with the exception of rodents (Ruskell 1978; Alvarado Mallart

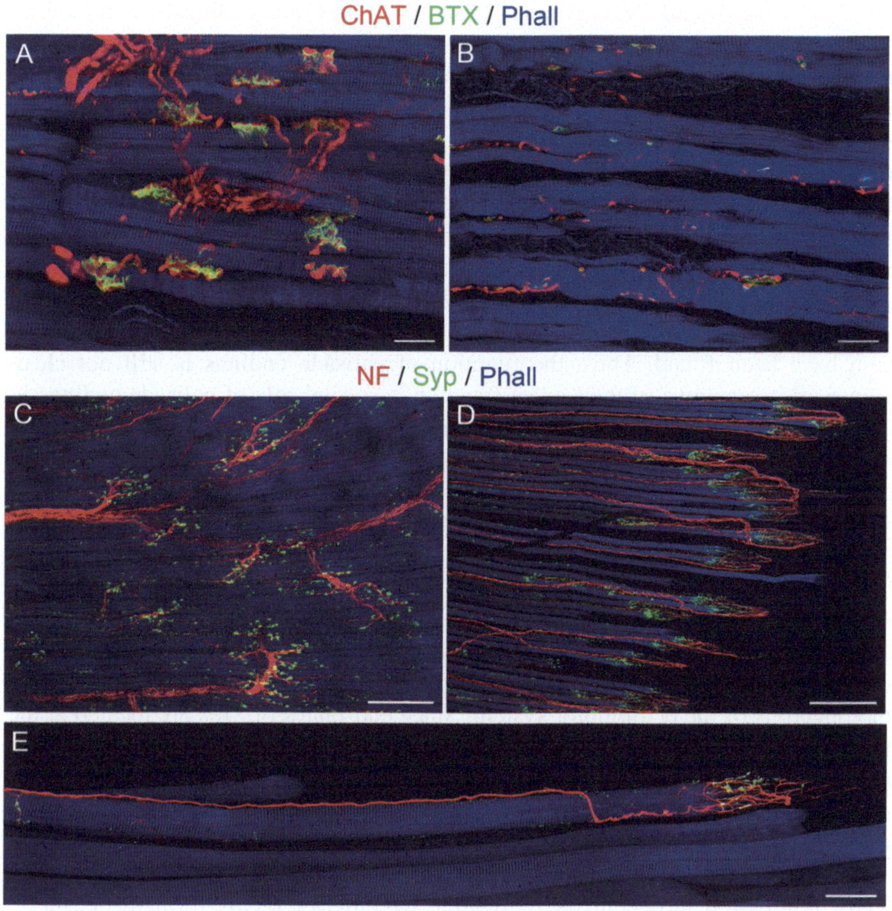

Fig. 4 **Innervation of lateral rectus muscle**. Projection images from the confocal laser scanning microscope. Images (**a–d**) are from the medial rectus muscle and **E** from the lateral rectus muscle of the cat. (**a, b**) Nerve fibers are labeled with anti-choline acetyltransferase (ChAT, red), motor terminals with α-bungarotoxin (BTX, green) and muscle fibers with phalloidin (Phall, blue). Showing *en plaque* motor terminals of singly innervated muscle fibers (**a**) and multiple *en grappe* motor terminals of multiply innervated muscle fibers (**b**). (**c, d, e**) Nerve fibers are labeled with anti-neurofilament (NF, red), nerve terminals with anti-synaptophysin (Syp, green) and muscle fibers with phalloidin (Phall, blue). (**c**) Showing *en plaque* motor terminals in the muscle belly. (**d**) Showing palisade endings at the muscle tendon junction of multiply-innervated muscle fiber. Terminal varicosities of palisade endings and *en grappe* motor terminal alongside multiply innervated muscle fibers exhibit synaptophysin immunoreactivity. The tendon, unlabeled, continues the muscle fibers to the right of the panel. (**e**) Showing a palisade ending which is formed by a nerve fiber that establishes multiple *en grappe* terminals alongside the muscle fiber. Terminal varicosities and en grappe terminals are synaptophysin-positive. The tendon, not labeled, continues the muscle fibers to the right of the image. Scale bars: 20 μm (**a, b**), 200 μm (**c, d**), and 50 μm (**e**)

and Pincon Raymond 1979; Blumer et al. 2016). Palisade endings are formed by axons that come from the muscle and extend into the tendon where they return towards the muscle-tendon junction to establish terminal varicosities around single muscle fiber tips (Fig. 4d, e). Molecular studies have shown that palisade endings are cholinergic structures (Konakci et al. 2005; Blumer et al. 2009) and have an exocytotic machinery for neurotransmitter release, most likely acetylcholine because of their cholinergic phenotype (Blumer et al. 2020). Additionally, neuronal tracing studies indicate that palisade endings originate from the oculomotor nuclei, most likely from MIF motoneurons (Lienbacher et al. 2011; Zimmermann et al. 2013). Despite clear motor features, palisade endings have sensory-like terminal varicosities that contact the tendon and so far no receptors for cholinergic transmission have been found. Thus, the function of palisade endings is still not clear. However, irrespective of their exact function, a special role of palisade endings in convergence is suggested due to their high number and accelerated postnatal development in the medial rectus of frontal-eyes species (Blumer et al. 2016, 2017).

Anatomical Evidence

The selective injection of a retrograde tracer at the level of the distal portion of extraocular muscles (the myotendinous zone) has allowed to selectively label MIF motoneurons of the oculomotor system. Since end-plates of SIFs are located at the muscle belly, SIF motoneurons remain unlabeled by this procedure. In general, MIF motoneurons are located preferentially in the periphery of the abducens, trochlear and oculomotor nuclei, whereas SIF motoneurons are distributed within the boundaries of the extraoculomotor nuclei. In the monkey, the anatomical segregation within the IIIrd nucleus between MIF and SIF motoneurons is very remarkable. Thus, non-twitch motoneurons of medial and inferior rectus form the C group located dorsomedial to the nucleus, and those of the inferior oblique and superior rectus lie near the midline in the S group. SIF medial rectus motoneurons in monkeys lie in the A- and B-groups within the oculomotor nucleus (Büttner-Ennever et al. 2001; Wasicky et al. 2004; Büttner-Ennever 2006; Erichsen et al. 2014; Tang et al. 2015). Similar findings have been obtained in humans (Horn et al. 2018). In the rat there is less anatomical segregation between MIF and SIF motoneurons, because MIF motoneurons, although found in the periphery, are also intermixed with SIF motoneurons within the nucleus, especially regarding abducens motoneurons (Eberhorn et al. 2006). In the cat, only MIF motoneurons of the medial rectus and lateral rectus muscles have been labeled. MIF medial rectus motoneurons appear located ventrolaterally and rostrally within the cat oculomotor nucleus, so that, as in primates, cats show some segregation of MIF and SIF medial rectus motoneurons, although with a different pattern (Bohlen et al. 2017b). In contrast, MIF lateral rectus motoneurons of the cat appear intermingled with SIF motoneurons in the abducens nucleus, without any particular distribution pattern (Hernández et al. 2019).

MIF and SIF motoneurons can also be differentiated by their somatic size and by the percentage of synaptic boutons terminating on their soma. Several studies have demonstrated, in all species studied so far, that the cell bodies of MIF motoneurons are smaller than those of SIF motoneurons (Büttner-Ennever et al. 2001; Wasicky et al. 2004; Eberhorn et al. 2005; Erichsen et al. 2014; Bohlen et al. 2017b; Hernández et al. 2019). This difference is in congruence with the previously reported smaller diameter of MIF axons and fibers (Namba et al. 1968; Browne 1976; Nelson et al. 1986) and also with the longer antidromic activation latencies of MIF abducens motoneurons, as compared with SIF motoneurons, in the cat abducens nucleus following the electrical stimulation of the VIth nerve (Hernández et al. 2019). Moreover, SIF motoneurons have been described as receiving a larger density of somatic synaptic boutons as compared with MIF motoneurons. This difference has been shown both in monkeys, between the groups A and B (SIF) in contrast with the C group (MIF) of medial rectus motoneurons (Erichsen et al. 2014), and in cats between MIF and SIF abducens motoneurons (Hernández et al. 2019).

The study of the afferent inputs to MIF and SIF motoneurons has revealed that these two types of motoneuron receive, in general, different projections. In primates, anterograde labeling of afferents to the twitch and non-twitch subgroups of the IIIrd nucleus has demonstrated that, except for the abducens nucleus input, the vestibular projections to twitch and non-twitch oculomotor motoneurons arise from different pools of vestibular neurons, and that pretectal inputs terminate only over non-twitch motoneurons (Wasicky et al. 2004). The authors propose that twitch motoneurons primarily would drive eye movements and non-twitch motoneurons would be involved in tonic muscle activity, such as gaze holding and vergence. In another study, by means of retrograde transneuronal tracing with rabies virus, remarkable differences in innervation of "fast" and "slow" motoneurons have been revealed in the abducens nucleus of primates. MIF abducens motoneurons receive inputs from the supraoculomotor area, the central mesencephalic reticular formation and subgroups of neurons located in the medial vestibular and prepositus hypoglossi nuclei. On the other hand, SIF abducens motoneurons are innervated by all known sources of afferents to this nucleus (Ugolini et al. 2006). Based on these results, the authors conclude that MIF motoneurons would be involved exclusively in fixations and slow eye movements and would not participate in fast eye movements like saccades. Along the same lines, other authors have demonstrated in monkeys that the central mesencephalic reticular formation projects densely to the medial rectus motoneurons of the C group (MIF) whereas it scarcely terminates on the A and B groups of medial rectus motoneurons (SIF) suggesting different roles for MIF and SIF motoneurons. In particular, the authors suggest that the C group of medial rectus motoneurons (MIF) might participate in accommodation-related vergence (Bohlen et al. 2017a).

Physiological Evidence

All the anatomical evidence stated above points to a clear distinction between SIF and MIF motoneurons. Based mainly on the works that have reported that MIF motoneurons receive mostly tonic afferents whereas SIF motoneurons are driven by all known sources of inputs to extraocular motoneurons, a functional segregation between the two types of extraocular motoneurons has been suggested, so that they might participate in different types of eye movement. MIF motoneurons would contribute to slow eye movements (such as vergence) and gaze holding (fixations), but not to fast eye movements (saccades, the vestibulo-ocular reflex), whereas SIF motoneurons would participate in the entire repertoire of eye movements (Wasicky et al. 2004; Ugolini et al. 2006; Bohlen et al. 2017a, 2017b). However, works reporting the activity of extraocular motoneurons, recorded along with eye movements under alert conditions, have demonstrated that all motoneurons display a tonic-phasic discharge pattern and participate in all different types of eye movement. (Robinson 1970; Schiller 1970; Fuchs and Luschei 1971; Delgado-García et al. 1986; de la Cruz et al. 1990; Evinger and Baker 1991; Pastor et al. 1991; Stahl and Simpson 1995; Davis-López de Carrizosa et al. 2011; Hernández et al. 2017). Therefore, these neurophysiological works have indicated that individual extraocular motoneurons do not specialize in producing any of the different types of eye movements (see however Henn and Cohen 1972).

The controversy then is whether MIF and SIF motoneurons subserve different roles. Recently, it has been possible to identify electrophysiologically MIF *versus* SIF motoneurons in the cat abducens nucleus, by placing a stimulating electrode close to the myotendinous junction of the lateral rectus muscle to selectively activate MIF motoneurons, whereas the stimulating electrode placed in the VIth nerve activated both types of motoneuron (Hernández et al. 2019). Extracellular single-unit recordings in the alert behaving cat of identified MIF and SIF abducens motoneurons have demonstrated that both types of motoneuron participate in fixations, saccades, vergence and the vestibulo-ocular reflex, so that there is no functional segregation depending on whether the eye movement is slow or fast. However, significant differences were found between MIF and SIF motoneurons with respect to their sensitivities to eye position (k) and eye velocity (r) in the oculomotor repertoire (Fig. 5). In particular, MIF motoneurons show smaller gains to eye position (k; Fig. 5b, c) and eye velocity (r; Fig. 5b, d) and discharge at lower firing frequencies compared with SIF motoneurons (Fig. 5a, b). We proposed that a functional segregation is present between MIF and SIF motoneurons, but not based on the type of eye movement in which they participate, but in their relative contribution to the generation of muscle tension (Hernández et al. 2019). It is needed more research to determine whether MIF motoneurons contribute to the fine adjustment of eye movements.

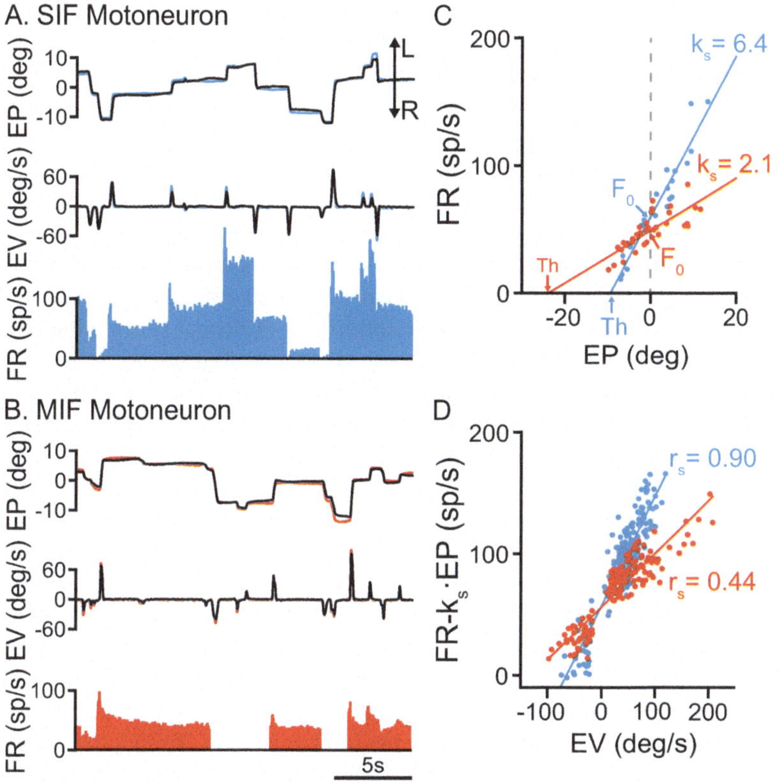

Fig. 5 Discharge characteristics of MIF and SIF motoneurons
(**a**) Firing rate (FR, in spikes/s) of a SIF abducens motoneuron during spontaneous eye movements (EP, eye position in degrees; EV, eye velocity in degrees/s). (**b**) Same for a MIF abducens motoneuron. Note its lower FR. (**c**) Correlation between FR and EP for the SIF and MIF motoneurons illustrated in (**a**) and (**b**). Linear regression lines represent the neuronal eye position sensitivity (k_s), F_0 represents the FR at straight ahead gaze, and Th is eye position threshold for recruitment. Note lower k_s, F_0 and Th for the MIF motoneuron. (**d**) Correlation between FR and EV for the motoneurons in (**a**) and (**b**). The linear regression line between FR and EV represents the neuronal eye velocity sensitivity (r_s), which is lower for the MIF motoneuron. (Reproduced from Hernández et al. 2019)

6 Response of Extraocular Motoneurons to Injury and Administration of Neurotrophic Factors

Neurons are highly dependent on their target cells for survival. The mediators of this trophic interactions are the so-called neurotrophic factors, which are produced mainly by target cells (Purves 1990; Castrén 2013; Lewin and Carter 2014) although other sources can also contribute (Korsching 1993). As neurons mature, they lose dependence on target-derived factors for survival, but these molecules are essential in the adult CNS for the adequate maintenance of neuronal physiological and structural properties, as well as for neuronal plasticity (Purves 1990; Sofroniew et al. 1990, 1993; Castrén 2013; Lewin and Carter 2014). Extraocular motoneurons have been widely used as a model to study the plastic changes that occur after target disconnection induced by axotomy, as well as the effects of exogenously provided neurotrophic factors in the axotomy state. In particular, the motoneurons of the abducens nucleus have been the main focus of these studies because they offer several advantages: (i) they can be recorded in the chronic alert animal preparation, which allows the correlation of motoneuron firing with behavior (motor in this case) during the performance of different types of eye movement; (ii) extracellular single-unit recordings of abducens motoneurons can be carried out after their electrophysiological identification by means of their antidromic electrical stimulation from the VIth nerve and the subsequent collision test; (iii) their afferents are well characterized: they comprise a threefold system of reciprocal connections mainly involving vestibular neurons (ipsilateral inhibitory, contralateral excitatory) arising in the medial vestibular nucleus (McCrea et al. 1987; Escudero et al. 1992), ponto-medullary projections (ipsilateral excitatory, contralateral inhibitory) originating in the reticular formation (Hikosaka et al. 1978; Igusa et al. 1980; Strassman et al. 1986a, b), and prepositus hypoglossi neurons (ipsilateral excitatory, contralateral inhibitory) (McCrea and Baker 1985; Escudero et al. 1992); (iv). These different inputs are mainly responsible for particular types of eye movement, thus, vestibular neurons provide the signals for the vestibular-ocular reflex, ponto-medullary neurons drive the response of abducens motoneurons during on- and off-directed saccades, and the prepositus hypoglossi input is largely responsible of the tonic firing of these motoneurons (Escudero and Delgado-García 1988; Escudero et al. 1992). Therefore, alterations in the firing pattern of abducens motoneurons during a particular type of eye movement can be related to the loss of the input signal arising from a specific afferent source.

Axotomy of Extraocular Motoneurons and Administration of Neurotrophic Factors During Postnatal Development

Axotomy of all extraocular motoneurons has been carried out in neonatal rats (postnatal day 0, P0) by unilateral enucleation. This procedure leads to the death of a significant proportion of motoneurons in the three extraocular motor nuclei, i.e., abducens, trochlear and oculomotor (Morcuende et al. 2013). Approximately by P10 only 40% of the initial population of extraocular motoneurons survive the axotomy. The section of the VIth nerve in kittens also produces a similar percentage of cell death in the abducens nucleus as 2 months after axotomy (Pásaro et al. 1985). Cell death induced by axotomy in neonatal animals is a common phenomenon that also happens in non-extraocular motoneurons, such as facial and spinal motoneurons (Sendtner et al. 1992; Yan et al. 1992; Koliatsos et al. 1993; Clatterbuck et al. 1994) indicating that during early postnatal development neurons depend on target connection for survival. By contrast, in the adult CNS, axotomy is not followed by cell death, as demonstrated for abducens motoneurons in adult cats, although adult axotomized motoneurons exhibit notorious changes in their firing and synaptic properties (Delgado-García et al. 1988). Alterations at the physiological and biochemical levels have also been described in other skeletal motoneurons after axotomy (Titmus and Faber 1990; Moran and Graeber 2004; Olmstead et al. 2015) and are restored when they reinnervate their original muscle (Delgado-García et al. 1988; Titmus and Faber 1990; Navarro et al. 2007).

The fact that during postnatal development motoneurons depend on their target muscle for survival has been explained on the lack of target-derived neurotrophic support, since when postnatal axotomized motoneurons are exogenously provided with neurotrophic factors they are rescued from cell death. This has been extensively demonstrated in the oculomotor system (Morcuende et al. 2013; Benítez-Temiño et al. 2016). Thus, after monocular enucleation at P0 in rats, the supply of different neurotrophic factors, applied intraorbitally by means of a Gelfoam implant, has resulted in a significant percentage of axotomized extraocular motoneurons surviving as compared with saline administration. These neurotrophic factors include nerve growth factor (NGF), brain-derived neurotrophic factor (BDNF), neurotrophin-3 (NT-3) and glial cell-line-derived neurotrophic factor (GDNF). For the dose used, NGF and GDNF represent the most potent survival factors for these motoneurons, followed by BDNF and lastly by NT-3. The response to the administration of these different trophic molecules in neonatal animals after axotomy is similar between the different populations of extraocular motoneurons (i.e, oculomotor, trochlear and abducens). However, It is unknown whether MIF and SIF motoneurons respond differentially to applied neurotrophins.

As compared with other brainstem and spinal motoneurons, BDNF and NT-3 yield similar degrees of cell survival after neonatal axotomy, with BDNF being more efficient than NT-3 (Sendtner et al. 1992; Yan et al. 1992; Koliatsos et al. 1993; Clatterbuck et al. 1994). Moreover, and similar to extraocular motoneurons, GDNF stands also as the most potent survival factor for other developing motoneuronal types following axonal injury (Oppenheim et al. 1995; Yuan et al. 2000; Chen et al. 2010). NGF is a survival factor as powerful as GDNF to rescue injured extraocular motoneurons from cell death in new-borns; however, this is not the case for other brainstem and spinal motoneurons. Thus, NGF is unable to rescue developing injured facial or spinal motoneurons in either neonatal rats or cultures (Sendtner et al. 1992; Houenou et al. 1994; Vejsada et al. 1995). These data are in congruence with the peculiar expression of trkA (the high-affinity receptor for NGF) on adult extraocular motoneurons which, by contrast, is not present in other brainstem and spinal adult motoneurons (Henderson et al. 1993; Koliatsos et al. 1993; Benítez-Temiño et al. 2004; Morcuende et al. 2011).

Another striking difference of axotomized neonatal extraocular motoneurons treated with neurotrophic factors is that the survival effect of these molecules-applied as a single dose- is long-lasting (Morcuende et al. 2013), in contrast to other motoneuronal types in which the neuroprotective effects of neurotrophic factors are transient, and so they eventually die at longer survival times (Eriksson et al. 1994; Vejsada et al. 1995, 1998).

Effects of Axotomy on the Discharge and Synaptic Properties of Extraocular Motoneurons in Adult Mammals

Axotomy of extraocular motoneurons has been extensively studied for abducens motoneurons following the section of the VIth nerve. In contrast to immature motoneurons, axotomy of adult motoneurons is not followed by cell death. However, injured motoneurons present several physiological and structural alterations that recover with reinnervation. In alert chronic cats, the discharge of axotomized abducens motoneurons has been recorded and correlated with the horizontal movement of the control eye. Due to the high degree of conjugacy of cat's horizontal eye movements, it has been demonstrated in control animals that similar motoneuronal firing parameters are obtained when correlation of the discharge is performed with one eye or another. (Delgado-García et al. 1986; Calvo et al. 2018). After axotomy, the firing behavior of the motoneuron is greatly reduced. Thus, during eye fixations, axotomized motoneurons fire at low rate and, frequently, their discharge shows an exponential decay during the course of the fixation. This means that the motoneuron is unable to maintain the tonic discharge for prolonged period of time, thereby remaining silent over wide parts of the oculomotor range. This alteration produces low values in neuronal eye position sensitivity (k), which are significantly lower after axotomy as compared to control. Similarly, axotomized abducens motoneurons show small and delayed bursts of discharge during on-directed saccades, but pauses during off-directed saccades. This behavior also leads to a significant decrease in neuronal eye velocity sensitivity (r). Firing frequency is also reduced during the vestibulo-ocular reflex. Consequently, there is a significant reduction, as compared to control, in neuronal eye position and velocity sensitivities during the slow phases of the reflex (Delgado-García et al. 1988; Davis-López de Carrizosa et al. 2009, 2010; Calvo et al. 2018).

All these changes have been explained based on a decreased excitability of axotomized abducens motoneurons and a loss of synaptic afferents (Delgado-García et al. 1988). The low excitability of abducens motoneurons after axotomy contrasts sharply with the increased excitability shown in axotomized spinal motoneurons. For example, axotomized abducens motoneurons do not exhibit dendritic spikes and/or partial responses as happens in axotomized spinal motoneurons (Kuno and Llinás 1970a, b). Neither there is an increase in input resistance, membrane time constant or changes in afterhyperpolarization (Gustafsson 1979; Gustafsson and Pinter 1984a, b; Titmus and Faber 1990), demonstrating again singularities in extraocular motoneurons compared to spinal ones.

Following axotomy, a general response that has been described in different types of brainstem (including extraocular) and spinal motoneurons is the retraction of afferent boutons contacting with the injured cell (Hamberger et al. 1970; Sumner and Sutherland 1973; Sumner 1975a, b; Delgado-García et al. 1988; Davis-López de Carrizosa et al. 2009, 2010; Alvarez et al. 2011; Calvo et al. 2018). In the case of axotomized abducens motoneurons, the retraction of synaptic inputs arising from the prepositus hypoglossi nucleus could explain the partial loss of the tonic signal in these injured motoneurons, the withdrawal of boutons originating in reticular neurons is likely the cause of the decreased bursts during on-directed saccades, and synaptic stripping from vestibular afferents might be responsible of the decreased firing of injured abducens motoneurons during the vestibulo-ocular reflex (Delgado-García et al. 1988).

Nonetheless, a striking difference between extraocular and spinal motoneurons after nerve section is that inhibitory synapses are preferentially lost from the cell body of axotomized extraocular motoneurons, whereas in injured spinal motoneurons inhibitory synapses are largely retained in contrast to excitatory synapses which are mostly lost on the cell body (Delgado-García et al. 1988; Lindå et al. 1992; Brännström and Kellerth 1998; Lindå et al. 2000; Alvarez et al. 2011; Rotterman et al. 2019).

Another interesting aspect of synaptic stripping after a lesion, common to all neuronal types analyzed so far (e.g., axotomized brainstem and spinal motoneurons), is that it occurs associated with an increase in glial processes, which in many cases appear interposed between the bouton and the somatic membrane of the injured neuron. The so-called distal glial reaction is characterized by the hypertrophy of astrocytes, which express increased levels of glial fibrillary acidic protein (GFAP), and by the proliferation of microglial cells (Svensson et al. 1993; Davis-López de Carrizosa et al. 2009, 2010; Calvo et al. 2018). Reactive glial cells have been implicated in the removal of presynaptic terminals from axotomized neurons (Blinzinger and Kreutzberg 1968; Graeber and Kreutzberg 1988; Kreutzberg et al. 1989; Svensson et al. 1993; Yamada et al. 2011; Kettenmann et al. 2013; Spejo and Oliveira 2015).

BDNF and NT-3: Two Neurotrophins with Complementary Actions on the Function/Structure of Extraocular Motoneurons

Extraocular motoneurons are responsive to several neurotrophic factors, such as neurotrophins and vascular endothelial growth factor (VEGF, see below). It has been demonstrated that these motoneurons express the high-affinity receptors for NGF (TrkA), BDNF (TrkB) and NT-3 (TrkC) and that extraocular muscles express NGF, BDNF and NT-3, and in consequence these factors may act as target-derived trophic molecules for extraocular motoneurons (Benítez-Temiño et al. 2004, 2016; Davis-López de Carrizosa et al. 2010; Morcuende et al. 2011; Hernández et al. 2017). Neurotrophic factors are essential for survival during development but as neurons mature they lose dependence on these molecules, which are mainly provided by target cells. However, although in the adult state, the lack of target-derived neurotrophic factors (e.g., by axotomy) is not followed by retrograde cell death, it does produce remarkable alterations of numerous functional and structural aspects of target-deprived neurons, including all types of skeletal motoneurons (Purves 1990).

Interestingly, the exogenous administration of different neurotrophic factors to axotomized extraocular motoneurons produce distinctive and peculiar effects, in all cases in the direction of preventing or recovering the axotomy-induced alterations, as has been demonstrated clearly for abducens motoneurons (Davis-López de Carrizosa et al. 2009, 2010; Benítez-Temiño et al. 2016; Calvo et al. 2018).

A striking situation occurs after the administration of the neurotrophins BDNF and/or NT-3 to axotomized abducens motoneurons (Davis-López de Carrizosa et al. 2009; Benítez-Temiño et al. 2016). BDNF recovers the tonic firing of injured abducens motoneurons, so that motoneurons can sustain a tonic discharge that is proportional to the position of the eye in the orbit. This yields neuronal eye position sensitivities in BDNF-treated axotomized motoneurons similar to control, whereas untreated axotomized motoneurons have significantly lower position gain. However, BDNF is unable to recover the loss of the phasic signal in axotomized abducens motoneurons, so that these neurons discharge, as during axotomy, small bursts of spikes during on-directed saccades. Therefore, the reduction in neuronal eye velocity sensitivity observed in axotomized motoneurons is not restored by BDNF. The tonic component of abducens motoneuron firing arises from the output of a velocity-to-position integrator (Aksay et al. 2001) whose main output cells are the burst-tonic neurons of the prepositus hypoglossi nucleus (Escudero et al. 1992). So this input is likely retained by BDNF treatment. Indeed, confocal analysis shows, when BDNF is delivered to axotomized motoneurons, a partial recovery of the synaptic loss due to axotomy (Fig. 6).

NT-3 administration to axotomized abducens motoneurons leads to the complementary effects to those observed after BDNF treatment. Thus, NT-3 recovers the phasic behavior of axotomized abducens motoneurons, so that high frequency bursts of spikes appear, as in control, preceding on-directed saccades, but this neurotrophin is unable to recover the loss of the tonic firing in axotomized motoneurons that occurs during eye fixations. As a consequence, there is a recovery in neuronal eye velocity sensitivity, with values similar to control, but neuronal eye position sensitivity is not restored after axotomy and NT-3 treatment, showing values significantly lower than control (Fig. 6). The phasic component of abducens motoneurons is provided by burst reticular neurons (excitatory and inhibitory, EBN, IBN) located in the paramedian pontine reticular formation (Highstein et al. 1976; Hikosaka et al. 1978; Igusa et al. 1980) which convey a strong burst during on-directed saccades and a pause during saccades in the opposite direction (Büttner-Ennever 2006). Therefore, it is likely that NT-3 exerts a trophic support on these afferents. This is in consonance with the finding, at the confocal level, of a partial recovery of synaptic boutons observed in NT-3-treated axotomized motoneurons, in comparison with the large synaptic detachment that occurs in untreated axotomized motoneurons.

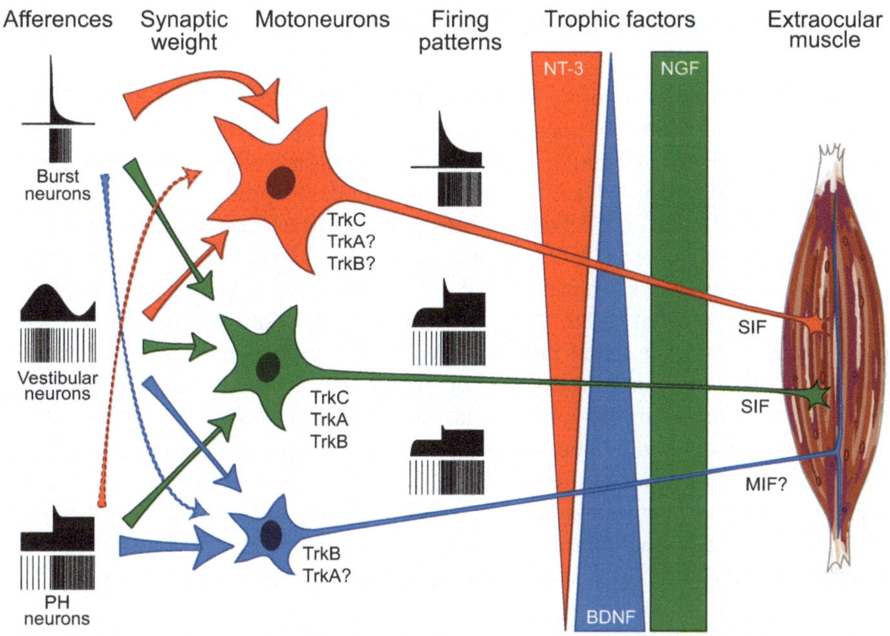

Fig. 6 A hypothetical link between motoneuronal type and neurotrophin delivery from muscle. SIF and MIF muscle fibers would supply motoneurons with different combinations of neurotrophins leading to variations in motoneuronal discharge. Thus, SIF muscle fibers would produce mainly BDNF whereas MIF muscle fibers would synthetize mainly NT-3. This different neurotrophic supply could affect motoneuronal synaptic inputs (illustrated to the left in the figure) in a critical way, so that BDNF would enhance synaptic transmission from tonic afferents (leading to a more tonic firing in MIF motoneurons) whereas NT-3 would improve synaptic strength from phasic afferents (leading to a more tonic-phasic discharge in SIF motoneurons). BDNF and NT-3 would be secreted as a gradient in a complementary way, whereas the release of NGF would be more homogeneous between muscle fibers. In turn, SIF and MIF motoneurons should contain the different Trk receptors for the three neurotrophins, as indicated. (Reproduced from Benítez-Temiño et al. 2016)

Remarkably, when BDNF and NT-3 are applied together to axotomized abducens motoneurons, the complementary effects of each neurotrophic add and the result is the complete recovery of the typical tonic-phasic firing pattern of abducens motoneurons. Also, the density of synaptic afferents impinging on these axotomized motoneurons treated with the two factors recovered to values similar to control.

It is important to emphasize that both neurotrophins require a continuous supply in order to recover abducens motoneurons from the axotomy state. When administration ceases, the effects of trophic support wanes and motoneuron firing resembles again the axotomy state (Davis-López de Carrizosa et al. 2009).

NGF Activity in Extraocular Motoneurons

Unlike other cranial and spinal motoneurons (Lindsay 1994; Ferri et al. 2002), extraocular motoneurons are peculiar in that they express the high affinity receptor for NGF, TrkA (Benítez-Temiño et al. 2004). In contrast, other cranial and spinal motoneurons only express TrkA transiently during development so that only TrkB (for BDNF) and TrkC (for NT-3) are present in adults (Merlio et al. 1992; Henderson et al. 1993; Koliatsos et al. 1993; Piehl et al. 1994). During development, spinal motoneurons contain TrkA receptors but their expression in downregulated during maturation (Ernfors et al. 1989, 1991). Interestingly, the expression of TrkA receptors is induced in adult spinal motoneurons by axotomy (Omura et al. 2005) but disappears after the reinnervation of the target muscle (Koliatsos et al. 1993). Motoneurons affected by amyotrophic lateral sclerosis (ALS), a disease that does not affect extraocular motoneurons, shift their expression of Trk receptors from TrkB and Trk C to TrkA (Nishio et al. 1998). The reasons for the continuous responsiveness of extraocular motoneurons to NGF (due to constitutive expression of TrkA), in contrast to other motoneuronal types are unclear, but it might be related to their higher resistivity to some motoneuronal degenerative diseases, like ALS (Hernández et al. 2017).

NGF has also been administered to injured abducens motoneurons (Davis-López de Carrizosa et al. 2010; Benítez-Temiño et al. 2016). NGF recovers the loss of the tonic and phasic components of the firing pattern of axotomized abducens motoneurons, so that neuronal eye position and velocity sensitivities do not fall to the low levels typical of the axotomy situation. Strikingly, these two parameters (k and r) increase even above control values, demonstrating an increased excitability in axotomized motoneurons when provided with this factor. The recovery of the tonic-phasic discharge produced by NGF administration is in congruence with the restoration of a normal synaptic complement in NGF-treated injured motoneurons, as observed by confocal microscopy.

NGF treatment produces two singular responses in axotomized cells, not observed in control. First, motoneurons exhibit a notorious increase in discharge variability during ocular fixations, so that the coefficient of variation in NGF-treated injured motoneurons (i.e., the percentage ratio of the standard deviation to the mean firing during stationary eye positions) raises significantly in relation to both, control and axotomy. NGF has been shown to regulate excitability through several ionic currents (Zhang and Nicol 2004; Luther and Birren 2009), one of them present in oculomotor motoneurons (Nieto-Gonzalez et al. 2009). There are several mechanisms proposed to explain an increase in firing irregularity: (i) fluctuations produced by synaptic noise after any depolarization that carries membrane potential near spike threshold (Calvin and Stevens 1968); (ii) the location of inhibitory synapses near the axon hillock (Shadlen and Newsome 1998); and (iii) the synchronization of inputs (Lampl et al. 1999). The mechanism by which NGF increases the irregularity of firing in axotomized abducens motoneurons has not yet been revealed.

The second singularity of NGF treatment is the increase that produces in eye position threshold (i.e., recruitment threshold), in comparison with both axotomy and control. The elevation in recruitment threshold is accompanied by an increase in the ratio of inhibition to excitation (González-Forero et al. 2002, 2004; Salama-Cohen et al. 2006; Davis-López de Carrizosa et al. 2010).

By means of selective inhibitors of either the high-affinity receptor (TrkA) or the low-affinity receptor (p75) of NGF, it has been demonstrated that the different effects of NGF on axotomized abducens motoneurons are signaled selectively by the activation of one or the other receptor. In particular, NGF acting through TrkA receptors mostly regulates the tonic-phasic discharge pattern related to eye movements and maintains the synaptic inputs on these motoneurons. On the other hand, p75 signaling is mainly involved in the increase of both firing irregularity and recruitment threshold (Davis-López de Carrizosa et al. 2010).

VEGF: A Powerful Neurotrophic Factor for Extraocular Motoneurons

VEGF was first discovered by its angiogenic activity (Senger et al. 1983; Ferrara and Henzel 2012) and later confirmed as a molecule that also acts in the CNS as a neuroprotective factor (Storkebaum et al. 2004; Lambrechts and Carmeliet 2006; Lange et al. 2016). Indeed, during the course of evolution, VEGF appeared first as a neurotrophic factor essential for the development of the nervous system in invertebrates lacking or having a rudimentary circulatory system, and later on it acquired its vascular promoting activity in animals with a well-established vascular system (Zacchigna et al. 2008). VEGF has been shown to be neuroprotective in diverse types of neurons and following different types of insults (Carmeliet and Storkebaum 2002; Storkebaum et al. 2004; Nicoletti et al. 2008; Lange et al. 2016). However, it should be emphasized that it plays a special role as a neuroprotective factor in motoneurons. Thus: (i) low levels of VEGF in the mutant mice VEGF$^{\delta/\delta}$ lead to motoneuron degeneration resembling ALS (Oosthuyse et al. 2001); (ii) its delivery to animal models of ALS slows the progression of motoneuronal degeneration and increases life expectancy (Azzouz et al. 2004; Wang et al. 2016); (iii) its administration in the spinal cord of rats exposed to excitotoxic motoneuron degeneration prevents paralysis and motoneuronal death (Tovar-y-Romo et al. 2007); and (iv) it improves motor behavior in different models of motoneuronal degeneration (Azzouz et al. 2004; Storkebaum et al. 2005; Tovar-y-Romo et al. 2007).

Extraocular motoneurons, as spinal motoneurons, express the two tyrosine kinase receptors of VEGF, VEGFR-1 and VEGFR-2, and are therefore responsive to this factor (Calvo et al. 2018). Moreover, VEGF can act as a retrograde neurotrophic support for these motoneurons since they express this factor (Calvo et al. 2018) Following the section of the VIth nerve, the administration of VEGF prevents and recovers (depending on the time of VEGF administration in relation to lesion) the axotomy-induced alterations in abducens motoneurons, both physiologically and morphologically. VEGF-treated axotomized motoneurons show a normal discharge activity. Thus, they fire at higher tonic rates for eye fixations more in the on (temporal) direction exhibiting a normal firing regularity. They also show during saccades the behavior typical of control, that is, a high-frequency burst of spikes for on-directed saccades and a pause or abrupt decay in firing rate for off-directed saccades, and finally they modulate properly during the vestibulo-ocular reflex. Consequently, VEGF treatment produces in axotomized abducens motoneurons eye-related parameters similar to control and significantly different to axotomy (Calvo et al. 2018).

Since the different signals encoded by abducens motoneurons are provided preferentially by particular afferents as stated above (Escudero and Delgado-García 1988; Büttner-Ennever 2006), these physiological findings indicate that VEGF is also acting to maintain or recover synaptic inputs arising from different sources, which are lost following axotomy. Indeed, immunocytochemical procedures viewed at the confocal level have revealed, in agreement with the physiological data, a normal density of afferent synaptic boutons in VEGF-treated axotomized motoneurons demonstrating that VEGF blocks the synaptic stripping that occurs in injured neurons. In parallel, the astroglial reaction around motoneurons that ensues nerve section, which is associated with synaptic detachment, is not present when VEGF is delivered (Calvo et al. 2018).

7 Higher Resistance of Extraocular Motoneurons to Degeneration in ALS

ALS is a devastating disease characterized by the progressive degeneration of motoneurons causing muscle weakness and later on motor paralysis that ultimately leads to the loss of respiratory function, thereby resulting in death in 3–5 years (Cleveland and Rothstein 2001; Peters et al. 2015). However, not all motoneurons are equally affected by the disease. Thus, whereas most motoneuronal types degenerate, extraocular motoneurons are resistant to neurodegeneration in ALS. At early stages of the disease, a severe neuronal loss is observed in facial, trigeminal, hypoglossal and spinal motoneuronal pools. In contrast, the oculomotor, trochlear and abducens nuclei remain unaffected, as observed both in animal models of the disease and in human ALS (DePaul et al. 1988; Reiner et al. 1995; Nimchinsky et al. 2000; Haenggeli and Kato 2002). The mechanisms of the selective resistance of extraocular motoneurons in ALS are still unclear, although several hypotheses have been proposed.

Extraocular motoneurons are enriched in calcium-binding proteins, especially parvalbumin, as compared with other motoneuronal types, which might confer them with a neuroprotective mechanism against high intracellular concentrations of free, extremely cytotoxic Ca^{++} ions due to the buffering capacity of these calcium-binding proteins (Alexianu et al. 1994; Elliott and Snider 1995; Reiner et al. 1995; de la Cruz et al. 1998; Dekkers et al. 2004). The high concentration of calcium buffering proteins in extraocular motoneurons together with the expression of different types of glutamate receptors between resistant *versus* vulnerable motoneurons might prevent motoneuronal degeneration in extraoculomotor nuclei produced by glutamate overstimulation (Medina et al. 1996; Laslo et al. 2001). Along these lines, it is important to note that one of the main mechanisms proposed to explain the etiology of ALS is glutamate excitotoxicity. Indeed, the only FDA-approved therapy in ALS, Riluzole, functions by decreasing glutamate toxicity (Bruijn et al. 2004). Motoneurons are especially vulnerable to excitotoxicity in comparison with other neuronal types because they contain a high number of AMPA receptors permeable to Ca^{++}. The Ca^{++} entry via these AMPA receptors has been proposed as the cause of the selective motoneuron death. Ca^{++}-permeable AMPA receptors are characterized by the absence of the GluR2 subunit, and motoneurons are relatively deficient in GluR2. Thus, it has been hypothesized that the selective vulnerability of motoneurons to excitotoxicity is due to their relative deficiency in GluR2, that increases the risk of Ca^{++} entry in these neurons (Van Damme et al. 2002).

Other authors have proposed that the presence of the neuropeptide calcitonin gene-related peptide (CGRP) might predict the degree of motoneuron vulnerability in ALS animal models. Thus, oculomotor, trochlear and abducens nuclei (III, IV and VI) contain mostly non-CGRP motoneurons, whereas motor nuclei V, VII and XII (trigeminal, facial and hypoglossal) contain equal numbers of high-CGRP, low-CGRP, and non-CGRP motoneurons, and the spinomedullary group (ambiguus nucleus and lumbar spinal cord) contains mainly motoneurons with high levels of CGRP. At end-stages of mutant mice showing ALS-like disease, the extent of motoneuron loss correlates with the expression level of CGRP: neurons with high CGRP die by 80%, those with low CGRP by 50% and those with non CGRP (i.e., extraocular motoneurons) are not affected in any of the three motor nuclei (Ringer et al. 2011).

An alternative mechanism proposed to explain the pathogenesis of ALS lies in a low availability of neurotrophic factors (Bruijn et al. 2004). It is well known that neurotrophic factors selectively regulate the growth and survival of certain populations of neurons in the central and peripheral nervous systems and that they play an essential role in neuronal development and in the maintenance of differentiated neurons in the adult (Purves 1990). In this sense, it is important to highlight that extraocular motoneurons contain a greater amount of neurotrophins (NGF, BDNF, NT-3) than other motoneurons vulnerable to ALS (e.g., those of the facial and hypoglossal nuclei), which either contain neurotrophins to a lesser extent or are virtually devoid of these proteins, depending on each neurotrophin. Target muscles are the main source of neurotrophic supply for motoneurons, so it is also important to test the presence of neurotrophins in muscles. It has been shown that extraocular muscles contain neurotrophins to a greater extent compared to facial (buccinator) and tongue muscles. Therefore, it has been suggested that the differences in neurotrophin availability could be related to the particular resistance of extraocular motoneurons to neurodegeneration (Hernández et al. 2017b). In relation to neurotrophins, another aspect that characterized extraocular motoneurons as compared with other motoneuronal types is that only extraocular motoneurons express the high affinity receptor of NGF, TrkA, in the adult CNS (Benítez-Temiño et al. 2004; Davis-López de Carrizosa et al. 2010; Morcuende et al. 2011). Other brainstem or spinal motoneurons express the high affinity receptors for BDNF, TrkB, and NT-3, TrkC, but lack TrkA (Koliatsos et al. 1991; Merlio et al. 1992; Henderson et al. 1993; Piehl et al. 1994). Extraocular motoneurons also express TrkB and TrkC (Benítez-Temiño et al. 2004). Moreover, abducens motoneurons respond to NGF exogenous administration, so that they recover after axotomy the typical tonic-phasic discharge pattern (although with certain peculiarities) which is lost by injury (Davis-López de Carrizosa et al. 2010). Therefore, the particular responsiveness of extraocular motoneurons to NGF, in contrast to other motoneuronal types, might contribute to their higher resistance to neurodegeneration in diseases such as ALS.

Another neurotrophic factor which stands out as highly neuroprotective for unhealthy motoneurons is VEGF. Low levels of VEGF in mutant mice lead to adult-onset motoneuron degeneration resembling ALS (Oosthuyse et al. 2001), and its delivery to animal models of ALS delays the onset of motoneuron degeneration, preserves neuromuscular junctions, and prolongs the survival of these animals (Azzouz et al. 2004; Storkebaum et al. 2005). Therefore, it may be possible that VEGF acts as an essential factor to maintain brainstem and spinal motoneurons in a healthy state. It has been shown recently that whereas virtually all extraocular motoneurons of the oculomotor, trochlear and abducens nuclei express VEGF and its main receptor, VEGFR-2 (also named Flk-1), the majority of facial and hypoglossal motoneurons lack both VEGF and its receptor Flk-1 (Silva-Hucha et al. 2017). These data have led to the hypothesis that differences in VEGF availability and signaling could contribute to the different susceptibility of extraocular motoneurons, as compared with other cranial and spinal motoneurons, to neurodegeneration in ALS (Silva-Hucha et al. 2017). Moreover, it has also been demonstrated that VEGF protects motoneurons against excitotoxicity by upregulation of the subunit GluR2 of the AMPA receptor since AMPA receptors containing GluR2 are not permeable to Ca^{++} ions (Bogaert et al. 2010). Altogether, the presence of higher amounts of VEGF in extraocular motoneurons might allow these neurons to express AMPA receptors containing the subunit GluR2 and, in this way, rendering extraocular motoneurons highly impermeable to the entry of Ca^{++} ions, which gives them neuroprotection against excitotoxicity. Since other motoneurons have a lesser amount or lack VEGF, they could be more susceptible to excitotoxicity, which is one of the main hypothesis explaining the pathogenesis of ALS (Bruijn et al. 2004; Silva-Hucha et al. 2017).

8 Conclusions and Perspectives

The complementary effects of BDNF and NT-3 in regulating the tonic *versus* the phasic firing are highly remarkable in view of the existence of functional diversity in the oculomotor system. These findings could indicate that different classes of motoneurons might be defined with respect to their trophic influence from the muscle fibers they innervate. Thus, an interesting topic to investigate would be to analyze the possibility of a link between neurotrophic factors and the structural/functional differences between MIF and SIF extraocular motoneurons. In this line, it could be hypothesized that MIF motoneurons would respond preferentially to BDNF, whereas SIF motoneurons might be regulated mostly by NT-3 (Fig. 3). This suggestion remains, however, to be explored.

VEGF is a potent neurotrophic factor for extraocular motoneurons, able to restore both components of the discharge of these motoneurons, the tonic and the phasic, along with a normal synaptic complement, restoring all axotomy-induced morphophysiological alterations to control values. Moreover, extraocular motoneurons are enriched in this molecule and its main receptor in comparison with other motoneurons, suggesting that it might play an important role in the higher resistance of extraocular motoneurons to degeneration in motoneuronal diseases such as ALS.

Acknowledgments This research was funded by MINISTERIO DE CIENCIA, INNOVACIÓN Y UNIVERSIDADES (SPAIN), grant PGC2018-094654-B-100 and PID2021-124300NB-I00 (MCI/AEI/FEDER, UE) "A way of making Europe". Supported also by project P20_00529 Consejería de Transformación Económica, Industria y Conocimiento, Junta de Andalucía-FEDER to AMP and RRC and Fonds zur Förderung der wissenschaftlichen Forschung (Austria), Grant P32463-B to RB.

References

Aksay E, Gamkrelidze G, Seung HS, Baker R, Tank DW (2001) In vivo intracellular recording and perturbation of persistent activity in a neural integrator. Nat Neurosci 4:184–193. https://doi.org/10.1038/84023

Alexianu ME, Ho BK, Mohamed AH, La Bella V, Smith RG, Appel SH (1994) The role of calcium-binding proteins in selective motoneuron vulnerability in amyotrophic lateral sclerosis. Ann Neurol 36:846–858. https://doi.org/10.1002/ana.410360608

Alvarado Mallart RM, Pincon Raymond M (1979) The palisade endings of cat extraocular muscles: a light and electron microscope study. Tissue Cell 11:567–584. https://doi.org/10.1016/0040-8166(79)90063-6

Alvarez FJ, Titus-Mitchell HE, Bullinger KL, Kraszpulski M, Nardelli P, Cope TC (2011) Permanent central synaptic disconnection of proprioceptors after nerve injury and regeneration. I. Loss of VGLUT1/IA synapses on motoneurons. J Neurophysiol 106:2450–2470. https://doi.org/10.1152/jn.01095.2010

Azzouz M, Ralph GS, Storkebaum E, Walmsley LE, Mitrophanous KA, Kingsman SM, Carmeliet P, Mazarakis ND (2004) VEGF delivery with retrogradely transported lentivector prolongs survival in a mouse ALS model. Nature 429:413–417. https://doi.org/10.1038/nature02544

Benítez-Temiño B, Morcuende S, Mentis GZ, de la Cruz RR, Pastor AM (2004) Expression of Trk receptors in the oculomotor system of the adult cat. J Comp Neurol 473:538–552. https://doi.org/10.1002/cne.20095

Benítez-Temiño B, Davis-López de Carrizosa MA, Morcuende S, Matarredona ER, de la Cruz RR, Pastor AM (2016) Functional diversity of Neurotrophin actions on the oculomotor system. Int J Mol Sci 17(12). https://doi.org/10.3390/ijms17122016

Blinzinger K, Kreutzberg G (1968) Displacement of synaptic terminals from regenerating motoneurons by microglial cells. Z Zellforsch Mikrosk Anat 85:145–157. https://doi.org/10.1007/bf00325030

Blumer R, Konakci KZ, Pomikal C, Wieczorek G, Lukas JR, Streicher J (2009) Palisade endings: cholinergic sensory organs or effector organs? Invest Ophthalmol Vis Sci 50:1176–1186. https://doi.org/10.1167/iovs.08-2748

Blumer R, Maurer-Gesek B, Gesslbauer B, Blumer M, Pechriggl E, Davis-López de Carrizosa MA, Horn AK, May PJ, Streicher J, de la Cruz RR, Pastor AM (2016) Palisade endings are a constant feature in the extraocular muscles of frontal-eyed, but not lateral-eyed, animals. Invest Ophthalmol Vis Sci 57:320–331. https://doi.org/10.1167/iovs.15-18716

Blumer R, Streicher J, Davis-Lopez de Carrizosa MA, de la Cruz RR, Pastor AM (2017) Palisade endings of extraocular muscles develop postnatally following different time courses. Invest Ophthalmol Vis Sci 58:5105–5121. https://doi.org/10.1167/iovs.17-22643

Blumer R, Streicher J, Carrero-Rojas G, Calvo PM, de la Cruz RR, Pastor AM (2020) Palisade Endings Have an Exocytotic Machinery But Lack Acetylcholine Receptors and Distinct Acetylcholinesterase Activity. Invest Ophthalmol Vis Sci. Dec 1;61(14):31. https://doi.org/10.1167/iovs.61.14.31

Bogaert E, Van Damme P, Poesen K, Dhondt J, Hersmus N, Kiraly D, Scheveneels W, Robberecht W, Van Den Bosch L (2010) VEGF protects motor neurons against excitotoxicity by upregulation of GluR2. Neurobiol Aging 31:2185–2191. https://doi.org/10.1016/j.neurobiolaging.2008.12.007

Bohlen MO, Warren S, May PJ (2017a) A central mesencephalic reticular formation projection to medial rectus motoneurons supplying singly and multiply innervated extraocular muscle fibers. J Comp Neurol 525:2000–2018. https://doi.org/10.1002/cne.24187

Bohlen MO, Warren S, Mustari MJ, May PJ (2017b) Examination of feline extraocular motoneuron pools as a function of muscle fiber innervation type and muscle layer. J Comp Neurol 525:919–935. https://doi.org/10.1002/cne.24111

Bondi AY, Chiarandini DJ (1979) Ionic basis for electrical properties of tonic fibres in rat extraocular muscles. J Physiol 295:273–281. https://doi.org/10.1113/jphysiol.1979.sp012968

Brännström T, Kellerth JO (1998) Changes in synaptology of adult cat spinal alpha-motoneurons after axotomy. Exp Brain Res 118:1–13. doi: https://doi.org/10.1007/s002210050249

Browne JS (1976) The contractile properties of slow muscle fibres in sheep extraocular muscle. J Physiol 254:535–555. https://doi.org/10.1113/jphysiol.1976.sp011245

Bruijn LI, Miller TM, Cleveland DW (2004) Unraveling the mechanisms involved in motor neuron degeneration in ALS. Annu Rev Neurosci 27:723–749. https://doi.org/10.1146/annurev.neuro.27.070203.144244

Büttner-Ennever JA (ed) (2006) Neuroanatomy of the oculomotor system. Elsevier, Amsterdam

Büttner-Ennever JA, Horn AK, Scherberger H, D'Ascanio P (2001) Motoneurons of twitch and nontwitch extraocular muscle fibers in the abducens, trochlear, and oculomotor nuclei of monkeys. J Comp Neurol 438:318–335. https://doi.org/10.1002/cne.1318

Calvin WH, Stevens CF (1968) Synaptic noise and other sources of randomness in motoneuron interspike intervals. J Neurophysiol 31:574–587. https://doi.org/10.1152/jn.1968.31.4.574

Calvo PM, de la Cruz RR, Pastor AM (2018) Synaptic loss and firing alterations in Axotomized Motoneurons are restored by vascular endothelial growth factor (VEGF) and VEGF-B. Exp Neurol 304:67–81. https://doi.org/10.1016/j.expneurol.2018.03.004

Carmeliet P, Storkebaum E (2002) Vascular and neuronal effects of VEGF in the nervous system: implications for neurological disorders. Semin Cell Dev Biol 13:39–53. https://doi.org/10.1006/scdb.2001.0290

Castrén E (2013) Trophic factors: neurotrophic factors. In: Pfaff DW (ed) Neuroscience in the 21st century. Springer, New York, pp 1555–1589

Chen J, Chu YF, Chen JM, Li BC (2010) Synergistic effects of NGF, CNTF and GDNF on functional recovery following sciatic nerve injury in rats. Adv Med Sci 55:32–42. https://doi.org/10.2478/v10039-010-0020-9

Chiarandini DJ (1976) Activation of two types of fibres in rat extraocular muscles. J Physiol 259:199–212. https://doi.org/10.1113/jphysiol.1976.sp011461

Chiarandini DJ, Stefani E (1979) Electrophysiological identification of two types of fibres in rat extraocular muscles. J Physiol 290:453–465. https://doi.org/10.1113/jphysiol.1979.sp012783

Clatterbuck RE, Price DL, Koliatsos VE (1994) Further characterization of the effects of brain-derived neurotrophic factor and ciliary neurotrophic factor on axotomized neonatal and adult mammalian motor neurons. J Comp Neurol 342:45–56. https://doi.org/10.1002/cne.903420106

Cleveland DW, Rothstein JD (2001) From Charcot to Lou Gehrig: deciphering selective motor neuron death in ALS. Nat Rev Neurosci 2:806–819. https://doi.org/10.1038/35097565

Consi TR, Macagno ER, Necles N (1987) The oculomotor system of Daphnia magna. The eye muscles and their motor neurons. Cell Tissue Res 247:515–523. https://doi.org/10.1007/BF00215744

Davidowitz J, Rubinson K, Jacoby J, Philips G (1996) Myofibril size variation along the length of extraocular muscle in rabbit and rat. I: orbital layer. Tissue Cell 28:63–76. https://doi.org/10.1016/s0040-8166(96)80045-0

Davis-López de Carrizosa MA, Morado-Díaz CJ, Tena JJ, Benítez-Temiño B, Pecero ML, Morcuende SR, de la Cruz RR, Pastor AM (2009) Complementary actions of BDNF and neurotrophin-3 on the firing patterns and synaptic composition of motoneurons. J Neurosci 29:575–587. https://doi.org/10.1523/JNEUROSCI.5312-08.2009

Davis-López de Carrizosa MA, Morado-Díaz CJ, Morcuende S, de la Cruz RR, Pastor AM (2010) Nerve growth factor regulates the firing patterns and synaptic composition of motoneurons. J Neurosci 30:8308–8319. https://doi.org/10.1523/JNEUROSCI.0719-10.2010

Davis-López de Carrizosa MA, Morado-Díaz CJ, Miller JM, de la Cruz RR, Pastor AM (2011) Dual encoding of muscle tension and eye position by abducens motoneurons. J Neurosci 31:2271–2279. https://doi.org/10.1523/JNEUROSCI.5416-10.2011

de la Cruz RR, Escudero M, Delgado-García JM (1990) Behaviour of medial rectus Motoneurons in the alert cat. Eur J Neurosci 1:288–295

de la Cruz RR, Pastor AM, Martínez-Guijarro FJ, López-García C, Delgado-García JM (1998) Localization of parvalbumin, calretinin, and calbindin D-28k in identified extraocular motoneurons and internuclear neurons of the cat. J Comp Neurol 390:377–391. https://doi.org/10.1002/(sici)1096-9861(19980119)390:3%3C377::aid-cne6%3E3.0.co;2-z

Dekkers J, Bayley P, Dick JR, Schwaller B, Berchtold MW, Greensmith L (2004) Over-expression of parvalbumin in transgenic mice rescues motoneurons from injury-induced cell death. Neuroscience 123:459–466. https://doi.org/10.1016/j.neuroscience.2003.07.013

Delgado-García JM, del Pozo F, Baker R (1986) Behavior of neurons in the abducens nucleus of the alert cat--I. Motoneurons. Neuroscience 17:929–952. https://doi.org/10.1016/0306-4522(86)90072-2

Delgado-García JM, Del Pozo F, Spencer RF, Baker R (1988) Behavior of neurons in the abducens nucleus of the alert cat—III. Axotomized motoneurons. Neuroscience 24:143–160. https://doi.org/10.1016/0306-4522(88)90319-3

DePaul R, Abbs JH, Caligiuri M, Gracco VL, Brooks BR (1988) Hypoglossal, trigeminal, and facial motoneuron involvement in amyotrophic lateral sclerosis. Neurology 38:281–283. https://doi.org/10.1212/wnl.38.2.281

Dieringer N, Precht W (1986) Functional organization of eye velocity and eye position signals in abducens motoneurons of the frog. J Comp Physiol A 158:179–194. https://doi.org/10.1007/BF01338561

Dietrich D, Glasauer S, Straka H (2017) Functional organization of vestibulo-ocular responses in abducens motoneurons. J Neurosci 37:4032–4045. https://doi.org/10.1523/JNEUROSCI.2626-16.2017

Eberhorn AC, Ardeleanu P, Büttner-Ennever JA, Horn AK (2005) Histochemical differences between motoneurons supplying multiply and singly innervated extraocular muscle fibers. J Comp Neurol 491:352–366. https://doi.org/10.1002/cne.20715

Eberhorn AC, Büttner-Ennever JA, Horn AK (2006) Identification of motoneurons supplying multiply- or singly-innervated extraocular muscle fibers in the rat. Neuroscience 137:891–903. https://doi.org/10.1016/j.neuroscience.2005.10.038

Elliott JL, Snider WD (1995) Parvalbumin is a marker of ALS-resistant motor neurons. Neuroreport 6:449–452. https://doi.org/10.1097/00001756-199502000-00011

Erichsen JT, Wright NF, May PJ (2014) Morphology and ultrastructure of medial rectus subgroup motoneurons in the macaque monkey. J Comp Neurol 522:626–641. https://doi.org/10.1002/cne.23437

Eriksson NP, Lindsay RM, Aldskogius H (1994) BDNF and NT-3 rescue sensory but not motoneurones following axotomy in the neonate. Neuroreport 5:1445–1448

Ernfors P, Henschen A, Olson L, Persson H (1989) Expression of nerve growth factor receptor mRNA is developmentally regulated and increased after axotomy in rat spinal cord motoneurons. Neuron 2:1605–1613. https://doi.org/10.1016/0896-6273(89)90049-4ç

Ernfors P, Wetmore C, Eriksdotter-Nilsson M, Bygdeman M, Strömberg I, Olson L, Persson H (1991) The nerve growth factor receptor gene is expressed in both neuronal and non-neuronal tissues in the human fetus. Int J Dev Neurosci 9:57–66. https://doi.org/10.1016/0736-5748(91)90073-u

Escudero M, Delgado-García JM (1988) Behavior of reticular, vestibular and prepositus neurons terminating in the abducens nucleus of the alert cat. Exp Brain Res 71:218–222. https://doi.org/10.1007/bf00247538

Escudero M, de la Cruz RR, Delgado-García JM (1992) A physiological study of vestibular and prepositus hypoglossi neurones projecting to the abducens nucleus in the alert cat. J Physiol 458:539–560. https://doi.org/10.1113/jphysiol.1992.sp019433

Evinger C, Baker R (1991) Are there subdivisions of extraocular motoneuronal pools that can be controlled selectively? In: Humphrey DR, Freund H-J (eds) Motor control: concepts and issues. Wiley, pp 23–31

Ferrara N, Henzel WJ (2012) Pituitary follicular cells secrete a novel heparin-binding growth factor specific for vascular endothelial cells. 1989. Biochem Biophys Res Commun 425:540–547. https://doi.org/10.1016/j.bbrc.2012.08.021

Ferri CC, Ghasemlou N, Bisby MA, Kawaja MD (2002) Nerve growth factor alters p75 neurotrophin receptor-induced effects in mouse facial motoneurons following axotomy. Brain Res 950:180–185. https://doi.org/10.1016/s0006-8993(02)03035-4

Fuchs AF, Luschei ES (1970) Firing patterns of abducens neurons of alert monkeys in relationship to horizontal eye movement. J Neurophysiol 33:382–392. https://doi.org/10.1152/jn.1970.33.3.382

Fuchs AF, Luschei ES (1971) The activity of single trochlear nerve fibers during eye movements in the alert monkey. Exp Brain Res 13:78–89. https://doi.org/10.1007/BF00236431

Fuchs AF, Robinson DA (1966) A method for measuring horizontal and vertical eye movement chronically in the monkey. J Appl Physiol 21:1068–1070. https://doi.org/10.1152/jappl.1966.21.3.1068

Fuchs AF, Scudder CA, Kaneko CR (1988) Discharge patterns and recruitment order of identified motoneurons and internuclear neurons in the monkey abducens nucleus. J Neurophysiol 60:1874–1895. https://doi.org/10.1152/jn.1988.60.6.1874

Ghasia FF, Angelaki D (1995) Do motoneurons encode the noncommutativity of ocular rotations? Neuron 47:281–293. https://doi.org/10.1016/j.neuron.2005.05.031

Goldberg J (1980) Activity of abducens nucleus units in the alert cat. PhD thesis. University of California, Berkeley

Goldberg SJ, Meredith MAM, Shall MS (1998) ExtraocularMotor unit and whole-muscle responses in the lateral rectus muscle of the squirrel monkey. J Neurosci 18:10629–10639. https://doi.org/10.1523/JNEUROSCI.18-24-10629.1998

Goldstein HP, Robinson DA (1986) Hysteresis and slow drift in abducens unit activity. J Neurophysiol 55:1044–1056. https://doi.org/10.1152/jn.1986.55.5.1044

González-Forero D, Alvarez FJ, de la Cruz RR, Delgado-García JM, Pastor AM (2002) Influence of afferent synaptic innervation on the discharge variability of cat abducens motoneurones. J Physiol 541:283–299. https://doi.org/10.1113/jphysiol.2001.013405

González-Forero D, Pastor AM, Delgado-García JM, de la Cruz RR, Alvarez FJ (2004) Synaptic structural modification following changes in activity induced by tetanus neurotoxin in cat abducens neurons. J Comp Neurol 471:201–218. https://doi.org/10.1002/cne.20039

Graeber MB, Kreutzberg GW (1988) Delayed astrocyte reaction following facial nerve axotomy. J Neurocytol 17:209–220. https://doi.org/10.1007/bf01674208

Graft W, Simpson JI (1981) Relations between the semicircular canals, the optic axis, and the extraocular muscles in lateral-eyed and frontal-eyed animals. In: Fuchs A, Becker (eds) Progress in oculomotor research, Elsevier North Holland, p 409–417

Gustafsson B (1979) Changes in motoneurone electrical properties following axotomy. J Physiol 293:197–215. https://doi.org/10.1113/jphysiol.1979.sp012885

Gustafsson B, Pinter MJ (1984a) Relations among passive electrical properties of lumbar alpha-motoneurones of the cat. J Physiol 356:401–431. https://doi.org/10.1113/jphysiol.1984.sp015473

Gustafsson B, Pinter MJ (1984b) Effects of axotomy on the distribution of passive electrical properties of cat motoneurones. J Physiol 356:433–442. https://doi.org/10.1113/jphysiol.1984.sp015474

Haenggeli C, Kato AC (2002) Differential vulnerability of cranial motoneurons in mouse models with motor neuron degeneration. Neurosci Lett 335:39–43. https://doi.org/10.1016/s0304-3940(02)01140-0

Hamberger A, Hansson HA, Sjöstrand J (1970) Surface structure of isolated neurons: detachment of nerve terminals during axon regeneration. J Cell Biol 47:319–331. https://doi.org/10.1083/jcb.47.2.319

Henderson CE, Camu W, Mettling C, Gouin A, Poulsen K, Karihaloo M, Rullamas J, Evans T, McMahon SB, Armanini MP, Berkemeler L, Phillips HS, Rosenthal A (1993) Neurotrophins promote motor neuron survival and are present in embryonic limb bud. Nature 363:266–270. https://doi.org/10.1038/363266a0

Henn V, Cohen B (1972) Eye muscle motor neurons with different functional characteristics. Brain Res 45:561–568. https://doi.org/10.1016/0006-8993(72)90483-0

Henneman E (1981) Recruitment of motoneurons: the size principle. In: Desmett JE (ed) Motor unit types, recruitment and plasticity in health and disease, vol 9. Prog Clin Neurophysiol, Basel, pp 26–60

Henneman E, Somjen G, Carpenter DO (1965a) Functional significance of cell size in spinal motoneurons. J Neurophysiol 28:560–580. https://doi.org/10.1152/jn.1965.28.3.560

Henneman E, Somjen G, Carpenter DO (1965b) Excitability and inhibitability of motoneurons of different sizes. J Neurophysiol 28:599–620. https://doi.org/10.1152/jn.1965.28.3.599

Hernández RG, Benítez-Temiño B, Morado-Díaz CJ, Davis-López de Carrizosa MA, de la Cruz RR, Pastor AM (2017a) Effects of selective deafferentation on the discharge characteristics of medial rectus motoneurons. J Neurosci 37:9172–9188. https://doi.org/10.1523/JNEUROSCI.1391-17.2017

Hernández RG, Silva-Hucha S, Morcuende S, de la Cruz RR, Pastor AM, Benítez-Temiño B (2017b) Extraocular motor system exhibits a higher expression of neurotrophins when compared with other brainstem motor systems. Front Neurosci 11:399. https://doi.org/10.3389/fnins.2017.00399

Hernández RG, Calvo PM, Blumer R, de la Cruz RR, Pastor AM (2019) Functional diversity of motoneurons in the oculomotor system. Proc Natl Acad Sci U S A 116:3837–3846. https://doi.org/10.1073/pnas.1818524116

Hess A, Pilar G (1963) Slow fibres in the extraocular muscles of the cat. J Physiol 169:780–798. https://doi.org/10.1113/jphysiol.1963.sp007296

Highstein SM, Holstein (2006) The anatomy of the vestibular nuclei. In: Büttner-Ennever JA (ed) Neuroanatomy of the oculomotor system, 1st edn. Elsevier, Amsterdam, pp 157–203

Highstein SM, Maekawa K, Steinacker A, Cohen B (1976) Synaptic input from the pontine reticular nuclei to abducens motoneurons and internuclear neurons in the cat. Brain Res 112:162–167. https://doi.org/10.1016/0006-8993(76)90344-9

Hikosaka O, Igusa Y, Nakao S, Shimazu H (1978) Direct inhibitory synaptic linkage of pontomedullary reticular burst neurons with abducens motoneurons in the cat. Exp Brain Res 33:337–352. https://doi.org/10.1007/bf00235558

Hoh JFY (2021) Myosin heavy chains in extraocular muscle fibres: distribution, regulation and function. Acta Physiol (Oxf) 231:e13535. https://doi.org/10.1111/apha.13535

Horn AKE, Horng A, Buresch N, Messoudi A, Härtig W (2018) Identification of functional cell groups in the abducens nucleus of monkey and human by Perineuronal nets and choline acetyltransferase immunolabeling. Front Neuroanat 12:45. https://doi.org/10.3389/fnana.2018.00045

Houenou LJ, Li L, Lo AC, Yan Q, Oppenheim RW (1994) Naturally occurring and axotomy-induced motoneuron death and its prevention by neurotrophic agents: a comparison between chick and mouse. Prog Brain Res 102:217–226. https://doi.org/10.1016/S0079-6123(08)60542-7

Igusa Y, Sasaki S, Shimazu H (1980) Excitatory premotor burst neurons in the cat pontine reticular formation related to the quick phase of vestibular nystagmus. Brain Res 182:451–456. https://doi.org/10.1016/0006-8993(80)91202-0

Isomura G (1981) Comparative anatomy of the extrinsic ocular muscles in vertebrates. Anat Anz 150:498.-515

Keller EL, Robinson DA (1972) Abducens unit behavior in the monkey during vergence movements. Vis Res 12:369–382. https://doi.org/10.1016/0042-6989(72)90082-x

Kettenmann H, Kirchhoff F, Verkhratsky A (2013) Microglia: new roles for the synaptic stripper. Neuron 77:10–18. https://doi.org/10.1016/j.neuron.2012.12.023

Koliatsos VE, Crawford TO, Price DL (1991) Axotomy induces nerve growth factor receptor immunoreactivity in spinal motor neurons. Brain Res 549:297–304. https://doi.org/10.1016/0006-8993(91)90471-7

Koliatsos VE, Clatterbuck RE, Winslow JW, Cayouette MH, Price DL (1993) Evidence that brain-derived neurotrophic factor is a trophic factor for motor neurons in vivo. Neuron 10:359–367

Konakci KZ, Streicher J, Hoetzenecker W, Blumer MJ, Lukas JR, Blumer R (2005) Molecular characteristics suggest an effector function of palisade endings in extraocular muscles. Invest Ophthalmol Vis Sci 46:155–165. https://doi.org/10.1167/iovs.04-

Korsching S (1993) The neurotrophic factor concept: a reexamination. J Neurosci 13:2739–2748. https://doi.org/10.1523/JNEUROSCI.13-07-02739.1993

Kreutzberg GW, Graeber MB, Streit WJ (1989) Neuron-glial relationship during regeneration of motorneurons. Metab Brain Dis 4:81–85. https://doi.org/10.1007/bf00999498

Kuno M, Llinás R (1970a) Enhancement of synaptic transmission by dendritic potentials in chromatolysed motoneurones of the cat. J Physiol 210:807–821. https://doi.org/10.1113/jphysiol.1970.sp009243

Kuno M, Llinás R (1970b) Alterations of synaptic action in chromatolysed motoneurones of the cat. J Physiol 210:823–838. https://doi.org/10.1113/jphysiol.1970.sp009244

Lambrechts D, Carmeliet P (2006) VEGF at the neurovascular interface: therapeutic implications for motor neuron disease. Biochim Biophys Acta 1762:1109–1121. https://doi.org/10.1016/j.bbadis.2006.04.005

Lampl I, Reichova I, Ferster D (1999) Synchronous membrane potential fluctuations in neurons of the cat visual cortex. Neuron 22:361–374. https://doi.org/10.1016/s0896-6273(00)81096-x

Land MF (1969) Movements of the retinae of jumping spiders (salticidae dedryphantinae) in response to visual stimuli. J Exp Biol 51:471–493

Lange C, Storkebaum E, de Almodóvar CR, Dewerchin M, Carmeliet P (2016) Vascular endothelial growth factor: a neurovascular target in neurological diseases. Nat Rev Neurol 12:439–454. https://doi.org/10.1038/nrneurol.2016.88

Laslo P, Lipski J, Funk GD (2001) Differential expression of group I metabotropic glutamate receptors in motoneurons at low and high risk for degeneration in ALS. Neuroreport 12:1903–1908. https://doi.org/10.1097/00001756-200107030-00027

Lewin GR, Carter BD (2014) Neurotrophic factors. Springer, Berlin, p 1087
Lienbacher K, Mustari M, Ying HS, Buttner-Ennever JA, Horn AK (2011) Do palisade endings in extraocular muscles arise from neurons in the motor nuclei? Invest Ophthalmol Vis Sci 52:2510–2519. https://doi.org/10.1167/iovs.10-6008
Lindå H, Cullheim S, Risling M (1992) A light and electron microscopic study of intracellularly HRP-labeled lumbar motoneurons after intramedullary axotomy in the adult cat. J Comp Neurol 318:188–208. https://doi.org/10.1002/cne.903180205
Lindå H, Shupliakov O, Ornung G, Ottersen OP, Storm-Mathisen J, Risling M, Cullheim S (2000) Ultrastructural evidence for a preferential elimination of glutamate-immunoreactive synaptic terminals from spinal motoneurons after intramedullary axotomy. J Comp Neurol 425:10–23
Lindsay RM (1994) Trophic protection of motor neurons: clinical potential in motor neuron diseases. J Neurol 242(1 Suppl 1):S8–S11. https://doi.org/10.1007/bf00939232
Lorente de No R (1933) Vestibulo-ocular reflex arc. Arch Neurol Psychiatr 20:245–291
Luther JA, Birren SJ (2009) p75 and TrkA signaling regulates sympathetic neuronal firing patterns via differential modulation of voltage-gated currents. J Neurosci 29:5411–5424. https://doi.org/10.1523/JNEUROSCI.3503-08.2009
McCrea RA, Baker R (1985) Anatomical connections of the nucleus prepositus of the cat. J Comp Neurol 237:377–407. https://doi.org/10.1002/cne.902370308
McCrea RA, Strassman A, May E, Highstein SM (1987) Anatomical and physiological characteristics of vestibular neurons mediating the horizontal vestibulo-ocular reflex of the squirrel monkey. J Comp Neurol 264:547–570. https://doi.org/10.1002/cne.902640408
Medina L, Figueredo-Cardenas G, Rothstein JD, Reiner A (1996) Differential abundance of glutamate transporter subtypes in amyotrophic lateral sclerosis (ALS)-vulnerable versus ALS-resistant brain stem motor cell groups. Exp Neurol 142:287–295. https://doi.org/10.1006/exnr.1996.0198
Merlio JP, Ernfors P, Jaber M, Persson H (1992) Molecular cloning of rat trkC and distribution of cells expressing messenger RNAs for members of the trk family in the rat central nervous system. Neuroscience 51:513–532. https://doi.org/10.1016/0306-4522(92)90292-a
Miller JM, Bockisch CJ, Pavlovski DS (2002) Missing lateral rectus force and absence of nedial rectus cocontraction in ocular convergence. 87:2421–2433. https://doi.org/10.1152/jn.00566.2001
Moran LB, Graeber MB (2004) The facial nerve axotomy model. Brain Res Brain Res Rev 44:154–178. https://doi.org/10.1016/j.brainresrev.2003.11.004
Morcuende S, Matarredona ER, Benítez-Temiño B, Muñoz-Hernández R, Pastor AM, de la Cruz RR (2011) Differential regulation of the expression of neurotrophin receptors in rat extraocular motoneurons after lesion. J Comp Neurol 519:2335–2352. https://doi.org/10.1002/cne.22630
Morcuende S, Muñoz-Hernández R, Benítez-Temiño B, Pastor AM, de la Cruz RR (2013) Neuroprotective effects of NGF, BDNF, NT-3 and GDNF on axotomized extraocular motoneurons in neonatal rats. Neuroscience 250:31–48. https://doi.org/10.1016/j.neuroscience.2013.06.050
Morgan DL, Proske U (1984) Vertebrate slow muscle: its structure, pattern of innervation, and mechanical properties. Physiol Rev 64:103–169. https://doi.org/10.1152/physrev.1984.64.1.103
Namba T, Nakamura T, Takahashi A, Grob D (1968) Motor nerve endings in extraocular muscles. J Comp Neurol 134:385–396. https://doi.org/10.1002/cne.901340402
Navarro X, Vivó M, Valero-Cabré A (2007) Neural plasticity after peripheral nerve injury and regeneration. Prog Neurobiol 82:163–201. https://doi.org/10.1016/j.pneurobio.2007.06.005
Nelson JS, Goldberg SJ, McClung JR (1986) Motoneuron electrophysiological and muscle contractile properties of superior oblique motor units in cat. J Neurophysiol 55:715–726. https://doi.org/10.1152/jn.1986.55.4.715
Nicoletti JN, Shah SK, McCloskey DP, Goodman JH, Elkady A, Atassi H, Hylton D, Rudge JS, Scharfman HE, Croll SD (2008) Vascular endothelial growth factor is up-regulated after status epilepticus and protects against seizure-induced neuronal loss in hippocampus. Neuroscience 151:232–241. https://doi.org/10.1016/j.neuroscience.2007.09.083

Nieto-Gonzalez JL, Carrascal L, Nunez-Abades P, Torres B (2009) Muscarinic modulation of recruitment threshold and firing rate in rat oculomotor nucleus motoneurons. J Neurophysiol 101:100–111. https://doi.org/10.1152/jn.90239.2008

Nimchinsky EA, Young WG, Yeung G, Shah RA, Gordon JW, Bloom FE, Morrison JH, Hof PR (2000) Differential vulnerability of oculomotor, facial, and hypoglossal nuclei in G86R superoxide dismutase transgenic mice. J Comp Neurol 416:112–125. https://doi.org/10.1002/(sici)1096-9861(20000103)416:1%3C112::aid-cne9%3E3.0.co;2-k

Nishio T, Sunohara N, Furukawa S (1998) Neutrophin switching in spinal motoneurons of amyotrophic lateral sclerosis. Neuroreport 9:1661–1665. https://doi.org/10.1097/00001756-199805110-00073

Olmstead DN, Mesnard-Hoaglin NA, Batka RJ, Haulcomb MM, Miller WM, Jones KJ (2015) Facial nerve axotomy in mice: a model to study motoneuron response to injury. J Vis Exp 23:e52382. https://doi.org/10.3791/52382

Omura T, Sano M, Omura K, Hasegawa T, Doi M, Sawada T, Nagano A (2005) Different expressions of BDNF, NT3, and NT4 in muscle and nerve after various types of peripheral nerve injuries. J Peripher Nerv Syst 10(293–300). https://doi.org/10.1111/j.1085-9489.2005.10307.x

Oosthuyse B, Moons L, Storkebaum E, Beck H, Nuyens D, Brusselmans K, Van Dorpe J, Hellings P, Gorselink M, Heymans S, Theilmeier G, Dewerchin M, Laudenbach V, Vermylen P, Raat H, Acker T, Vleminckx V, Van Den Bosch L, Cashman N, Fujisawa H, Drost MR, Sciot R, Bruyninckx F, Hicklin DJ, Ince C, Gressens P, Lupu F, Plate KH, Robberecht W, Herbert JM, Collen D, Carmeliet P (2001) Deletion of the hypoxia-response element in the vascular endothelial growth factor promoter causes motor neuron degeneration. Nat Genet 28:131–138. https://doi.org/10.1038/88842

Oppenheim RW, Houenou LJ, Johnson JE, Lin LF, Li L, Lo AC, Newsome AL, Prevette DM, Wang S (1995) Developing motor neurons rescued from programmed and axotomy-induced cell death by GDNF. Nature 373:344–346. https://doi.org/10.1038/373344a0

Pásaro R, Torres B, Delgado-García JM (1985) Morphological effects of VIth nerve section in the kitten as revealed by horseradish peroxidase. Neurosci Lett 58:207–211. https://doi.org/10.1016/0304-3940(85)90165-x

Pastor AM, González-Forero D (2003) Recruitment order of cat abducens motoneurons and internuclear neurons. J Neurophysiol 90:2240–2252. https://doi.org/10.1152/jn.00402.2003

Pastor AM, Torres B, Delgado-Garcia JM, Baker R (1991) Discharge characteristics of medial rectus and abducens motoneurons in the goldfish. J Neurophysiol 66:2125–2140. https://doi.org/10.1152/jn.1991.66.6.2125

Pastor AM, Calvo PM, de la Cruz RR, Baker R, Straka H (2019) Discharge properties of morphologically identified vestibular neurons recorded during horizontal eye movements in the goldfish. J Neurophysiol 121:1865–1878. https://doi.org/10.1152/jn.00772.2018

Peters OM, Ghasemi M, Brown RH Jr (2015) Emerging mechanisms of molecular pathology in ALS. J Clin Invest 125:1767–1779. https://doi.org/10.1172/JCI71601

Piehl F, Frisén J, Risling M, Hökfelt T, Cullheim S (1994) Increased trkB mRNA expression by axotomized motoneurones. Neuroreport 5:697–700. https://doi.org/10.1097/00001756-199402000-00009

Powers RK, Binder MD (2000) Relationship between the time course of the afterhyperpolarization and discharge variability in cat spinal motoneurones. J Physiol 528:131–150. https://doi.org/10.1111/j.1469-7793.2000.t01-1-00131.x

Precht W, Grippo J, Richter A (1967) Effects of horizontal angular acceleration on neurons in the abducens nucleus. 5:527–531. https://doi.org/10.1016/0006-8993(67)90028-5

Purves D (1990) Body and brain. A trophic theory of neural connections. Harvard University Press, Cambridge. https://doi.org/10.1126/science.244.4907.993

Reiner A, Medina L, Figueredo-Cardenas G, Anfinson S (1995) Brainstem motoneuron pools that are selectively resistant in amyotrophic lateral sclerosis are preferentially enriched in parvalbumin: evidence from monkey brainstem for a calcium-mediated mechanism in sporadic ALS. Exp Neurol 131:239–250. https://doi.org/10.1016/0014-4886(95)90046-2

Ringer C, Weihe E, Schütz B (2011) Calcitonin gene-related peptide expression levels predict motor neuron vulnerability in the superoxide dismutase 1-G93A mouse model of amyotrophic lateral sclerosis. Neurobiol Dis 45:547–554. https://doi.org/10.1016/j.nbd.2011.09.011

Robinson DA (1970) Oculomotor unit behavior in the monkey. J Neurophysiol 33:393–403. https://doi.org/10.1152/jn.1970.33.3.393

Robinson DA (1981) The use of control systems analysis in the neurophysiology of eye movements. Annu Rev Neurosci 4:463–503. https://doi.org/10.1146/annurev.ne.04.030181.002335

Rotterman TM, Akhter ET, Lane AR, MacPherson KP, García VV, Tansey MG, Alvarez FJ (2019) Spinal motor circuit synaptic plasticity after peripheral nerve injury depends on microglia activation and a CCR2 mechanism. J Neurosci 39:3412–3433. https://doi.org/10.1523/JNEUROSCI.2945-17.2019

Ruskell G (1978) The fine structure of innervated myotendinous cylinders in monkey. J Neurocytol 7:693–708. https://doi.org/10.1007/BF01205145

Salama-Cohen P, Arévalo MA, Grantyn R, Rodríguez-Tébar A (2006) Notch and NGF/p75NTR control dendrite morphology and the balance of excitatory/inhibitory synaptic input to hippocampal neurones through Neurogenin 3. J Neurochem 97:1269–1278. https://doi.org/10.1111/j.1471-4159.2006.03783.x

Schiller PH (1970) The discharge characteristics of single units in the oculomotor and abducens nuclei of the unanesthetized monkey. Exp Brain Res 10:347–362. https://doi.org/10.1007/bf02324764

Sendtner M, Holtmann B, Kolbeck R, Thoenen H, Barde YA (1992) Brain-derived neurotrophic factor prevents the death of motoneurons in newborn rats after nerve section. Nature 360:757–759. https://doi.org/10.1038/360757a0

Senger DR, Galli SJ, Dvorak AM, Perruzzi CA, Harvey VS, Dvorak HF (1983) Tumor cells secrete a vascular permeability factor that promotes accumulation of ascites fluid. Science 219:983–985. https://doi.org/10.1126/science.6823562

Shadlen MN, Newsome WT (1998) The variable discharge of cortical neurons: implications for connectivity, computation, and information coding. J Neurosci 18:3870–3896. https://doi.org/10.1523/JNEUROSCI.18-10-03870.1998

Silva-Hucha S, Hernández RG, Benítez-Temiño B, Pastor ÁM, de la Cruz RR, Morcuende (2017) Extraocular motoneurons of the adult rat show higher levels of vascular endothelial growth factor and its receptor Flk-1 than other cranial motoneurons. PLoS One 12:e0178616. https://doi.org/10.1371/journal.pone.0178616

Skavenski AA, Robinson DA (1973) Role of abducens neurons in vestibuloocular reflex. J Neurophysiol 36:724–738. https://doi.org/10.1152/jn.1973.36.4.724

Sofroniew MV, Galletly NP, Isacson O, Svendsen CN (1990) Survival of adult basal forebrain cholinergic neurons after loss of target neurons. Science 247:338–342. https://doi.org/10.1126/science.1688664

Sofroniew MV, Cooper JD, Svendsen CN, Crossman P, Ip NY, Lindsay RM, Zafra F, Lindholm D (1993) Atrophy but not death of adult septal cholinergic neurons after ablation of target capacity to produce nRNAs for NGF, BDNF and NT3 J Neurosci 13:5263–5276. doi: 10.1523/JNEUROSCI.13-12-05263.1993.

Spejo AB, Oliveira AL (2015) Synaptic rearrangement following axonal injury: old and new players. Neuropharmacology 96:113–123. https://doi.org/10.1016/j.neuropharm.2014.11.002

Spencer RF, Porter JD (2006) Biological organization of the extraocular muscles. In: Büttner-Ennever JA (ed) Neuroanatomy of the oculomotor system, 1st edn. Elsevier, Amsterdam, pp 43–80

Stahl JS, Simpson JI (1995) Dynamics of abducens nucleus neurons in the awake rabbit. J Neurophysiol 73:1383–1395. https://doi.org/10.1152/jn.1995.73.4.1383

Storkebaum E, Lambrechts D, Carmeliet P (2004) VEGF: once regarded as a specific angiogenic factor, now implicated in neuroprotection. BioEssays 26:943–954. https://doi.org/10.1002/bies.20092

Storkebaum E, Lambrechts D, Dewerchin M, Moreno-Murciano MP, Appelmans S, Oh H, Van Damme P, Rutten B, Man WY, De Mol M, Wyns S, Manka D, Vermeulen K, Van Den Bosch L, Mertens N, Schmitz C, Robberecht W, Conway EM, Collen D, Moons L, Carmeliet P (2005) Treatment of motoneuron degeneration by intracerebroventricular delivery of VEGF in a rat model of ALS. Nat Neurosci 8:85–92. https://doi.org/10.1038/nn1360

Strassman A, Highstein SM, McCrea RA (1986a) Anatomy and physiology of saccadic burst neurons in the alert squirrel monkey. I. Excitatory burst neurons J Comp Neurol 249:337–357. https://doi.org/10.1002/cne.902490303

Strassman A, Highstein SM, McCrea RA (1986b) Anatomy and physiology of saccadic burst neurons in the alert squirrel monkey. II. Inhibitory burst neurons J Comp Neurol 249:358–380. https://doi.org/10.1002/cne.902490304-

Sumner BE (1975a) A quantitative analysis of the response of presynaptic boutons to postsynaptic motor neuron axotomy. Exp Neurol 46:605–615. https://doi.org/10.1016/0014-4886(75)90129-6

Sumner BE (1975b) A quantitative analysis of boutons with different types of synapse in normal and injured hypoglossal nuclei. Exp Neurol 49:406–417. https://doi.org/10.1016/0014-4886(75)90097-7

Sumner BE, Sutherland FI (1973) Quantitative electron microscopy on the injured hypoglossal nucleus in the rat. J Neurocytol 2:315–328. https://doi.org/10.1007/bf01104033

Svensson M, Eriksson P, Persson JK, Molander C, Arvidsson J, Aldskogius H (1993) The response of central glia to peripheral nerve injury. Brain Res Bull 30:499–506. https://doi.org/10.1016/0361-9230(93)90284-i

Tang X, Büttner-Ennever JA, Mustari MJ, Horn AK (2015) Internal organization of medial rectus and inferior rectus muscle neurons in the C group of the oculomotor nucleus in monkey. J Comp Neurol 523:1809–1823. https://doi.org/10.1002/cne.23760

Titmus MJ, Faber DS (1990) Axotomy-induced alterations in the electrophysiological characteristics of neurons. Prog Neurobiol 35:1–51. https://doi.org/10.1016/0301-0082(90)90039-j

Tovar-Y-Romo LB, Zepeda A, Tapia R (2007) Vascular endothelial growth factor prevents paralysis and motoneuron death in a rat model of excitotoxic spinal cord neurodegeneration. J Neuropathol Exp Neurol 66:913–922. https://doi.org/10.1097/nen.0b013e3181567c16

Ugolini G, Klam F, Doldan Dans M, Dubayle D, Brandi AM, Büttner-Ennever J, Graf W (2006) Horizontal eye movement networks in primates as revealed by retrograde transneuronal transfer of rabies virus: differences in monosynaptic input to "slow" and "fast" abducens motoneurons. J Comp Neurol 498:762–785. https://doi.org/10.1002/cne.21092

Van Damme P, Van Den Bosch L, Van Houtte E, Callewaert G, Robberecht W (2002) GluR2-dependent properties of AMPA receptors determine the selective vulnerability of motor neurons to excitotoxicity. J Neurophysiol 88:1279–1287. https://doi.org/10.1152/jn.2002.88.3.1279

Vejsada R, Sagot Y, Kato AC (1995) Quantitative comparison of the transient rescue effects of neurotrophic factors on axotomized motoneurons in vivo. Eur J Neurosci 7:108–115. https://doi.org/10.1111/j.1460-9568.1995.tb01025.x

Vejsada R, Tseng JL, Lindsay RM, Acheson A, Aebischer P, Kato AC (1998) Synergistic but transient rescue effects of BDNF and GDNF on axotomized neonatal motoneurons. Neuroscience 84:129–139. https://doi.org/10.1016/s0306-4522(97)00497-1

Walls GL (1962) The evolutionary history of eye movements. Vis Res 2:69–80. https://doi.org/10.1016/0042-6989(62)90064-0

Wang Y, Duan W, Wang W, Wen D, Liu Y, Liu Y, Li Z, Hu H, Lin H, Cui C, Li D, Dong H, Li C (2016) scAAV9-VEGF prolongs the survival of transgenic ALS mice by promoting activation of M2 microglia and the PI3K/Akt pathway. Brain Res 1648:1–10. https://doi.org/10.1016/j.brainres.2016.06.043

Wasicky R, Horn AK, Büttner-Ennever JA (2004) Twitch and nontwitch motoneuron subgroups in the oculomotor nucleus of monkeys receive different afferent projections. J Comp Neurol 479:117–129. https://doi.org/10.1002/cne.20296

Yamada J, Nakanishi H, Jinno S (2011) Differential involvement of perineuronal astrocytes and microglia in synaptic stripping after hypoglossal axotomy. Neuroscience 182:1–10. https://doi.org/10.1016/j.neuroscience.2011.03.030

Yamanaka Y, Bach-Y-Rita P (1970) Relations between extraocular muscle contaction and extension times in each phase of nystagmus. Exp Neurol 20:143–155. https://doi.org/10.1016/0014-4886(68)90130-1

Yan Q, Elliott J, Snider WD (1992) Brain-derived neurotrophic factor rescues spinal motor neurons from axotomy-induced cell death. Nature 360:753–755. https://doi.org/10.1038/360753a0

Yuan Q, Wu W, So KF, Cheung AL, Prevette DM, Oppenheim RW (2000) Effects of neurotrophic factors on motoneuron survival following axonal injury in newborn rats. Neuroreport 11:2237–2241

Zacchigna S, Lambrechts D, Carmeliet P (2008) Neurovascular signalling defects in neurodegeneration. Nat Rev Neurosci 9:169–181. https://doi.org/10.1038/nrn2336

Zhang YH, Nicol GD (2004) NGF-mediated sensitization of the excitability of rat sensory neurons is prevented by a blocking antibody to the p75 neurotrophin receptor. Neurosci Lett 366:187–192. https://doi.org/10.1016/j.neulet.2004.05.042

Zimmermann L, Morado-Diaz CJ, Davis-Lopez de Carrizosa MA, De La Cruz RR, May PJ, Streicher J, Pastor AM, Blumer R (2013) Axons giving rise to the palisade endings of feline extraocular muscles display motor features. J Neurosci 33(2784–2793):2013. https://doi.org/10.1523/JNEUROSCI.4116-12.2013

Part III
Motoneuron Disease

Motoneuron Diseases

Francesco Lotti and Serge Przedborski

Abstract Motoneuron diseases (MNDs) represent a heterogeneous group of progressive paralytic disorders, mainly characterized by the loss of upper (corticospinal) motoneurons, lower (spinal) motoneurons or, often both. MNDs can occur from birth to adulthood and have a highly variable clinical presentation, even within gene-positive forms, suggesting the existence of environmental and genetic modifiers. A combination of cell autonomous and non-cell autonomous mechanisms contributes to motoneuron degeneration in MNDs, suggesting multifactorial pathogenic processes.

Keywords Amyotrophic lateral sclerosis · Cell autonomy · Motor cortex · Spinal cord · C9ORF72 · FUS · Motoneuron diseases · Neurodegeneration · Superoxide dismutase-1 · Spinal muscular atrophy · TDP43

1 Introduction

Motoneuron diseases (MNDs) refers to a collection of neurological conditions grouped together on the basis of shared clinical and neuropathological hallmarks: muscle wasting and weakness attributed to the specific loss of upper motoneurons – giant multipolar, pyramidal neurons called Betz cells in layer V of the primary motor cortex of the cerebral cortex – and/or lower motoneurons – primarily multipolar alpha-motoneurons found in the brainstem motor nuclei and anterior horn of the spinal cord (Highley and Ince 2012; Rowland et al. 2010). Furthermore, it is increasingly recognized that while motoneurons are the main and shared target of the neurodegenerative processes in MNDs, non-motoneurons may also be affected (Highley and Ince 2012), suggesting that these paralytic disorders may be more multisystem neurodegenerative disorders than initially thought. This is why,

F. Lotti · S. Przedborski (✉)
Departments of Neurology, Pathology & Cell Biology, and Neuroscience, College of Physicians and Surgeons, Columbia University, New York, NY, USA
e-mail: sp30@columbia.edu

M. J. O'Donovan, M. Falgairolle (eds.), *Vertebrate Motoneurons*, Advances in Neurobiology 28, https://doi.org/10.1007/978-3-031-07167-6_13

now-a-day, reference to frontal temporal dementia is often part of the discussion about the clinical spectrum of MNDs, as mentioned below.

Most MNDs are *sporadic*, meaning that they seem to occur at random with no family history of the diseases. However, some cases are inherited, due to mutations in a variety of genes (Kim et al. 2020a; Shribman et al. 2019; Tisdale and Pellizzoni 2015) or caused by viral infections such as poliovirus, West Nile virus or enterovirus D68 (Fatemi and Chakraborty 2019), biochemical defects such as hexosaminidase deficiency (Johnson et al. 1982) and even, presumably, by heavy metal poisoning provoked by lead or mercury exposures (Campbell et al. 1970; Mitchell 1987).

Typically, MNDs include diseases such as amyotrophic lateral sclerosis (ALS), progressive bulbar palsy, primary lateral sclerosis, progressive muscular atrophy, spinal muscular atrophy (SMA), Kennedy's disease, and post-polio syndrome. However, rather than discussing each MND, this chapter will focus mainly on ALS (also known as Lou Gehrig's disease or motoneurone disease), as the prototypical example of adult-onset MND, and on SMA, as the prototypical example of childhood-onset MND. ALS, which is the most common MND, is typically a mid-life progressive paralytic disorder that presents as sporadic in 90% and familial in 10% of all cases characterized neuropathologically by the loss of both upper motoneurons and lower motor neurons associated with gliosis and a variety of proteinaceous aggregates such as ubiquitinated, transactive response DNA-binding protein of 43 kDa (TDP43) neuronal cytoplasmic inclusions (Highley and Ince 2012; Ince and Wharton 2007). As for SMA, it is a pure lower MND due to autosomal recessive mutations in the *SMN1* gene that encodes for the survival motor neuron protein (SMN) (Ince and Wharton 2007; Tisdale and Pellizzoni 2015). We recognize that this selection introduces a bias, yet these two MNDs encompass the vast majority of patients afflicted with disorders of the motoneurons, and thus should provide a faithful picture of the state of affairs regarding these neurodegenerative paralytic conditions. Reader interested in other kind of MNDs is directed to the following references (Fatemi and Chakraborty 2019; Ince and Wharton 2007; Shribman et al. 2019).

2 What Some Numbers Tell About MNDs

Logroscino et al. 2018 while this figure is dramatic, relative to other common neurodegenerative disorders such as Parkinson's disease and Alzheimer's disease, it is more than 12- and 84-fold smaller, respectively (Feigin et al. 2019). Although this relatively low number of affected individuals reflects the MNDs low worldwide all-age prevalence (4·5 per 100 000 people) and incidence (0·78 per 100 000 person-years) (Logroscino et al. 2018), it fails to capture the dramatic impact of MNDs at the public health level, in part, due to the fact that these paralytic disorders cause disproportionally higher disability and fatality rates compared to other neurodegenerative disorders. For instance, based on the data reported in two large epidemiological studies (Feigin et al. 2019; Logroscino et al. 2018), if one calculates ratios of the disability-adjusted life year divided by the prevalence for MNDs, Parkinson's disease and Alzheimer's disease in 2016, it is found that the overall disease burden among individuals with MNDs is 4–5 times higher than that among individuals with Parkinson's disease or Alzheimer's disease. Likewise, if one calculates the worldwide annual case fatality rate in individuals with MNDs for 2016, it is more than 10%, which is 2- and 3-times higher than that in individuals with Parkinson's disease and Alzheimer's disease, respectively.

Other striking epidemiological findings worth mentioning include sex and geographical differences in the occurrence of MNDs. For instance, the prevalence of MNDs was found to be consistently higher in males than females across all age groups, a difference thought to be attributed to distinct exposure of females to environmental factors, such as smoking (GBD 2017). However, the evidence for an association of smoking with MNDs was found to be insufficient (Logroscino et al. 2018). Moreover, Logroscino et al. (2018) reported that the highest incidence and prevalence for MNDs are found in high-income countries such as North America, Australia/New Zeeland and western Europe whereas the lowest incidence and prevalence are found in sub-Saharan Africa, a geographical heterogeneity that persists even after controlling for differences in age structure among different countries. Although one may argue that such epidemiological differences are rooted in health care disparity among countries, Logroscino et al. (2018) also showed that the age-standardized prevalence and incidence for MNDs are much lower than expected in high-income Asia Pacific countries such as Japan, Singapore or South Korea. Thus, sociodemographic development inequalities are unlikely the sole factor responsible for geographic variation in MNDs occurrence and burden of disease. Cogent to this intriguing observation, Hardiman (2018) pointed out that, while the incidence variation of SMA across populations, which, as indicated above, is monogenic form of MND, might be a function of different carrier rates of the disease-causing variants of the *SMN1* gene across different ancestral populations, the reasons for the geographic heterogeneity of largely sporadic MNDs such as 90% of all cases of ALS is unclear. Over the years, some examples of geographic clusters of ALS have been described, supporting a role of lifestyles, occupations, sports and environment in the occurrence of ALS (Ingre et al. 2015). One such example includes the Western Pacific form of ALS, mainly in Guam and the Kii Peninsula of Honshu Island, Japan, where the disease had a prevalence 50–100 times higher than in other parts of the world (Kuzuhara et al. 2001; Plato et al. 2003). Likewise, higher occurrence of ALS was found in Italy among professional football players (Chio et al. 2005, 2009) and in the USA among deployed members of the army (IOM 2006). Geographic heterogeneity is also well documented for a rare MND called monomelic amyotrophy, which is marked by an insidious onset of focal weakness and muscle wasting most commonly affecting an arm and a hand (Brannagan III 2010). Indeed, while this particular MND has a global distribution, it is essentially seen in individuals from Japan, China and India (Brannagan III 2010). These clusters while quite enlightening must be regarded as the exceptions rather than the rules. Moreover, a careful review of the long list of occupational and environmental factors did not find any definitive association with MNDs (Logroscino et al. 2018), including ALS (Ingre et al. 2015). Yet, before discounting a contribution of the environment in MNDs, it would be important to study these non-genetic factors, no longer individually but in combination and via their potential interaction with genetic factors. Here, the latter do not refer to pathogenic mutations in genes such as *C9ORF72*, superoxide dismutase 1 (*SOD1*), fused in sarcoma/translocated in liposarcoma (*FUS/TLS*), TAR DNA Binding Protein 43 (*TARDBP*) or *SMN1*, which are known genetic causes of MNDs (Kim et al. 2020a; Shribman et al. 2019; Tisdale and Pellizzoni 2015), but rather genetic variants associated with an increased risk of developing MNDs. Along this line, it must be mentioned that the gene encoding for *ATAXIN-2*, which is a known cause of spinocerebellar ataxia when its polyglutamine expansion is greater than 34 glutamines, increases the risk of developing ALS when its polyglutamine expansion is between 27 and 33 (Elden et al. 2010).

Lastly, it is important to emphasize that the occurrence of MNDs is rare before the age of 50 but, thereafter, it increases dramatically with a peak at around 85 years (Logroscino et al. 2018). This observation supports the notion that, like in other neurodegenerative disorders such as Parkinson's disease and Alzheimer's disease (Tanner et al. 2014), older age is a risk factor for developing MNDs, a fact that may have far reaching implications for the generations to come. Indeed, as previously noted (Przedborski 2017), over the past century, the growth rate of the population ages 65 and above in industrialized countries has far exceeded that of the population as a whole. Therefore, it can be anticipated that, over the next generations, the proportion of elderly citizens will double and consequently, the number of persons suffering from MNDs will rise dramatically as well.

3 The Different Facets of Genetics in MNDs

Over the past two decades, advances in genetics have been transformative for our understanding of the possible causes and mechanisms of MNDs. Not only has genetics led to the discovery of a long list of gene mutations that cause MNDs, but it also led to the emergence of new concepts that have modified our approach to these paralytic disorders such as *oligogenic* inheritance (i.e. a trait influenced by a few genes, which represents an intermediate between monogenic inheritance in which a trait is determined by a single causative gene, and polygenic inheritance, in which a trait is influenced by many genes and often environmental factors) and *pleiotropy* (i.e. a phenomenon by which one gene controls the expression of several phenotypic traits).

Sporadic Does Not Equate Non-genetic in MNDs

An absence of family history for the disease in a patient with a MND is necessary to refer to this case as sporadic, but it is not sufficient to rule out a genetic basis. Indeed, a single, isolated occurrence of a genetic MND in a family can results from wrong family histories related to, for example, non-paternity, adoption, or failure to recognize MNDs in affected family members as well as monogenic causes with reduced penetrance or *de novo* mutations. Relevant to this, roughly 11% of cases of sporadic ALS in populations of European ancestry were found to harbor mutations in a known ALS-related gene, of which more than half were hexanucleotide repeat expansion in *C9ORF72* and, to a lesser extent, mutations in *SOD1* (Renton et al. 2014); of note, this relatively high proportion of pathogenic *C9ORF72* repeat expansion in sporadic ALS is true for white populations with high Scandinavian admixture (Majounie et al. 2012), but is very low in other populations.

Moreover, even in absence of known pathogenic mutations, evidence of heritability has been well documented, for example, in sporadic ALS by parent-offspring (Ryan et al. 2019), twin (Al-Chalabi et al. 2010; Graham et al. 1997), pedigree (Wingo et al. 2011) and genome wide single-nucleotide polymorphism data (Fogh et al. 2014; Keller et al. 2014; McLaughlin et al. 2017; van Rheenen et al. 2016). These studies have used different populations, samples sizes, and methodologies, hence, have generated different estimates of heritability of ALS ranging from as high as 85% to as low as 7.2%. However, the main point here is not the large variability, but the fact that all of these studies converge toward the same conclusion that sporadic ALS carries some significant heritability. Incidentally, since ALS is more frequent in men than in women (Rowland et al. 2010), it is remarkable to note that its heritability estimates were reported to be the highest in mother-daughter pairings (Ryan et al. 2019), suggesting the possibility that sex influences the risk of the disease differently between related and unrelated individuals.

These findings indicate that a small fraction of sporadic ALS heritability can be explained by mutations in known ALS genes, but a much larger portion remains unsolved albeit possible sources include mutations in new genes (Cirulli et al. 2015; Gelfman et al. 2019), epigenetic alterations (Young et al. 2017), and oligogenic/polygenic effects (van Blitterswijk et al. 2012a). Relevant to the latter, 60% of sporadic ALS patients were found to have rare or novel variants in 169 known and candidate ALS disease genes (Couthouis et al. 2014). Likewise, in 46 sporadic ALS patients living in Hong Kong, 67% had at least one rare variant in the exon of 40 ALS genes and 22% had two or more rare variants (Pang et al. 2017). In aggregate, these findings suggest that a significant proportion of sporadic patients may harbor rare variants in ALS risk genes, which may help to identify the missing heritability of sporadic ALS.

Same Phenotype Due to Mutations in Different Genes

Among MNDs, ALS stands out from the rest on the basis that both gene-positive and gene-negative forms occur in substantial proportions, and strikingly these two distinct forms of ALS, while showing population differences are indistinguishable at the patient level both clinically and neuropathologically, implying that they might share common underlying mechanisms. Over the years, it has been thought that this phenotypic similarity justifies the analysis of rare gene-positive MNDs, as it could well illuminate the pathogenesis of both. However, the gene-positive counterparts of gene-negative ALS are typically due to mutations in ostensibly dissimilar genes, with mutations in *SOD1*, *TARDBP*, *FUS*, *VCP*, *C9ORF72*, and *PFN1* accounting for about 60–70% of all familial ALS cases (Nguyen et al. 2018). The genetic heterogeneity associated with specific MNDs is to be expected since thus far the taxonomy of MNDs rests on clinical criteria, lumping under the same label disorders that merely look alike. Nonetheless, this striking situation raises the possibility that no matter how disparate these mutated genes appear to be, the functions of the respective gene products might intersect in common pathways, a view that is in keeping with the demonstrations of shared neuropathology that spans across genes (Highley and Ince 2012). Along this line, overlapping cellular and molecular pathways have begun to emerge. In the case of ALS, such converging themes include disturbances in RNA metabolism, impaired proteostasis (including trafficking and degradation defects), axonal transport disruption, aberrant liquid-liquid phase transition and neuroinflammation (Beers and Appel 2019; Cook and Petrucelli 2019; Kim et al. 2020a; Schon and Przedborski 2011; Taylor et al. 2016; Tisdale and Pellizzoni 2015). This suggests that there may be more commonality in the cellular and molecular mechanisms of neurodegeneration among the different causes of a given MND phenotype than among different MND phenotypes. Along this line, we can further speculate that pathway analyses in genetic MNDs such as X-linked recessive spinobulbar muscular atrophy (Kennedy's disease), hereditary spastic paraplegia and SMA may shed light on the mechanisms of bulbar, spinal or cortical motoneuron degeneration in, respectively, progressive bulbar palsy, progressive muscular atrophy and primary lateral sclerosis (Chio et al. 2011).

Different Phenotypes Due to Mutations in the Same Gene

Just as similar MND phenotypes can be caused by mutations in different genes, mutations in the same gene can give rise to more than one clinical phenotype of MNDs (e.g. the same mutation causing both bulbar or spinal onset, or frontotemporal dementia without ALS). This observation suggests that while the disease classification scheme has clinical significance, it may be equally helpful to view different MNDs as reflecting different, and perhaps more nuanced, expressions of shared, fundamental underlying problems. Among MNDs, this pleotropic phenomenon come in different flavors: same gene, different mutations with different severity; same gene, different mutations with different diseases; and same gene, same mutations, different manifestations.

Let us start with the case of the same mutated gene associated with distinct severity of MND, a situation well illustrated by *SOD1* and *SMN* mutations. As already mentioned, some autosomal dominant gain-of-function *SOD1* modulations such as p.A5V (also called A4V) give rise to a rapidly progressive and thus lethal form of ALS while others such as p.G94A (also called G93A) are associated with much more protracted but otherwise identical clinical phenotype (Cudkowicz et al. 1997), despite having comparable pathobiochemical properties (Tiwari and Hayward 2005). In the case of SMA, the exact same autosomal recessive loss-of-function mutation in the *SMN1* gene is associated with a clinical spectrum in terms of age of onset and severity (Tisdale and Pellizzoni 2015). This spectrum goes from a prenatal form of the disease that is uniformly fatal *in utero* or a few months after birth to an adult-onset form of the disease with mild muscle weakness and normal life expectancy (Tisdale and Pellizzoni 2015). While the basis of this pleiotropism in *SOD1*-related ALS remains unknown, in SMA it is attributed to the number of copies of the *SMN1* hypomorphic paralogue *SMN2,* which correlates inversely with SMA severity (McAndrew et al. 1997; Wirth et al. 2006). As illustrated with mutations in *BSCL2,* the same mutated gene can give rise to different neurological disorders of the motor system, even within a common pedigree (Ito and Suzuki 2009). Indeed, while *BSCL2* loss-of-function mutations are known to cause a form of congenital generalized lipodystrophy, two gain-of-toxic-function mutations, i.e. N88S and S90L, have been identified in autosomal dominant cases of ALS, hereditary spastic paraplegia as well as distal hereditary motor neuropathy (Ito and Suzuki 2009). Another example of pleiotropism is seen in *IGHMBP2* mutations (Yuan et al. 2017). *IGHMBP2* mutations are usually linked to a form of SMA with respiratory distress type 1 or SMARD1, where most infants die before age of one (Grohmann et al. 2001). However, compound heterozygous mutations in *IGHMBP2* were found in individuals with recessive Charcot-Marie Tooth disease type 2, a slowly progressive peripheral axonal neuropathy characterized by distal weakness, atrophy, sensory loss and variable foot deformity, but no significant respiratory problems (Cottenie et al. 2014). Likewise, patients with *C9ORF72* repeat expansion also show marked phenotypic variability, both among and within families (Hsiung et al. 2012; Van Mossevelde et al. 2017b). Although several different neurodegenerative disorders have been described in *C9ORF72* repeat expansion carriers, the majority of these patients are diagnosed with either frontotemporal dementia, ALS, or both (Van Mossevelde et al. 2017a). This pleiotropism led many researchers to propose that the sizes of the *C9ORF72* repeat expansion might predict the evolution toward frontotemporal dementia or ALS in *C9ORF72* carriers, but, up to now, this hypothesis remains to be proven (Van Mossevelde et al. 2017a). It should also be mentioned that pathological expansion in *HTT* genes known to be linked to the fatal, dementing, movement disorder Huntington's disease, was reported to be associated, in some cases, with clinical hallmarks of ALS or frontal temporal dementia (Dewan et al. 2021). Lastly, mutations in *OPTN* and *VCP* genes are even more striking with respect to pleiotropism in that they were found to cause broad ranges of diseases such as ALS and glaucoma (Toth and Atkin 2018) and inclusion body myopathy, Paget's disease of the bone, frontotemporal dementia and ALS (Meyer and Weihl 2014), respectively.

4 The Devil Is in the Details

Despite that all MNDs have in common being paralytic neurodegenerative disorders, the clinical and the neuropathological presentations vary widely among patients not only with different, but also with the same neurological condition. This striking clinical heterogeneity is well illustrated in ALS where it is attributed to variation in the anatomic location, extent, and proportion of lower motoneuron, upper motoneuron, and non-motoneuron involvement among patients (Harms and Baloh 2013; Swinnen and Robberecht 2014). In keeping with this view, these authors further propose that at one end of a MND clinical spectrum are patients with progressive muscular atrophy, a form of adult-onset MND with only lower motoneurons involvement, whereas at the other end are patients with primary lateral sclerosis, a form of adult-onset MND with primarily upper motoneurons involvement. Of note, most cases of both progressive muscular atrophy (Kim et al. 2009) and primary lateral sclerosis (Gordon et al. 2006a) eventually progress to meet criteria for ALS (i.e. presence of upper motor neuron and lower motor neuron signs, progression of disease, and the absence of an alternative explanation), suggesting that they likely represent phenotypic variants of ALS rather than distinct forms of MNDs, a view supported by the finding that mutation frequency in for example *SOD1*, *FUS* or *TARDBP* was similar in progressive muscular atrophy and ALS patients (van Blitterswijk et al. 2012b), that rare primary lateral sclerosis patients harbor mutations in ALS genes (Silani et al. 2020), and that cases share TDP43 aggregate pathology on neuropathological evaluation (Kosaka et al. 2012; Mackenzie and Briemberg 2020; Pamphlett 2010). In addition, ALS and frontotemporal dementia is increasingly regarded as two ends of another MND clinical spectrum which, this time, revolves around the proportion of motoneuron versus non-motoneuron involvement based on shared neuropathological findings and genetic causes (Lillo and Hodges 2009).

The age at onset of clinical manifestations of ALS and its rate of progression are also highly variable among patients (Swinnen and Robberecht 2014), even though the mean survival from the time of diagnosis is about 4 years, with less than 20% surviving more than 5 years and only 10% surviving more than a decade (Rowland et al. 2010). Among key factors influencing the course of the disease is the anatomical site first affected. Indeed, patients with bulbar onset, where muscles controlling speech and swallowing are affected first, progress and die faster than those with spinal onset ALS (Rowland et al. 2010). Moreover, in the case of gene-positive ALS, it can be speculated that some gene mutations are more *aggressive* than others, insofar as they are associated with earlier onset and/or faster progression.

Consistent with this view is the observation that mutation in specific ALS genes such as *FUS* P525L is particularly malignant, striking individuals at very young age and progressing very rapidly whereas *FUS* mutations at the R521 site are associated with later onset and slower progression (Naumann et al. 2019). Likewise and as already alluded to, patients carrying a p.A5V mutation in *SOD1* typically die after about 1 year from diagnosis, while those carrying the p.G94A mutation in the same gene have a mean survival of 10.5 years (Cudkowicz et al. 1997).

Lastly, one must know that while a common first manifestation of ALS is a painless weakness in a single body region (e.g. hand, foot, arm, leg, tongue or pharynx), over time, the motor deficit not only worsens at the site of symptoms onset, but also extends to contiguous regions of the body (Ravits and La Spada 2009); of note, such contiguous progression is observed in most but not in all patients (Ravits and La Spada 2009). Relevant to the observation of contiguous progression is the increased attention paid to the possible self-propagation and spread, in a prion-like manner (i.e. biochemical phenomenon by which a protein transfer folding characteristics to surrounding proteins with identical or similar amino acid sequences) of several ALS-related proteins including TDP43, SOD1, and FUS (McAlary et al. 2019). In keeping with this speculation, both *in vitro* and *in vivo* experiments have demonstrated that mutations, posttranslational modifications or simply high concentration of TDP43, SOD1, and FUS promote the misfolding and subsequent aggregation of these proteins. Once misfolded, TDP43, SOD1, and FUS can transmit their misfolded conformation onto normal variants of the same proteins (McAlary et al. 2019). Of note, SOD1, TDP-43, and FUS are all supersaturated in spinal motoneurons (Ciryam et al. 2017), suggesting that these neurons may provide a fertile environment for ALS-associated prion-like seeds to form and propagate. While this mechanism of disease spreading could provide a conceptual framework for the progression of the neurodegenerative process in MNDs like ALS, how misfolded proteins could travel from cell-to-cell and cause the demise of the recipient cells remains elusive. Perhaps, a hint about the latter may be found in the discovery that disease mutations in ALS proteins such as FUS can promote aberrant liquid-liquid phase transition within liquid-like compartments (Murakami et al. 2015; Patel et al. 2015), thereby perturbing many vital functions of the cell (Alberti 2017). Furthermore, since not all motoneurons are equally affected in MNDs like ALS (Kanning et al. 2010), one may wonder whether the spatiotemporal pattern of progression in these diseases rests solely on the hypothesized spread of misfolded protein or merely reflects the interaction between this ubiquitous pathological stressor and the known differential susceptibility of motoneurons (Kanning et al. 2010). Accordingly, different subpopulations of motoneurons would succumb within different areas of the nervous system and at different times according to their differential vulnerability in an ordered spatiotemporal pattern. For instance, it is well documented in transgenic mice expressing mutant *SOD1* that spinal motoneuron innervating larger, pale muscle fibers that generate more force (called fast fatigable motor units) degenerate consistently sooner than their contiguous counterparts innervating small muscle fibers rich in myoglobin content, mitochondria, and blood capillaries that contract slowly and generate relatively small forces (called slow motor units) (Frey et al. 2000; Pun et al. 2006). In addition, other motoneuron subtypes, such as those that innervate the extraocular muscles of the eyes or the striated muscles of the rectum and urethral sphincter are spared until the end stage of disease (Kanning et al. 2010; Nijssen et al. 2017). The same pattern of motoneuron vulnerability is observed in a transgenic mouse model of TDP43 proteinopathy (Spiller et al. 2016). In this other model of MNDs, despite widespread neuronal expression of cytoplasmic TDP43, only motoneurons in the spinal cord and in the hypoglossal nucleus are lost, whereas those in the oculomotor, trigeminal, and facial nuclei are less affected (Spiller et al. 2016). As discussed in Ragagnin et al. (2019) distinct intrinsic properties of motoneurons, ranging from metabolic to transcriptomic, are likely underpinning their differential propensity to degenerate in MNDs. For instance, our group used a computational model in which detailed morphology and ion conductance were paired with intracellular ATP production and consumption (Le Masson et al. 2014). This work showed that the occurrence of an irreversible, dramatic ionic instability could be initiated in our model of fast fatigable motor unit-innervating motoneurons for much smaller bioenergetic defects than in our model of slow motor unit-innervating motoneurons (Le Masson et al. 2014). Moreover, several transcriptomic analyses revealed a number of interesting genes that were differentially expressed between vulnerable and susceptible motoneurons (Brockington et al. 2013; Kaplan et al. 2014; Kline et al. 2017). Among these, matrix metalloproteinase 9 (MMP9) has been identified as highly expressed in vulnerable and much less or not expressed at all in resistant motoneurons (Kaplan et al. 2014), and to correlate with the expression of death receptor 6 in these cells (Mishra et al. 2020). Interestingly, Mentis et al. (2011) have shown in a severe mouse model of SMA that motoneurons innervating proximal muscles are more affected than those innervating distal muscles, which is in keeping with the clinical picture of greater proximal than distal weakness in SMA patients (Kaufmann and De Vivo 2010). Given this very different profile of motoneuron susceptibility compared to the one observed in ALS, it would be valuable to determine whether some of the identified makers of motoneuron vulnerability in ALS are conserved in SMA.

5 Onset and Neuromuscular Junction Remodeling in MNDs

As mentioned above, most MNDs are sporadic and with the absence of presymptomatic markers for virtually all of them, patients are seeking medical attention only when the first signs of the disease emerge. Because of the compensatory mechanisms discussed below, the onset of manifestations (e.g. loss of finger dexterity) does not equate with the onset of the disease. Instead, the beginning of clinical signs merely corresponds to a neurodegenerative stage at which the extent of muscle innervation is no longer sufficient to maintain a normal motor activity. It is thus clear that the onset of the disease occurs at some unknown preceding time, which, depending on how fast the neurodegenerative process evolves, it can range from a few months to several years. This interpretation is well illustrated by transgenic mutant SOD1 mice, which express copious amount of the pathogenic protein and show several motoneuronal abnormalities prenatally (Branchereau et al. 2019; Martin et al. 2020) and yet, they develop the first signs of motor deficit only several months later (Nagai et al. 2006). Our ability to determine the actual onset of the disease is at this point undermined by the lack of biomarkers and the little knowledge about the true kinetics of motoneuron loss, even if levels of neurofilaments in both blood and CSF of patients with MNDs show some promise (Poesen and Van Damme 2018).

Although MNDs are identified as diseases of motoneurons, there have been attempts to redefine them as a *distal axonopathy* since overt neuropathological changes occur at the neuromuscular junction in both patients with ALS and in animals models at early stages of the diseases and even before any significant motoneuron cell body loss (Fischer et al. 2004; Vinsant et al. 2013) and prior to the onset of motor deficit (Clark et al. 2016; Fischer et al. 2004; Martineau et al. 2018, 2020; Pun et al. 2006; Tallon et al. 2016; Vinsant et al. 2013). As discussed below, changes at the level of the neuromuscular junctions are key determinants in the onset and progression of the paralytic phenotype in MNDs. For instance, using transgenic mutant $SOD1^{G93A}$ mice, it has been reported (Fischer et al. 2004; Fischer and Glass 2007) that neuromuscular junction denervation and muscle wasting are already detected in pre-symptomatic animals at postnatal day P47 whereas overt spinal motoneuron cell body loss is observed only after P80. The early occurrence of neuromuscular junction denervation in transgenic mutant SOD1 mice was corroborated by other teams (Clark et al. 2016; Frey et al. 2000; Killian et al. 1994; Martineau et al. 2018, 2020; Pun et al. 2006; Schaefer et al. 2005; Tallon et al. 2016; Tremblay et al. 2017; Vinsant et al. 2013). This pre-symptomatic phase is attributed to the formation of new collateral axonal branches, reinnervating other nearby denervated neuromuscular junctions (Martineau et al. 2018). As shown by Martineau et al. (2018), this period of neuromuscular junction plasticity involved a constant axonal terminal retraction and growth, a synaptic plasticity that was reported to be more frequent in female than male mice (Martineau et al. 2020) and particularly intense in slow-type, disease-resistant synapses compared to fast fatigable-type synapses with the highest vulnerability (Frey et al. 2000). Thus, it can be inferred that early muscle denervation is a consequence of nerve terminal degeneration where disease-resistant motoneurons that have a high sprouting capacity compensate for the loss of nerve terminals by disease-susceptible motoneurons that have a low sprouting capacity. However, as the disease progresses, even the disease-resistant motoneurons become affected and the sprouting compensatory mechanisms eventually fail, hence leading to clinical onset. This striking early distal axonopathy is not restricted to the transgenic mutant SOD1 mice, as other models of MNDs show similar early neuromuscular junction pathology as reviewed in (Fischer and Glass 2007), including in mouse models of SMA (Kariya et al. 2008; Kong et al. 2009). For instance, transgenic mice expressing the P525L mutant of *FUS* exhibit an age-dependent muscle denervation phenotype that begins prior to spinal motoneuron loss (Sharma et al. 2016). Likewise, in those *C9orf72* BAC transgenic mice that develop an ALS-like phenotype, prominent neuromuscular junction denervation is observed in both tibialis anterior and diaphragm muscles (Liu et al. 2016). Collectively, these findings, in both human and mouse models, also provide solid ground for the notion that nerve terminal degeneration and the consequent neuromuscular junction denervation precede spinal motoneuron death. They also raise the significant prospect that therapeutic strategies, aimed at blocking/slowing nerve terminal degeneration, may extend muscle function and thus greatly improve the quality of life for patients with MNDs. Yet, a structurally intact neuromuscular junction does not necessary imply normal function. To this end, it has been shown that anatomically intact neuromuscular junctions in a severe model of SMA may indeed be dysfunctional (Fletcher et al. 2017). However, as long as the nerve terminal is preserved and connected to muscle fibers, strategies aimed at boosting/recovering synaptic function can be envisioned.

Remarkably, while the temporal dissociation between neuromuscular junction denervation and motoneuron death in MND is well recognized, it remains unclear whether this is due to: (i) a unique process that affects the whole motoneuron at once, but with greater impact on nerve terminals than on motoneuron soma; (ii) a process that specifically damages the nerve terminals, which, in turn, triggers a secondary dying-back process that kills the motoneurons; or (iii) a situation in which nerve terminals and soma destruction are governed by simultaneous but distinct molecular pathways with a faster time-course of the former. Irrespective of these three pathogenic scenarios, preventing motoneuron death may be necessary but not sufficient to improve the quality of life and likely survival of patients with MNDs. Indeed, deletion of the pro-death gene *Bax* nullified motoneuron loss with limited effect on neuromuscular junction denervation in end-stage transgenic mutant SOD1 mice (Gould et al. 2006). A similar situation is also true for SMA where the activation of the tumor suppressor p53 drives motoneuron death in the SMNΔ7 mouse model of the disease (Simon et al. 2017). Genetic or pharmacological inhibition of p53 prevents motoneuron loss without any effect on neuromuscular junction denervation, consistent with synaptic dysfunction and motoneuron survival being two independent events in SMA (Simon et al. 2017).

Mechanisms of Axon Pathology in MNDs

Given the increase attention paid to neuromuscular junction and axon degeneration in MNDs, it is worth mentioning that over the past decade, significant strides have been made toward unraveling mechanisms of axon degeneration, especially after mechanical lesion (Conforti et al. 2014). One of the most striking discoveries regarding axon biology pertains to the demonstration that the gain-of-function mutation, *slow Wallerian degeneration* (Wld^S), delays axon degeneration after nerve injury (Neukomm and Freeman 2014). This Wallerian degeneration, which refers to degenerative changes in the distal segment of an axon when its continuity with its cell body is interrupted by a focal lesion, requires the pro-degenerative molecules sterile alpha and TIR motif-containing protein-1 (SARM1) and PHR1

(PAM-Highwire-Rpm-1) E3 ubiquitin ligase. Conversely, the nicotinamide mononucleotide adenylyltransferase-2 (NMNAT2) is essential for axon growth and survival (Conforti et al. 2014). While studies in lower organisms suggest a role for at least SARM1 in ALS axonopathy (Veriepe et al. 2015), a significant involvement of the Wallerian machinery in axon pathobiology in MNDs has not been convincingly demonstrated in vertebrate models of the diseases (Fischer et al. 2005; Vande Velde et al. 2004). For instance, genetic experiments in which Wld^{S} mice were crossed to SMA mice have ruled out a contribution of Wallerian degeneration to the motor axon loss observed in SMA (Kariya et al. 2009).

The lack of molecular insights into the intrinsic mechanisms underlying axon degeneration in MNDs led the performance of a cell-based high-throughput screen for small molecules that could protect against axon pathology in MNDs and shed light into its molecular basis. Of the >50,000 small molecules tested for their ability to promote axon outgrowth on inhibitory substrata, the most potent hits were the antagonists of the HMG-CoA reductase, statins (Bahia El Idrissi et al. 2016). In studying the metabolic pathway downstream to the HMG-CoA reductase pathway, these investigators found that the observed stimulatory effects on axon growth, in both mouse- and human-derived neurons, were mediated not by inhibiting coenzyme Q or cholesterol synthesis but protein prenylation – the covalent attachment of a lipid moiety to a cysteine residue at or near the C-terminus of a protein (Wang and Casey 2016). At this time, the type of prenylation and the targets of prenylation implicated in MND axonopathy remains to be elucidated.

6 Non-cell Autonomous Drivers of Neurodegeneration in MNDs

Traditionally, MNDs, like other common neurodegenerative disorders, have been approached in a *neurocentric* fashion, leading to the emergence of a host of *cell autonomous* pathogenic hypotheses including defects in RNA metabolism, protein quality control mechanisms, mitochondrial functions, axonal transport and aberrant liquid-liquid phase transition to cite only a few. Many of these proposed mechanisms of neurodegeneration derive from studies of MND-associated genes and the reader is referred to reviews on these topics for detail (Cook and Petrucelli 2019; Kim et al. 2020a; Schon and Przedborski 2011; Taylor et al. 2016; Tisdale and Pellizzoni 2015).

Aside from these cell autonomous processes, growing attention has been paid over the past recent years to the potential role of non-neuronal cells to the dysfunction and/or death of motoneurons in MNDs. Support to a non-cell autonomous contribution to MNDs goes back to the seminal work of Hamburger, Levi-Montalcini and Cohen (Levi-Montalcini 1987) demonstrating that the motoneuron developmental death was regulated by peripheral tissues in the chick embryo. It should thus not come as a surprise that in MNDs, a great deal of attention has been given to the question of the role of skeletal muscles in the demise of their innervating motoneurons. In SMA, while there is disagreement about whether low levels of wild-type SMN protein (i.e. mimicking the situation seen in SMA) cause muscle pathology and contribute the SMA phenotype (Gavrilina et al. 2008; Iyer et al. 2015; Kim et al. 2020b), there seems to be a consensus around the following two conclusions: (i) low level of Smn in muscles does not cause spinal motoneuron loss in mice with normal level of Smn everywhere else; and, (ii) overexpression of Smn in muscles does not mitigate spinal motoneuron loss in SMA mouse models. A similar picture appear to exist for ALS where it was found that a partial reduction of the toxic mutant SOD1 in muscle failed to modify the paralytic phenotype seen in these transgenic mice (Towne et al. 2008) and overexpression of mutant SOD1 in skeletal muscles, while associated with a host of local alterations including metabolic and morphological changes (Loeffler et al. 2016), did not cause consistent motoneuron degeneration (Dobrowolny et al. 2008; Martin and Wong 2020; Towne et al. 2008). Based on these studies it appears that a role of skeletal muscles in motoneuron degeneration remains unconvincing. In contrast, more compelling *in vivo* and *in vitro* studies support the contribution of a variety of other non-motoneuron cells in the expression of the disease phenotypes in models of MNDs, ranging from sensory proprioceptors (Fletcher et al. 2017; Mentis et al. 2011), Renshaw cells (Wootz et al. 2013) and other inhibitory interneurons (Schutz 2005) as well as a range of glial and immune cells such as microglia (Beers et al. 2006; Boillee et al. 2006), astrocytes (Barbeito et al. 2004; Cassina et al. 2008; Di Giorgio et al. 2007, 2008; Fritz et al. 2013; Haidet-Phillips et al. 2011; Ikiz et al. 2015; Marchetto et al. 2008; Meyer et al. 2014; Nagai et al. 2007; Pehar et al. 2004, 2005; Re et al. 2014; Yamanaka et al. 2008b), oligodendroglia (Kang et al. 2013), Schwann cells (Lobsiger et al. 2009) and T cells (Banerjee et al. 2008; Beers et al. 2008). To date, however, despite the growing number of non-cell autonomous hypotheses centered on different glial and immune cells, the lion's share of attention has been paid to microglia and astrocytes and to a lesser extent to T cells, which is illustrated in the next two subsections. However, the focus of this chapter on these specific non-neuronal cell types should not be misconstrued as suggesting that other glial cells such oligodendrocytes (Kang et al. 2013; Saez-Atienzar et al. 2021; Yamanaka et al. 2008a) or blood-borne cells including macrophages (Beers and Appel 2019; Chiot et al. 2020) are not contributing to the non-cell autonomous pathogenic hypothesis of MNDs. For instance, TDP43 inclusions were detected in both neurons and oligodendrocytes in spinal cord samples from patients with ALS and, many oligodendrocytes in spinal cords of transgenic mutant SOD1 mice had a reduced expression of myelin basic protein and monocarboxylate transporter-1 suggesting that these cells may have myelination and metabolic alterations (Philips et al. 2013), which could affect the function and survival of motoneurons.

Glial Cells as Motoneuron Death Drivers in MNDs

Over the past two decades, the potential contribution of glial cells to the degeneration of motoneurons has been increasingly recognized (Van Harten et al. 2021). Thus far, two broad hypotheses, which are not mutually exclusive, are proposed for the mechanisms by which glial cells, and in particular microglia and astrocytes might play a pathogenic role in MNDs. The first hypothesis, rests on the idea that the disease process, while not necessarily killing glial cells, corrupt their normal homeostatic functions (e.g. supplying nutrients and neurotrophic factors, buffering ions), thus impairing their supportive/protective effects on neighboring cells including motoneurons. Relevant to this first hypothesis are the investigations in neuron-astrocyte co-culture systems (Phatnani et al. 2013) that revealed downregulations in mutant SOD1-expressing astrocytes of about 34 genes including *Osteopontin, Sparcl1, Sepp1, Apolipoprotein D and Pleiotrophin* consistent with the idea that ALS glia have a compromised molecular response to injury. Likewise, Tyzack et al. (2017) showed that the neuroprotective EphB1-ephrin-B1 cross-talk between motoneurons and astrocytes is disrupted in human stem cell-derived ALS astrocytes and transgenic mutant SOD1 mice. Disruption of glial cell supportive/protective functions in MNDs can also be indirect as proposed by Gerbino et al. (2020) in light of their observation that the conditional deletion of in *TANK-binding kinase 1* (*TBK1*) in motoneurons reduces the responsiveness of glial cells to pathological stimuli.

Although these studies, as discussed by Van Harten et al. (2021), suggest that part of the non-cell autonomous arm of ALS pathogenesis is related to a loss of beneficial properties, data from other studies argue that ALS astrocytes kill motoneurons by a gain of toxic function. Thus, the second hypothesis posits that many, if not all the of the glial cells studied so far have the capacity to mount a neuroinflammatory response, which, for example, via the production and release of cytokines and chemokines as well as other cytotoxic molecules may exert deleterious effects on neighboring cells such as motoneurons. We review below, the available data often cited in support of a toxic role of specifically microglia and astrocytes in MND pathogenesis.

Microglia are the resident immune cells of the CNS and there is morphological and molecular evidence of microglial activation in areas of motoneuron degeneration in post-mortem samples (Henkel et al. 2009; Moisse and Strong 2006). Similar neuroinflammatory changes that are seen in patients with MNDs are also seen in mouse models of these diseases (Alexianu et al. 2001; Almer et al. 2002; Butovsky et al. 2012; Chiu et al. 2009, 2013; Elliott 2001; Hall et al. 1998). It is also important to note that haploinsufficiency of *C9orf72* in mice alters systemic immune response and induces a mild age-related neuroinflammation, but without evidence of neurodegeneration (Lall and Baloh 2017). In addition, the discovery that loss-of-function mutations in *TBK1* -a key regulator of innate immunity- can cause ALS further supports the idea that immune dysregulation might be a common feature of the disease (Cirulli et al. 2015; Freischmidt et al. 2015; Gerbino et al. 2020). These findings raised the idea that microglia, once activated, initiate an inflammatory process which contributes to the degeneration of neighboring motoneurons. Although the mechanisms by which microglia might kill motoneurons remain to be established, evidence from cultured microglia supports the idea that overexpression of mutant SOD1 or deletion of TDP43 suffices to trigger molecular alterations in microglia promoting their neurotoxic properties (Weydt et al. 2004; Xiao et al. 2007). One key change observed in microglia is the selective activation of nuclear factor κB (NF-κB) that can induce gliosis and motoneuron death. Inhibition of NF-κB signaling suppresses microglia-mediated neuroinflammatory toxicity when mutant SOD1 microglia are cocultured with motoneurons, suggesting that microglia-induced motoneuron death in ALS is executed via the classical NF-κB pathway (Frakes et al. 2014; Zhao et al. 2015).

As discussed above electrophysiological and neuropathological changes occur in, respectively, skeletal muscle and neuromuscular junction at early stages of MNDs (Fischer and Glass 2007) and here again non-cell autonomous mechanisms revolving around other types of microglial, namely monocytes/macrophages which, unlike microglia reside outside the brain, have been found to progressively infiltrate peripheral nerves in a mouse model of ALS (Chiu et al. 2009; Graber et al. 2010; Kano et al. 2012). Although these findings raise the possibility that the infiltrating immune cells in peripheral nerves contribute to the early denervation of the neuromuscular junctions, the available data fail to inform about the phenotypes of these cells, which is necessary to address properly this important question. Nonetheless, some clues about the possible underpinning mechanism by which microglia and related cell types may contribute to degeneration of nerve terminals in MNDs start to emerge. Indeed, the work of Vukojicic et al. (2019) shows that microglial-specific ablation or pharmacological inhibition of the complement component 1q (C1q) restored vulnerable synapses in a mouse model of SMA, consistent with the classic complement activation being a main pathway involved in the synaptic elimination in this MND. Furthermore, using flow cytometry and a bone marrow chimera, it was found that activated macrophages in sciatic nerves differ from activated microglia in spinal cords (Chiu et al. 2009); hence, suggesting that distinct non-cell autonomous molecular mechanisms may contribute to the destruction of components of the motor pathway localized in the central versus the peripheral nervous systems in MNDs. Also worth mentioning is the demonstration that replacing peripheral mutant SOD1 macrophages with wild-type macrophages had marginal effects in transgenic mutant SOD1 mice, unless wild-type macrophages were also engineered to produce less reactive oxygen species and were grafted not at pre-symptomatic but at early symptomatic stages (Chiot et al. 2020).

However, despite the large amount of data supporting a pathogenic role of microglia in MNDs, immunosuppressive and anti-inflammatory drugs including celecoxib (Cudkowicz et al. 2006), glatiramer acetate (Gordon et al. 2006b), minocycline (Gordon et al. 2004), NP001 (Miller et al. 2014), pioglitazone (Dupuis et al. 2012), cyclosporine (Appel et al. 1988) and the combination of basiliximab, mycophenolate, tacrolimus with steroids (Fournier et al. 2018) produced either no or only marginal clinical improvement. Immune cells such as microglia are highly plastic in that they adapt their phenotype and function based on the signals that are received constantly from their environment (Gordon 2003). Until now, studies on microglia as those cited above have relied on general surface markers that ignore microglial heterogeneity, thereby underestimating the complexity of the microglial response in neurodegenerative disorders like MNDs. To circumvent this shortcoming, strategies aimed at capturing microglial multidimensional signature rather than assessing the state of selected inflammatory factors are emerging. Studies using whole spinal cord samples and, more recently, acutely purified microglia, have consistently found evidence of neuroinflammatory genes and/or gene product changes in ALS (Butovsky et al. 2012; Chiu et al. 2013; D'Erchia et al. 2017; Jiang et al. 2005; Lerman et al. 2012; Nardo et al. 2013; Nikodemova et al. 2014; Noristani et al. 2015). However, all of these studies, even those using acutely purified microglia, utilized a bulk RNA sequencing approach. A major caveat of such an approach is that changes in bulk-sequencing data, especially in pathological tissues, may primarily reflect changes in the abundance of cell types rather than in gene transcription per cell (Srinivasan et al. 2016). Thus, meaningful signatures may be masked by mere differences in the type and number of cells present in the studied tissue. Consistent with this view, studies of neuroinflammation in models of Alzheimer's disease at single-cell resolution have revealed previously unobserved heterogeneity in microglial populations, identified disease stage-specific microglial states, and determined the trajectory of the cellular reprogramming of microglia in response to neurodegeneration (Keren-Shaul et al. 2017; Mathys et al. 2017). Microglial heterogeneity was also shown to be a relevant component of the progression of the paralytic phenotype in mouse models of ALS (Haukedal and Freude 2019; Keren-Shaul et al. 2017). Another critical question about microglial heterogeneity that begins to be testable is the anatomical distribution of different phenotypically-defined clusters of microglia within diseased tissues. For instance, while resolution of the technique may have to improve, spatial transcriptomics (Maniatis et al. 2019) offers an unprecedented opportunity to shed light into this question.

Aside from microglia, mounting evidence indicates that astrocytes can also contribute to neuroinflammation in MNDs. For example, in ALS patients, reactive astrocytes have been found throughout the cerebral gray matter and the spinal cord (Nagy et al. 1994; Schiffer et al. 1996). In transgenic mutant SOD1 animals, signs of astrogliosis are detected in the spinal cord with the increase in the number of astrocytes occurring concomitantly to the loss of motoneurons (Almer et al. 2002; Hall et al. 1998; Levine et al. 1999). Astrocytes may shift from a quiescent to reactive state in response to a variety of pathological stimuli and it has been proposed that there might be two major subtypes of reactive astrocytes: LPS-induced neurotoxic A1 astrocytes and ischemic penumbra-associated neuroprotective A2 astrocytes (Liddelow et al. 2017). In this *in vitro* model, microglia induce the astrocyte A1 phenotype through the release of IL-1α/TNF-α/C1q, resulting in the death of neurons and oligodendrocytes. From a pathogenic point of view, while this A1/A2 model is quite appealing, whether astrocytes contribute to motoneuron death in MNDs via an A1 mechanism (Liddelow et al. 2017) remains to be demonstrated albeit knocking out IL-1α/TNF-α/C1q was reported to markedly extend survival of transgenic mutant SOD1 mice (Guttenplan et al. 2020). Yet, we believe that there may be more nuances to astrocyte phenotypes than this strict dichotomy (Liddelow et al. 2017) given the transcriptomic and proteomic heterogeneity of these cells that are starting to be reported in different human diseases of the nervous system. Nonetheless, once astrocytes become reactive, they contribute to neuroinflammation not only directly, but also indirectly, by facilitating the entry of blood-borne immune components into the brain parenchyma through the blood-brain barrier (Farina et al. 2007; Johann et al. 2015; Philips and Robberecht 2011).

In addition to possibly contributing to motoneuron degeneration via neuroinflammation, astrocytes, as mentioned above, can also directly mediate motoneuron death via the release of neurotoxic factors. In particular, astrocytes expressing mutant SOD1 trigger the selective loss of spinal motoneurons, by releasing soluble toxic factors, while leaving undamaged spinal GABAergic or dorsal root ganglion neurons or embryonic stem cell-derived interneurons (Nagai et al. 2007). Although the precise nature of the factors released by ALS astrocytes that are toxic to motoneurons is still under investigation, our group has reported that ALS astrocytes generate an aberrant fragment of amyloid precursor protein (and perhaps of amyloid-like protein 1), which by activating the death receptor 6 expressed by motoneurons trigger neurodegeneration (Mishra et al. 2020). Nonetheless, since deletion of death receptor 6 only provided partial protection against neurodegeneration in transgenic mutant SOD1 mice (Mishra et al. 2020), it is likely that additional non-cell autonomous mechanisms are at play. A similar role for astrocytes in the pathology of SMA has been proposed, although a clear link with toxicity has not been established (Abati et al. 2020).

Other Immune Cells as Modulators of Neurodegeneration in MNDs

More recently, major interest in exploring whether immune cells such as T cells also contribute to the non-cell autonomous pathogenesis of MNDs has developed (Henkel et al. 2009; Moisse and Strong 2006). In ALS, emerging evidence supports the involvement of the peripheral immune system in balancing glial neurotrophic and neurotoxic functions. For instance, lack of $CD4^+$ T cells in mutant SOD1 mice skews the glia phenotype towards toxicity due to reduced expression of neurotrophic factors such as IGF-1, GDNF, BDNF and anti-inflammatory cytokines such as IL-4, TGF-β and to a concomitant increase in the expression of proinflammatory molecules such as TNF-α and NOX2 (Beers et al. 2008). In addition, passive transfer of *ex vivo* activated $CD4^+$ T lymphocytes into transgenic mutant SOD1 mice delayed onset of the paralytic phenotype and extended their survival (Banerjee et al. 2008). The protective effects of $CD4^+$ T cells were attributed to the presence of regulatory T lymphocytes (Tregs), a subgroup of $CD4^+$ cells that modulates the immune system by maintaining self-tolerance (Rajabinejad et al. 2020). Suggestive of their protective effect, passive transfer of endogenous Tregs from ALS mice in the early stage of disease into ALS mice lacking functional T lymphocytes prolonged disease duration and survival of recipient mice (Beers et al. 2011). The authors attributed the Tregs-mediated neuroprotection to the increased release of IL-4, which can directly promote an anti-inflammatory microglial phenotype that includes the production of neurotrophic factors (Beers et al. 2011). However, as neurodegeneration progresses, it seems that there is a shift from a supportive to a toxic Tregs response (Beers and Appel 2019). Extending these observations to ALS patients, Beers et al. (2011) also reported that the number of Tregs in blood correlates with the rate of progression of the motor deficit, and a clinical trial is ongoing to determine whether the enhancement of Treg numbers and function can slow disease progression (https://clinicaltrials.gov/ct2/show/NCT04055623).

While $CD4^+$ T cells are observed in the spinal cord during all stages of the disease, $CD8^+$ T cells are present only at the end stage and made up only a small fraction of the total T-cell population (Beers et al. 2008; Chiu et al. 2008). Because of their small numbers, the role of $CD8^+$ T cells has been largely ignored, until recently. Infiltrating $CD8^+$ T cells have been historically considered as being detrimental to motoneurons. In keeping with this view, the microglial-specific depletion of MHC-I and concomitant lack of $CD8^+$ T cell infiltration in the spinal cord were associated with delayed paralysis as well as prolonged survival in mutant SOD1 mice (Nardo et al. 2018). Likewise, Coque et al. (2019) reported an increased number of surviving motoneurons in $CD8^+$ T cells-depleted transgenic mutant SOD1 mice. Interestingly, in this study, the authors also showed that purified mutant SOD1-expressing $CD8^+$ T cells supposedly trigger the death of primary motoneurons in a MHC-I-dependent manner through Fas and granzyme death pathways (Coque et al. 2019).

Also interesting is the developing story about the potential role of the natural killer (NK) cells in MNDs. NK cells are a type of cytotoxic lymphocyte critical to the innate immune system, which provide rapid responses to virus-infected cells. It has been proposed that during ALS progression, NK cells play a major role in the death of motoneurons by invading the central nervous system. It started, in part, from a longitudinal cohort study reporting an increased number of NK cells in peripheral blood from ALS patients (Murdock et al. 2017) and the claim that a high NK cell frequency is observed in the spinal cord of end-stage mutant SOD1 mice (Finkelstein et al. 2011). Building on these published findings, Garofalo et al. (2020) reported that NK cells infiltrate the CNS affected areas of transgenic mutant SOD1 mice and sporadic ALS patients. In addition, these authors reported that neuronal cells from the spinal cord of transgenic $SOD1^{G93A}$ mice express increased levels of the chemokine CCL2, and that CCL2 neutralization reduced NK cells infiltration. Moreover, NK cell depletion increased survival in transgenic $SOD1^{G93A}$ and $TDP43^{A315T}$ mice and modulated the activation of microglia and the infiltration of Treg cells in transgenic $SOD1^{G93A}$ mice (Garofalo et al. 2020). Finally, they found that motoneurons of both ALS patients and transgenic mutant SOD1 mice express high levels of NK2GD ligands, important for NK cell cytotoxic activity (Garofalo et al. 2020).

Acknowledgements We thank Neil Shneider, Matthew Harms and George Mentis for their insightful comments on this chapter and Jennifer Heredia for her critical editing of this manuscript. S.P. is supported by the Department of Defense (W81XWH-13-0416), the National Institute of Health (NS107442, NS117583, NS111176, AG064596), and Project-ALS. F.L. is supported by grants from Cure SMA, Thompson Family Foundation Initiative (TFFI), Project-ALS and National Institute of Health (R21NS101575).

References

Abati E, Citterio G, Bresolin N, Comi GP, Corti S (2020) Glial cells involvement in spinal muscular atrophy: could SMA be a neuroinflammatory disease? Neurobiol Dis 140:104870. https://doi.org/10.1016/j.nbd.2020.104870

Alberti S (2017) Phase separation in biology. Curr Biol 27:R1097–R1102. https://doi.org/10.1016/j.cub.2017.08.069

Al-Chalabi A, Fang F, Hanby MF, Leigh PN, Shaw CE, Ye W, Rijsdijk F (2010) An estimate of amyotrophic lateral sclerosis heritability using twin data. J Neurol Neurosurg Psychiatry 81:1324–1326. https://doi.org/10.1136/jnnp.2010.207464

Alexianu ME, Kozovska M, Appel SH (2001) Immune reactivity in a mouse model of familial ALS correlates with disease progression. Neurology 57:1282–1289

Almer G, Vukosavic S, Romero N, Przedborski S (2002) Inducible nitric oxide synthase upregulation in a transgenic mouse model of familial amyotrophic lateral sclerosis. J Neurochem 72:2415–2425. https://doi.org/10.1046/j.1471-4159.1999.0722415.x

Appel SH, Stewart SS, Appel V, Harati Y, Mietlowski W, Weiss W, Belendiuk GW (1988) A double-blind study of the effectiveness of cyclosporine in amyotrophic lateral sclerosis. Arch Neurol 45:381–386

Bahia El Idrissi N, Bosch S, Ramaglia V, Aronica E, Baas F, Troost D (2016) Complement activation at the motor end-plates in amyotrophic lateral sclerosis. J Neuroinflammation 13:72. https://doi.org/10.1186/s12974-016-0538-2

Banerjee R et al (2008) Adaptive immune neuroprotection in G93A-SOD1 amyotrophic lateral sclerosis mice. PLoS One 3:e2740

Barbeito LH et al (2004) A role for astrocytes in motor neuron loss in amyotrophic lateral sclerosis. Brain Res Brain Res Rev 47:263–274

Beers DR, Appel SH (2019) Immune dysregulation in amyotrophic lateral sclerosis: mechanisms and emerging therapies. Lancet Neurol 18:211–220. https://doi.org/10.1016/S1474-4422(18)30394-6

Beers DR et al (2006) Wild-type microglia extend survival in PU.1 knockout mice with familial amyotrophic lateral sclerosis. Proc Natl Acad Sci U S A 103:16021–16026. https://doi.org/10.1073/pnas.0607423103

Beers DR, Henkel JS, Zhao W, Wang J, Appel SH (2008) CD4+ T cells support glial neuroprotection, slow disease progression, and modify glial morphology in an animal model of inherited ALS. Proc Natl Acad Sci U S A 105:15558–15563. https://doi.org/10.1073/pnas.0807419105

Beers DR et al (2011) Endogenous regulatory T lymphocytes ameliorate amyotrophic lateral sclerosis in mice and correlate with disease progression in patients with amyotrophic lateral sclerosis. Brain 134:1293–1314. https://doi.org/10.1093/brain/awr074

Boillee S et al (2006) Onset and progression in inherited ALS determined by motor neurons and microglia. Science 312:1389–1392

Branchereau P et al (2019) Relaxation of synaptic inhibitory events as a compensatory mechanism in fetal SOD spinal motor networks. elife 8:e51402. https://doi.org/10.7554/eLife.51402

Brannagan TH III (2010) Monomelic muscular atrophy. In: Rowland LP, Pedley TA (eds) Merritt's neurology, 12th edn. Lippincott, Williams & Wilkins, Philadelphia, p 812

Brockington A et al (2013) Unravelling the enigma of selective vulnerability in neurodegeneration: motor neurons resistant to degeneration in ALS show distinct gene expression characteristics and decreased susceptibility to excitotoxicity. Acta Neuropathol 125:95–109. https://doi.org/10.1007/s00401-012-1058-5

Butovsky O et al (2012) Modulating inflammatory monocytes with a unique microRNA gene signature ameliorates murine ALS. J Clin Invest 122:3063–3087. https://doi.org/10.1172/JCI62636

Campbell AM, Williams ER, Barltrop D (1970) Motor neurone disease and exposure to lead. J Neurol Neurosurg Psychiatry 33:877–885. https://doi.org/10.1136/jnnp.33.6.877

Cassina P et al (2008) Mitochondrial dysfunction in S0D1(G93A)-bearing astrocytes promotes motor neuron degeneration: prevention by mitochondrial-targeted antioxidants. J Neurosci 28:4115–4122

Chio A, Benzi G, Dossena M, Mutani R, Mora G (2005) Severely increased risk of amyotrophic lateral sclerosis among Italian professional football players. Brain 128:472–476. https://doi.org/10.1093/brain/awh373

Chio A, Calvo A, Dossena M, Ghiglione P, Mutani R, Mora G (2009) ALS in Italian professional soccer players: the risk is still present and could be soccer-specific. Amyotroph Lateral Scler 10:205–209. https://doi.org/10.1080/17482960902721634

Chio A, Calvo A, Moglia C, Mazzini L, Mora G, Group Ps (2011) Phenotypic heterogeneity of amyotrophic lateral sclerosis: a population based study. J Neurol Neurosurg Psychiatry 82:740–746. https://doi.org/10.1136/jnnp.2010.235952

Chiot A et al (2020) Modifying macrophages at the periphery has the capacity to change microglial reactivity and to extend ALS survival. Nat Neurosci 23:1339–1351. https://doi.org/10.1038/s41593-020-00718-z

Chiu IM et al (2008) T lymphocytes potentiate endogenous neuroprotective inflammation in a mouse model of ALS. Proc Natl Acad Sci U S A 105:17913–17918. https://doi.org/10.1073/pnas.0804610105

Chiu IM et al (2009) Activation of innate and humoral immunity in the peripheral nervous system of ALS transgenic mice. Proc Natl Acad Sci U S A 106:20960–20965. https://doi.org/10.1073/pnas.0911405106

Chiu IM et al (2013) A neurodegeneration-specific gene-expression signature of acutely isolated microglia from an amyotrophic lateral sclerosis mouse model. Cell Rep 4:385–401. https://doi.org/10.1016/j.celrep.2013.06.018

Cirulli ET et al (2015) Exome sequencing in amyotrophic lateral sclerosis identifies risk genes and pathways. Science 347:1436–1441. https://doi.org/10.1126/science.aaa3650

Ciryam P et al (2017) Spinal motor neuron protein supersaturation patterns are associated with inclusion body formation in ALS. Proc Natl Acad Sci U S A 114:E3935–E3943. https://doi.org/10.1073/pnas.1613854114

Clark JA, Southam KA, Blizzard CA, King AE, Dickson TC (2016) Axonal degeneration, distal collateral branching and neuromuscular junction architecture alterations occur prior to symptom onset in the SOD1(G93A) mouse model of amyotrophic lateral sclerosis. J Chem Neuroanat 76:35–47. https://doi.org/10.1016/j.jchemneu.2016.03.003

Conforti L, Gilley J, Coleman MP (2014) Wallerian degeneration: an emerging axon death pathway linking injury and disease. Nat Rev Neurosci 15:394–409. https://doi.org/10.1038/nrn3680

Cook C, Petrucelli L (2019) Genetic convergence brings clarity to the enigmatic red line in ALS. Neuron 101:1057–1069. https://doi.org/10.1016/j.neuron.2019.02.032

Coque E et al (2019) Cytotoxic CD8(+) T lymphocytes expressing ALS-causing SOD1 mutant selectively trigger death of spinal motoneurons. Proc Natl Acad Sci U S A 116:2312–2317. https://doi.org/10.1073/pnas.1815961116

Cottenie E et al (2014) Truncating and missense mutations in IGHMBP2 cause Charcot-Marie Tooth disease type 2. Am J Hum Genet 95:590–601. https://doi.org/10.1016/j.ajhg.2014.10.002

Couthouis J, Raphael AR, Daneshjou R, Gitler AD (2014) Targeted exon capture and sequencing in sporadic amyotrophic lateral sclerosis. PLoS Genet 10:e1004704. https://doi.org/10.1371/journal.pgen.1004704
Cudkowicz ME et al (1997) Epidemiology of mutations in superoxide dismutase in amyotrophic lateral sclerosis. Ann Neurol 41:210–221. https://doi.org/10.1002/ana.410410212
Cudkowicz ME, Shefner JM, Schoenfeld DA, Zhang H, Andreasson KI, Rothstein JD, Drachman DB (2006) Trial of celecoxib in amyotrophic lateral sclerosis. Ann Neurol 60:22–31. https://doi.org/10.1002/ana.20903
D'Erchia AM et al (2017) Massive transcriptome sequencing of human spinal cord tissues provides new insights into motor neuron degeneration in ALS. Sci Rep 7:10046. https://doi.org/10.1038/s41598-017-10488-7
Dewan R et al (2021) Pathogenic huntingtin repeat expansions in patients with frontotemporal dementia and amyotrophic lateral sclerosis. Neuron 109:448–460 e444. https://doi.org/10.1016/j.neuron.2020.11.005
Di Giorgio FP, Carrasco MA, Siao MC, Maniatis T, Eggan K (2007) Non-cell autonomous effect of glia on motor neurons in an embryonic stem cell-based ALS model. Nat Neurosci 10:608–614. https://doi.org/10.1038/nn1885
Di Giorgio FP, Boulting GL, Bobrowicz S, Eggan KC (2008) Human embryonic stem cell-derived motor neurons are sensitive to the toxic effect of glial cells carrying an ALS-causing mutation. Cell Stem Cell 3:637–648. https://doi.org/10.1016/j.stem.2008.09.017
Dobrowolny G et al (2008) Skeletal muscle is a primary target of SOD1G93A-mediated toxicity. Cell Metab 8:425–436. https://doi.org/10.1016/j.cmet.2008.09.002
Dupuis L et al (2012) A randomized, double blind, placebo-controlled trial of pioglitazone in combination with riluzole in amyotrophic lateral sclerosis. PLoS One 7:e37885. https://doi.org/10.1371/journal.pone.0037885
Elden AC et al (2010) Ataxin-2 intermediate-length polyglutamine expansions are associated with increased risk for ALS. Nature 466:1069–1075. https://doi.org/10.1038/nature09320
Elliott JL (2001) Cytokine upregulation in a murine model of familial amyotrophic lateral sclerosis. Brain Res Mol Brain Res 95:172–178
Farina C, Aloisi F, Meinl E (2007) Astrocytes are active players in cerebral innate immunity. Trends Immunol 28:138–145. https://doi.org/10.1016/j.it.2007.01.005
Fatemi Y, Chakraborty R (2019) Acute flaccid myelitis: a clinical overview for 2019. Mayo Clin Proc 94:875–881. https://doi.org/10.1016/j.mayocp.2019.03.011
Feigin VL et al (2019) Global, regional, and national burden of neurological disorders, 1990–2016: a systematic analysis for the Global Burden of Disease Study 2016. Lancet Neurol 18:459–480. https://doi.org/10.1016/S1474-4422(18)30499-X
Finkelstein A et al (2011) Abnormal changes in NKT cells, the IGF-1 axis, and liver pathology in an animal model of ALS. PLoS One 6:e22374. https://doi.org/10.1371/journal.pone.0022374
Fischer LR, Glass JD (2007) Axonal degeneration in motor neuron disease. Neurodegener Dis 4:431–442. https://doi.org/10.1159/000107704
Fischer LR et al (2004) Amyotrophic lateral sclerosis is a distal axonopathy: evidence in mice and man. Exp Neurol 185:232–240. https://doi.org/10.1016/j.expneurol.2003.10.004
Fischer LR et al (2005) The WldS gene modestly prolongs survival in the SOD1G93A fALS mouse. Neurobiol Dis 19:293–300. https://doi.org/10.1016/j.nbd.2005.01.008
Fletcher EV et al (2017) Reduced sensory synaptic excitation impairs motor neuron function via Kv2.1 in spinal muscular atrophy. Nat Neurosci 20:905–916. https://doi.org/10.1038/nn.4561
Fogh I et al (2014) A genome-wide association meta-analysis identifies a novel locus at 17q11.2 associated with sporadic amyotrophic lateral sclerosis. Hum Mol Genet 23:2220–2231. https://doi.org/10.1093/hmg/ddt587
Fournier CN et al (2018) An open label study of a novel immunosuppression intervention for the treatment of amyotrophic lateral sclerosis. Amyotroph Lateral Scler Frontotemporal Degener 19:242–249. https://doi.org/10.1080/21678421.2017.1421666

Frakes AE et al (2014) Microglia induce motor neuron death via the classical NF-kappaB pathway in amyotrophic lateral sclerosis. Neuron 81:1009–1023. https://doi.org/10.1016/j.neuron.2014.01.013

Freischmidt A et al (2015) Haploinsufficiency of TBK1 causes familial ALS and fronto-temporal dementia. Nat Neurosci 18:631–636. https://doi.org/10.1038/nn.4000

Frey D, Schneider C, Xu L, Borg J, Spooren W, Caroni P (2000) Early and selective loss of neuromuscular synapse subtypes with low sprouting competence in motoneuron diseases. J Neurosci 20:2534–2542

Fritz E et al (2013) Mutant SOD1-expressing astrocytes release toxic factors that trigger motoneuron death by inducing hyperexcitability. J Neurophysiol 109:2803–2814. https://doi.org/10.1152/jn.00500.2012

Garofalo S et al (2020) Natural killer cells modulate motor neuron-immune cell cross talk in models of Amyotrophic Lateral Sclerosis. Nat Commun 11:1773. https://doi.org/10.1038/s41467-020-15644-8

Gavrilina TO et al (2008) Neuronal SMN expression corrects spinal muscular atrophy in severe SMA mice while muscle-specific SMN expression has no phenotypic effect. Hum Mol Genet 17:1063–1075. https://doi.org/10.1093/hmg/ddm379

GBD (2017) Global, regional, and national comparative risk assessment of 84 behavioural, environmental and occupational, and metabolic risks or clusters of risks, 1990-2016: a systematic analysis for the Global Burden of Disease Study 2016. Lancet 390:1345–1422. https://doi.org/10.1016/S0140-6736(17)32366-8

Gelfman S et al (2019) A new approach for rare variation collapsing on functional protein domains implicates specific genic regions in ALS. Genome Res 29:809–818. https://doi.org/10.1101/gr.243592.118

Gerbino V et al (2020) The loss of TBK1 kinase activity in motor neurons or in all cell types differentially impacts ALS disease progression in SOD1 mice. Neuron 106:789–805 e785. https://doi.org/10.1016/j.neuron.2020.03.005

Gordon S (2003) Alternative activation of macrophages. Nat Rev Immunol 3:23–35. https://doi.org/10.1038/nri978

Gordon PH et al (2004) Placebo-controlled phase I/II studies of minocycline in amyotrophic lateral sclerosis. Neurology 62:1845–1847

Gordon PH, Cheng B, Katz IB, Pinto M, Hays AP, Mitsumoto H, Rowland LP (2006a) The natural history of primary lateral sclerosis. Neurology 66:647–653. https://doi.org/10.1212/01.wnl.0000200962.94777.71

Gordon PH et al (2006b) Randomized controlled phase II trial of glatiramer acetate in ALS. Neurology 66:1117–1119

Gould TW et al (2006) Complete dissociation of motor neuron death from motor dysfunction by Bax deletion in a mouse model of ALS. J Neurosci 26:8774–8786. https://doi.org/10.1523/JNEUROSCI.2315-06.2006

Graber DJ, Hickey WF, Harris BT (2010) Progressive changes in microglia and macrophages in spinal cord and peripheral nerve in the transgenic rat model of amyotrophic lateral sclerosis. J Neuroinflammation 7:8–8. https://doi.org/10.1186/1742-2094-7-8

Graham AJ, Macdonald AM, Hawkes CH (1997) British motor neuron disease twin study. J Neurol Neurosurg Psychiatry 62:562–569. https://doi.org/10.1136/jnnp.62.6.562

Grohmann K et al (2001) Mutations in the gene encoding immunoglobulin mu-binding protein 2 cause spinal muscular atrophy with respiratory distress type 1. Nat Genet 29:75–77

Guttenplan KA, Weigel MK, Adler DI, Couthouis J, Liddelow SA, Gitler AD, Barres BA (2020) Knockout of reactive astrocyte activating factors slows disease progression in an ALS mouse model. Nat Commun 11:3753. https://doi.org/10.1038/s41467-020-17514-9

Haidet-Phillips AM et al (2011) Astrocytes from familial and sporadic ALS patients are toxic to motor neurons. Nat Biotechnol 29:824–828. https://doi.org/10.1038/nbt.1957

Hall ED, Oostveen JA, Gurney ME (1998) Relationship of microglial and astrocytic activation to disease onset and progression in a transgenic model of familial ALS. Glia 23:249–256

Hardiman O (2018) Global burden of motor neuron diseases: mind the gaps. Lancet Neurol 17:1030–1031. https://doi.org/10.1016/S1474-4422(18)30398-3
Harms MB, Baloh RH (2013) Clinical neurogenetics: amyotrophic lateral sclerosis. Neurol Clin 31:929–950. https://doi.org/10.1016/j.ncl.2013.05.003
Haukedal H, Freude K (2019) Implications of microglia in amyotrophic lateral sclerosis and frontotemporal dementia. J Mol Biol 431:1818–1829. https://doi.org/10.1016/j.jmb.2019.02.004
Henkel JS, Beers DR, Zhao W, Appel SH (2009) Microglia in ALS: the good, the bad, and the resting. J NeuroImmune Pharmacol 4:389–398. https://doi.org/10.1007/s11481-009-9171-5
Highley JR, Ince PG (2012) The neuropathology of the motor neuron diseases. In: Strong MJ (ed) Amyotrophic lateral sclerosis and the frontotemporal dementias, 1st edn. Oxford University Press, Oxford, pp xii, 413 p. https://doi.org/10.1093/med/9780199590674.003.0019
Hsiung GY et al (2012) Clinical and pathological features of familial frontotemporal dementia caused by C9ORF72 mutation on chromosome 9p. Brain 135:709–722. https://doi.org/10.1093/brain/awr354
Ikiz B et al (2015) The regulatory machinery of neurodegeneration in in vitro models of amyotrophic lateral sclerosis. Cell Rep 12:335–345. https://doi.org/10.1016/j.celrep.2015.06.019
Ince PG, Wharton SB (2007) Chapter 5: cytopathology of the motor neuron. Handb Clin Neurol 82:89–119. https://doi.org/10.1016/S0072-9752(07)80008-X
Ingre C, Roos PM, Piehl F, Kamel F, Fang F (2015) Risk factors for amyotrophic lateral sclerosis. Clin Epidemiol 7:181–193. https://doi.org/10.2147/CLEP.S37505
IOM (2006) Amyotrophic lateral sclerosis in veterans: review of the scientific literature. The National Academies Press, Washington, DC. https://doi.org/10.17226/11757
Ito D, Suzuki N (2009) Seipinopathy: a novel endoplasmic reticulum stress-associated disease. Brain 132:8–15. https://doi.org/10.1093/brain/awn216
Iyer CC et al (2015) Low levels of Survival Motor Neuron protein are sufficient for normal muscle function in the SMNDelta7 mouse model of SMA. Hum Mol Genet 24:6160–6173. https://doi.org/10.1093/hmg/ddv332
Jiang YM et al (2005) Gene expression profile of spinal motor neurons in sporadic amyotrophic lateral sclerosis. Ann Neurol 57:236–251. https://doi.org/10.1002/ana.20379
Johann S et al (2015) NLRP3 inflammasome is expressed by astrocytes in the SOD1 mouse model of ALS and in human sporadic ALS patients. Glia 63:2260–2273. https://doi.org/10.1002/glia.22891
Johnson WG, Wigger HJ, Karp HR, Glaubiger LM, Rowland LP (1982) Juvenile spinal muscular atrophy: a new hexosaminidase deficiency phenotype. Ann Neurol 11:11–16. https://doi.org/10.1002/ana.410110103
Kang SH et al (2013) Degeneration and impaired regeneration of gray matter oligodendrocytes in amyotrophic lateral sclerosis. Nat Neurosci 16:571–579. https://doi.org/10.1038/nn.3357
Kanning KC, Kaplan A, Henderson CE (2010) Motor neuron diversity in development and disease. Annu Rev Neurosci 33:409–440. https://doi.org/10.1146/annurev.neuro.051508.135722
Kano O, Beers DR, Henkel JS, Appel SH (2012) Peripheral nerve inflammation in ALS mice: cause or consequence. Neurology 78:833–835. https://doi.org/10.1212/WNL.0b013e318249f776
Kaplan A et al (2014) Neuronal matrix metalloproteinase-9 is a determinant of selective neurodegeneration. Neuron 81:333–348. https://doi.org/10.1016/j.neuron.2013.12.009
Kariya S et al (2008) Reduced SMN protein impairs maturation of the neuromuscular junctions in mouse models of spinal muscular atrophy. Hum Mol Genet 17:2552–2569
Kariya S, Mauricio R, Dai Y, Monani UR (2009) The neuroprotective factor Wld(s) fails to mitigate distal axonal and neuromuscular junction (NMJ) defects in mouse models of spinal muscular atrophy. Neurosci Lett 449:246–251. https://doi.org/10.1016/j.neulet.2008.10.107
Kaufmann P, De Vivo DC (2010) Spinal musclular atrophies of childhood. In: Rowland LP, Pedley TA (eds) Merrit's neurology, 12th edn. Wolters Kluwer, Philadelphia, pp 810–811
Keller MF et al (2014) Genome-wide analysis of the heritability of amyotrophic lateral sclerosis. JAMA Neurol 71:1123–1134. https://doi.org/10.1001/jamaneurol.2014.1184

Keren-Shaul H et al (2017) A unique microglia type associated with restricting development of Alzheimer's disease. Cell 169:1276–1290 e1217. https://doi.org/10.1016/j.cell.2017.05.018

Killian JM, Wilfong AA, Burnett L, Appel SH, Boland D (1994) Decremental motor responses to repetitive nerve stimulation in ALS. Muscle Nerve 17:747–754. https://doi.org/10.1002/mus.880170708

Kim WK, Liu X, Sandner J, Pasmantier M, Andrews J, Rowland LP, Mitsumoto H (2009) Study of 962 patients indicates progressive muscular atrophy is a form of ALS. Neurology 73:1686–1692. https://doi.org/10.1212/WNL.0b013e3181c1dea3

Kim G, Gautier O, Tassoni-Tsuchida E, Ma XR, Gitler AD (2020a) ALS genetics: gains, losses, and implications for future therapies. Neuron 108:822–842. https://doi.org/10.1016/j.neuron.2020.08.022

Kim JK et al (2020b) Muscle-specific SMN reduction reveals motor neuron-independent disease in spinal muscular atrophy models. J Clin Invest 130:1271–1287. https://doi.org/10.1172/JCI131989

Kline RA et al (2017) Comparison of independent screens on differentially vulnerable motor neurons reveals alpha-synuclein as a common modifier in motor neuron diseases. PLoS Genet 13:e1006680. https://doi.org/10.1371/journal.pgen.1006680

Kong L et al (2009) Impaired synaptic vesicle release and immaturity of neuromuscular junctions in spinal muscular atrophy mice. J Neurosci 29:842–851

Kosaka T et al (2012) Primary lateral sclerosis: upper-motor-predominant amyotrophic lateral sclerosis with frontotemporal lobar degeneration – immunohistochemical and biochemical analyses of TDP-43. Neuropathology 32:373–384. https://doi.org/10.1111/j.1440-1789.2011.01271.x

Kuzuhara S, Kokubo Y, Sasaki R, Narita Y, Yabana T, Hasegawa M, Iwatsubo T (2001) Familial amyotrophic lateral sclerosis and parkinsonism-dementia complex of the Kii Peninsula of Japan: clinical and neuropathological study and tau analysis. Ann Neurol 49:501–511

Lall D, Baloh RH (2017) Microglia and C9orf72 in neuroinflammation and ALS and frontotemporal dementia. J Clin Invest 127:3250–3258. https://doi.org/10.1172/JCI90607

Le Masson G, Przedborski S, Abbott LF (2014) A computational model of motor neuron degeneration. Neuron 83:975–988. https://doi.org/10.1016/j.neuron.2014.07.001

Lerman BJ et al (2012) Deletion of galectin-3 exacerbates microglial activation and accelerates disease progression and demise in a SOD1(G93A) mouse model of amyotrophic lateral sclerosis. Brain Behav 2:563–575. https://doi.org/10.1002/brb3.75

Levi-Montalcini R (1987) The nerve growth factor 35 years later. Science 237:1154–1162. https://doi.org/10.1126/science.3306916

Levine JB, Kong J, Nadler M, Xu Z (1999) Astrocytes interact intimately with degenerating motor neurons in mouse amyotrophic lateral sclerosis (ALS). Glia 28:215–224

Liddelow SA et al (2017) Neurotoxic reactive astrocytes are induced by activated microglia. Nature 541:481–487. https://doi.org/10.1038/nature21029

Lillo P, Hodges JR (2009) Frontotemporal dementia and motor neurone disease: overlapping clinic-pathological disorders. J Clin Neurosci 16:1131–1135. https://doi.org/10.1016/j.jocn.2009.03.005

Liu Y et al (2016) C9orf72 BAC mouse model with motor deficits and neurodegenerative features of ALS/FTD. Neuron 90:521–534. https://doi.org/10.1016/j.neuron.2016.04.005

Lobsiger CS, Boillee S, McAlonis-Downes M, Khan AM, Feltri ML, Yamanaka K, Cleveland DW (2009) Schwann cells expressing dismutase active mutant SOD1 unexpectedly slow disease progression in ALS mice. Proc Natl Acad Sci U S A 106:4465–4470. https://doi.org/10.1073/pnas.0813339106

Loeffler JP, Picchiarelli G, Dupuis L, Gonzalez De Aguilar JL (2016) The role of skeletal muscle in amyotrophic lateral sclerosis. Brain Pathol 26:227–236. https://doi.org/10.1111/bpa.12350

Logroscino G et al (2018) Global, regional, and national burden of motor neuron diseases 1990–2016: a systematic analysis for the Global Burden of Disease Study 2016. Lancet Neurol 17:1083–1097. https://doi.org/10.1016/S1474-4422(18)30404-6

Mackenzie IRA, Briemberg H (2020) TDP-43 pathology in primary lateral sclerosis. Amyotroph Lateral Scler Frontotemporal Degener 21:52–58. https://doi.org/10.1080/21678421.2020.1790607

Majounie E et al (2012) Frequency of the C9orf72 hexanucleotide repeat expansion in patients with amyotrophic lateral sclerosis and frontotemporal dementia: a cross-sectional study. Lancet Neurol 11:323–330. https://doi.org/10.1016/S1474-4422(12)70043-1

Maniatis S et al (2019) Spatiotemporal dynamics of molecular pathology in amyotrophic lateral sclerosis. Science 364:89–93. https://doi.org/10.1126/science.aav9776

Marchetto MC, Muotri AR, Mu Y, Smith AM, Cezar GG, Gage FH (2008) Non-cell-autonomous effect of human SOD1 G37R astrocytes on motor neurons derived from human embryonic stem cells. Cell Stem Cell 3:649–657. https://doi.org/10.1016/j.stem.2008.10.001

Martin LJ, Wong M (2020) Skeletal muscle-restricted expression of human SOD1 in transgenic mice causes a fatal ALS-like syndrome. Front Neurol 11:592851. https://doi.org/10.3389/fneur.2020.592851

Martin E, Cazenave W, Allain AE, Cattaert D, Branchereau P (2020) Implication of 5-HT in the dysregulation of chloride homeostasis in prenatal spinal motoneurons from the G93A mouse model of amyotrophic lateral sclerosis. Int J Mol Sci 21(3):1107. https://doi.org/10.3390/ijms21031107

Martineau E, Di Polo A, Vande Velde C, Robitaille R (2018) Dynamic neuromuscular remodeling precedes motor-unit loss in a mouse model of ALS. elife 7:e41973. https://doi.org/10.7554/eLife.41973

Martineau E, Di Polo A, Vande Velde C, Robitaille R (2020) Sex-specific differences in motor-unit remodeling in a mouse model of ALS. eNeuro 7:ENEURO.0388-0319.2020. https://doi.org/10.1523/ENEURO.0388-19.2020

Mathys H et al (2017) Temporal tracking of microglia activation in neurodegeneration at single-cell resolution. Cell Rep 21:366–380. https://doi.org/10.1016/j.celrep.2017.09.039

McAlary L, Plotkin SS, Yerbury JJ, Cashman NR (2019) Prion-like propagation of protein misfolding and aggregation in amyotrophic lateral sclerosis. Front Mol Neurosci 12:262. https://doi.org/10.3389/fnmol.2019.00262

McAndrew PE et al (1997) Identification of proximal spinal muscular atrophy carriers and patients by analysis of SMNT and SMNC gene copy number. Am J Hum Genet 60:1411–1422

McLaughlin RL et al (2017) Genetic correlation between amyotrophic lateral sclerosis and schizophrenia. Nat Commun 8:14774. https://doi.org/10.1038/ncomms14774

Mentis GZ et al (2011) Early functional impairment of sensory-motor connectivity in a mouse model of spinal muscular atrophy. Neuron 69:453–467. https://doi.org/10.1016/j.neuron.2010.12.032

Meyer H, Weihl CC (2014) The VCP/p97 system at a glance: connecting cellular function to disease pathogenesis. J Cell Sci 127:3877–3883. https://doi.org/10.1242/jcs.093831

Meyer K et al (2014) Direct conversion of patient fibroblasts demonstrates non-cell autonomous toxicity of astrocytes to motor neurons in familial and sporadic ALS. Proc Natl Acad Sci U S A 111:829–832. https://doi.org/10.1073/pnas.1314085111

Miller RG et al (2014) NP001 regulation of macrophage activation markers in ALS: a phase I clinical and biomarker study. Amyotroph Lateral Scler Frontotemporal Degener 15:601–609. https://doi.org/10.3109/21678421.2014.951940

Mishra V et al (2020) Systematic elucidation of neuron-astrocyte interaction in models of amyotrophic lateral sclerosis using multi-modal integrated bioinformatics workflow. Nat Commun 11:5579. https://doi.org/10.1038/s41467-020-19177-y

Mitchell JD (1987) Heavy metals and trace elements in amyotrophic lateral sclerosis. Neurol Clin 5:43–60

Moisse K, Strong MJ (2006) Innate immunity in amyotrophic lateral sclerosis. Biochim Biophys Acta 1762:1083–1093. https://doi.org/10.1016/j.bbadis.2006.03.001

Murakami T et al (2015) ALS/FTD mutation-induced phase transition of FUS liquid droplets and reversible hydrogels into irreversible hydrogels impairs RNP granule function. Neuron 88:678–690. https://doi.org/10.1016/j.neuron.2015.10.030

Murdock BJ, Zhou T, Kashlan SR, Little RJ, Goutman SA, Feldman EL (2017) Correlation of peripheral immunity with rapid amyotrophic lateral sclerosis progression. JAMA Neurol 74:1446–1454. https://doi.org/10.1001/jamaneurol.2017.2255

Nagai M, Kikuchi H, Przedborski S (2006) Experimental models of motor neuron diseases. In: Mitsumoto H, Przedborski S, Gordon PH (eds) Amyotrophic lateral sclerosis, Neurological disease and therapy, vol 78. Taylor & Francis, New York, pp 525–549

Nagai M, Re DB, Nagata T, Chalazonitis A, Jessell TM, Wichterle H, Przedborski S (2007) Astrocytes expressing ALS-linked mutated SOD1 release factors selectively toxic to motor neurons. Nat Neurosci 10:615–622. https://doi.org/10.1038/nn1876

Nagy D, Kato T, Kushner PD (1994) Reactive astrocytes are widespread in the cortical gray matter of amyotrophic lateral sclerosis. J Neurosci Res 38:336–347. https://doi.org/10.1002/jnr.490380312

Nardo G et al (2013) Transcriptomic indices of fast and slow disease progression in two mouse models of amyotrophic lateral sclerosis. Brain 136:3305–3332. https://doi.org/10.1093/brain/awt250

Nardo G et al (2018) Counteracting roles of MHCI and CD8(+) T cells in the peripheral and central nervous system of ALS SOD1(G93A) mice. Mol Neurodegener 13:42. https://doi.org/10.1186/s13024-018-0271-7

Naumann M et al (2019) Phenotypes and malignancy risk of different FUS mutations in genetic amyotrophic lateral sclerosis. Ann Clin Transl Neurol 6:2384–2394. https://doi.org/10.1002/acn3.50930

Neukomm LJ, Freeman MR (2014) Diverse cellular and molecular modes of axon degeneration. Trends Cell Biol 24:515–523. https://doi.org/10.1016/j.tcb.2014.04.003

Nguyen HP, Van Broeckhoven C, van der Zee J (2018) ALS genes in the genomic era and their implications for FTD. Trends Genet 34:404–423. https://doi.org/10.1016/j.tig.2018.03.001

Nijssen J, Comley LH, Hedlund E (2017) Motor neuron vulnerability and resistance in amyotrophic lateral sclerosis. Acta Neuropathol 133:863–885. https://doi.org/10.1007/s00401-017-1708-8

Nikodemova M, Small AL, Smith SM, Mitchell GS, Watters JJ (2014) Spinal but not cortical microglia acquire an atypical phenotype with high VEGF, galectin-3 and osteopontin, and blunted inflammatory responses in ALS rats. Neurobiol Dis 69:43–53. https://doi.org/10.1016/j.nbd.2013.11.009

Noristani HN et al (2015) Brca1 is expressed in human microglia and is dysregulated in human and animal model of ALS. Mol Neurodegener 10:34. https://doi.org/10.1186/s13024-015-0023-x

Pamphlett R (2010) Study of 962 patients indicates progressive muscular atrophy is a form of ALS. Neurology 74:1926. Author reply 1926-1927. https://doi.org/10.1212/WNL.0b013e3181e03b59

Pang SY et al (2017) Burden of rare variants in ALS genes influences survival in familial and sporadic ALS. Neurobiol Aging 58:238 e239–238 e215. https://doi.org/10.1016/j.neurobiolaging.2017.06.007

Patel A et al (2015) A liquid-to-solid phase transition of the ALS protein FUS accelerated by disease mutation. Cell 162:1066–1077. https://doi.org/10.1016/j.cell.2015.07.047

Pehar M et al (2004) Astrocytic production of nerve growth factor in motor neuron apoptosis: implications for amyotrophic lateral sclerosis. J Neurochem 89:464–473. https://doi.org/10.1111/j.1471-4159.2004.02357.x

Pehar M, Vargas MR, Cassina P, Barbeito AG, Beckman JS, Barbeito L (2005) Complexity of astrocyte-motor neuron interactions in amyotrophic lateral sclerosis. Neurodegener Dis 2:139–146

Phatnani HP et al (2013) Intricate interplay between astrocytes and motor neurons in ALS. Proc Natl Acad Sci U S A 110:E756–E765. https://doi.org/10.1073/pnas.1222361110

Philips T, Robberecht W (2011) Neuroinflammation in amyotrophic lateral sclerosis: role of glial activation in motor neuron disease. Lancet Neurol 10:253–263. https://doi.org/10.1016/S1474-4422(11)70015-1

Philips T et al (2013) Oligodendrocyte dysfunction in the pathogenesis of amyotrophic lateral sclerosis. Brain 136:471–482. https://doi.org/10.1093/brain/aws339

Plato CC et al (2003) Amyotrophic lateral sclerosis and parkinsonism-dementia complex of Guam: changing incidence rates during the past 60 years. Am J Epidemiol 157:149–157. https://doi.org/10.1093/aje/kwf175

Poesen K, Van Damme P (2018) Diagnostic and prognostic performance of neurofilaments in ALS. Front Neurol 9:1167. https://doi.org/10.3389/fneur.2018.01167

Przedborski S (2017) Neurodegeneration. In: Ikezu T, Gendelman HE (eds) Neuroimmune pharmacology, 2nd edn. Springer, Cham, pp 345–354. https://doi.org/10.1007/978-3-319-44022-4_22

Pun S, Santos AF, Saxena S, Xu L, Caroni P (2006) Selective vulnerability and pruning of phasic motoneuron axons in motoneuron disease alleviated by CNTF. Nat Neurosci 9:408–419. https://doi.org/10.1038/nn1653

Ragagnin AMG, Shadfar S, Vidal M, Jamali MS, Atkin JD (2019) Motor neuron susceptibility in ALS/FTD. Front Neurosci 13:532. https://doi.org/10.3389/fnins.2019.00532

Rajabinejad M, Ranjbar S, Afshar Hezarkhani L, Salari F, Gorgin Karaji A, Rezaiemanesh A (2020) Regulatory T cells for amyotrophic lateral sclerosis/motor neuron disease: a clinical and preclinical systematic review. J Cell Physiol 235:5030–5040. https://doi.org/10.1002/jcp.29401

Ravits JM, La Spada AR (2009) ALS motor phenotype heterogeneity, focality, and spread: deconstructing motor neuron degeneration. Neurology 73:805–811. https://doi.org/10.1212/WNL.0b013e3181b6bbbd

Re DB et al (2014) Necroptosis drives motor neuron death in models of both sporadic and familial ALS. Neuron 81:1001–1008. https://doi.org/10.1016/j.neuron.2014.01.011

Renton AE, Chio A, Traynor BJ (2014) State of play in amyotrophic lateral sclerosis genetics. Nat Neurosci 17:17–23. https://doi.org/10.1038/nn.3584

Rowland LP, Mitsumoto H, Przedborski S (2010) Amyotrophic lateral sclerosis, progressive muscular atrophy and primary lateral sclerosis. In: Rowland LP, Pedley TA (eds) Merritt's neurology, 12th edn. Lippincott, Williams & Wilkins, Philadelphia, pp 802–808

Ryan MA, Heverin M, McLaughlin RL, Hardiman O (2019) Lifetime risk and heritability of amyotrophic lateral sclerosis. JAMA Neurol 76:1367–1374. https://doi.org/10.1001/jamaneurol.2019.2044

Saez-Atienzar S et al (2021) Genetic analysis of amyotrophic lateral sclerosis identifies contributing pathways and cell types. Sci Adv 7:eabd9036. https://doi.org/10.1126/sciadv.abd9036

Schaefer AM, Sanes JR, Lichtman JW (2005) A compensatory subpopulation of motor neurons in a mouse model of amyotrophic lateral sclerosis. J Comp Neurol 490:209–219. https://doi.org/10.1002/cne.20620

Schiffer D, Cordera S, Cavalla P, Migheli A (1996) Reactive astrogliosis of the spinal cord in amyotrophic lateral sclerosis. J Neurol Sci 139:27–33

Schon EA, Przedborski S (2011) Mitochondria: the next (neurode)generation. Neuron 70:1033–1053. https://doi.org/10.1016/j.neuron.2011.06.003

Schutz B (2005) Imbalanced excitatory to inhibitory synaptic input precedes motor neuron degeneration in an animal model of amyotrophic lateral sclerosis. Neurobiol Dis 20:131–140. https://doi.org/10.1016/j.nbd.2005.02.006

Sharma A et al (2016) ALS-associated mutant FUS induces selective motor neuron degeneration through toxic gain of function. Nat Commun 7:10465. https://doi.org/10.1038/ncomms10465

Shribman S, Reid E, Crosby AH, Houlden H, Warner TT (2019) Hereditary spastic paraplegia: from diagnosis to emerging therapeutic approaches. Lancet Neurol 18:1136–1146. https://doi.org/10.1016/S1474-4422(19)30235-2

Silani V, Corcia P, Harms MB, Rouleau G, Siddique T, Ticozzi N (2020) Genetics of primary lateral sclerosis. Amyotroph Lateral Scler Frontotemporal Degener 21:28–34. https://doi.org/10.1080/21678421.2020.1837177

Simon CM et al (2017) Converging mechanisms of p53 activation drive motor neuron degeneration in spinal muscular atrophy. Cell Rep 21:3767–3780. https://doi.org/10.1016/j.celrep.2017.12.003

Spiller KJ, Cheung CJ, Restrepo CR, Kwong LK, Stieber AM, Trojanowski JQ, Lee VM (2016) Selective motor neuron resistance and recovery in a new inducible mouse model of TDP-43 proteinopathy. J Neurosci 36:7707–7717. https://doi.org/10.1523/JNEUROSCI.1457-16.2016
Srinivasan K et al (2016) Untangling the brain's neuroinflammatory and neurodegenerative transcriptional responses. Nat Commun 7:11295. https://doi.org/10.1038/ncomms11295
Swinnen B, Robberecht W (2014) The phenotypic variability of amyotrophic lateral sclerosis. Nat Rev Neurol 10:661–670. https://doi.org/10.1038/nrneurol.2014.184
Tallon C, Russell KA, Sakhalkar S, Andrapallayal N, Farah MH (2016) Length-dependent axo-terminal degeneration at the neuromuscular synapses of type II muscle in SOD1 mice. Neuroscience 312:179–189. https://doi.org/10.1016/j.neuroscience.2015.11.018
Tanner CM, Goldman SM, Ross GW, Grate SJ (2014) The disease intersection of susceptibility and exposure: chemical exposures and neurodegenerative disease risk Alzheimers. Dement 10:S213–S225. https://doi.org/10.1016/j.jalz.2014.04.014
Taylor JP, Brown RH Jr, Cleveland DW (2016) Decoding ALS: from genes to mechanism. Nature 539:197–206. https://doi.org/10.1038/nature20413
Tisdale S, Pellizzoni L (2015) Disease mechanisms and therapeutic approaches in spinal muscular atrophy. J Neurosci 35:8691–8700. https://doi.org/10.1523/JNEUROSCI.0417-15.2015
Tiwari A, Hayward LJ (2005) Mutant SOD1 instability: implications for toxicity in amyotrophic lateral sclerosis. Neurodegener Dis 2:115–127. https://doi.org/10.1159/000089616
Toth RP, Atkin JD (2018) Dysfunction of optineurin in amyotrophic lateral sclerosis and glaucoma. Front Immunol 9:1017. https://doi.org/10.3389/fimmu.2018.01017
Towne C, Raoul C, Schneider BL, Aebischer P (2008) Systemic AAV6 delivery mediating RNA interference against SOD1: neuromuscular transduction does not alter disease progression in fALS mice. Mol Ther 16:1018–1025. https://doi.org/10.1038/mt.2008.73
Tremblay E, Martineau E, Robitaille R (2017) Opposite synaptic alterations at the neuromuscular junction in an ALS mouse model: when motor units matter. J Neurosci 37:8901–8918. https://doi.org/10.1523/JNEUROSCI.3090-16.2017
Tyzack GE et al (2017) A neuroprotective astrocyte state is induced by neuronal signal EphB1 but fails in ALS models. Nat Commun 8:1164. https://doi.org/10.1038/s41467-017-01283-z
van Blitterswijk M et al (2012a) Evidence for an oligogenic basis of amyotrophic lateral sclerosis. Hum Mol Genet 21:3776–3784. https://doi.org/10.1093/hmg/dds199
van Blitterswijk M et al (2012b) Genetic overlap between apparently sporadic motor neuron diseases. PLoS One 7:e48983. https://doi.org/10.1371/journal.pone.0048983
Van Harten ACM, Phatnani H, Przedborski S (2021) Non-cell autonomous pathogenic mechanisms in amyotrophic lateral sclerosis. Trends Neurosci 44(8):658–668
Van Mossevelde S, van der Zee J, Cruts M, Van Broeckhoven C (2017a) Relationship between C9orf72 repeat size and clinical phenotype. Curr Opin Genet Dev 44:117–124. https://doi.org/10.1016/j.gde.2017.02.008
Van Mossevelde S et al (2017b) Clinical evidence of disease anticipation in families segregating a C9orf72 repeat expansion. JAMA Neurol 74:445–452. https://doi.org/10.1001/jamaneurol.2016.4847
van Rheenen W et al (2016) Genome-wide association analyses identify new risk variants and the genetic architecture of amyotrophic lateral sclerosis. Nat Genet 48:1043–1048. https://doi.org/10.1038/ng.3622
Vande Velde C, Garcia ML, Yin X, Trapp BD, Cleveland DW (2004) The neuroprotective factor Wlds does not attenuate mutant SOD1-mediated motor neuron disease. NeuroMolecular Med 5:193–203. https://doi.org/10.1385/NMM:5:3:193
Veriepe J, Fossouo L, Parker JA (2015) Neurodegeneration in C. elegans models of ALS requires TIR-1/Sarm1 immune pathway activation in neurons. Nat Commun 6:7319. https://doi.org/10.1038/ncomms8319
Vinsant S et al (2013) Characterization of early pathogenesis in the SOD1(G93A) mouse model of ALS: part II, results and discussion. Brain Behav 3:431–457. https://doi.org/10.1002/brb3.142

Vukojicic A et al (2019) The classical complement pathway mediates microglia-dependent remodeling of spinal motor circuits during development and in SMA. Cell Rep 29:3087–3100 e3087. https://doi.org/10.1016/j.celrep.2019.11.013

Wang M, Casey PJ (2016) Protein prenylation: unique fats make their mark on biology. Nat Rev Mol Cell Biol 17:110–122. https://doi.org/10.1038/nrm.2015.11

Weydt P, Yuen EC, Ransom BR, Moller T (2004) Increased cytotoxic potential of microglia from ALS-transgenic mice. Glia 48:179–182. https://doi.org/10.1002/glia.20062

Wingo TS, Cutler DJ, Yarab N, Kelly CM, Glass JD (2011) The heritability of amyotrophic lateral sclerosis in a clinically ascertained United States research registry. PLoS One 6:e27985. https://doi.org/10.1371/journal.pone.0027985

Wirth B, Brichta L, Schrank B, Lochmuller H, Blick S, Baasner A, Heller R (2006) Mildly affected patients with spinal muscular atrophy are partially protected by an increased SMN2 copy number. Hum Genet 119:422–428. https://doi.org/10.1007/s00439-006-0156-7

Wootz H et al (2013) Alterations in the motor neuron-renshaw cell circuit in the Sod1(G93A) mouse model. J Comp Neurol 521:1449–1469. https://doi.org/10.1002/cne.23266

Xiao Q, Zhao W, Beers DR, Yen AA, Xie W, Henkel JS, Appel SH (2007) Mutant SOD1(G93A) microglia are more neurotoxic relative to wild-type microglia. J Neurochem 102:2008–2019. https://doi.org/10.1111/j.1471-4159.2007.04677.x

Yamanaka K et al (2008a) Mutant SOD1 in cell types other than motor neurons and oligodendrocytes accelerates onset of disease in ALS mice. Proc Natl Acad Sci U S A 105:7594–7599. https://doi.org/10.1073/pnas.0802556105

Yamanaka K et al (2008b) Astrocytes as determinants of disease progression in inherited amyotrophic lateral sclerosis. Nat Neurosci 11:251–253. https://doi.org/10.1038/nn2047

Young PE, Kum Jew S, Buckland ME, Pamphlett R, Suter CM (2017) Epigenetic differences between monozygotic twins discordant for amyotrophic lateral sclerosis (ALS) provide clues to disease pathogenesis. PLoS One 12:e0182638. https://doi.org/10.1371/journal.pone.0182638

Yuan JH et al (2017) Clinical diversity caused by novel IGHMBP2 variants. J Hum Genet 62:599–604. https://doi.org/10.1038/jhg.2017.15

Zhao W, Beers DR, Bell S, Wang J, Wen S, Baloh RH, Appel SH (2015) TDP-43 activates microglia through NF-kappaB and NLRP3 inflammasome. Exp Neurol 273:24–35. https://doi.org/10.1016/j.expneurol.2015.07.019

Electrical and Morphological Properties of Developing Motoneurons in Postnatal Mice and Early Abnormalities in SOD1 Transgenic Mice

Jacques Durand and Anton Filipchuk

Abstract In this chapter, we review electrical and morphological properties of lumbar motoneurons during postnatal development in wild-type (WT) and transgenic superoxide dismutase 1 (SOD1) mice, models of amyotrophic lateral sclerosis. First we showed that sensorimotor reflexes do not develop normally in transgenic $SOD1^{G85R}$ pups. Fictive locomotor activity recorded in *in vitro* whole brainstem/spinal cord preparations was not induced in these transgenic $SOD1^{G85R}$ mice using NMDA and 5HT in contrast to WT mice. Further, abnormal electrical properties were detected as early as the second postnatal week in lumbar motoneurons of SOD1 mice while they develop clinical symptoms several months after birth. We compared two different strains of mice (G85R and G93A) at the same postnatal period using intracellular recordings and patch clamp recordings of WT and SOD1 motoneurons. We defined three types of motoneurons according to their discharge firing pattern (transient, sustained and delayed onset firing) when motor units are not yet mature. The delayed-onset firing motoneurons had the higher rheobase compared to the transient and sustained firing groups in the WT mice. We demonstrated hypoexcitability in the delayed onset-firing motoneurons of SOD1 mice. Intracellular staining of motoneurons revealed dendritic overbranching in SOD1 lumbar motoneurons that was more pronounced in the sustained firing motoneurons. We suggested that motoneuronal hypoexcitability is an early pathological sign affecting a subset of lumbar motoneurons in the spinal cord of SOD1 mice.

Keywords Spinal cord · ALS · SOD1 transgenic mice · Hypoexcitability · Motor neuron

J. Durand (✉)
Institut de Neurosciences de la Timone (INT) P3M team, Aix Marseille Université, Marseille, cedex 05, France
e-mail: Jacques.Durand@univ-amu.fr

A. Filipchuk
Department for Integrative and Computational Neuroscience (ICN), Paris-Saclay Institute of Neuroscience (NeuroPSI), Gif-sur-Yvette, France
e-mail: anton.filipchuk@cnrs.fr

M. J. O'Donovan, M. Falgairolle (eds.), *Vertebrate Motoneurons*, Advances in Neurobiology 28, https://doi.org/10.1007/978-3-031-07167-6_14

1 Introduction

Motoneurons of different classes and subtypes (fast/slow, alpha/gamma) are grouped together into motor pools each of which innervates a single muscle in vertebrates (Kanning et al. 2010). An adult motoneuron innervates a group of muscular fibers, both forming a motor unit. Most muscles are composed of different motor units with different ratios between slow, fast-resistant and fast fatigable depending on the muscle (Burke et al. 1971; Burke 1981). In neonatal animals, a single muscle fiber is innervated by several motoneurons (polyinnervation) up to the third postnatal week and motor units achieve a complete maturation depending on the muscle a month after birth in rodents (Navarrete and Vrbova 1983, 1993; Balice-Gordon and Thompson 1988). Gap junctions are present during the two first postnatal weeks in rodents facilitating synchronization of motor discharge firing between motoneurons. At adulthood, an orderly recruitment from slow to fast motor units is a general rule for most movements (Henneman 1957; Henneman et al. 1965). Motoneuron discharge firing is modulated and permanently adjusted during the muscle contraction by proprioceptive afferents from muscle spindles (Matthews 1982; Rossi-Durand 2006, for review).

During the neonatal period, the motoneuron is the main target for numerous abnormalities caused by neurological diseases. Early abnormalities have been described in motoneurons from animal models of neurodegenerative diseases during embryonic and postnatal periods in disorders such as spinal muscular atrophy (Mentis et al. 2011; Vrbova 2007), Charcot–Marie–Tooth diseases (Irobi et al. 2010) and amyotrophic lateral sclerosis (Amendola and Durand 2008; Amendola et al. 2004; Bories et al. 2007; Durand et al. 2006; Raoul et al. 2002; Van Zundert et al. 2012; Vinsant et al. 2013). Amyotrophic lateral sclerosis (ALS) is a fatal neurodegenerative disease mainly affecting motor neurons and is one of the most complex and dramatic diseases with no available curative therapy (Leigh and Swash 1996; Kanning et al. 2010). It is known that large motoneurons are affected first in humans. In transgenic mouse model of ALS, the motoneuron subtypes are not equally vulnerable (Pun et al. 2006; Hadzipasic et al. 2014). A number of sporadic and familial ALS patients have mutations in the gene coding the superoxide dismutase 1 (SOD1) (Rosen et al. 1993; Boillee et al. 2006; Robberecht and Philips 2013). Most mechanisms leading to the disease have been described in the SOD1 mice, (Gurney et al. 1994; Dal Canto and Gurney 1997; Bruijn et al. 1997, 1998, 2004). Early changes in excitability were described in the motoneurons of SOD1 mice including both increased and decreased excitability (Pieri et al. 2003; Kuo et al. 2004, 2005; Bories et al. 2007; Van Zundert et al. 2008; Pambo-Pambo et al. 2009; Martin et al. 2013; Elbasiouny et al. 2010). Perturbed intracellular trafficking in SOD1 mice was indicated by the reduced transport of selective cargoes described months before disease onset (Williamson and Cleveland 1999), the alterations in fast axonal transport (Kieran et al. 2005) and neurofilament accumulation (Bruijn et al. 2004). A direct link between retrograde axonal transport and mutant SOD1 was made with dynein interaction (Zhang et al. 2007).

We described morphological and electrical pathological signs in motoneurons during the postnatal period of SOD1 mice (Amendola et al. 2004, 2007; Durand et al. 2006; Bories et al. 2007; Amendola and Durand 2008; Filipchuk and Durand 2012; Filipchuk et al. 2008, 2021). The time course of the disease may differ depending on the mutations and the number of copies of mutated proteins expressed in the mutant animals. In our experiments, we have used low expressor lines of SOD1 mice (G85R and G93A-low) that become paralyzed several months after birth (6 to 8 months) (Dal Canto and Gurney 1997; Bruijn et al. 1997, 1998; Amendola et al. 2004; Pambo-Pambo et al. 2009) in contrast to high expressor strains (SOD1 G93A-high) that are paralyzed at 4 months (Gurney et al. 1994; Vinsant et al. 2013). In the high expressor SOD1-G93A strain, neuromuscular junctions start to disconnect between 14 and 30 days after birth depending on the muscles (Vinsant et al. 2013). The advantage of the former models (low expressor lines) is the longer presymptomatic time allowing us to decorticate the earlier pathological mechanisms (Amendola et al. 2004; Bories et al. 2007; Pambo-Pambo et al. 2009). The homozygous $SOD1^{G85R}$ strain of mice was first chosen for several reasons. First, all the $SOD1^{G85R}$ homozygous pups are viable and carries the mutation causing the disease; thus no genetic screening was needed to study sensorimotor reflexes using Fox battery tests (Fox 1965) and spinal networks using electrophysiological recordings in the whole brainstem spinal cord preparation (Amendola et al. 2004). Second, the $SOD1^{G85R}$ strain was found to be a good model because the disease-onset did not depend on the enzyme activity and astrocytes were involved in the pathology (Bruijn et al. 1997, 1998). Homozygous and heterozygous $SOD1^{G85R}$ mice develop clinical symptoms at 6 and 12 months of age, respectively (Bruijn et al. 1997, 1998; Amendola et al. 2004, 2007; Amendola 2008). We studied the electrical and morphological properties of motoneurons in both low expressor models (G85R and G93A-low) with comparable time courses (Amendola et al. 2007; Pambo-Pambo et al. 2009; Filipchuk et al. 2021). We also studied the electrical properties of lumbar motoneurons from the high expressor line of SOD1 G93A mice (G93A-high) in the whole brainstem-spinal cord preparation to compare with most studies performed in slice or in cultured neurons from this transgenic strain (Pieri et al. 2003; Kuo et al. 2004, 2005; Quinlan et al. 2011; Leroy et al. 2014).

2 Altered Sensorimotor Development in SOD1 Mice

In earlier work, we addressed the question of whether the SOD1 mutation has an impact on spinal motor networks in postnatal mutant mice (Amendola et al. 2004). We detected subtle functional alterations in postnatal $SOD1^{G85R}$ mice during sensorimotor development. Six sensorimotor responses appearing from P1 to P6 were measured daily in all male and female mice (Fox 1965; Amendola et al. 2004; Amendola 2008). When the appropriate responses were observed, the pup was given a score of 1 for the corresponding test. Tests were as follows. (i) Righting: the pup was placed on it back; it righted itself within10 s. (ii) Cliff-drop aversion: the pup was placed on the edge of a cliff (a table), the forepaws and the head over the edge; it turned and crawled away from the cliff. (iii) Forepaw and (iv) hind paw grasping: when the inside of one paw was gently stroked with an object, the paw flexed to grasp the object. (v) Forelimb and (vi) Vibrissae placing responses were also tested between P3 and P8 (Amendola et al. 2004). Righting and hind paw grasping responses were significantly delayed for SOD1^{G85R} pups up to P7 (Amendola et al. 2004). Our behavioral testing from WT and SOD1^{G85R} showed a transient motor deficit with a motor impairment of only the hind limbs. Behavioral deficits were also found in postnatal SOD1^{G93A} mice (Van Zundert et al. 2008).

Electrophysiological studies using extracellular recordings from the ventral roots in entire brainstem-spinal cord preparations *in vitro* during perfusion of NMDA and 5HT, showed a lack of alternating rhythmic activities in SOD1 mice, in contrast to WT preparations recorded in the same chamber. This lack of rhythmic responses suggested early hypoexcitability of lumbar networks probably due to changes in endogenous membrane properties and/or alterations in lumbar network connectivity.

A delayed maturation of supraspinal descending pathways has been suggested (Amendola et al. 2004; Amendola 2008). Recent results show a delay in the development of 5HT immunoreactive descending fibers in SOD1^{G93A} mice (Martin et al. 2020). Further, we recently brought some arguments in support of a lower density of boutons in motoneuronal pools of SOD1 G85R mice (Filipchuk et al. 2021). In transgenic SOD1^{G93A} mice, it was shown that NMA and 5HT could generate fictive locomotor activity just after birth at P0-P3 (Milan et al. 2014). This was not in contrast with our results since they were obtained in younger animals and in a different strain. Further, in previous experiments we could get alternating rhythmic activities in SOD1^{G85R} in *in vitro* preparations when dopamine was added to the cocktail NMDA, 5HT (Amendola 2008). Gross deficits in locomotion were found in SOD1 juvenile mice (Wooley et al. 2005; Van Zundert et al. 2008; Mead et al. 2011; Vinsant et al. 2013).

3 Electrical Properties of Developing Motoneurons in Postnatal Mice

Electrical properties of developing spinal motoneurons in neonatal mice have been investigated in several studies (Mynlieff and Beam 1992; Bories 2004; Bories et al. 2007; Pambo-Pambo et al. 2009; Pambo-Pambo 2010; Nakanishi and Whelan 2010; Quinlan et al. 2011; Durand et al. 2015).

Early Changes in Electrical Properties of Lumbar Motoneurons from WT Mice

Here we focus on the different subgroups of motoneurons defined by their pattern of discharge during the postnatal period (Pambo-Pambo et al. 2009; Durand et al. 2015). Using intracellular current pulses, we found three subgroups of postnatal motoneurons according to their discharge firing pattern (Fig. 1).

The delayed-onset firing motoneurons have the highest rheobase and, consequently, are the least excitable motoneurons. We suggested they might represent the future fast motoneurons in adult animals (Durand et al. 2015). A transient firing pattern with a single or a short train of spike (Fig. 1a) was the most preponderant between P0 and P3 as already described in neonatal rat and these motoneurons are considered as the least mature (Vinay et al. 2000a, b, 2002; Mentis et al. 2007). The sustained firing pattern (Fig. 1b) was the most frequent in older animals and was expressed by 100% of the motoneurons at P9 when recorded in an entire brainstem-spinal cord preparation (Durand et al. 2015). The distribution of the different

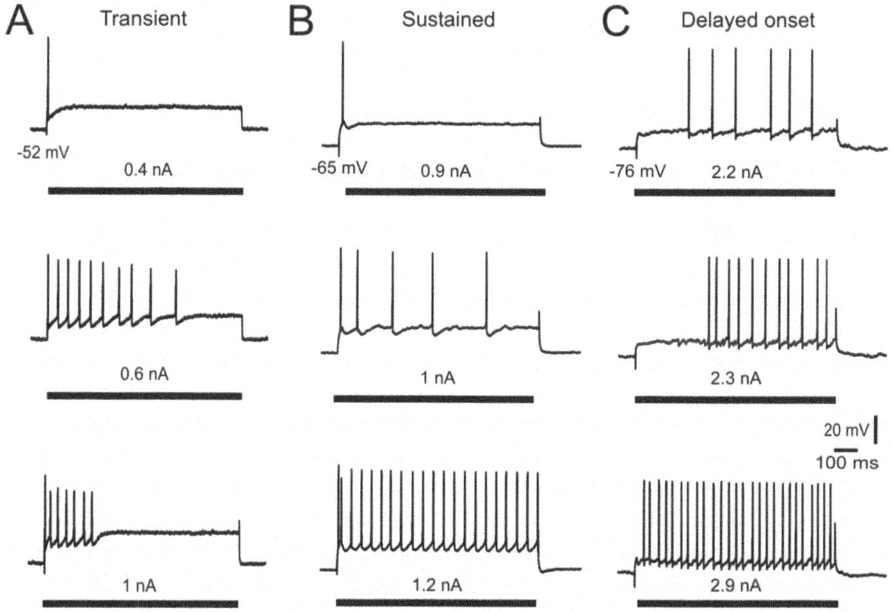

Fig. 1 Three subgroups of lumbar motoneurons defined according to their discharge firing pattern in postnatal mice. (**a**) The transient firing is present in 32% of motoneurons at postnatal days P3-P5 (Durand et al. 2015). (**b**) The sustained firing is the most frequent after P7-P8. (**c**) The delayed onset firing is observed in 20–30% of cases when recorded in the whole brainstem-spinal cord preparation. Intracellular recordings of three lumbar motoneurons are illustrated with voltage deflections in response to current pulse injections of increasing intensities indicated above each black bar (nA). Membrane potentials are shown below first upper traces for each motoneuron. Same calibration bars for all recordings in the lower right panel

patterns depends on the postnatal age of the mouse (Durand et al. 2015). The delayed-onset firing pattern (Fig. 1c) was found in 20–30% of lumbar motoneurons in mice aged between P3 and P10. However, delayed-onset firing pattern (also called late spiking) was recorded in 65–70% motoneurons in spinal cord slices (Pambo-Pambo et al. 2009; Leroy et al. 2014). We suggested that the absence of neuromodulatory controls by supraspinal pathways in slice preparations was responsible for this different ratio (Durand et al. 2015). Several electrophysiological properties indicate that the three different patterns of discharge firing induced by intracellular current pulses injection in motoneurons are reflecting different states of maturation during this postnatal period (Durand et al. 2015). We also described five types of neonatal motoneurons using current ramp injection named transient, linear, linear with sustained firing, clockwise and counter-clockwise hysteresis (Amendola et al. 2007; Durand et al. 2015). Except for the transient groups, we found no correlation between the subgroups defined by their discharge pattern induced by current pulses and ramps (Pambo-Pambo et al. 2009; Durand et al. 2015).

Early Alterations in Electrical Properties of Lumbar Motoneurons from SOD1 Mice

We found differences in electrical properties of WT and $SOD1^{G85R}$ motoneurons as soon as the beginning of the second postnatal week (Bories 2004; Durand et al. 2006; Bories et al. 2007; Pambo-Pambo et al. 2009; Elbasiouny et al. 2010). We measured a lower gain indicated by a lower slope of the frequency (F)– intensity (I) relationship in $SOD1^{G85R}$ motoneurons, a lower input resistance and a higher rheobase indicating hypoexcitability of lumbar motoneurons (Bories et al. 2007). The slope of the frequency-intensity curve (F-I) measured during the steady state of the discharge firing induced by intracellular injection of current pulses is the gain of a neuron. It has been measured in lumbar motoneurons (n = 32) and neighbor neurons (n = 20) in WT and SOD1 mice (Fig. 2). Such changes in excitability were not present in P3-P6 motoneurons (but only in P7-P10) nor in spinal neurons, suggesting specific alterations of the motoneurons at that particular postnatal period (Bories et al. 2007; Amendola 2008).

Recently we confirmed hypoexcitability of lumbar motoneurons from $SOD1^{G93A}$ $_{low}$ mice (Filipchuk et al. 2021). Further, alterations in the electrical properties including a lower gain and higher voltage threshold (signs of hypoexcitability) were specific to the delayed-onset firing SOD1 postnatal motoneurons. We concluded that the delayed-onset firing motoneurons are the first functionally affected motoneurons in the second postnatal week in the mouse model of ALS (Filipchuk et al. 2021).

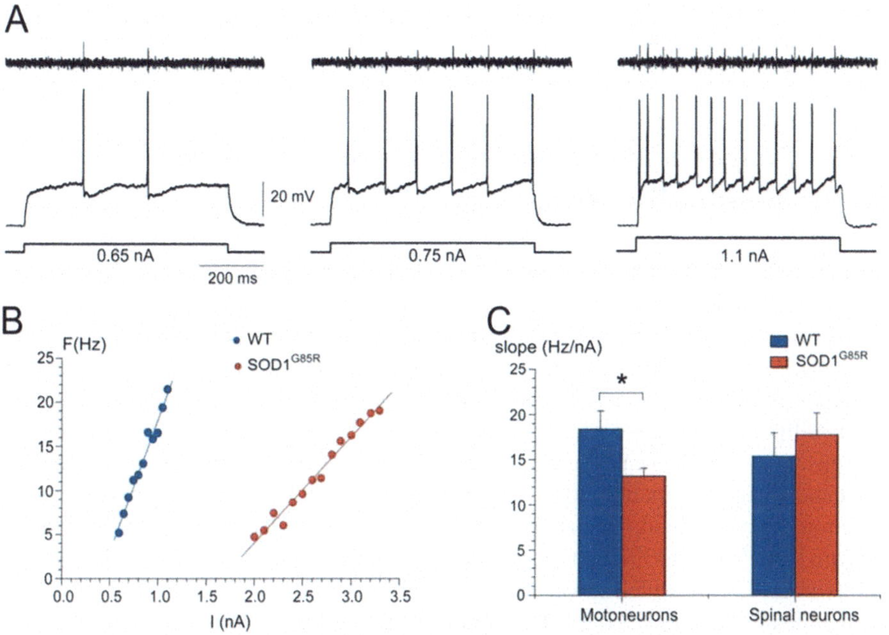

Fig. 2 Lower gain in SOD1 motoneurons. The frequency-intensity curve (F-I) measured in motoneurons and spinal neurons during the steady state of discharge firing induced by intracellular injection of current pulses with increasing intensities. In (**a**) the upper trace shows the extracellular recording from the L5 ventral root of the orthodromic spike evoked in the intracellularly recorded motoneuron (lower trace) confirming the identification of the motoneuron. In the lower traces, the voltage deflections of a WT motoneuron intracellularly recorded at P7 in response to injected pulses at three different current intensities (Em = −65 mV) illustrate the increased discharge firing frequency of action potentials. (**b**) Steady-state F-I curves of two motoneurons distinguished by different colors (WT in blue and SOD1-G85R in red). (**c**) Bar plots of the slopes of F-I relationships of the motoneurons (n = 32) and spinal neurons (n = 20). The slope of the F-I relationship of the SOD1 motoneurons was less than that of the WT motoneurons (P = 0.02; permutation exact test with general scores) whereas no difference was observed between the spinal neurons. (Adapted from Bories et al. 2007)

4 Morphological Properties of Developing Motoneurons in Postnatal Mice

Because the spatial pattern of the dendritic arborizations influences neuronal functions (Korogod and Tyc-Dumont 2009), it is important to determine how dendrites acquire their characteristic size and morphology during development, when neuronal branching patterns expand in space by increasing the number and length of segments. In cats and rodents dendritic maturation is completed at different times, it takes the first 6 postnatal weeks in the kitten (Krishnan 1980; Ulfhake et al. 1988; Ulfhake and Cullheim 1988) and the first 3 weeks after birth in the rat (Altman and Sudarshan 1975; Navarrete and Vrbova 1983, 1993). The complexity of dendritic arborizations depend on many intrinsic and extrinsic factors (Libersat and Duch 2004; Parrish et al. 2007).

Early Changes in Morphological Properties of Lumbar Motoneurons in WT Mice

In contrast to cat and rat motoneurons in which the branching structure (topology) of motoneuron dendrites exhibits some simplification during postnatal development (Cameron et al. 1991; Kalb 1994; Nunez-Abades et al. 1994; Núñez-Abades and Cameron 1995), it seems that motoneuronal complexity in the mouse (number of branches per neuron) does not change during the two first postnatal weeks (Li et al. 2005; Filipchuk and Durand 2012). Well branched dendritic trees of spinal motoneurons were found as early as P3 (average 75 ± 20 branching points per neuron) in postnatal mouse (Filipchuk 2011; Filipchuk and Durand 2012). In embryonic lumbar motoneurons, dendritic arborizations are already well developed in WT mice (Martin et al. 2013) although the total dendritic length and the number of branching points at E17.5 are much lower than that measured at P3-P4 (Martin et al. 2013; Filipchuk and Durand 2012). The number of branching points per motoneuron ranges between 34 and 100 and remains stable on average during the first two postnatal weeks in WT mice (Amendola and Durand 2008; Filipchuk and Durand 2012). The number of dendrites per motoneuron, number of terminals per motoneuron and the other parameters related to tree topology remain relatively constant (Amendola and Durand 2008; Li et al. 2005; Filipchuk and Durand 2012; Filipchuk et al. 2021). Lumbar motoneurons develop in mice by elongation of their terminal segments between P3 and P8 (Filipchuk 2011; Filipchuk and Durand 2012).

The complete morphologies of single WT motoneurons have been described in detail in Amendola and Durand (2008), Filipchuk (2011) and Filipchuk and Durand, (2012).

Early Changes in Morphological Properties of Lumbar Motoneurons in SOD1 Mice

We compared the detailed morphology of lumbar motoneurons from WT and SOD1 mice at the age P3-P4 and correlated it with the geometry of WT and SOD1 motoneurons at P8-P9. By measuring the length of intermediate and terminal dendritic segments, we determined how dendrites grow and we detected the first morphological anomalies in postnatal SOD1 motoneurons (Filipchuk and Durand 2012). Before the formation of supernumerary branches, SOD1 motoneurons exhibit longer terminal segments compared to WT motoneurons at P3-P4 (Fig. 3).

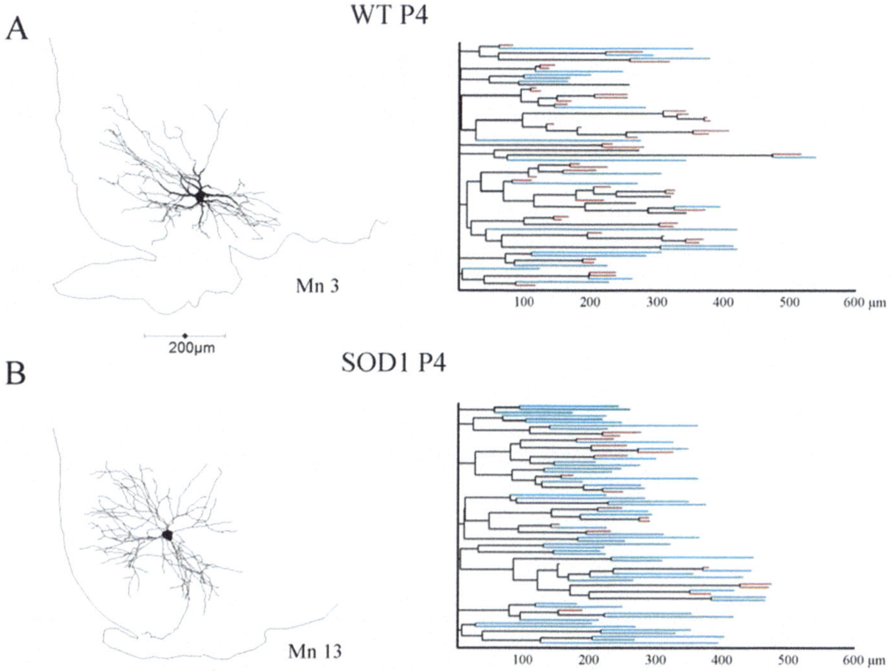

Fig. 3 Elongated distal dendrites in SOD1 motoneurons. Early morphological abnormalities in SOD1 motoneurons at P3-P4 concern the terminal dendrites. Dendrograms illustrate high percentage of short terminals (in red) in the WT motoneurons. Intermediate segments are shown in black; short terminal segments (<35 pm) are highlighted in red and longer ones (>35 pm) in blue. (Adapted from Filipchuk and Durand 2012)

Excessive dendritic overbranching was first measured in $SOD1^{G85R}$ lumbar motoneurons at postnatal days P8-P9 (Allene 2006; Amendola et al. 2007; Amendola and Durand 2008; Filipchuk and Durand 2012) and then confirmed in the SOD1-G93A-low expressor line (Filipchuk et al. 2010, 2021). Abnormal proliferation of dendritic arborizations of SOD1 lumbar motoneurons was reported at the beginning of the second postnatal week in $SOD1^{G85R}$ mice, concomitant with changes in input resistance, rheobase current and membrane capacitance measured in motoneurons recorded using the brainstem/spinal cord preparation (Bories et al. 2007).

The number of branches was twofold higher than that in the WT motoneurons at postnatal days P8 to P9. However, the rostrocaudal extension was found to be similar between WT and SOD1 motoneurons as well as the transverse extension. In the examples of Fig. 4, digitized full reconstructions showing the complete dendritic trees of two WT and two SOD1 motoneurons are illustrated in the transverse view. The motoneurons with the lowest number of branching points (upper two) and with the highest number of branching points (lower two) were compared. In both cases, the SOD1 motoneurons have a higher number of branching points (see Amendola and Durand 2008. Filipchuk et al. 2021).

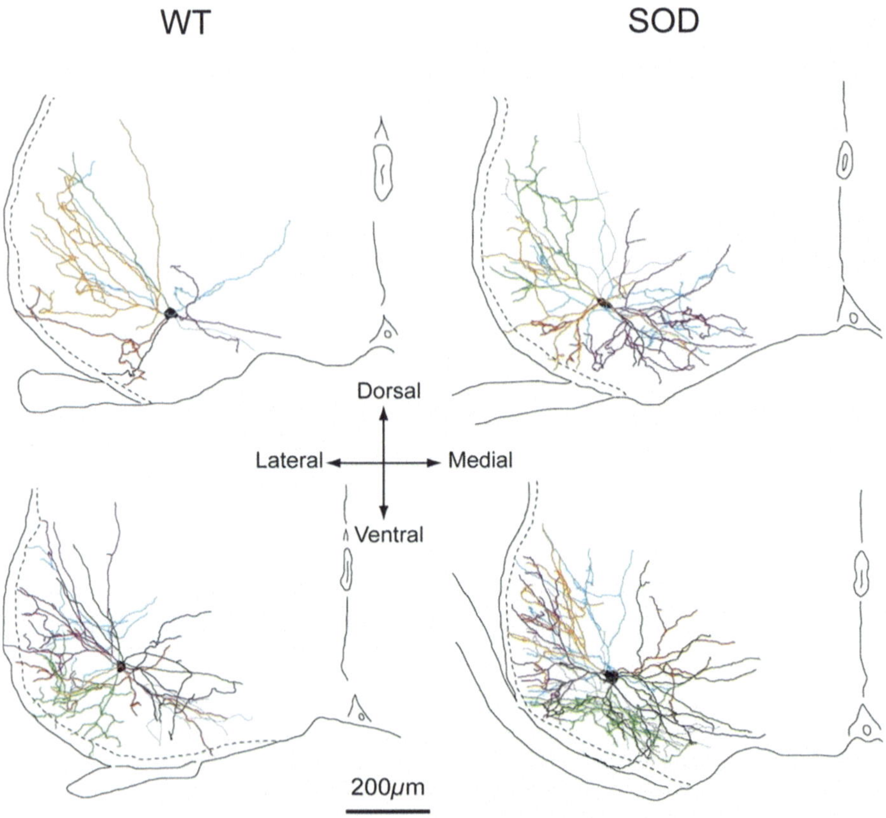

Fig. 4 Examples of full 3D reconstructions showing the dendritic trees of two WT and two SOD1 motoneurons (P8-P9) in transverse view. SOD1 motoneurons have a higher number of branching points in all cases. Dendritic overbranching in SOD1 motoneurons has been found originally in SOD1-G85R at postnatal days P8-P9. (Adapted from Amendola and Durand 2008)

The shape of dendritic arborizations depends on many intrinsic and extrinsic factors during development (Libersat and Duch 2004). The location of cell bodies in the ventral horn of the spinal cord also influences the orientation of dendrites. The dendrites projected mainly in 3 directions (dorsal, dorsolateral and medial) when the soma was in close proximity to the ventral horn boundary (Fig. 5a, c, e). Dendrites extended also ventrally when the soma was located centrally (Fig. 5b, d, f).

In many instances, multiple staining was observed following a single intracellular injection of neurobiotin (see also Amendola and Durand 2008). This could be due to gap junctions precluding the full reconstructions of motoneurons (Amendola et al. 2007; Amendola and Durand 2008; Filipchuk et al. 2021). In the example illustrated in Fig. 6, a second stained soma is visible in the background of the image.

During the course of these experiments, we could observe on rare occasions trifurcations in dendrites of SOD1 motoneurons (Fig. 6d). At higher magnification

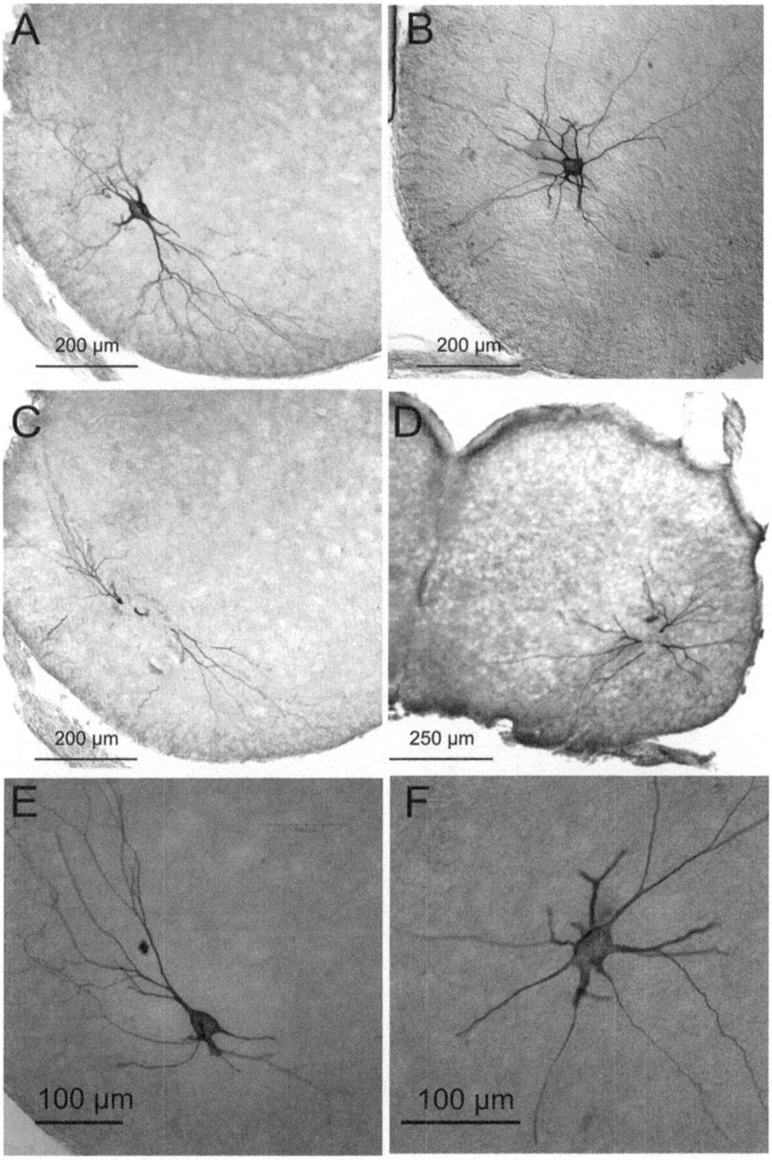

Fig. 5 Examples of intracellularly stained lumbar motoneurons from SOD1 mice at P8-P9: (**a**, **b**) The shape of the dendritic arborizations depends on the location of cell bodies in the ventral horn of the spinal cord. Digital images of lumbar motoneurons in sections (75 pm) containing soma and proximal dendrites. Note the high number of primary dendrites and branching points, the abnormal and opposite directions taken by some proximal dendrites direction in other branches. (**c**, **d**) Digital images of the sections n−1 or n+1 adjacent to the section containing the soma and showing the dendritic extension of SOD1 G93A motoneurons in both cases

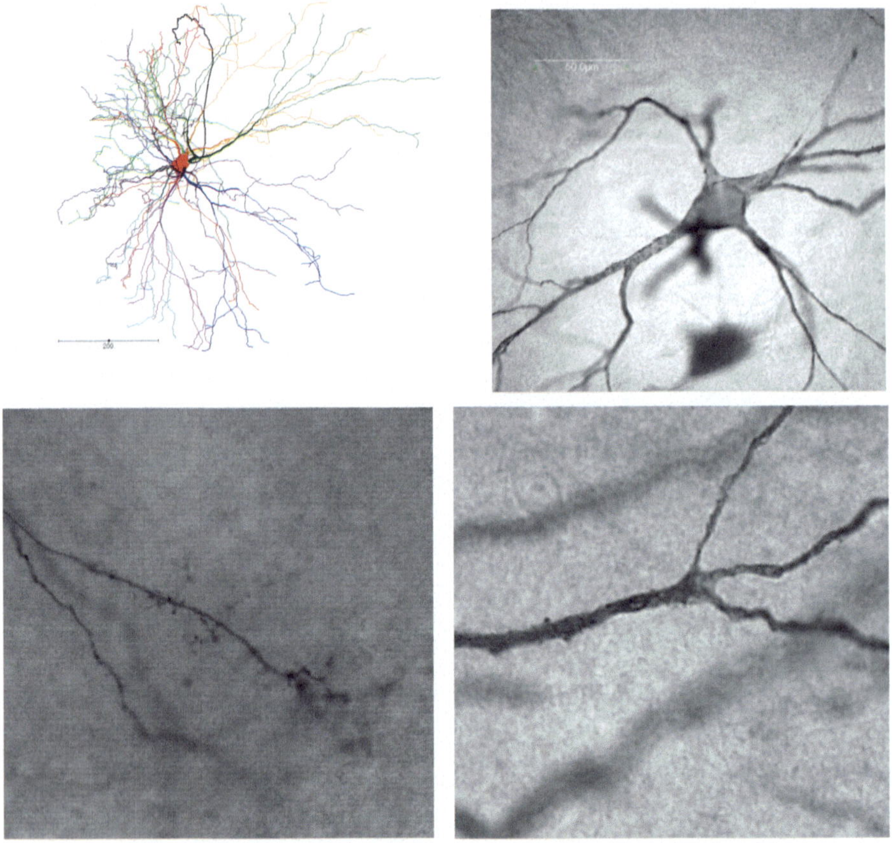

Fig. 6 Digitized full 3D reconstruction of a P8 lumbar SOD1 G93A-low motoneuron (from Filipchuk 2011). Upper right: Example of multiple staining following a single intracellular staining during the postnatal period; a second somatic region is visible in the background. Lower images: Small spines are detected in distal dendrites. Notice their very small size (<0.7 pm). lower right: an example of trifurcation in dendrite of a SOD1 motoneuron. Rare spines are present on thicker dendrite

only a few and very small spines (<0.5 μm) are detected 100 μm from the soma (Fig. 6c) and in distal dendrites, C). This is in contrast with golgi-cox staining where many big spines are observed in motoneuron dendrites at that postnatal age (Kanjhan et al. 2016).

5 Discussion

Several decades ago, Swash and Schwartz (1995) wrote about ALS "*The onset of the disease is difficult to delineate since the clinical manifestations develop gradually; retrospective analysis may suggest that the disease began several years before presentation with muscular weakness or wasting), perhaps with a long preclinical phase during which the disease, although active is not symptomatic*". This long pre-symptomatic period has now been confirmed in the transgenic SOD1 mice, models of ALS. We recently suggested that structural alterations perturbing intracellular trafficking might provoke an early dysfunction such as hypoexcitability of motoneurons (Durand et al. 2021). We will discuss here the possible causes of pathological alterations of dendritic trees in SOD1 motoneurons, the possible link between toxic misfolded SOD1 protein (Saxena et al. 2013) and the specific hypoexcitabilty in the delayed onset firing motoneurons.

Possible Mechanisms Involved in Pathological Alterations of Dendritic Tree

In $SOD1^{G85R}$ mutant mice, we showed that dendrites of SOD1 motoneurons had significantly longer terminal segments than the WT motoneurons at P3-P4 (Filipchuk and Durand 2012). This excessive elongation was then followed by pathological ramification of most individual dendrites between P4 and P8. During this short time, WT motoneurons grew only by elongation of their terminal segments, the number of branches being stable. In both WT and $SOD1^{G85R}$ motoneurons, the average length of intermediate segments remained invariant. An algorithm of $SOD1^{G85R}$ dendritic growth between P4-P8 was built, based on the detailed morphometric analysis, to outline the dynamics of abnormal dendritic development (Filipchuk and Durand 2012). The dendritic elongation could be a sign of acceleration of dendritic growth at P3-P4 which is consistent with reported results demonstrating accelerated maturation of electrical properties in SOD1 motoneurons (Quinlan et al. 2011; Van Zundert et al. 2008). On the other hand, excessively elongated terminals segments might indicate some extrinsic mechanisms influencing dendritic morphology. Immunofluorescent labeling of synaptic terminals revealed significantly lower densities of synaptophysin in $SOD1^{G85R}$ motoneuronal pools at postnatal day P3 compared with matched WT pools (Filipchuk et al. 2010, 2021). This indicated that $SOD1^{G85R}$ motoneurons may have lower synaptic input than WT motoneurons at that early age P3-P4. It's known that dendrites extend into rich synaptogenic regions in normal conditions (Vaughn et al. 1988; Vaughn 1989) and that dendritic density is proportional to the number of synapses which supports the idea that formation of a synaptic contact could stabilize particular filopodia (Vaughn et al. 1988). In the absence of synaptic activity, dendrites elongate until they encounter active presynaptic elements.

In cultured Purkinje cells, inhibition of electrical activity by tetrodotoxin (TTX) decreasing intracellular calcium level stops dendritic branching and provokes dendritic elongation (Schilling et al. 1991). It was hypothesized that for branching to occur, not only are synaptic contacts necessary but they must be functional, both in terms of neurotransmission and the generation of postsynaptic activity. It might be suggested that at P3-P4, dendrites elongate "looking for functional presynaptic partners" and likely between P4 and P8 they manage to establish contacts which trigger branching. A study on adult Drosophila motoneurons demonstrated that decreased excitability leads to increased dendritic branch elongation, whereas increased excitability causes branching (Duch et al. 2008). In chicken motoneurons, it was shown that blocking of Ca+ permeable AMPA (α-amino-3-hydroxy-5-methyl-4-isoxazolepropionic acid) receptors at different embryonic ages has different effect on dendritic morphology (elongation or branching) (Ni and Martin-Caraballo 2010).

Dendrite elongation and branching may involve the same mechanisms based on the microtubule-associated proteins (MAP2) phosphorylation (Hely et al. 2001). Perturbation of microtubule dynamism and impairment of axonal transport occur in ALS (Fanara et al. 2007; Farah et al. 2003; Kieran et al. 2005; Williamson and Cleveland 1999). Alterations in the integrity of microtubules have been observed very early in SOD1 mice with decreased levels of microtubule-associated proteins (Farah et al. 2003) as in human ALS patients (Kikuchi et al. 1999). Dendritic branching depends on microtubule assembly and disassembly under the control of the microtubule-associated proteins MAP2 phosphorylation (Audesirk et al. 1997; Friedrich and Aszodi 1991; Hely et al. 2001). MAP regulates many factors of microtubule dynamics, including polymerization (depolymerization), bundling, spacing, and interaction with actin filaments (Maccioni and Cambiazo 1995). MAP2 was found predominantly in dendrites. The phosphorylation state of MAP2 modulates its interaction with microtubules (Audesirk et al. 1997). Increased phosphorylation of MAP2 protein stimulates neurite branching. Dephosphorylated MAP2 favors elongation by promoting microtubule polymerization. It may play a key role in switching neurite development between elongation and branching. MAP2 phosphorylation depends on the intracellular calcium concentration (Maccioni and Cambiazo 1995).

Synaptic Activity and NMDA

The timing of afferent innervation and synapse formation coincides with the period of maximum growth and dendritic remodeling (Cline and Haas 2008; Curfs et al. 1993). In the spinal cord the first postnatal weeks are an important period for the establishment of the connections of descending pathways (Vinay et al. 2000b). In the activity-dependent regulation of dendrite development, NMDA-type glutamate receptors play a central role (Kalb 1994; Lee et al. 2005). Their activation during development promotes motoneurons dendrite growth. Incorporation of NR2B dramatically increases the number of secondary and tertiary dendritic branches of ventral spinal cord neurons (Sepulveda et al. 2010). The reduction of dendritic branching is related to the reduction of NR2A subunit in the cortex of SOD1 mice (Spalloni et al. 2011). We have already some preliminary indications that NR2A subunit expression is changed in the developing SOD1 mouse lumbar spinal cord (unpublished data).

Maturation of Motoneurons

In the newborn kitten, motoneurons innervating the short plantar muscles of the foot show less mature characteristics than those innervating triceps surae muscles (Ulfhake and Cullheim 1988) suggesting that dendritic maturation may be completed at different times in different motoneuron pools. It may correspond to different maturation of electrical properties of functionally distinct motoneurons (Russier et al. 2003; Cotel et al. 2009; Vinay et al. 2000a). Thus changes in excitability and distribution of firing patterns depend on the motoneuron populations under study during the maturation period. Postnatal maturation of dendrites of extensor motoneurons within cervical in rat spinal cord is delayed compared to flexor motoneurons (Curfs et al. 1993) which can be related to the fact that flexor muscles are innervated at an earlier stage than extensor ones. The development of repetitive firing of ankle extensor motoneurons is also delayed compared to the flexor ones (Vinay et al. 2000a). Sustained repetitive and transient firings were observed for neonatal rat brain stem and spinal motoneurons (Gao and Ziskind-Conhaim 1998; Viana et al. 1995; Vinay et al. 2000a, 2002). Transient features are presumably related to different maturation states of underlying slow Na^+ and Ca^{2+}currents in lumbar motoneurons.

Signs of denervation appear very early in certain muscles of the SOD1 G93A high expressor (Vinsant et al. 2013; Hadzipasic et al. 2014). We hypothesized that the early electrical alterations in motoneurons are pathological signs in SOD1 mice as they might be at the origin of the peripheral disconnection at the neuromuscular junction (Durand et al. 2021). Hypoexcitability was found in the delayed-onset firing motoneurons in postnatal SOD1 mice (Filipchuk et al. 2021). It was also described in adult transgenic SOD1 mice (Martínez-Silva et al. 2018), in iPSC derived from human patients (Sances et al. 2016) and also in ALS patients (Marchand-Pauvert et al. 2019). A direct consequence of early motoneuronal hypoexcitability may be weakening of motor units.

6 Conclusion

We studied in detail the earliest modifications in dendrites, their morphological evolution during a crucial period of maturation and attempted to establish a structure-functional correlation in the context of synaptic input integration in $SOD1^{G85R}$ and $SOD1^{G93A\text{-}low}$ mouse models of ALS. We have shown that excessive elongation of terminal segments represents the earliest signs of abnormal dendritic development in SOD1 motoneurons and is followed by intensive overbranching, starting between P4 and P8. The possible scenario of dendritic growth during this period has been elaborated (Filipchuk and Durand 2012). We have found that dendritic overbranching is a common feature in both $SOD1^{G85R}$ and $SOD1^{G93A}$ model mice at the beginning of the second postnatal week (Filipchuk et al. 2021).

We suggested that the sustained firing motoneurons are more resistant to pathological mechanisms involving the toxic accumulation of mutated proteins (Saxena et al. 2013).

Our data support the fact that pathological changes begin significantly earlier than overt clinical symptoms and suggest that treatment should start much earlier to be effective.

It seems that the denervation of muscles precipitates the clinical symptoms, but denervation starts very much earlier that the obvious motor deficits. Our data support the central origin of the denervation following motoneuronal hypoexcitability in ALS.

The transgenic SOD1 animals compensate for a long period during which clinical symptoms are not detected even when a number of motoneurons have been lost. Such compensatory mechanisms could explain a long asymptomatic period in ALS patients (Leigh and Swash 1996).

References

Allene C (2006) Etude de la morphologie des motoneurones lombaires d'un modèle murin de sclérose latérale amyotrophique. Master Neuroscience, Aix Marseille Université, 28 pp

Altman J, Sudarshan K (1975) Postnatal development of locomotion in the laboratory rat. Anim Behav 23:896–920

Amendola J (2008) Developpement postnatal d'un modèle murin de Sclérose Latérale Amyotrophique. PhD thesis, Aix Marseille Université, Faculté de Médecine, 234 pp

Amendola J, Durand J (2008) Morphological differences between wild-type and transgenic superoxide dismutase 1 lumbar motoneurons in postnatal mice. J Comp Neurol 511:329–341

Amendola J, Verrier B, Roubertoux P, Durand J (2004) Altered sensorimotor development in a transgenic mouse model of amyotrophic lateral sclerosis. Eur J Neurosci 20:2822–2826

Amendola J, Gueritaud JP, Lamotte d'Incamps B et al (2007) Postnatal electrical and morphological abnormalities in lumbar motoneurons from transgenic mouse models of amyotrophic lateral sclerosis. Arch Ital Biol 145:311–323

Audesirk G, Cabell L, Kern M (1997) Modulation of neurite branching by protein phosphorylation in cultured rat hippocampal neurons. Brain Res Dev Brain Res 102:247–260

Balice-Gordon RJ, Thompson WJ (1988) Synaptic rearrangements and alterations in motor unit properties in neonatal rat extensor digitorum longus muscle. J Physiol Lond 398:191–210

Boillee S, Vande Velde C, Cleveland DW (2006) ALS: a disease of motor neurons and their nonneuronal neighbors. Neuron 52:39–59

Bories C (2004) Propriétés électriques des motoneurones lombaires au cours du développement postnatal chez la souris transgénique SOD1-G85R, un modèle de sclérose latérale amyotrophique. Master Neuroscience, Aix Marseille Université, p 30

Bories C, Amendola J, Lamotte d'Incamps B, Durand J (2007) Early electrophysiological abnormalities in lumbar motoneurons in a transgenic mouse model of amyotrophic lateral sclerosis. Eur J Neurosci 25:451–459

Bruijn LI, Becher MW, Lee MK et al (1997) ALS-linked SOD1 mutant G85R mediates damage to astrocytes and promotes rapidly progressive disease with SOD1-containing inclusions. Neuron 18:327–338

Bruijn LI, Houseweart MK, Kato S et al (1998) Aggregation and motor neuron toxicity of an ALS-linked SOD1 mutant independent from wild-type SOD1. Science 281:1851–1854

Bruijn LI, Miller TM, Cleveland DW (2004) Unraveling the mechanisms involved in motor neuron degeneration in ALS. Annu Rev Neurosci 27:723–749

Burke RE (1981) Motor units: anatomy, physiology and functional organization. In: Brookhart JM, Mountcastle VB (eds) Handbook of physiology. section I, The nervous system, vol. II, motor control, part 1. American Physiological Society, Bethesda, pp 345–422

Burke RE, Levine DN, Zajac FE 3rd. (1971) Mammalian motor units: physiological-histochemical correlation in three types in cat gastrocnemius. Science 174:709–712. https://doi.org/10.1126/science.174.4010.709

Cameron WE, He F, Kalipatnapu P, Jodkowski JS, Guthrie RD (1991) Morphometric analysis of phrenic motoneurons in the cat during postnatal development. J Comp Neurol 314:763–776

Cline H, Haas K (2008) The regulation of dendritic arbor development and plasticity by glutamatergic synaptic input: a review of the synaptotrophic hypothesis. J Physiol 586:1509–1517

Cotel F, Antri M, Barthe JY, Orsal D (2009) Identified ankle extensor and flexor motoneurons display different firing profiles in the neonatal rat. J Neurosci 29:2748–2753

Curfs MH, Gribnau AA, Dederen PJ (1993) Postnatal maturation of the dendritic fields of motoneuron pools supplying flexor and extensor muscles of the distal forelimb in the rat. Development 117(2):535–541

Dal Canto MC, Gurney ME (1997) A low expressor line of transgenic mice carrying a mutant human Cu,Zn superoxide dismutase (SOD1) gene develops pathological changes that most closely resemble those in human amyotrophic lateral sclerosis. Acta Neuropathol 93:537–550

Duch C, Vonhoff F, Ryglewski S (2008) Dendrite elongation and dendritic branching are affected separately by different forms of intrinsic motoneuron excitability. J Neurophysiol 100:2525–2536

Durand J, Amendola J, Bories C, Lamotte d'Incamps B (2006) Early abnormalities in transgenic mouse models of amyotrophic lateral sclerosis. J Physiol Paris 99:211–220

Durand J, Filipchuk A, Pambo-Pambo A et al (2015) Developing electrical properties of postnatal mouse lumbar motoneurons. Front Cell Neurosci 9:349

Durand J, Filipchuk A, Pambo-Pambo A et al (2021) Hypoexcitability of motoneurons: an early pathological sign in ALS. Neuroscience 465:233–234

Elbasiouny SM, Amendola J, Durand J, Heckman CJ (2010) Evidence from computer simulations for alterations in the membrane biophysical properties and dendritic processing of synaptic inputs in mutant superoxide dismutase-1 motoneurons. J Neurosci 30:5544–5558

Fanara P, Banerjee J, Hueck RV et al (2007) Stabilization of hyperdynamic microtubules is neuroprotective in amyotrophic lateral sclerosis. J Biol Chem 282:23465–23472

Farah CA, Nguyen MD, Julien JP, Leclerc N (2003) Altered levels and distribution of microtubule-associated proteins before disease onset in a mouse model of amyotrophic lateral sclerosis. J Neurochem 84:77–86

Filipchuk A (2011) Morphometrical and electrical features of motoneurons affected by amyotrophic lateral sclerosis. PhD thesis Aix Marseille Université, Faculté de Médecine, p 207

Filipchuk AA, Durand J (2012) Postnatal dendritic development in lumbar motoneurons in mutant superoxide dismutase 1 mouse model of amyotrophic lateral sclerosis. Neuroscience 209:144–154

Filipchuk AA, Durand J, Korogod SM (2008) Charge transfer effectiveness as electrotonic indicator of the structural differences between samples of dendritic morphology. Neurophysiology 40:497–501

Filipchuk AA, Pambo-Pambo A, Liabeuf S et al (2010) Evidence of early somato-dendritic alterations in lumbar motoneurons of SOD1 juvenile mice. International Symposium on ALS/MND, Orlando, USA. Amyotroph Lateral Scler 11:39–40

Filipchuk AA, Pambo-Pambo A, Gaudel F et al (2021) Early hypoexcitability in a subgroup of spinal motoneurons in superoxide dismutase 1 transgenic mice, model of amyotrophic lateral sclerosis. Neuroscience 463:337–353. https://doi.org/10.1016/j.neuroscience.2021.01.39

Fox WM (1965) Reflex ontogeny and behavioral development of the mouse. Anim Behav 13:234–241

Friedrich P, Aszodi A (1991) MAP2: a sensitive cross-linker and adjustable spacer in dendritic architecture. FEBS Lett 295(1-3):5–9

Gao BX, Ziskind-Conhaim L (1998) Development of ionic currents underlying changes in action potential waveforms in rat spinal motoneurons. J Neurophysiol 80:3047–3061

Gurney ME, Pu H, Chiu AY, Dal Canto MC et al (1994) Motor neuron degeneration in mice that express a human Cu,Zn superoxide dismutase mutation. Science 264(5166):1772–1775

Hadzipasic M, Tahvildari B, Nagy M, Bianc M, Horwich AL, McCormick DA (2014) Selective degeneration of a physiological subtype of spinal motor neuron in mice with SOD1-linked ALS. PNAS 111:16883–16888

Hely TA, Graham B, Ooyen AV (2001) A computational model of dendrite elongation and branching based on MAP2 phosphorylation. J Theor Biol 210(3):375–384

Henneman E (1957) Relation between size of neurons and their susceptibility to discharge. Science 126:1345–1347

Henneman E, Somjen G, Carpenter DO (1965) Functional significance of cell size in spinal motoneurons. J Neurophysiol 28:560–580

Irobi J, Almeida-Souza L, Asselbergh B et al (2010) Mutant HSPB8 causes motor neuron- specific neurite degeneration. Hum Mol Genet 19:3254–3265

Kalb RG (1994) Regulation of motor neuron dendrite growth by NMDA receptor activation. Development 120:3063–3071

Kanjhan R, Noakes PG, Bellingham MC (2016) Emerging roles of Filopodia and Dendritic spines in motoneuron plasticity during development and disease. Neural Plasticity, Article ID 3423267, 31 pages. https://doi.org/10.1155/2016/3423267

Kanning KC, Kaplan A, Henderson CE (2010) Motor neuron diversity in development and disease. Annu Rev Neurosci 33:409–440

Kieran D, Hafezparast M, Bohnert S, Dick JR, Martin J, Schiavo G, Fisher EM, Greensmith L (2005) A mutation in dynein rescues axonal transport defects and extends the life span of ALS mice. J Cell Biol 169:561–567

Kikuchi H, Doh-ura K, Kawashima T et al (1999) Immunohistochemical analysis of spinal cord lesions in amyotrophic lateral sclerosis using microtubule-associated protein 2 (MAP2) antibodies. Acta Neuropathol 97:13–21

Korogod S, Tyc-Dumont S (2009) Electrical dynamics of the dendritic space. Cambridge University Press, Cambridge, p 211

Krishnan RV (1980) Cortico-spinal tract interaction in locomotor development and motor units formation: an hypothesis. Int J Neurosci 10:89–93

Kuo JJ, Schonewille M, Siddique T et al (2004) Hyperexcitability of cultured spinal motoneurons from presymptomatic ALS mice. J Neurophysiol 91:571–575

Kuo JJ, Siddique T, Fu R, Heckman CJ (2005) Increased persistent Na(+) current and its effect on excitability in motoneurones cultured from mutant SOD1 mice. J Physiol 563:843–854

Lee LJ, Lo FS, Erzurumlu RS (2005) NMDA receptor-dependent regulation of axonal and dendritic branching. J Neurosci 25:2304–2311

Leigh PN, Swash M (1996) Motor neuron disease. Springer, London, p 468

Leroy F, Lamotte d'Incamps B, Imhoff-Manuel RD et al (2014) Early intrinsic hyperexcitability does not contribute to motoneuron degeneration in amyotrophic lateral sclerosis. elife 3:E04046

Li Y, Brewer D, Burke RE, Ascoli GA (2005) Developmental changes in spinal motoneuron dendrites in neonatal mice. J Comp Neurol 483(3):304–317

Libersat F, Duch C (2004) Mechanisms of dendritic maturation. Mol Neurobiol 29(3):303–320

Maccioni RB, Cambiazo V (1995) Role of microtubule-associated proteins in the control of microtubule assembly. Physiol Rev 75(4):835–864

Marchand-Pauvert V, Peyre I, Lackmy-Vallee A et al (2019) Absence of hyperexcitability of spinal motoneurons in patients with amyotrophic lateral sclerosis. J Physiol 597:5445–5467

Martin E, Cazenave W, Cattaert D, Branchereau P (2013) Embryonic alteration of motoneuronal morphology induces hyperexcitability in the mouse model of amyotrophic lateral sclerosis. Neurobiol Dis 54:116–126

Martin E, Cazenave W, Allain AE, Cattaert D, Branchereau P (2020) Implication of 5-HT in the dysregulation of chloride homeostasis in prenatal spinal motoneurons from the G93A Mouse model of Amyotrophic lateral sclerosis. Int J Mol Sci 21(3):1107. https://doi.org/10.3390/ijms21031107

Martínez-Silva ML, Imhoff-Manuel RD, Sharma A et al (2018) Hypoexcitability precedes denervation in the large fast-contracting motor units in two unrelated mouse models of ALS. elife 7:e30955

Matthews PBC (1982) Where does Sherrington's "muscular sense" originate – muscles, joints, corollary discharges. Annu Rev Neurosci 5:189–218

Mead RJ, Bennett EJ, Kennerley AJ et al (2011) Optimised and rapid pre-clinical screening in the SOD1(G93A) transgenic mouse model of amyotrophic lateral sclerosis (ALS). PLoS One 6:e23244. https://doi.org/10.1371/journal.pone.0023244

Mentis GZ, Diaz E, Moran LB, Navarrete R (2007) Early alterations in the electrophysiological properties of rat spinal motoneurones following neonatal axotomy. J Physiol 582(Pt 3):1141–1161

Mentis GZ, Blivis D, Liu W, Drobac E, Crowder ME, Kong L, Alvarez FJ, Sumner CJ, O'Donovan MJ (2011) Early functional impairment of sensory-motor connectivity in a mouse model of spinal muscular atrophy. Neuron 69:453–467

Milan L, Barrière G, De Deurwaerdère P et al (2014) Monoaminergic control of spinal locomotor networks in SOD1G93A newborn mice. Front Neural Circuits 8:77

Mynlieff M, Beam KG (1992) Characterization of voltage-dependent calcium currents in mouse motoneurons. J Neurophysiol 68:85–92

Nakanishi ST, Whelan PJ (2010) Diversification of intrinsic motoneuron electrical properties during normal development and botulinum toxin-induced muscle paralysis in early postnatal mice. J Neurophysiol 103:2833–2845

Navarrete R, Vrbova G (1983) Changes of activity patterns in slow and fast muscles during postnatal-development. Dev Brain Res 8(1):11–19

Navarrete R, Vrbova G (1993) Activity-dependent interactions between motoneurones and muscles: their role in the development of the motor unit. Prog Neurobiol 41:93–124

Ni X, Martin-Caraballo M (2010) Differential effect of glutamate receptor blockade on dendritic outgrowth in chicken lumbar motoneurons. Neuropharmacology 58:593–604

Núñez-Abades PA, Cameron WE (1995) Morphology of developing rat genioglossal motoneurons studied in vitro: relative changes in diameter and surface area of somata and dendrites. J Comp Neurol 353:129–142

Nunez-Abades PA, He F, Barrionuevo G, Cameron WE (1994) Morphology of developing rat genioglossal motoneurons studied in vitro: changes in length, branching pattern, and spatial distribution of dendrites. J Comp Neurol 339:401–420

Pambo-Pambo AB (2010) Etude du développement postnatal des motoneurones lombaires de souches de souris transgéniques, modèles de la sclérose latérale amyotrophique. PhD thesis Aix Marseille Université, Faculté de Médecine, p 258

Pambo-Pambo A, Durand J, Gueritaud JP (2009) Early excitability changes in lumbar motoneurons of transgenic SOD1G85R and SOD1G(93A-Low) mice. J Neurophysiol 102:3627–3642

Parrish JZ, Emoto K, Kim MD, Jan YN (2007) Mechanisms that regulate establishment, maintenance, and remodeling of dendritic fields. Annu Rev Neurosci 30:399–423

Pieri M, Albo F, Gaetti C, Spalloni A, Bengtson CP, Longone P, Cavalcanti S, Zona C (2003) Altered excitability of motor neurons in a transgenic mouse model of familial amyotrophic lateral sclerosis. Neurosci Lett 351:153–156

Pun S, Santos AF, Saxena S, Xu L, Caroni P (2006) Selective vulnerability and pruning of phasic motoneuron axons in motoneuron disease alleviated by CNTF. Nat Neurosci 9:408–419

Quinlan KA, Schuster JE, Fu R, Siddique T, Heckman CJ (2011) Altered postnatal maturation of electrical properties in spinal motoneurons in an ALS mouse model. J Physiol 589:2245–2260

Raoul C, Estevez AG, Nishimune H et al (2002) Motoneuron death triggered by a specific pathway downstream of Fas. Potentiation by ALS-linked SOD1 mutations. Neuron 35:1067–1083

Robberecht W, Philips T (2013) The changing scene of amyotrophic lateral sclerosis. Nat Rev Neurosci 14:248–264

Rosen DR, Siddique T, Patterson D et al (1993) Mutations in Cu/Zn superoxide dismutase gene are associated with familial amyotrophic lateral sclerosis. Nature 362:59–62

Rossi-Durand C (2006) Proprioception and myoclonus. Neurophysiol Clin 36:299–308

Russier M, Carlier E, Ankri N et al (2003) A-, T-, and H-type currents shape intrinsic firing of developing rat abducens motoneurons. J Physiol 549:21–36

Sances S, Bruijn L, Chandran S, Eggan K, Ho R, Klim J, Livesey MR, Lowry E, Macklis JD, Rushton D, Sadegh C, Sareen D, Wichterle H, Zhang SC, Svendsen CN (2016) Modeling ALS using motor neurons derived from human induced pluripotent stem cells. Nat Neurosci 19:542–553. https://doi.org/10.1038/nn.4273

Saxena S, Roselli F, Singh K et al (2013) Neuroprotection through excitability and mTOR required in ALS motoneurons to delay disease and extend survival. Neuron 80:80–96

Schilling K, Dickinson MH, Connor JA, Morgan JI (1991) Electrical activity in cerebellar cultures determines Purkinje cell dendritic growth patterns. Neuron 7(6):891–902

Sepulveda FJ, Bustos FJ, Inostroza E et al (2010) Differential roles of NMDA receptor subtypes NR2A and NR2B in dendritic branch development and requirement of RasGRF1. J Neurophysiol 103:1758–1770

Spalloni A, Origlia N, Sgobio C et al (2011) Postsynaptic alteration of NR2A subunit and defective autophosphorylation of alphaCaMKII at Threonine-286 contribute to abnormal plasticity and morphology of upper motor neurons in presymptomatic SOD1G93A mice, a murine model for amyotrophic lateral sclerosis. Cereb Cortex 21(4):796–805

Swash M, Schwartz MS (1995) Motoneuron disease: the clinical syndrome in motor neuron disease. Leigh and Swash M (eds) Springer, London, pp 1–17

Ulfhake B, Cullheim S (1988) Postnatal development of cat hind limb motoneurons. III: changes in size of motoneurons supplying the triceps surae muscle. J Comp Neurol 278:103–120

Ulfhake B, Cullheim S, Franson P (1988) Postnatal development of cat hind limb motoneurons. I: changes in length, branching structure, and spatial distribution of dendrites of cat triceps surae motoneurons. J Comp Neurol 278:69–87

Van Zundert B, Peuscher MH, Hynynen M et al (2008) Neonatal neuronal circuitry shows hyperexcitable disturbance in a Mouse model of the adult-onset neurodegenerative disease amyotrophic lateral sclerosis. J Neurosci 28:10864–10874

Van Zundert B, Izaurieta P, Fritz E, Alvarez FJ (2012) Early pathogenesis in the adult-onset neurodegenerative disease amyotrophic lateral sclerosis. J Cell Biochem 113:3301–3312

Vaughn JE (1989) Fine structure of synaptogenesis in the vertebrate central nervous system. Synapse 3(3):255–285

Vaughn JE, Barber RP, Sims TJ (1988) Dendritic development and preferential growth into synaptogenic fields: a quantitative study of Golgi-impregnated spinal motor neurons. Synapse 2(1):69–78

Viana F, Bayliss DA, Berger AJ (1995) Repetitive firing properties of developing rat brainstem motoneurones. J Physiol 486:745–761

Vinay L, Brocard F, Clarac F (2000a) Differential maturation of motoneurons innervating ankle flexor and extensor muscles in the neonatal rat. Eur J Neurosci 12(12):4562–4566

Vinay L, Brocard F, Pflieger JF, Simeoni-Alias J, Clarac F (2000b) Perinatal development of lumbar motoneurons and their inputs in the rat. Brain Res Bull 53:635–647

Vinay L, Brocard F, Clarac F, Norreel JC, Pearlstein E, Pflieger JF (2002) Development of posture and locomotion: an interplay of endogenously generated activities and neurotrophic actions by descending pathways. Brain Res Rev 40:118–129

Vinsant S, Mansfield C, Jimenez-Moreno R et al (2013) Characterization of early pathogenesis in the SOD1(G93A) mouse model of ALS: part II, results and discussion. Brain Behav 3:431–457

Vrbova G (2007) Understanding motoneurone development explains spinal muscular atrophy. Arch Ital Biol 145:325–335

Williamson TL, Cleveland D (1999) Slowing of axonal transport is a very early event in the toxicity of ALS-linked SOD1 mutants to motor neurons. Nat Neurosci 2:50–56

Wooley CM, Sher RB, Kale A, Frankel WN, Cox GA, Seburn KL (2005) Gait analysis detects early changes in transgenic SOD1(G93A) mice. Muscle Nerve 32:43–50

Zhang F, Strom AL, Fukada K, Lee S, Hayward LJ, Zhu H (2007) Interaction between familial amyotrophic lateral sclerosis (ALS)-linked SOD1 mutants and the dynein complex. J Biol Chem 282:16691–16699

From Physiological Properties to Selective Vulnerability of Motor Units in Amyotrophic Lateral Sclerosis

Marcin Bączyk, Marin Manuel, Francesco Roselli, and Daniel Zytnicki

Abstract Spinal alpha-motoneurons are classified in several types depending on the contractile properties of the innervated muscle fibers. This diversity is further displayed in different levels of vulnerability of distinct motor units to neurodegenerative diseases such as Amyotrophic Lateral Sclerosis (ALS). We summarize recent data suggesting that, contrary to the excitotoxicity hypothesis, the most vulnerable motor units are hypoexcitable and experience a reduction in their firing prior to symptoms onset in ALS. We suggest that a dysregulation of activity-dependent transcriptional programs in these motoneurons alter crucial cellular functions such as mitochondrial biogenesis, autophagy, axonal sprouting capability and re-innervation of neuromuscular junctions.

Keywords Motoneuron · Motor unit · Physiological type · Electrophysiology · ALS · Neurodegeneration · Activity-dependent transcription

---

Authors Marcin Bączyk, Marin Manuel, Francesco Roselli, and Daniel Zytnicki have equally contributed to this chapter.

M. Bączyk
Department of Neurobiology, Poznań University of Physical Education, Poznań, Poland

M. Manuel (✉) · D. Zytnicki
SPPIN – Saints-Pères Paris Institute for the Neurosciences, CNRS, Université de Paris, Paris, France
e-mail: marin.manuel@neurobio.org

F. Roselli
Department of Neurology, Ulm University, Ulm, Germany

Institute of Anatomy and Cell Biology, Ulm University, Ulm, Germany

German Center for Neurodegenerative Diseases (DZNE)-Ulm, Ulm, Germany

Neurozentrum Ulm, Ulm, Germany

M. J. O'Donovan, M. Falgairolle (eds.), *Vertebrate Motoneurons*, Advances in Neurobiology 28, https://doi.org/10.1007/978-3-031-07167-6_15

1 Introduction

Neurodegenerative diseases generally affect preferentially defined populations of neurons ("selective vulnerability", Roselli and Caroni 2015). For example, in Amyotrophic Lateral Sclerosis (ALS), the most prominent neurodegenerative disease of motoneurons, some motor pools are more vulnerable than others (Kaplan et al. 2014), but even in a given motor pool, the order of motoneuron degeneration depends on their physiological type (Fig. 1) (the reader may refer to the chapter "Diversity of mammalian motoneurons and motor units" for details on the properties of motor units and their classification): motoneurons innervating the fast-contracting and fatigable (FF) motor units degenerate first, followed by those innervating the fast-contracting fatigue resistant (FR) motor units, whereas the motoneurons innervating the slow-contracting (S) motor units are the most resistant (Pun et al. 2006; Hegedus et al. 2008). Moreover, gamma-motoneurons, which innervate intrafusal muscle fibers, are spared in this disease (Lalancette-Hebert et al. 2016). Despite more than 20 years of intensive research, the pathophysiological mechanisms that lead to motoneuron degeneration in ALS are still largely unknown. Among many others, the glutamate excitotoxic hypothesis has been proposed (Van Den Bosch et al. 2006; Ilieva et al. 2009) but recent studies have challenged this hypothesis (Saxena et al. 2013; Delestrée et al. 2014; Kim et al. 2017; Martínez-Silva et al. 2018). In this chapter, we will review recent data that challenge the excitotoxic hypothesis, and which shed new light on the possible role of intrinsic excitability as an essential determinant of motoneuron differential vulnerability in ALS. Finally, we will propose an alternative hypothesis based on a dysregulation of activity-dependent transcriptional programs to account for the selective vulnerability.

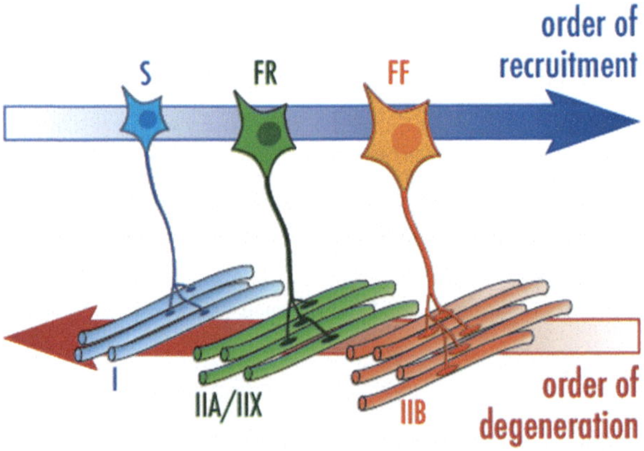

Fig. 1 Schema illustrating the order of degeneration of the main types of motor units during the course of ALS. NMJs on type IIB fibers (belonging to FF motor units) degenerate first, followed by NMJs on type IIA fibers (FR motor units), and the NMJs on type I fibers (S motor units) degenerate last. Note that the degeneration order is opposite to the recruitment order of motor units (S > FR > FF)

2 Selective Vulnerability of Motor Units to ALS

Studies in animal models of ALS and human patients have shown that motor units degenerate at different time points according to their type (Frey et al. 2000; Pun et al. 2006; Hegedus et al. 2008). Motor unit degeneration starts by the denervation of its constituent muscle fibers (i.e. the degeneration of neuromuscular junctions), a process that happens quite fast in the SOD1(G93A) mouse model (Schaefer et al. 2005; Pun et al. 2006). However, the remaining axonal branches are able to sprout and reinnervate vacant motor endplates for a substantial amount of time (Martineau et al. 2018), a process that seems to be also dependent on the type of motor unit (Pun et al. 2006). Following the degeneration of the neuromuscular junctions, the motor axon and then the cell body degenerate (Pun et al. 2006; Kanning et al. 2009). Anatomical observations in post-mortem ALS patients have shown that small diameter axons in the ventral horn (putatively gamma and S-type motor axons) are spared, while large diameter axons have degenerated (Kawamura et al. 1981; Sobue et al. 1981). Examination of cell bodies in the ventral horn of post-mortem ALS patient spinal cords similarly revealed a selective degeneration of the largest cell bodies (McIlwain 1991). These observations are corroborated by electrophysiological explorations in human patients, showing a decline in motor unit force specifically in the highest threshold (F type) motor units (Dengler et al. 1990), and the emergence of "giant" motor unit potentials, reflecting the sprouting of motor axons and the reinnervation of muscle fibers (Pinelli et al. 1991; Duleep and Shefner 2013).

Pun et al. (2006) have used the sparse labeling of axons in Thy1-GFP mice to map the distribution of motor units in the lateral gastrocnemius muscle and other muscles. They identified subcompartments in the hindlimb muscles innervated by separate motor unit populations, and further demonstrated that some of those subcompartments are constituted of a single muscle fiber type. For instance, the most lateral (l1) and the most medial (m3) compartments of the mouse LG are devoid of Type IIA and Type I fibers (Frey et al. 2000; Pun et al. 2006), and therefore constituted solely of FF motor units. Based on these results, the authors went on to show that FF motor units are the first to denervate, starting around P50 in the SOD1(G93A) mouse model (Fig. 2). Motoneurons innervating the FR motor units degenerate later (the denervation of their muscle fibers starts at P80-P90, a time at which the first motor symptoms start to be noticeable) whereas motoneurons innervating the S type motor units are the most resistant in ALS and usually do not show any sign of degeneration before end-stage (Pun et al. 2006). This sequence of degeneration has also been independently confirmed using force recordings and motor unit number estimation (Hegedus et al. 2007, 2008). In the SOD1(G37R) mouse model of ALS, at both presymptomatic and symptomatic disease stage, the NMJs of S motor units showed more nerve sprouting, poly-innervation, and perisynaptic Schwann cell process extension than those of FF or FR motor units indicating their high capability for muscle reinnervation (Tremblay et al. 2017) which may play an important role in their resistance to degeneration.

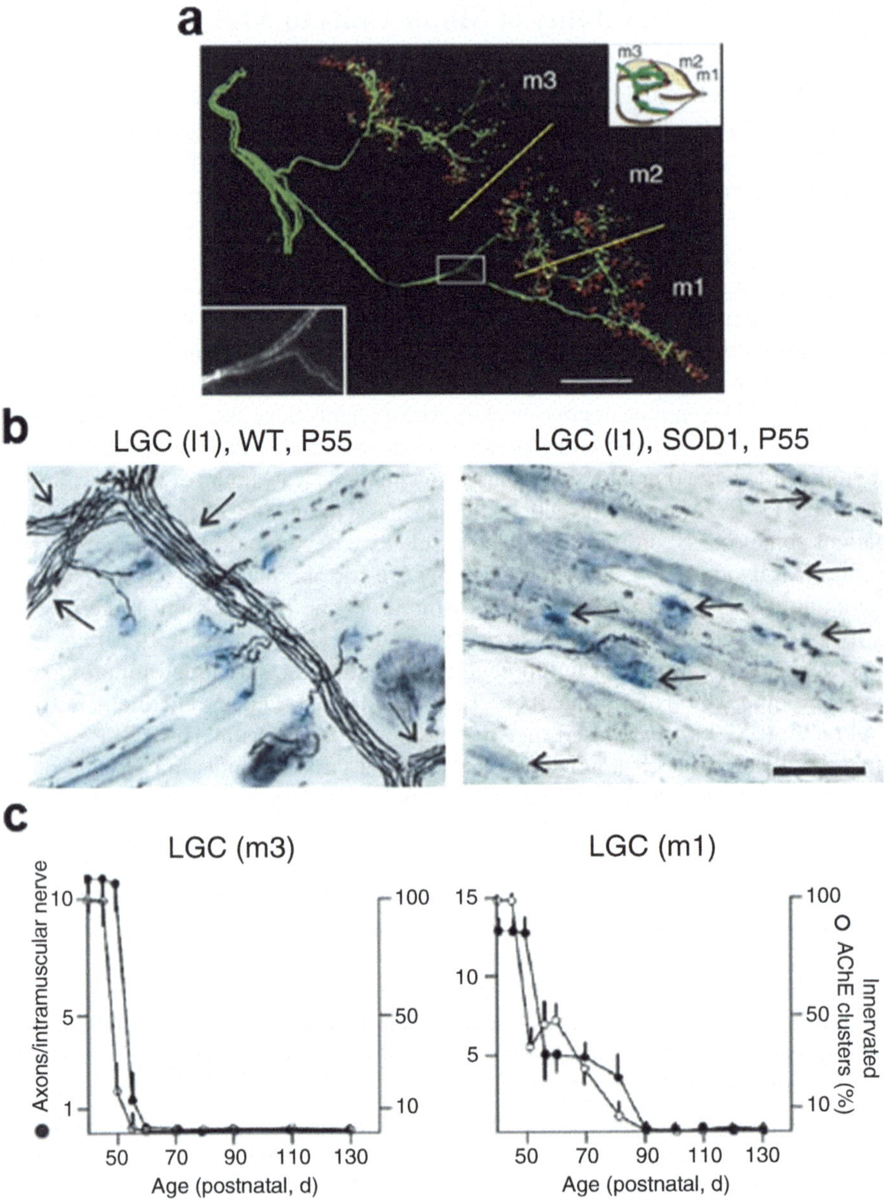

Fig. 2 NMJ denervation happens at different time points depending on motor unit type in the SOD1(G93A) mouse model of ALS. (**a**) Arborization pattern of axons innervating the medial compartment of LGC. Whole-mount preparation from Thy1-GFP transgenic mouse expressing mGFP (green) in few motoneurons and counterstained with RITC-α-bungarotoxin to visualize NMJs. Axon bundles branched out from LGC and SOL nerve (left) to innervate subcompartments m3, m2 and m1 (boundary regions indicated by yellow bars). Individual axons projected to only one subcompartment. (**b**) Intramuscular nerves (black axon bundles, arrows at left) and postsynaptic sites (acetylcholine esterase, blue) in LGC subcompartment l1 of wild-type and SOD1(G93A)

3 Excitability Changes in Motoneurons Is a Determinant of Their Vulnerability in ALS

The firing of any neuron depends both on the balance of excitatory and inhibitory inputs it receives, and on its intrinsic electrical properties that set its excitability. The long-lasting excitotoxicity hypothesis relies on the assumption that excessive excitatory glutamatergic inputs may lead to an overload of cytosolic calcium, which, in turn, would trigger apoptosis. In addition, calcium overload could also be triggered by intrinsic hyperexcitability of the motoneurons, as increased firing would over-activate voltage-dependent calcium channels (Van Den Bosch et al. 2006; Grosskreutz et al. 2010). Indeed, alterations of intrinsic excitability or alterations of excitatory inputs have been reported in several neurodegenerative diseases (reviewed in Roselli and Caroni 2015).

In mouse models of ALS, alterations in motoneuron intrinsic excitability depend on the time point (embryos, neonates, presymptomatic adults, symptomatic adults, endpoint), the motor pool (some motor pools are selectively spared in ALS), and also on the motor unit subtype (which is linked to the motoneuron vulnerability). Alterations in motoneuron electrical properties are detected before the motoneurons begin to degenerate. Indeed, already at the embryonic stage, input resistance is increased, rheobase is decreased and slope of the frequency-current relationship (F-I gain) is increased in spinal motoneurons (recorded either in cultures or in whole spinal cords) indicating that motoneurons are hyperexcitable at this stage (Pieri et al. 2003; Kuo et al. 2004; Martin et al. 2013). Shortly after birth (P4–P10), alterations of intrinsic properties have also been reported but results are somehow contradictory. Hypoglossal motoneurons were reported to be hyperexcitable (F-I gain is increased, (van Zundert et al. 2008)). However, Pambo-Pambo et al. (2009) did not observe any change in input resistance, rheobase, or stationary gain of spinal motoneurons suggesting that their excitability was unchanged. In the same vein, Quinlan et al. (2011) found that the excitability of spinal motoneurons is homeostatically maintained despite an increase in their input conductance (recruitment current and F-I gain unchanged). In contrast, Bories et al. (2007) reported a decrease in input resistance causing spinal motoneurons to be hypoexcitable. There are many confounding factors that could account for these discrepancies. First, not all studies use the same animal model. Bories et al. (2007) used SOD1(G85R) mice, Pambo-Pambo et al. (2009) used the SOD1(G85R) and the SOD1(G93A)-low expressor line, van Zundert et al. (2008) and Quinlan et al. (2011) used the SOD1(G93A)-high

Fig. 2 (continued) mice at P55. Note complete denervation (arrows at right) and absence of thick motor axons in the mutant mouse. Combined silver-esterase labeling. (**c**) NMJs in compartments containing exclusively FF motor units (such as LGC subcomp. m3) are denervated around P50 and remain so for the duration of the life of the animal. NMJs in compartments containing a mix of FF and FR motor units (such as LGC subcomp. m1) experience two waves of denervation, a first happening around P50 (presumably corresponding to the denervation of the FF motor units), and a second around P80 (presumably corresponding to the denervation of the FR motor units). (Adapted from Pun et al. (2006) with permission from Springer Nature) expressor line. Since the time course of the disease is different in these models, the same age range does not necessarily correspond to the same point in the evolution of the disease process. The criteria for defining hyper- or hypo-excitability is also not always consistent between studies. In addition, in these studies, motoneurons were pooled together and were not separated between the different physiological subtypes. Therefore, type-specific changes in motoneuron electrical properties, if any, could have been overlooked. Conclusions at the population level may be dependent on the proportion of each subtypes in the motoneuron samples investigated in mutant and control mice. It is then critically important to try separating the motoneurons according to their physiological subtype, i.e. to their vulnerability.

For this reason, Leroy et al. (2014) used the difference in firing observed in S and F-type lumbar motoneurons at P6–P10 (immediate vs. delayed-firing, see section "Motoneuron electrophysiological properties depend on their motor unit type", Fig. 3) to test whether the different types of motor units exhibit differential changes of their excitability in the SOD1(G93A) mice.

Most interestingly, it was found that the rheobase specifically decreases in immediate firing motoneurons of mutant SOD1 mice compared with wild-type (WT) mice (their voltage threshold for firing is lower than in immediate-firing controls) (Fig. 3d). This indicates that this motoneuron subpopulation (S-type) is hyperexcitable in neonatal SOD1(G93A) mice. In sharp contrast, the rheobase remained unchanged in the delayed firing motoneurons (Fig. 3c) indicating that the excitability of F-type motoneurons is unchanged. It then appears that the S-type motoneurons, which are the less vulnerable in ALS, display an hyperexcitability in the SOD1(G93A) mice at P6–P10 whereas the F-type motoneurons (i.e. more vulnerable than the S-type) have normal excitability (Leroy et al. 2014). These results represent a significant shift with the current hypothesis that the more hyperexcitable a motoneuron, the more vulnerable it is (Van Den Bosch et al. 2006; Grosskreutz et al. 2010). However, it is interesting to note that Venugopal et al. (2015) have found, at the same postnatal age and in the same mice, that among the trigeminal motoneurons (classified on their electrophysiological properties: rheobase, input resistance and membrane capacitance) only the "predicted fast types" motoneurons display a hyperexcitable shift marked by reduced rheobase and increased input resistance whereas the "predicted slow type" do not present significant alterations. These results are different from those obtained in lumbar motoneurons by Leroy et al. (2014) suggesting that excitability changes of motoneurons may not be parallel in the brainstem and in the spinal cord.

ALS is an adult-onset disease, such that there is ample time between the neonatal stage and the onset of symptoms for the properties of motoneurons to change and adapt. In adult presymptomatic ALS mice, in vivo intracellular recordings of motoneurons in SOD1(G93A) mice revealed that a fraction of motoneurons lose their ability to fire repetitively in response to a slow ramp of current (Delestrée et al. 2014). Indeed, this loss of repetitive firing is a manifestation of hypoexcitability. Moreover, the refinement of in vivo electrophysiological methods in adult mice allowed to intracellularly record or stimulate a single motoneuron while the force output of the motor unit was simultaneously recorded at tendon (Manuel and

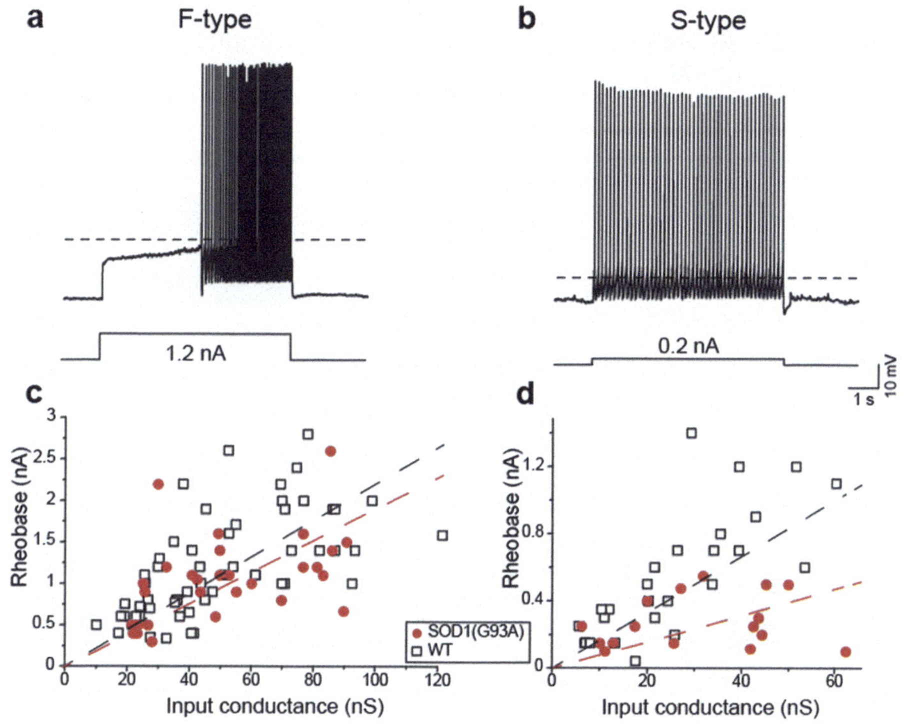

Fig. 3 Differential excitability changes of F- and S-type motoneurons in SOD1(G93A) mice at P6-P10. (**a**) The discharge of F-type motoneurons is delayed in response to a long depolarizing pulse at an intensity just above rheobase. Note that the motoneuron is slowly depolarizing during the first half of the pulse before reaching the spiking threshold. (**b**) In contrast, the S-type motoneurons discharge immediately at the onset of a long depolarizing pulse just above the rheobase. (**c**) Rheobase of F-type motoneurons is unchanged in SOD1(G93A) mice (red filled circles) with respect to WT mice (open squares). Rheobase is plotted against input conductance. (**d**) Rheobase of S-type motoneurons is significantly reduced in S-type motoneurons of SOD1(G93A) mice indicating that they are hyperexcitable. Same arrangement as in (**c**). (Adapted from Leroy et al. (2014) used under CC-BY)

Heckman 2011). This methodological development made possible in vivo type-identification of motoneurons in FF, FR and S subtypes on the basis of the contractile properties of their motor units (Martínez-Silva et al. 2018). This work showed that at P45–P50, most FF motor units and many FR motor units among the largest ones display a loss of repetitive firing (hypoexcitability) upon a slow ramp of depolarizing current, while the motoneurons of the smaller FR motor units and the S-type motor units display normal excitability (they discharge repetitively, their rheobase and the gain of the discharge frequency-injected current function are unaltered) (Fig. 4). This loss of repetitive firing occurs despite the fact that the motoneuron can still elicit a single spike in response to a current transient. It also occurs before the degeneration of the neuromuscular junction since the single spike is still able to elicit an electromyographic response and a twitch of the motor unit. Thus the loss of

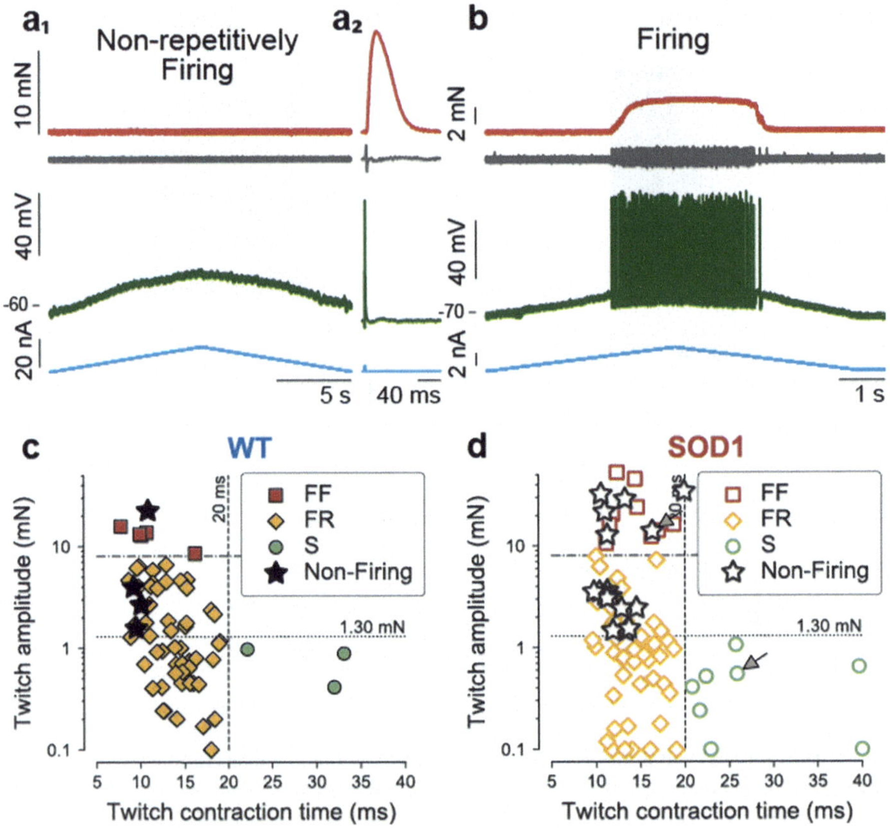

Fig. 4 Loss of repetitive firing in the most vulnerable motor units in presymptomatic SOD1(G93A) mice. (**a**) Example of a motoneuron which lost the property to fire repetitively (a1, green trace) in response to a slow triangular ramp of current (a1, blue trace) in a SOD1(G93A) mouse. However this motoneuron is still able to fire a single spike (a2, green trace) in response to a transient depolarizing pulse (a2, blue trace). Note that when the motoneuron is firing a motor unit action potential is recorded (a2, the grey trace shows the electromyographic response), and the motor unit displays a twitch force (a2, red trace) indicating that the neuromuscular junctions have not degenerated yet. (**b**) Example of a motoneuron that elicits a repetitive discharge in response to a slow triangular ramp of current (blue trace) in a SOD1(G93A) mouse (same arrangement as in **a**). (**c**–**d**) The different motor unit subtypes are displayed in function of their twitch amplitude and of their twitch contraction time. Non-firing motoneurons in response to the triangular ramp are highlighted with a star (filled stars in **c** and open stars in **d**). There are more non-firing motoneurons in the SOD1(G93A) mice (**d**) than in WT mice (21% vs. 6%). Remarkably the non-firing motoneurons exclusively clustered in the FF motor units and in the largest FR motoneurons (those that develop a twitch force larger than 1.3 mN) indicating that the non-firing property is observed only among the most vulnerable motoneurons. Arrows point to the two examples in panels **a** and **b**. (Adapted from Martínez-Silva et al. (2018) used under CC-BY)

repetitive firing precedes the degeneration onset. Disease markers (p-eIF2α and p62 aggregates) confirm that non-repetitively-firing motoneurons are at a more advanced disease stage (i.e., they are the most vulnerable) than those that can still discharge normally (Martínez-Silva et al. 2018). In addition, the same selective loss of repetitive firing also occurs in an unrelated ALS model, the FUS(P525L) mice (Sharma et al. 2016), indicating that this feature is common to both phenotypes. Motoneuron type-specific hypoexcitability was also suggested in spinal motoneurons in adult slices (Hadzipasic et al. 2014). Since functional identification of the motoneuron types based on the motor units contractile properties was not possible in these conditions, motoneurons were classified on their firing properties and retrograde labeling was used to tentatively correlate those clusters with the known composition in motor unit types in specific muscles. Recordings in SOD1(G85R) mice at 2–4 months show an hyperpolarization of the resting membrane potential, with no change in input resistance, making them potentially hypoexcitable in several of the putative vulnerable clusters. The same group has performed in vivo extracellular recordings of putative motoneurons during treadmill walking in symptomatic animals, and observed an overall decrease in motoneuron firing rates which is compatible with the hypoexcitable state (Hadzipasic et al. 2016). It should be noted, however, that the Meehan group has not reported such hypoexcitability when recording from SOD1(G93A) mice (Jensen et al. 2020), or from a mouse model with a different mutation of the SOD1 gene SOD1(G127X) (Meehan et al. 2010; Bonnevie et al. 2020). The origin of this discrepancy is still contentious, but it may be due to the fact that they performed their recordings in older mice (in which degeneration of the most vulnerable motoneurons is already under way) than in the previous studies (Delestrée et al. 2014; Martínez-Silva et al. 2018), as well as suboptimal discontinuous current-clamp switching rate, which may distort the firing of those cells (Manuel 2020).

Most of the data on alterations of spinal motoneuron excitability in ALS mice reported so far are compatible with the model illustrated in Fig. 5. At a very early stage (embryos), spinal motoneurons are hyperexcitable but the excitability declines with time at different speeds according to the motoneuron vulnerability. In neonates, the most resistant motoneurons are still hyperexcitable whereas the vulnerable motoneurons display a normal excitability. Later on the excitability of the most resistant motoneurons declines to become normal in presymptomatic adults at the time the most vulnerable motoneurons already display hypoexcitability, which precedes their degeneration. We do not know so far how the excitability of the most resistant motoneurons evolves at a time-point close to the disease end-stage. Interestingly, recordings in the cortex of cortico-spinal neurons, which are also known to degenerate in ALS, show that these cells are hyperexcitable early on (Pieri et al. 2009; Kim et al. 2017), returning to normal excitability in young adults, and then becoming hyperexcitable again at end stage (Kim et al. 2017). Furthermore, the timeline of excitability changes illustrated in Fig. 5 is corroborated by studies in induced pluripotent stem cell (iPSC)-derived motoneurons from human patients with different mutations. Recordings in these cells have shown an initial

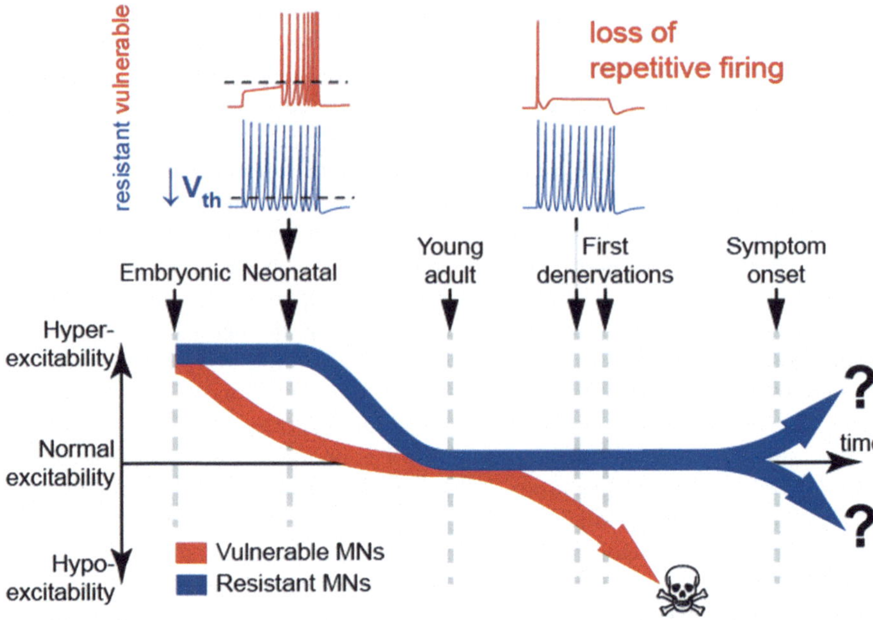

Fig. 5 Schematic timeline of the changes in excitability of vulnerable and resistant motoneurons during ALS. Top traces are cartoons showing the representative firing pattern of vulnerable and resistant motoneurons at neonatal stages, and just prior to the onset of NMJ denervation

hyperexcitability (Wainger et al. 2014; Devlin et al. 2015), which then turns into hypoexcitability as the cells mature (Sareen et al. 2014; Devlin et al. 2015; Naujock et al. 2016). Studies in IPSCs-derived motoneurons point to alterations in sodium or potassium currents which account for their excitability changes (Wainger et al. 2014; Devlin et al. 2015; Naujock et al. 2016). Most interestingly, a recent work demonstrated that the loss of repetitive discharge and the alterations in sodium and potassium currents in IPSC-derived motoneurons occurs only in presence of mutant astrocytes indicating that these processes are non-cell autonomous (Zhao et al. 2020).

Can the changes of intrinsic excitability contribute to the motoneuron vulnerability in ALS? In young IPSCs-derived motoneurons, a Kv7 potassium channel opener, Retigabine, was shown to reduce hyperexcitability to control levels and to increase the survival of the cells (Wainger et al. 2014). In older IPSCs-derived motoneurons, a potassium channel blocker, 4-Aminopyridine, was shown to reduce the hypoexcitability phenotype and also the stress of the reticulum endoplasmic as well as apoptosis, suggesting a neuroprotective action (Naujock et al. 2016). In adult presymptomatic double transgenic SOD1(G93A)/ChAT-cre mice, Saxena et al. (2013) performed in vivo chemogenetic manipulation of the excitability of lumbar motoneurons using an AAV9 viral vector that carried the floxed pharmacologically selective actuator module (PSAM) either coupled to 5HT3-receptor (PSAM-Act) for neuronal depolarization or to glycine-receptor (PSAM-Inh) for neuronal hyperpolarization. The expressed channels were then activated with the specific orthogonal ligand PSEM. The virus was delivered by intraspinal injection at the lumbar L3–L5 levels, thus avoiding autonomic pre-ganglionic cholinergic neurons, which are located more rostrally (Barber et al. 1984; Cabot 1996). During the injection, the tip of the capillary was located deep in the ventral horn, minimizing the infection of the cholinergic interneurons (so-called partition cells); nevertheless, control experiments were performed in which the injection only affected partition cells (injection site close to the ependymal canal). Enhancing motoneuron excitability with the PSAM-Act specifically reduced the amount of misfolded SOD1 protein, the endoplasmic reticulum stress in the FF-type motoneurons (i.e., the most vulnerable ones) and delayed the denervation of the neuromuscular junctions in the corresponding motor units. Conversely, reducing motoneuron excitability with the PSAM-Inh has opposite effects (Saxena et al. 2013). On the other hand, activating or inhibiting the partition cells did not result in any significant effect on disease markers in motoneuron. Of note, a beneficial effect similar to those obtained by direct motoneuron activation was also achieved by inactivation of Parvalbumin interneurons (Saxena et al. 2013), strengthening the concept that enhancement of motoneuron excitation is critical to obtain beneficial effects on disease markers. Altogether, these experiments indicate that there is a clear link between the excitability changes of motoneurons and their vulnerability. The next section presents some activity-dependent intracellular pathways that may be involved in such a link.

4 Molecular Pathways Linking Excitation/Excitability to Vulnerability

The alterations in motoneuronal excitability and firing properties are not isolated events but have far-reaching implications in terms of the cell biology of motoneurons. Indeed, different properties in the recruitment (i.e., how often motoneurons are used during physical activity), firing patterns (i.e., whether motoneurons display tonic or phasic firing patterns, and the characteristics of the burst in the phasic firing) and firing rates (i.e., the average frequency of action potentials as well as the frequency within each burst) are translated into distinct biological features that vary across motoneuron subtypes, in terms of transcriptional profiles, structural and functional plasticity and, in turn, to vulnerability to disease. In pathological conditions, changes in firing are expected to ricochet through signaling cascades down to widespread alterations in cell biological processes and transcriptional landscapes.

In this respect, transcription factors act as converging points of the transduction cascades (Fig. 6). Well-known activity-modulated transcription factors (such as CREB and SRF) as well as so-called Immediate-Early genes (so named because their expression is triggered by external stimuli within a few minutes of stimulation) are expressed in motoneurons, indicating that motoneurons display a robust activity-regulated transcription (Bączyk et al. 2020; Alami et al. 2020). To date, the precise

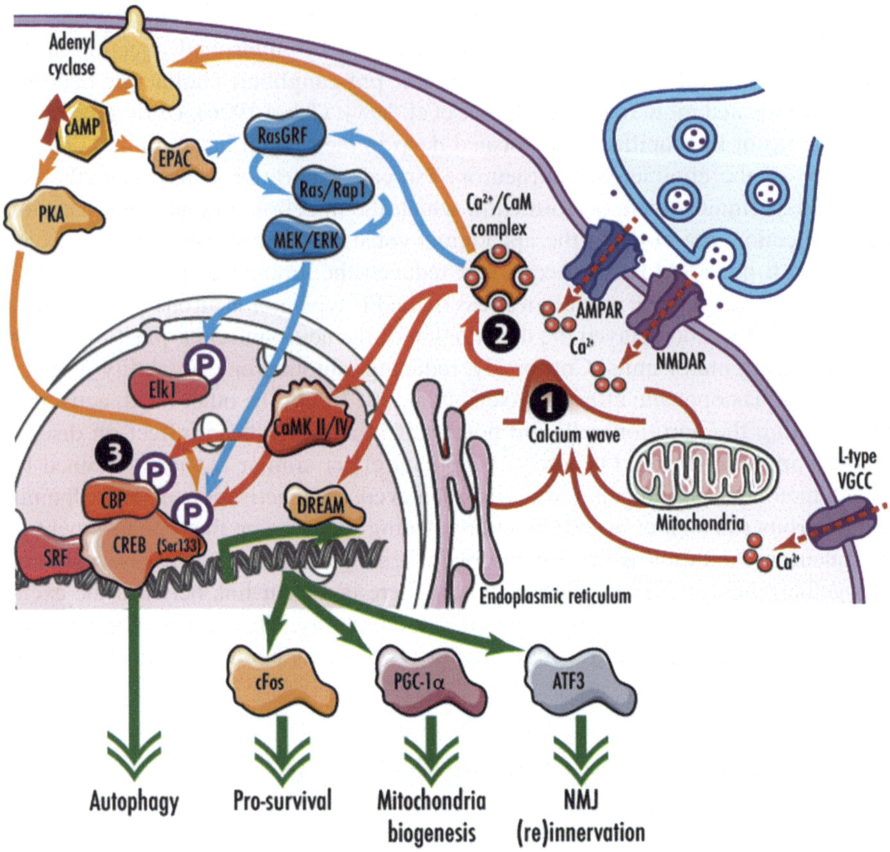

Fig. 6 Some of the pathways mediating activity-dependent transcriptional regulation in motoneurons. ① Neuronal activity is translated into a transient peak of cytoplasmic calcium levels due to the influx of calcium through synaptic receptors, Voltage-Dependent Calcium Channel and release of calcium from cellular stores. The elevated calcium levels trigger the formation of active calcium/Calmodulin complexes ②, which in turn activate adenylyl cyclases (and the downstream PKA pathway), MEK/ERK, and CaMKIV (and some isoforms of CaMKII). These kinases converge to phosphorylate CREB on serine 133, as well as CREB co-activator (CBP) and other transcription factors (Elk1, SRF) ③. Some transcription factors are activated directly by the elevation of the nuclear calcium level (such as DREAM). Downstream of CREB, multiple secondary transcription factors are activated (such as cFos, PGC-1α, ATF3), which bring about cellular responses in terms of promotion of cell survival, mitochondrial biogenesis, NMJ innervation, and autophagy (directly CREB dependent), thereby promoting cell survival. Loss of pCREB levels observed in ALS vulnerable motoneurons points toward the dysfunction of some or all of the upstream activation mechanisms and imply the dysfunction of multiple downstream cell biological pathways

contribution of these transcription factors to the control of the biology of motoneurons remains to be investigated. Nevertheless, the basic biology of these transcription factors is apparently conserved in motoneurons as much as in other neuronal subtypes. CREB is a transcription factor which constitutes the convergence-point for multiple signaling cascades: the phosphorylation of Serine 133, which enables CREB transcriptional activation, is brought about by PKA, MAPK (in particular by RSK2) (not shown), and CaMKIV (Mayr and Montminy 2001). Upon S133 phosphorylation, CREB binds the CBP/p300 transcriptional co-activator, which, in turn, upon phosphorylation by CaMKIV is responsible for recruiting the Type-II RNA polymerase and for starting the transcription of CREB-regulated genes (Saura and Cardinaux 2017). Notably, recruitment of CBP is aided by a second activity-regulated transcription factor named SRF (Serum Response Factor); SRF is activated upon phosphorylation by ERK and GSK-3beta and it is necessary for the transcription of a number of immediate-early genes (Lösing et al. 2017). It must be stressed that CREB and SRF are the best known but not necessarily the only transcription factors involved in translating neuronal activity into transcriptional programs. Downstream of the CREB and SRF responses lie a number of other transcriptional regulators, such as ATF3, NPAS4, Egr1-3, GADD45b and the Fos family. Thus, when considering the genes directly regulated by CREB and SRF (Zhang et al. 2005) and those indirectly regulated through second-order transcription factors, a large fraction of the genome expressed in neurons is under the regulation of neuronal activity. CREB and CREB-related proteins are necessary for the survival of several subpopulations of neurons (Mantamadiotis et al. 2002), but to date we do not have a clear understanding of their role in motoneuron survival.

Notably, the pattern of CREB phosphorylation in motoneurons reflects intrinsic differences between subpopulations, being the highest in FF- and the lowest in S-motoneurons (Saxena et al. 2013). Although the biochemical basis for such an effect are unknown, one may speculate that burst firing is more effective in inducing CREB phosphorylation than tonic firing. Notably, FF-motoneuron display high levels of pCREB even if their recruitment is anticipated to be infrequent, suggesting that once triggered, CREB may remain in a phosphorylated state for longer periods of time; nevertheless, a certain degree of motoneuron recruitment would be necessary to sustain CREB phosphorylation over time and thus guarantee neuroprotective transcriptional programs. Interestingly, epidemiological evidence (Pupillo et al. 2014; Fang et al. 2016; Gallo et al. 2016) suggests, not without controversy (Huisman et al. 2013; Visser et al. 2018) that moderate physical activity may decrease the risk of ALS (although intense physical activity may actually increase it). Furthermore, the differential phosphorylation level of CREB suggests that FF-motoneurons may be more sensitive than S-motoneurons to reduced excitability, since this would affect pCREB levels and therefore affect more severely FF-motoneurons than S-motoneurons. In fact, CREB phosphorylation is distinctively reduced in FF-motoneurons in the SOD1(G93A) ALS mouse model (Saxena et al. 2013). Which cellular processes controlled by CREB (and/or SRF) may be involved in bringing about motoneuron vulnerability differences, i.e. may sensitize motoneurons when pCREB levels are reduced? Among the CREB-induced transcription factors, ATF3 has been shown to prolong survival when overexpressed in the SOD1(G93A) mouse by enhancing the reinnervation sprouting (which takes places upon the loss of the FF-motoneurons) and maintaining muscle innervation; thus, loss of ATF3 may contribute to increased sensitivity. Moreover, CREB is a critical regulator of the expression of the transcription factor PGC-1α, which, in turn is the key activator of mitochondrial biogenesis (Li et al. 2017); in fact, PGC-1α levels are reduced in the spinal cord of ALS mouse models (Eschbach et al. 2013; Bayer et al. 2017) and mitochondrial abnormalities are considered a critical pathogenic step in ALS development (Carrì et al. 2017). Third, CREB controls directly the quantity of several factors involved in autophagy (at least in non-neuronal cells (Seok et al. 2014)) and it is possible that loss of CREB-dependent, activity-regulated transcription may generate or contribute to the autophagy overload seen in several ALS models and in patients (Rudnick et al. 2017; Catanese et al. 2019). Thus, loss of activity-dependent CREB phosphorylation would result in detrimental consequences in terms of mitochondrial biogenesis, autophagy and NMJ re-innervation. Such complex fall-out is in agreement with the large number of cell-biological abnormalities described in ALS mouse models.

Although CREB provides a prototypical example of activity-dependent transcription factor, it must be stressed it is not the only one (MEF family members, Elk-1, NPAS4 and MeCP2 are among the best understood regulators to date) and that a number of epigenetic mechanisms based on DNA methylation and chromatin remodeling may also be involved (Greer and Greenberg 2008; Kim et al. 2010; Ebert et al. 2013; Malik et al. 2014; Yap and Greenberg 2018).

CREB phosphorylation, and more broadly, the transcriptional responses linked to neuronal activity, are regulated by multiple signal transduction cascades (Fig. 6). Although the motoneuron-specific details of these signaling events are unknown, a few highly conserved signaling cascades are known to be active in motoneurons and may be involved not only in the regulation of activity-dependent transcription in healthy conditions but also in ALS-associated alterations linking disrupted activity to altered transcriptional responses.

In particular, the cAMP/PKA pathway is one of the most important regulators of the phosphorylation of CREB on Serine 133 either directly (with PKA interacting with CREB) or through the cAMP-dependent activation and stabilization of the active conformation of the small GTP binding protein Rap1 (Borland et al. 2009), which in turn activates the MEK/ERK cascade (which also phosphorylates CREB on S133; Takahashi et al. 2017). Moreover, the Ca^{2+}/Calmodulin-sensitive kinase CaMKIV has been shown to phosphorylate both CREB itself (Bito et al. 1996; Soderling 1999) and its co-activator CBP (Impey et al. 2002). CaMKIV is directly activated by CaM binding as well as by phosphorylation by the upstream CaMKK (Wayman et al. 2008) and it is highly enriched to the nucleus upon phosphorylation. Despite the apparent redundancy with PKA and ERK, CaMKIV has been identified as a critical mediator of large-scale adaptations of neurons to firing, in particular of homeostatic scaling, the global tuning of synaptic inputs aimed at stabilizing neuronal inputs and neuronal firing rates (Ibata et al. 2008; Joseph and Turrigiano 2017). More recently, the gamma isoform of CaMKII has been shown to be able to translocate from synapses to the nucleus, further contributing to activity-related CREB phosphorylation (Ma et al. 2014).

All the above-cited signaling cascades are linked to neuronal firing by their sensitivity to Ca^{2+} transients (i.e., they are sensitive to pulses of intracellular Ca^{2+}

concentration). The sudden and transient pulse in cytoplasmic Ca^{2+} concentration upon neuronal firing originates from multiple sources: (i) at the synaptic level, Ca^{2+}-permeable glutamate receptors such as NMDAR and GluR1/4-containing AMPA receptors; (ii) upon depolarization, voltage-dependent Ca^{2+} channels (VDCC) are open; (iii) Ca^{2+} is released by endoplasmic reticulum stores when DAG/InsP3 signaling is activated either by Ca^{2+} influx itself either by Gq-coupled GPCR as well as through Ca^{2+} -dependent Ca^{2+} release from mitochondria and the endoplasmic reticulum.

These Ca^{2+} ions are then bound to the Calcium-binding protein Calmodulin (Clapham 2007) which is then responsible for the activation of the neuronal isoforms of the cAMP-generating adenylyl cyclases (triggering PKA activation; Wang and Storm 2003) as well as of the Ca^{2+}/Calmodulin-dependent Ras-guanine-nucleotide-releasing factor, e.g. RasGRF1 and 2 (Krapivinsky et al. 2003), which leaves to MEK/ERK activation and, finally, to the activation of CaMKIV and CaMKII (Bito et al. 1996; Soderling 1999). It is currently object of investigation if the disturbance in pCREB levels in motoneurons observed in the SOD1(G93A) model of ALS results from the disturbances in the initiation of the signaling cascades (e.g., at the synaptic level or in the Ca^{2+} dynamics) or in downstream events (kinase activation).

5 Conclusion

Vulnerability of motoneurons in neurodegenerative diseases such as Amyotrophic Lateral Sclerosis depend on the physiological type of their motor unit. Here we suggest that reduced motoneuron firing and dysregulation of activity-dependent transcriptional programs induce this type-specific motoneuron vulnerability by altering crucial cellular functions such as mitochondrial biogenesis, autophagy, axonal sprouting capability and re-innervation of neuromuscular junctions. Further research is needed to validate this new conceptual framework in ALS, which might pave the way for innovative therapeutic approaches.

Acknowledgments MB is supported by National Science Centre 2017/26/D/NZ7/00728. MM and DZ are supported by NIH-NINDS R01NS110953, the Thierry Latran Foundation project "TRiALS", Association pour la Recherche sur la SLA et autres maladies du motoneurone (ARSLA), the Association Française contre les Myopathies (AFM) project "HYPERTOXIC", Radala Foundation for ALS Research, and Programme Hubert Curien "Polonium" for scientific exchanges. MM would like to thank Alexandra Elbakyan for her help with bibliography. FR is supported by the Thierry Latran Foundation (projects "Trials" and "Hypothals"), by the Radala Foundation, by the Deutsche Forschungsgemeinschaft (DFG) as part of the SFB1149 and with the individual grant no. 431995586 (RO-5004/8-1) and no. 443642953 (RO5004/9-1), by the Cellular and Molecular Mechanisms in Aging (CEMMA) Research Training Group and by BMBF (FKZ 01EW1705A, as member of the ERANET-NEURON consortium "MICRONET"). Clipart in Fig. 6 are adapted from Servier Medical Art by Servier, licensed under a Creative Commons Attribution 3.0 Unported License.

The authors declare no conflict of interest.

References

Alami NO, Tang L, Wiesner D et al (2020) Multiplexed chemogenetics in astrocytes and motoneurons restore blood–spinal cord barrier in ALS. Life Sci Alliance 3. https://doi.org/10.26508/lsa.201900571

Bączyk M, Alami NO, Delestrée N et al (2020) Synaptic restoration by cAMP/PKA drives activity-dependent neuroprotection to motoneurons in ALS. J Exp Med 217. https://doi.org/10.1084/jem.20191734

Barber RP, Phelps PE, Houser CR et al (1984) The morphology and distribution of neurons containing choline acetyltransferase in the adult rat spinal cord: an immunocytochemical study. J Comp Neurol 229:329–346. https://doi.org/10.1002/cne.902290305

Bayer H, Lang K, Buck E et al (2017) ALS-causing mutations differentially affect PGC-1α expression and function in the brain vs. peripheral tissues. Neurobiol Dis 97:36–45. https://doi.org/10.1016/j.nbd.2016.11.001

Bito H, Deisseroth K, Tsien RW (1996) CREB phosphorylation and dephosphorylation: a Ca(2+)- and stimulus duration-dependent switch for hippocampal gene expression. Cell 87:1203–1214. https://doi.org/10.1016/s0092-8674(00)81816-4

Bonnevie VS, Dimintiyanova KP, Hedegaard A et al (2020) Shorter axon initial segments do not cause repetitive firing impairments in the adult presymptomatic G127X SOD-1 Amyotrophic Lateral Sclerosis mouse. Sci Rep 10:1–16. https://doi.org/10.1038/s41598-019-57314-w

Bories C, Amendola J, Lamotte d'incamps B, Durand J (2007) Early electrophysiological abnormalities in lumbar motoneurons in a transgenic mouse model of amyotrophic lateral sclerosis. Eur J Neurosci 25:451–459. https://doi.org/10.1111/j.1460-9568.2007.05306.x

Borland G, Smith BO, Yarwood SJ (2009) EPAC proteins transduce diverse cellular actions of cAMP. Br J Pharmacol 158:70–86. https://doi.org/10.1111/j.1476-5381.2008.00087.x

Cabot JB (1996) Some principles of the spinal organization of the sympathetic preganglionic outflow. In: Progress in brain research. Elsevier, pp 29–42

Carrì MT, D'Ambrosi N, Cozzolino M (2017) Pathways to mitochondrial dysfunction in ALS pathogenesis. Biochem Biophys Res Commun 483:1187–1193. https://doi.org/10.1016/j.bbrc.2016.07.055

Catanese A, Olde Heuvel F, Mulaw M et al (2019) Retinoic acid worsens ATG10-dependent autophagy impairment in TBK1-mutant hiPSC-derived motoneurons through SQSTM1/p62 accumulation. Autophagy 15:1719–1737. https://doi.org/10.1080/15548627.2019.1589257

Clapham DE (2007) Calcium signaling. Cell 131:1047–1058. https://doi.org/10.1016/j.cell.2007.11.028

Delestrée N, Manuel M, Iglesias C et al (2014) Adult spinal motoneurones are not hyperexcitable in a mouse model of inherited amyotrophic lateral sclerosis. J Physiol 592:1687–1703. https://doi.org/10.1113/JPHYSIOL.2013.265843

Dengler R, Konstanzer A, Küther G et al (1990) Amyotrophic lateral sclerosis: macro–EMG and twitch forces of single motor units. Muscle Nerve 13:545–550. https://doi.org/10.1002/mus.880130612

Devlin A-C, Burr K, Borooah S et al (2015) Human iPSC-derived motoneurons harbouring TARDBP or C9ORF72 ALS mutations are dysfunctional despite maintaining viability. Nat Commun 6:1–12. https://doi.org/10.1038/ncomms6999

Duleep A, Shefner J (2013) Electrodiagnosis of motor neuron disease. Phys Med Rehabil Clin N Am 24:139–151. https://doi.org/10.1016/j.pmr.2012.08.022

Ebert DH, Gabel HW, Robinson ND et al (2013) Activity-dependent phosphorylation of MeCP2 threonine 308 regulates interaction with NCoR. Nature 499:341–345. https://doi.org/10.1038/nature12348

Eschbach J, Schwalenstöcker B, Soyal SM et al (2013) PGC-1α is a male-specific disease modifier of human and experimental amyotrophic lateral sclerosis. Hum Mol Genet 22:3477–3484. https://doi.org/10.1093/hmg/ddt202

Fang F, Hållmarker U, James S et al (2016) Amyotrophic lateral sclerosis among cross-country skiers in Sweden. Eur J Epidemiol 31:247–253. https://doi.org/10.1007/s10654-015-0077-7

Frey D, Schneider C, Xu L et al (2000) Early and selective loss of neuromuscular synapse subtypes with low sprouting competence in motoneuron diseases. J Neurosci 20:2534–2542. https://doi.org/10.1523/JNEUROSCI.20-07-02534.2000

Gallo V, Vanacore N, Bueno-de-Mesquita HB et al (2016) Physical activity and risk of Amyotrophic Lateral Sclerosis in a prospective cohort study. Eur J Epidemiol 31:255–266. https://doi.org/10.1007/s10654-016-0119-9

Greer PL, Greenberg ME (2008) From synapse to nucleus: calcium-dependent gene transcription in the control of synapse development and function. Neuron 59:846–860. https://doi.org/10.1016/j.neuron.2008.09.002

Grosskreutz J, van den Bosch L, Keller BU (2010) Calcium dysregulation in amyotrophic lateral sclerosis. Cell Calcium 47:165–174. https://doi.org/10.1016/j.ceca.2009.12.002

Hadzipasic M, Tahvildari B, Nagy M et al (2014) Selective degeneration of a physiological subtype of spinal motor neuron in mice with SOD1-linked ALS. Proc Natl Acad Sci U S A 111. https://doi.org/10.1073/pnas.1419497111

Hadzipasic M, Ni W, Nagy M et al (2016) Reduced high-frequency motor neuron firing, EMG fractionation, and gait variability in awake walking ALS mice. Proc Natl Acad Sci U S A 113:E7600–E7609. https://doi.org/10.1073/PNAS.1616832113

Hegedus J, Putman CT, Gordon T (2007) Time course of preferential motor unit loss in the SOD1 G93A mouse model of amyotrophic lateral sclerosis. Neurobiol Dis 28:154–164. https://doi.org/10.1016/j.nbd.2007.07.003

Hegedus J, Putman CT, Tyreman N, Gordon T (2008) Preferential motor unit loss in the SOD1 G93A transgenic mouse model of amyotrophic lateral sclerosis. J Physiol 586:3337–3351. https://doi.org/10.1113/jphysiol.2007.149286

Huisman MHB, Seelen M, de Jong SW et al (2013) Lifetime physical activity and the risk of amyotrophic lateral sclerosis. J Neurol Neurosurg Psychiatry 84:976–981. https://doi.org/10.1136/jnnp-2012-304724

Ibata K, Sun Q, Turrigiano GG (2008) Rapid synaptic scaling induced by changes in postsynaptic firing. Neuron 57:819–826. https://doi.org/10.1016/J.NEURON.2008.02.031

Ilieva H, Polymenidou M, Cleveland DW (2009) Non-cell autonomous toxicity in neurodegenerative disorders: ALS and beyond. J Cell Biol 187:761–772. https://doi.org/10.1083/jcb.200908164

Impey S, Fong AL, Wang Y et al (2002) Phosphorylation of CBP mediates transcriptional activation by neural activity and CaM kinase IV. Neuron 34:235–244. https://doi.org/10.1016/s0896-6273(02)00654-2

Jensen DB, Kadlecova M, Allodi I, Meehan CF (2020) Spinal motoneurones are intrinsically more responsive in the adult G93A SOD1 mouse model of Amyotrophic Lateral Sclerosis. J Physiol. https://doi.org/10.1113/JP280097

Joseph A, Turrigiano GG (2017) All for one but not one for all: excitatory synaptic scaling and intrinsic excitability are coregulated by CaMKIV, whereas inhibitory synaptic scaling is under independent control. J Neurosci 37:6778–6785. https://doi.org/10.1523/JNEUROSCI.0618-17.2017

Kanning KC, Kaplan A, Henderson CE (2009) Motor neuron diversity in development and disease. Annu Rev Neurosci 33:409–440

Kaplan A, Spiller KJ, Towne C et al (2014) Neuronal matrix Metalloproteinase-9 is a determinant of selective neurodegeneration. Neuron 81:333–348. https://doi.org/10.1016/J.NEURON.2013.12.009

Kawamura Y, Dyck PJ, Shimono M et al (1981) Morphometric comparison of the vulnerability of peripheral motor and sensory neurons in amyotrophic lateral sclerosis. J Neuropathol Exp Neurol 40:667–675. https://doi.org/10.1097/00005072-198111000-00008

Kim T-K, Hemberg M, Gray JM et al (2010) Widespread transcription at neuronal activity-regulated enhancers. Nature 465:182–187. https://doi.org/10.1038/nature09033

Kim J, Hughes EG, Shetty AS et al (2017) Changes in the excitability of neocortical neurons in a mouse model of amyotrophic lateral sclerosis are not specific to corticospinal neurons and are modulated by advancing disease. J Neurosci 37:9037–9053. https://doi.org/10.1523/JNEUROSCI.0811-17.2017

Krapivinsky G, Krapivinsky L, Manasian Y et al (2003) The NMDA receptor is coupled to the ERK pathway by a direct interaction between NR2B and RasGRF1. Neuron 40:775–784. https://doi.org/10.1016/s0896-6273(03)00645-7

Kuo JJ, Schonewille M, Siddique T et al (2004) Hyperexcitability of cultured spinal motoneurons from presymptomatic ALS mice. J Neurophysiol 91:571–575. https://doi.org/10.1152/jn.00665.2003

Lalancette-Hebert M, Sharma A, Lyashchenko AK, Shneider NA (2016) Gamma motor neurons survive and exacerbate alpha motor neuron degeneration in ALS. Proc Natl Acad Sci U S A 113:E8316–E8325. https://doi.org/10.1073/pnas.1605210113

Leroy F, Lamotte d'Incamps B, Imhoff-Manuel RD, Zytnicki D (2014) Early intrinsic hyperexcitability does not contribute to motoneuron degeneration in amyotrophic lateral sclerosis. elife 3. https://doi.org/10.7554/eLife.04046

Li PA, Hou X, Hao S (2017) Mitochondrial biogenesis in neurodegeneration. J Neurosci Res 95:2025–2029. https://doi.org/10.1002/jnr.24042

Lösing P, Niturad CE, Harrer M et al (2017) SRF modulates seizure occurrence, activity induced gene transcription and hippocampal circuit reorganization in the mouse pilocarpine epilepsy model. Mol Brain 10:30. https://doi.org/10.1186/s13041-017-0310-2

Ma H, Groth RD, Cohen SM et al (2014) γCaMKII shuttles Ca^{2+}/CaM to the nucleus to trigger CREB phosphorylation and gene expression. Cell 159:281–294. https://doi.org/10.1016/j.cell.2014.09.019

Malik AN, Vierbuchen T, Hemberg M et al (2014) Genome-wide identification and characterization of functional neuronal activity-dependent enhancers. Nat Neurosci 17:1330–1339. https://doi.org/10.1038/nn.3808

Mantamadiotis T, Lemberger T, Bleckmann SC et al (2002) Disruption of CREB function in brain leads to neurodegeneration. Nat Genet 31:47–54. https://doi.org/10.1038/ng882

Manuel M (2020) Sub-optimal discontinuous current-clamp switching rates lead to deceptive mouse neuronal firing. bioRxiv 20200813250134. https://doi.org/10.1101/2020.08.13.250134

Manuel M, Heckman CJ (2011) Adult mouse motor units develop almost all of their force in the subprimary range: a new all-or-none strategy for force recruitment? J Neurosci 31:15188–15194. https://doi.org/10.1523/JNEUROSCI.2893-11.2011

Martin E, Cazenave W, Cattaert D, Branchereau P (2013) Embryonic alteration of motoneuronal morphology induces hyperexcitability in the mouse model of amyotrophic lateral sclerosis. Neurobiol Dis 54:116–126. https://doi.org/10.1016/j.nbd.2013.02.011

Martineau É, Di Polo A, Vande Velde C, Robitaille R (2018) Dynamic neuromuscular remodeling precedes motor-unit loss in a mouse model of ALS. elife 7:e41973. https://doi.org/10.7554/eLife.41973

Martínez-Silva M de L, Imhoff-Manuel RD, Sharma A et al (2018) Hypoexcitability precedes denervation in the large fast-contracting motor units in two unrelated mouse models of ALS. elife 7:e30955. https://doi.org/10.7554/ELIFE.30955

Mayr B, Montminy M (2001) Transcriptional regulation by the phosphorylation-dependent factor CREB. Nat Rev Mol Cell Biol 2:599–609. https://doi.org/10.1038/35085068

McIlwain DL (1991) Nuclear and cell body size in spinal motor neurons. Adv Neurol 56:67–74

Meehan CF, Moldovan M, Marklund SL et al (2010) Intrinsic properties of lumbar motor neurones in the adult G127insTGGG superoxide dismutase-1 mutant mouse in vivo: evidence for increased persistent inward currents. Acta Physiol Oxf Engl 200:361–376. https://doi.org/10.1111/j.1748-1716.2010.02188.x

Naujock M, Stanslowsky N, Bufler S et al (2016) 4-aminopyridine induced activity rescues hypoexcitable motor neurons from amyotrophic lateral sclerosis patient-derived induced pluripotent stem cells. Stem Cells Dayt Ohio 34:1563–1575. https://doi.org/10.1002/STEM.2354

Pambo-Pambo A, Durand J, Gueritaud J-P (2009) Early excitability changes in lumbar motoneurons of transgenic SOD1G85R and SOD1G(93A-Low) mice. J Neurophysiol 102:3627–3642. https://doi.org/10.1152/jn.00482.2009

Pieri M, Albo F, Gaetti C et al (2003) Altered excitability of motor neurons in a transgenic mouse model of familial amyotrophic lateral sclerosis. Neurosci Lett 351:153–156. https://doi.org/10.1016/j.neulet.2003.07.010

Pieri M, Carunchio I, Curcio L et al (2009) Increased persistent sodium current determines cortical hyperexcitability in a genetic model of amyotrophic lateral sclerosis. Exp Neurol 215:368–379. https://doi.org/10.1016/j.expneurol.2008.11.002

Pinelli P, Pisano F, Ceriani F, Miscio G (1991) EMG evaluation of motor neuron sprouting in amyotrophic lateral sclerosis. Ital J Neurol Sci 12:359–367. https://doi.org/10.1007/BF02335775

Pun S, Santos AF, Saxena S et al (2006) Selective vulnerability and pruning of phasic motoneuron axons in motoneuron disease alleviated by CNTF. Nat Neurosci 9:408–419. https://doi.org/10.1038/nn1653

Pupillo E, Messina P, Giussani G et al (2014) Physical activity and amyotrophic lateral sclerosis: a European population-based case-control study. Ann Neurol 75:708–716. https://doi.org/10.1002/ana.24150

Quinlan KA, Schuster JE, Fu R et al (2011) Altered postnatal maturation of electrical properties in spinal motoneurons in a mouse model of amyotrophic lateral sclerosis. J Physiol 589:2245–2260

Roselli F, Caroni P (2015) From intrinsic firing properties to selective neuronal vulnerability in neurodegenerative diseases. Neuron 85:901–910. https://doi.org/10.1016/J.NEURON.2014.12.063

Rudnick ND, Griffey CJ, Guarnieri P et al (2017) Distinct roles for motor neuron autophagy early and late in the SOD1(G93A) mouse model of ALS. Proc Natl Acad Sci U S A 114:E8294–E8303. https://doi.org/10.1073/PNAS.1704294114

Sareen D, Gowing G, Sahabian A et al (2014) Human induced pluripotent stem cells are a novel source of neural progenitor cells (iNPCs) that migrate and integrate in the rodent spinal cord. J Comp Neurol 522:2707–2728. https://doi.org/10.1002/cne.23578

Saura CA, Cardinaux J-R (2017) Emerging roles of CREB-regulated transcription coactivators in brain physiology and pathology. Trends Neurosci 40:720–733. https://doi.org/10.1016/j.tins.2017.10.002

Saxena S, Roselli F, Singh K et al (2013) Neuroprotection through excitability and mTOR required in ALS motoneurons to delay disease and extend survival. Neuron 80:80–96. https://doi.org/10.1016/j.neuron.2013.07.027

Schaefer AM, Sanes JR, Lichtman JW (2005) A compensatory subpopulation of motor neurons in a mouse model of amyotrophic lateral sclerosis. J Comp Neurol 490:209–219. https://doi.org/10.1002/cne.20620

Seok S, Fu T, Choi S-E et al (2014) Transcriptional regulation of autophagy by an FXR-CREB axis. Nature 516:108–111. https://doi.org/10.1038/nature13949

Sharma A, Lyashchenko AK, Lu L et al (2016) ALS-associated mutant FUS induces selective motor neuron degeneration through toxic gain of function. Nat Commun 7:10465. https://doi.org/10.1038/NCOMMS10465

Sobue G, Matsuoka Y, Mukai E et al (1981) Pathology of myelinated fibers in cervical and lumbar ventral spinal roots in amyotrophic lateral sclerosis. J Neurol Sci 50:413–421. https://doi.org/10.1016/0022-510X(81)90153-2

Soderling TR (1999) The Ca-calmodulin-dependent protein kinase cascade. Trends Biochem Sci 24:232–236. https://doi.org/10.1016/s0968-0004(99)01383-3

Takahashi M, Li Y, Dillon TJ, Stork PJS (2017) Phosphorylation of Rap1 by cAMP-dependent protein kinase (PKA) creates a binding site for KSR to sustain ERK activation by cAMP. J Biol Chem 292:1449–1461. https://doi.org/10.1074/jbc.M116.768986

Tremblay E, Martineau É, Robitaille R (2017) Opposite synaptic alterations at the neuromuscular junction in an ALS mouse model: when motor units matter. J Neurosci 37:8901–8918. https://doi.org/10.1523/JNEUROSCI.3090-16.2017

Van Den Bosch L, Van Damme P, Bogaert E, Robberecht W (2006) The role of excitotoxicity in the pathogenesis of amyotrophic lateral sclerosis. Biochim Biophys Acta 1762:1068–1082. https://doi.org/10.1016/J.BBADIS.2006.05.002

van Zundert B, Peuscher MH, Hynynen M et al (2008) Neonatal neuronal circuitry shows hyperexcitable disturbance in a mouse model of the adult-onset neurodegenerative disease amyotrophic lateral sclerosis. J Neurosci 28:10864–10874. https://doi.org/10.1523/JNEUROSCI.1340-08.2008

Venugopal S, Hsiao C-F, Sonoda T et al (2015) Homeostatic dysregulation in membrane properties of masticatory motoneurons compared with oculomotor neurons in a mouse model for amyotrophic lateral sclerosis. J Neurosci 35:707–720. https://doi.org/10.1523/JNEUROSCI.1682-14.2015

Visser AE, Rooney JPK, D'Ovidio F et al (2018) Multicentre, cross-cultural, population-based, case-control study of physical activity as risk factor for amyotrophic lateral sclerosis. J Neurol Neurosurg Psychiatry 89:797–803. https://doi.org/10.1136/jnnp-2017-317724

Wainger BJ, Kiskinis E, Mellin C et al (2014) Intrinsic membrane hyperexcitability of amyotrophic lateral sclerosis patient-derived motor neurons. Cell Rep 7:1–11. https://doi.org/10.1016/j.celrep.2014.03.019

Wang H, Storm DR (2003) Calmodulin-regulated adenylyl cyclases: cross-talk and plasticity in the central nervous system. Mol Pharmacol 63:463–468. https://doi.org/10.1124/mol.63.3.463

Wayman GA, Lee Y-S, Tokumitsu H et al (2008) Calmodulin-kinases: modulators of neuronal development and plasticity. Neuron 59:914–931. https://doi.org/10.1016/j.neuron.2008.08.021

Yap E-L, Greenberg ME (2018) Activity-regulated transcription: bridging the gap between neural activity and behavior. Neuron 100:330–348. https://doi.org/10.1016/j.neuron.2018.10.013

Zhang X, Odom DT, Koo S-H et al (2005) Genome-wide analysis of cAMP-response element binding protein occupancy, phosphorylation, and target gene activation in human tissues. Proc Natl Acad Sci U S A 102:4459–4464. https://doi.org/10.1073/pnas.0501076102

Zhao C, Devlin A, Chouhan AK et al (2020) Mutant C9orf72 human iPSC-derived astrocytes cause non-cell autonomous motor neuron pathophysiology. Glia 68:1046–1064. https://doi.org/10.1002/glia.23761

Index

M. O'Donovan, M. Falgairolle (eds.), *Vertebrate Motoneurons*, Advances in Neurobiology 28, https://doi.org/10.1007/978-3-031-07167-6

N

O

The manufacturer's authorised representative in the EU is Springer Nature Customer Service Centre GmbH, Europaplatz 3, 69115 Heidelberg, Germany. If you have any concerns regarding our products, please contact ProductSafety@springernature.com

Printed and bound by CPI Group (UK) Ltd, Croydon, CR0 4YY
30/09/2026
02225462-0015